Semiconductor Pulse and Switching Circuits

Electronic Technology Series

SERIES EDITOR
Irving L. Kosow, Ph.D.

Herbert W. Richter
ELECTRICAL AND ELECTRONIC DRAFTING

Curtis Johnson
PROCESS CONTROL INSTRUMENTATION TECHNOLOGY

Luces M. Faulkenberry
AN INTRODUCTION TO OPERATIONAL AMPLIFIERS

Joseph Greenfield
PRACTICAL DIGITAL DESIGN USING ICs

Irving L. Kosow
STUDY GUIDE IN DIRECT CURRENT CIRCUITS: A PERSONALIZED SYSTEM OF INSTRUCTION

Irving L. Kosow
STUDY GUIDE IN ALTERNATING CURRENT CIRCUITS: A PERSONALIZED SYSTEM OF INSTRUCTION

Rodney B. Faber
APPLIED ELECTRICITY AND ELECTRONICS FOR TECHNOLOGY

Sol Lapatine
ELECTRONICS IN COMMUNICATION

Ulises M. Lopez and George E. Warrin
ELECTRONIC DRAWING AND TECHNOLOGY

Dan I. Porat and Arpad Barna
INTRODUCTION TO DIGITAL TECHNIQUES

Semiconductor Pulse and Switching Circuits

Santokh S. Basi

Saskatchewan Technical Institute
Moose Jaw, Canada

John Wiley and Sons

New York Chichester Brisbane Toronto

Library of Congress Cataloging in Publication Data:

Basi, Santokh, 1933-
Semiconductor pulse and switching circuits.

(Electronic technology series)
Bibliography: p.
Includes index.
1. Pulse circuits. 2. Switching circuits.
3. Semiconductors. I. Title.
TK7868.P8B33 621.3815'34 79-15379
ISBN 0-471-05539-5

Printed in the United States

10 9 8 7 6 5 4 3 2 1

Preface

Although several well-known books on pulse circuits are already in print, each with some strong points and some weak points, an author always strives to perfect the depth (contents), order of topics, and presentation (language) of the subject matter, even when using existing texts. It was with this thought in mind that this book was written.

In addition, the reader will find the following features that distinguish this book from its predecessors:

1. The topics discussed are only those that are pertinent to pulse circuits. The discussion is kept to the point. No redundant or "nice-to-know" material is included.
2. The underlying idea throughout the book is that the reader should be able to follow the material presented without any outside help. As an aid to understanding, many examples are given at every step along the way.
3. The content, and the depth to which it is discussed, is ideally suited for an introductory course in pulse circuits especially, for the training of technicians and technologists.
4. Wherever possible, mathematics, rather than elaborate descriptive prose, is used as a tool to explain circuit behavior.
5. Calculus is sometimes used to show the derivation of a useful equation. However, the reader does *not* need knowledge of calculus to follow the presentation or to solve problems.
6. For the most part, the operation of a circuit is described in *reduced-type* print and in orderly steps. This is done to spotlight it amid the rest of the presentation in that section.
7. Within the presentation, important terms and phrases are printed in *italic* script to emphasize their importance.
8. Self-Evaluation Questions, Summaries, and Glossaries of important terms serve to display the contents of the chapter. These sections indicate the important topics, or the "meat" of the chapter.
9. The text is outlined so that the material presented in one chapter follows that of the preceding chapter in the most logical

manner. For the most part, however, a chapter is an entity by itself. This makes it easier for the reader to arrange his/her own order of reading or to read only the necessary chapters.

10. The operation of circuits containing discrete components is explained first, and then the corresponding TTL IC circuits are discussed. This is done to show the direct link that exists between circuits having discrete components and the newer ICs.
11. Chapters 1 and 2 review the important topics in basic electrical circuit theory. It is assumed that the reader is already familiar with these and does not need to spend any appreciable amount of time on them. However, it is important that the reader understands them well enough to follow their use in other pulse circuits.
12. Chapter 11 is devoted to logic fundamentals and their use in pulse and switching circuits. Details are provided for the design of combinational switching networks, starting with the word problem and ending with the required logic gate circuit. Common IC families and their important characteristics also are discussed in this chapter.
13. The experiment provided at the end of each chapter is an integral part of the chapter. It is designed to complement the important material presented in the chapter. Each experiment has been laboratory- (student-) tested. An average student takes about three hours to perform the experiment and write a brief report on it. A slower student may have to do some calculations in advance in order to finish it in three hours. Sometimes it will take more than three hours of laboratory time to cover all the important topics presented in a chapter. In such cases, the experiment is divided into parts such that each part will take three hours of laboratory time. The equipment prescribed for each experiment consists of standard-value and readily-available components. The test equipment consists of a dual-beam (or dual-trace) oscilloscope, a multimeter, a multifunction generator, and a dual power supply—standard equipment in an electronics laboratory.

Chapters 1 and 2 review the important concepts of electric circuit theory. These are used to explain the operation of pulse circuits discussed in other chapters. It is assumed that the reader is already familiar with these and, if necessary, will use them as reference material.

The response of an RC circuit to a step input (pulse) is discussed in detail in Ch. 3. The role of an inductor as a waveshaping device is downplayed, because it is seldom used in practical circuits. The damping characteristics of an RLC circuit are discussed to explain the oscillations that are sometimes encountered in pulse circuits. The proper way to compensate an attenuator is also shown.

The basic characteristics of semiconductor devices such as diodes, bipolar junction transistors, and field effect transistors (FET) are reviewed in Chs. 4 and 5. Attention is focused on their use as a switch. Factors that control their turn-ON time and turn-OFF time are discussed in detail. The 7404 TTL IC hex-inverter is used to show the switching action of diodes and transistors, as well as the direct link that exists between ICs and circuits having discrete components.

Multivibrators are discussed in Chs. 6 and 7. In each case an analysis of a circuit using discrete components is followed by a discussion on the operation of the corresponding TTL IC. The use of F/Fs in binary registers and counters is also explained.

Voltage-sweep and current-sweep principles, as well as corresponding circuits, are discussed in Ch. 8. Methods of producing linear voltage sweep are emphasized.

Chapter 9 contains discussion on special pulse circuits not covered earlier. A detailed discussion on the Schmitt trigger, UJT relaxation oscillator, and the use of 555 timer in astable, monostable, and ramp-generator mode is presented.

A brief discussion on pulse transformers and pulse-delay techniques is presented in Ch. 10. Its importance in pulse circuits is rapidly diminishing and therefore only a short discussion is devoted to it.

Chapter 11 is devoted to logic fundamentals and logic circuits available in various IC families. The process involved in the design of combinational switching networks, starting with a word problem and ending up with the required logic-gate network, is explained in detail. The operation of common IC gates in various IC families is explained. An effort is made to show that IC circuits are really miniaturized semiconductor discrete-component circuits. Important characteristics and associated terms used to identify IC parameters are discussed briefly.

The importance of the place that an experiment holds in the whole technical learning process cannot be overemphasized. "Hands on" experience is a must if one is to have a real understanding of circuit behavior. To this end, an appended experiment forms an integral part of each chapter. Each experiment has been laboratory- (student-) tested for instruction, clarity, order of performance, time period, and its relationship to the material presented in the chapter. It has been found that:

1. The experiment is well-suited to the material presented in the chapter.
2. The experimenter is able to follow the instructions and perform the experiment without any additional help.
3. A three-hour laboratory period is more than enough to completely finish the experiment, including the report. Our students record all the information pertaining to the experiment on the experiment sheets. They are required to answer the questions and write the discussion on separate sheets that are attached to the report.
4. Whenever topics to be covered in an experiment require a longer laboratory time, the experiment is divided into parts. In these cases, each part of an experiment is an entity in itself, and requires about three hours of laboratory time.

My heartfelt thanks go to my editor, Dr. Irving L. Kosow, without whose able guidance this book could never have become a reality. I am also grateful to various reviewers, especially, Prof. David M. Hata (Portland Community College), Prof. Arthur Seidman (Pratt Institute), Prof. Arthur Del Giorno (City Univ. of N.Y.) and Dr. Stephen R. Chesier (Purdue University) for constructive comments that helped me to focus on the important topics in pulse circuits.

I am grateful, too, to my dear wife, Rajinder, and my three children, Rav Iqubal, Sanjinder, and Sanraj (my helper), for putting up with me while I pounded away on the typewriter.

Santokh S. Basi

Contents

Semiconductor Pulse and Switching Circuits

Chapter 1 Review of Basic Electrical Laws and Theorems

1-1 Objectives

The objectives of this chapter are to:

1. Review the units of electrical measurements and establish the conventional direction of current in a circuit.
2. Review the definition of resistance and briefly explain the differences between conductors and insulators.
3. Establish voltage-drop terminology as it relates to the conventional direction of current in a circuit.
4. Review Ohm's law and the units of resistance.
5. Review the concept of electrical power and the meaning of the power rating of a component.
6. Review the definition of capacitance.
7. Review the relationship between the charge on, the voltage across, and the capacitance of a capacitor.
8. Derive the equation for the instantaneous current in a capacitor.
9. Review the relationship that exists between the capacitance of, the voltage across, and the **dc** current charging a capacitor.
10. Review the definition of inductance.
11. Review capacitive reactance of a capacitor to a sinusoidal **ac** voltage.
12. State the equation for the instantaneous voltage across an inductor.
13. Review the conditions that occur when a **dc** current flows in an inductor.
14. Review inductive reactance of an inductor to a sinusoidal **ac** voltage.
15. Review Kirchhoff's current and voltage laws.
16. Review methods of calculating equivalent impedance when two or more impedances are connected in series and/or in parallel.
17. Review Superposition and Thévenin's theorems.

1-2 Self-Evaluation Questions

Watch for the answers to the following questions as you read the chapter. They should help you to understand the material presented. When you have finished the chapter, return to this section and be sure that you can answer all of these questions. Then test your understanding by working out solutions to the problems at the end of the chapter.

It is strongly recommended that the questions be satisfactorily answered before attempting to solve the Problems (Sec. 1-17).

1. Explain in your own words the following terms used in describing electric circuits:
   a. Potential.
   b. Current.
   c. Conventional direction of current.
2. Define resistance and outline the basic properties of a:
   a. Conductor.
   b. Insulator.
3. Show how the voltage at a point in a circuit with respect to another point in that circuit is commonly expressed.
4. State and explain Ohm's law.
5. Define electrical power and explain what is meant by a *watt*.
6. Show how the electrical power being dissipated in a circuit is calculated.
7. Define capacitance and explain what is meant by a *farad*.
8. State the relationship that exists between the potential difference across a charged capacitor and its capacitance.
9. State the equation for the instantaneous current in a capacitor.
10. State the equation for the instantaneous voltage across a capacitor.
11. Explain why the voltage across a capacitor increases as a constant (**dc**) current charges it.
12. State the equation used to calculate the capacitive reactance of a capacitor when the operating voltage is a sinusoidal.
13. Define *inductance* and explain what is meant by a *henry*.
14. State the equation for the instantaneous current in an inductor.
15. State the equation for the instantaneous voltage across an inductor.
16. Explain what happens to the voltage drop across the inductor if the current in it is **dc**.
17. State the equation used to calculate the inductive reactance of an inductor when the operating voltage is a sinusoidal.
18. State and explain the Kirchhoff's voltage (KVL) and current (KCL) laws.
19. Show how to calculate the equivalent impedance when two or more impedances are connected in **a.** series; **b.** parallel.
20. State and explain the Superposition theorem.
21. State and explain Thévenin's theorem.

1-3 Potential Difference and Current

Most electrical power sources (generators or batteries) have two output terminals. The basic function of a power source is to develop and (continuously) provide a surplus of electrons at one output terminal and a deficiency of electrons at the other output terminal.

This results in a difference of potential between the two terminals that is measured in *volts* (V). The output terminal having a surplus of electrons is called the *negative* terminal and the output terminal having a deficiency of electrons is called the *positive* terminal.

When an electric circuit is connected to (across) a power source (usually called the "source"), it provides a path for electrocharge flow. The rate of charge flow is known as *current* (I) in the circuit. The practical unit of current measurement is an *ampere* (A). A current of one ampere is said to be flowing at a point in a circuit if an electronic charge of one coulomb (C) is moving past that point every second. Stated in equation form:

$$I = \frac{Q}{t} \text{ amperes (A)} \tag{1-1}$$

where:

Q is the charge in coulombs (C)
t is the time in seconds (s)

The instantaneous value of current may be calculated by differentiating Eq. (1-1). Accordingly,

$$i = \frac{dq}{dt}, \text{ rate of charge flow in coulombs per second (C/s)} \tag{1-2}$$

where:

i is the instantaneous value of current (A)
dq/dt is the incremental flow of charge per incremental unit of time taken over an infinitesimally small increment of time (C/s)

The electrons in an electric circuit drift from the negative to the positive terminal. However, it is a standard practice to denote the direction of current from the positive to the negative terminal. This is called the *conventional direction.* The arrow symbol on all semiconductor devices shows the flow of current in this the conventional direction. *Conventional current* is used in this book as it is extensively used in most engineering literature throughout the world.

1-4 Resistance (R)

When a potential difference is applied to a closed electric circuit, a charge begins to drift. All circuit materials present a certain amount of opposition or *resistance* to the rate of charge flow or current. Some materials present more opposition to the flow of charge than others.

Resistance is that property of a material that *opposes* a current and dissipates electrical energy. The *electrical* energy is converted by the circuit property of resistance and is usually dissipated in the form of *heat* energy. The equation symbol for resistance is R.

Materials that present very little resistance to current are called *conductors*. Materials that present a high resistance to current are called *insulators*. Materials between these two extremes are called *semiconductors*.

The standard unit of resistance measurement is an ohm, usually abbreviated by a Greek letter omega (Ω). A material is said to possess a resistance of one Ω if a one-volt potential difference across it results in one ampere of current in it.

The standard symbol for a resistor is shown in Fig. 1-1. It shows a current of I amperes flowing through it. The potential at point a is V_a and at point b is V_b. The potential difference between points a and b is written as V_{ab}. V_{ab} means the potential at point a with respect to b and is equal to $V_a - V_b$. Obviously, then, V_{ba} means the potential at point b with respect to a. In Fig. 1-1, V_a is more positive than V_b. For this reason, the conventional direction of current is from a to b.

Ohm's law relates all three quantities (R, I, and V_{ab}). It states that the current in a resistor is directly proportional to the potential difference (*voltage drop*) across it. The constant of proportionality is the resistance (R) of the resistor. If the potential difference across a resistor is doubled, its current will also double. Mathematically, Ohm's law is:

$$R = \frac{V}{I} \text{ ohms } (\Omega) \tag{1-3a}$$

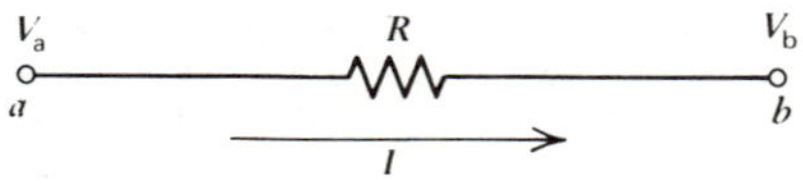

Figure 1-1 Current in and voltage-drop across a resistor.

or, rearranging terms,

$$V = I \cdot R \text{ volts (V)} \qquad \text{(1-3b)}$$

where

V is the voltage drop across the resistor in volts (V)
I is the current in the resistor in amperes (A)
R is the resistance of the resistor in ohms (Ω)

Example 1-1
Voltage drop across a 5-kΩ resistor is measured as 25-V. Calculate the current in the resistor.

Solution

$$I = \frac{V}{R} \qquad \text{(1-3)}$$

$$= \frac{25 \text{ V}}{5 \times 10^3 \ \Omega}$$

$$= 5 \times 10^{-3}$$

$$= \mathbf{5\ mA}$$

Note that Ohm's law (Eq. 1-3) holds true for **dc** as well as for **ac** circuits. However, when analyzing **ac** circuits, it is important to consider the magnitude as well as the direction of phasor quantities (see Sec. 1-11).

1-5 Power (*P*)

As mentioned earlier, the power source works to maintain a constant potential difference, also known as the *electromotive force* (emf), between its output terminals while supplying a steady current to an electric circuit. The electrical power supplied by the source is converted to other energy forms by the circuit. The practical unit of electrical power is the *watt* (W). Power as the rate of doing work is also measured in *joules per second* (J/s). One watt of power is equivalent to work done at the rate of one joule per second, that is, 1 W = 1 J/s.

Consider a power source with a terminal potential difference (voltage) of V volts, supplying a current of I amperes to a circuit.

The electrical power supplied by the source to the circuit is given by:

$$P = V \cdot I \text{ watts (W)} \tag{1-4a}$$

Equation (1-4a) is applicable to a circuit as well as to a component in a circuit. For example, if the current in a resistor is found to be 5 mA, while the voltage drop across it is 10 V, then it is dissipating 50 mW (10 × 5 = 50) of electrical power. The real significance of such calculations in practical circuits is to ensure that the actual power being dissipated in a component is below its maximum limit, given by the manufacturer. In this book, we are mainly concerned with the power ratings of the semiconductor devices and resistors, to assure that they are not exceeded during normal operation. Equation (1-4a) and its two forms given below are sufficient for that purpose. Active or reactive power is beyond the scope of this discussion.

Applying Ohm's law to Eq. (1-4a) yields its two other versions:

$$P = V \cdot I \tag{1-4a}$$

$$= V \cdot \left(\frac{V}{R}\right)$$

$$= \frac{V^2}{R} \text{ watts (W)} \tag{1-4b}$$

or

$$P = V \cdot I$$

$$= (I \cdot R) \cdot I$$

$$= I^2 \cdot R \text{ watts (W)}, \tag{1-4c}$$

where all terms have been defined above.

Example 1-2

Current in a 100 Ω-resistor is measured to be 500 mA. Calculate the power rating required for this resistor.

Solution

$$P = I^2 \cdot R \qquad \text{(1-4c)}$$

$$= (0.5)^2 \times 100$$

$$= \mathbf{25\ W}$$

1-6 Capacitance (*C*)

A capacitor is a circuit element that is encountered quite frequently in switching circuits. In its simplest form, a capacitor consists of two conductive plates separated by a dielectric or insulator. The *capacitance* of such a capacitor depends upon the surface area of the plates, the dielectric constant of the insulating material, and the distance between the two conductive plates. In fact, in practical circuits, a certain amount of capacitance exists between any two conductors.

The basic property of a capacitor is that it *opposes* any attempt to instantaneously *change* the potential difference across it. This property is called the capacitance (C) of the capacitor and is measured in farads (F).

The potential difference (voltage) across a charged capacitor is directly proportional to the electrical charge on its plates. The constant of proportionality is the capacitance of the capacitor. Stated in the equation form:

$$Q = C \cdot V \text{ coulombs (C)} \tag{1-5}$$

where:

C is the capacitance of the capacitor in farads (F)
V is the potential difference across the capacitor in volts (V)

While the basic unit of capacitance is the farad, only smaller capacitances usually are found in electronic circuits. These range from several microfarads (μF) down to several picofarads (pF).

Example 1-3

A 0.047-μF capacitor is shown in Fig. 1-2. The voltage across it is found to be 10 V. Find the charge on the capacitor.

Solution

$$Q = C \cdot V \tag{1-5}$$

$$= 0.047 \times 10^{-6} \times 10$$

$$= \mathbf{0.47\ \mu C}$$

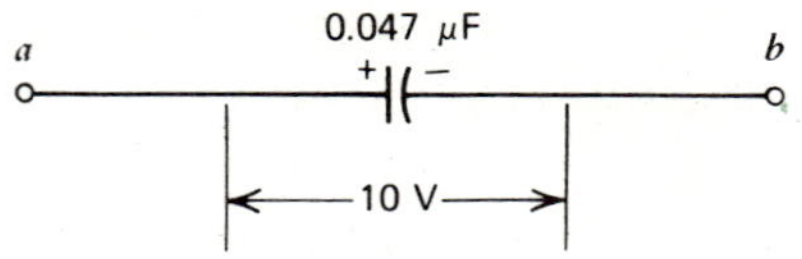

Figure 1-2 A charged capacitor, Ex. 1-3.

An expression for the instantaneous current in a capacitor is obtained by differentiating* Eq. (1-5) in the following manner:

$$Q = C \cdot V \tag{1-5}$$

$$\frac{dq}{dt} = C \cdot \frac{dv}{dt}$$

$$i = C \frac{dv}{dt} \text{ amperes (A)} \tag{1-6}$$

where:

i is the instantaneous current in the capacitor (A)
C is the capacitance of the capacitor (F)
dv/dt is the change in voltage across the capacitor with an infinitesimally small change in current charging it (V/s)

Equation (1-6) can be solved to find the instantaneous voltage across the capacitor in the following manner:

$$i = C \cdot \frac{dv}{dt} \tag{1-6}$$

$$dv = \frac{1}{C} \cdot i\,dt$$

$$v = \frac{1}{C} \int i\,dt \text{ volts (V)} \tag{1-7}$$

* Calculus is used merely to show the derivation of Eq. (1-8) only. No examples or problems require knowledge of calculus.

where:

i is the instantaneous current in the capacitor (A)
v is the instantaneous voltage across the capacitor (V)
C is the capacitance of the capacitor (F)

When the charging current does not change with time, that is, when it is a **dc** current, Eq. (1-7) reduces to:

$$V = \frac{1}{C}\int i\,dt \tag{1-7}$$

or

$$V = \frac{1}{C}\cdot I \cdot t$$

$$= \frac{I}{C}\cdot t = \frac{Q}{C} \text{ volts (V)} \tag{1-8}$$

where:

V is the voltage across the capacitor at time t (V)
I is the **dc** current charging the capacitor (A)
C is the capacitance of the capacitor (F)
t is the time duration for which the **dc** current has been charging the capacitor (s)

Example 1-4
A 5-μF capacitor is charged by a constant current of 8 mA **dc**. Find voltage across the capacitor after:

a. 1 ms.
b. 10 ms.
c. 1 s.

Solution

$$V = \frac{I}{C}\cdot t \tag{1-8}$$

a.

$$V = \frac{8 \times 10^{-3}}{5 \times 10^{-6}} \times 1 \times 10^{-3}$$

$$= \mathbf{1.6\ V}$$

b.

$$V = \frac{8 \times 10^{-3}}{5 \times 10^{-6}} \times 10 \times 10^{-3}$$

$$= \mathbf{16\ V}$$

c.

$$V = \frac{8 \times 10^{-3}}{5 \times 10^{-6}} \times 1$$

$$= \mathbf{1.6\ kV}$$

1-7 Capacitive Reactance (X_c)

Another instance of interest occurs when the operating voltage across (or current in) a capacitor is *sinusoidal* in nature. Under these circumstances, the **ac** voltage across the capacitor and the **ac** current in it are related in the following manner:

$$X_c = \frac{v_c}{i_c} \text{ ohms } (\Omega) \tag{1-9}$$

where:

X_c is the impedance (reactance) of the capacitor measured in ohms (Ω)
v_c is the voltage across the capacitor (V)
i_c is the sinusoidal **ac** current in the capacitor (A)

X_c (capacitive reactance) is a *frequency dependent* quantity. Knowing the frequency of the operating voltage, the value of X_c is calculated in the following manner:

$$X_c = \frac{1}{j\omega C} = -\frac{j}{2\pi f C} = \frac{1}{2\pi f C}\angle{-90^\circ} \text{ ohms } (\Omega) \tag{1-10}$$

where:

C is the capacitance of the capacitor in farads (F)
$\omega = 2\pi f$, which is measured in radians per second (rad/s)
f is the frequency of the operating voltage, measured in cycles per second or simply, hertz (Hz)
j is the imaginary axis of the complex plane

Example 1-5
Potential difference across a 0.01-μF capacitor is measured to be 110 V. The operating frequency of the power source is 60 Hz. Find:

a. Reactance of the capacitor.
b. Current in the capacitor.

Solution

a.
$$X_c = -j\frac{1}{2\pi fC} \tag{1-10}$$

$$= -j\frac{1}{2 \times 3.14 \times 60 \times 0.01 \times 10^{-6}}$$

$$= -j265 \times 10^3$$

or

$$= \mathbf{265 \angle -90^\circ \ k\Omega}$$

b.
$$i = \frac{v}{-jX_c}$$

$$= \frac{110\ 0^\circ}{265 \angle -90^\circ \ k\Omega}$$

$$= \mathbf{0.415 \angle 90^\circ \ mA}$$

Note from the solution of Ex. 1-5 that:

1. Pure capacitive reactance has a complex impedance which is represented at -90°.
2. The current in a capacitor leads the voltage across the capacitor by 90°.

Note also from Eq. (1-10) that the capacitive reactance X_c is small at high frequencies and large at low frequencies. Knowledge of Eqs. (1-9) and (1-10) is sufficient to follow the meaning of X_c as it occasionally appears in the following chapters.

In most switching circuits, the current charging a capacitor is neither a constant **dc** nor a sinusoidal **ac**, but instead, it is usually a continuously varying current. Let us examine Eq. (1-6) again.

$$i = C\frac{dv}{dt} \tag{1-6}$$

The term dv/dt gives the instantaneous rate at which the capacitor charges (or, slope of the capacitor charge curve). This rate of charge (dv/dt) is proportional to the instantaneous value of charging current. Therefore, the rate at which the capacitor charges depends upon the instantaneous value of the charging current. For the most part, we will be concerned with the manner in which a capacitor charges or discharges when a step (pulse) input is applied to the circuit. In such a case, it will be found that the capacitor charges exponentially from some initial value (volts) to a final value, sometimes the steady-state value. In general, the manner in which a capacitor charges/discharges from its initial value to its final value is known as its *transient* response. This and other important aspects of the capacitor charging/discharging process are discussed in detail in Ch. 3. For the time being it is important to note that:

1. The voltage across a capacitor cannot be changed instantaneously.
2. The rate at which voltage across a capacitor changes is directly proportional to the charging/discharging current and inversely proportional to its capacitance (Eq. 1-7).

1-8 Inductance (L)

When a circuit or an element in a circuit has the property or ability to *oppose* any *change in current*, it is said to possess self-inductance, or simply *inductance*. A circuit element that is specially designed to exhibit this property is called an inductor or a coil. The inductance of a coil depends partly on the geometry of its construction—the number of turns, the length and the crosssectional area of the core, as well as the material of the core. The inductance of a given coil (once formed) is constant and is independent of the current in it or voltage across it. The practical unit of inductance is the *henry* (H).

Figure 1-3 shows the practical symbol for a coil, L. It shows a current of i amperes in the coil and the voltage drop across it as v volts. According to Lenz's law, these three quantities (i, v, and L) are related in the following manner:

$$v = L\frac{di^*}{dt} \text{ volts (V)} \tag{1-11}$$

* Calculus is used to show the current-voltage relationship in an inductor. No example or problem requires the use of calculus.

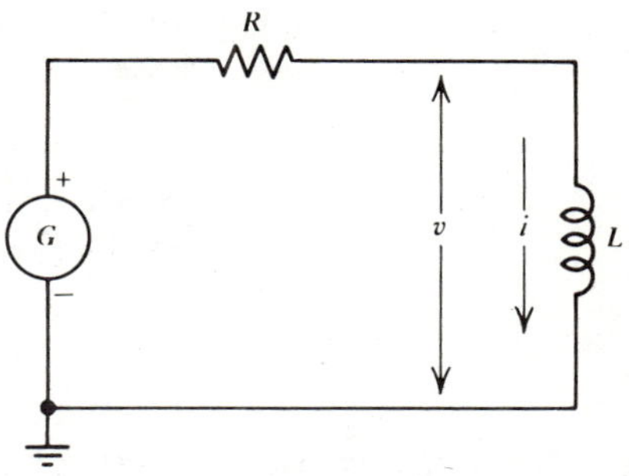

Figure 1-3 Current in an inductor.

where:

L is the inductance of the coil (H)

$\frac{di}{dt}$ is the rate of change of current in the inductor (A/s)

A closer look at Eq. (1-11) reveals:

1. Magnitude of the potential difference (v) across a coil is directly proportional to the rate of change of current that flows through it and its inductance L.
2. If a constant (steady) current (**dc**) exists in an inductor, then there is no voltage drop across it ($di/dt = 0$).
3. The polarity of the potential difference across a coil depends upon whether the current in it is increasing ($di/dt = +$) or decreasing ($di/dt = -$).

Rearranging and integrating Eq. (1-11) yields an equation for the instantaneous current in the coil.*

$$v = L\frac{di}{dt} \tag{1-11}$$

$$di = \frac{1}{L}v \cdot dt$$

$$i = \frac{1}{L}\int v \cdot dt \text{ amperes (A)} \tag{1-12}$$

where:

i is the instantaneous current in the coil (A)
v is the instantaneous voltage drop across the coil (V)
L is the inductance of the coil (H)

Example 1-6
Consider the coil shown in Fig. 1-3. The rate of current increase in the coil is measured to be 100 amperes per millisecond (A/ms). The potential difference across it at that instant is 50 V. Find inductance of the coil.

Solution

$$L = \frac{v}{di/dt} \qquad (1\text{-}11)$$

$$= \frac{50\ \text{V}}{100 \times 10^{+3}\ \text{A/s}}$$

$$= \mathbf{0.5\ mH}$$

1-9 Inductive Reactance (X_L)

When the operating current in an inductor is sinusoidal in nature, then the voltage across the inductor, the current through it, and its inductance are related in the following manner:

$$V_L = I \times jX_L,\ \text{volts (V)} \qquad (1\text{-}13)$$

where:

V_L is the voltage across the inductor (V)
I is the current flowing in the inductor (A)
X_L is the inductive reactance of the inductor (Ω)
j is the imaginary axis

Note that Eq. (1–13) is just another version of the Ohm's law (Eq. 1-4). The inductive reactance X_L is given by:

$$X_L = j2\pi fL = 2\pi fL\,\underline{/90^\circ}\ \text{ohms}\ (\Omega) \qquad (1\text{-}14)$$

where:

f is the frequency of the operating voltage (Hz)
L is the inductance of the coil (H)

Example 1-7
Voltage across an inductor is measured to be 30 volts, while the **ac** current in it is 15 mA. The operating voltage waveform is sinusoidal in nature. Find the reactance of the coil in kilohms.

Solution

$$X_L = \frac{V_L}{I_L} \tag{1-13}$$

$$= \frac{30\text{ V}}{15\text{ mA}}$$

$$= 2\text{ k}\Omega$$

or

$$jX_L = \mathbf{2\ \angle 90^\circ\ k\Omega}$$

Example 1-8
The coil of Ex. 1-7 has an inductance of 10 mH. Find the frequency of the operating voltage in kilohertz.

Solution

$$f = \frac{X_L}{2\pi L} \tag{1-14}$$

$$= \frac{2\text{ k}\Omega}{2\pi(10 \times 10^{-3})}$$

$$= \mathbf{31.83\ kHz}$$

Without going into detail, it should serve as a sufficient reminder that X_c and X_L have meanings only if the operating voltage (current) is sinusoidal in nature.

In practical switching circuits, use of an inductor is avoided whenever possible. However, in instances where an inductor must be used, the current in it is usually neither a constant **dc** nor a sinusoidal **ac**. Instead, it is a *time-varying* quantity. In a typical switching-circuit application, the current in an inductor starts from some initial value and then eventually reaches its final steady-state value. The manner in which current goes from its initial value to its final value is called its *transient* response. In switching applications, we are usually concerned with the initial current, the final current, and the transient response. These and other aspects of switching characteristics of an inductor are discussed in Ch. 3. For the present, it is important to note that:

1. Current in an inductor *cannot* be changed instantaneously.
2. Voltage across an inductor is directly proportional to its inductance and the rate of change of current in it (Eq. 1-11).

1-10 Kirchoff's Laws

Kirchhoff's voltage law (KVL) and Kirchhoff's current law (KCL) are equally important as extensions of Ohm's law. The KVL states that around a closed loop (path) in an electric circuit, the sum of potential *drops* across various components is equal to the sum of applied emf's (rises) in that loop. Consider the circuit shown in Fig. 1-4. The circuit contains three resistors (components), R_1, R_2, and R_3. The emf's applied to this circuit are E_1 and E_2. Note that E_1 and E_2 are the terminal voltages of the two power sources. Let us say that the voltage drop across R_1 is V_1, across R_2 is V_2, and across R_3 is V_3. Then, according to KVL:

$$V_{R_1} + V_{R_2} + V_{R_3} = E_1 + E_2 \text{ volts (V)} \qquad (1\text{-}15)$$

Note that if the two power sources are connected in series (aiding), then their overall combined voltage is equated to the *sum* of the

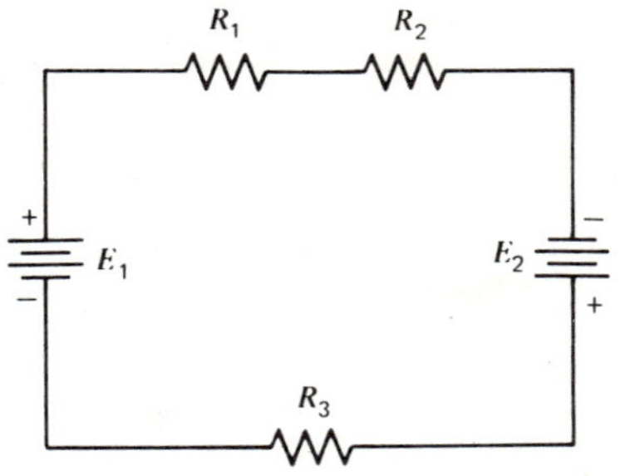

Figure 1-4 Multisource circuit illustrating KVL.

two. If the two power sources are connected such that their polarities *oppose* each other, then their overall combined voltage is equivalent to the *difference* between the two. The sources shown in Fig. 1-4 are connected in *series* (*aiding*) *and therefore are added.*

Example 1-9

Refer to the circuit of Fig. 1-4. Assume $V_{R_1} = 50$ V, $E_1 = 100$ V, $V_{R_2} = 80$ V, $E_2 = 50$ V. Find V_{R_3}.

Solution

Using KVL:

$$V_{R_1} + V_{R_2} + V_{R_3} = E_1 + E_2 \qquad (1\text{-}15)$$

$$50 + 80 + V_{R_3} = 100 + 50$$

$$V_{R_3} = \mathbf{20\ V}$$

Example 1-10

Figure 1-5 shows a simple electric circuit. Find V_a, V_b, and V_{ac}. Note: V_a means the voltage at point a with respect to ground. V_{ac} means the voltage at point a with respect to the voltage at point c.

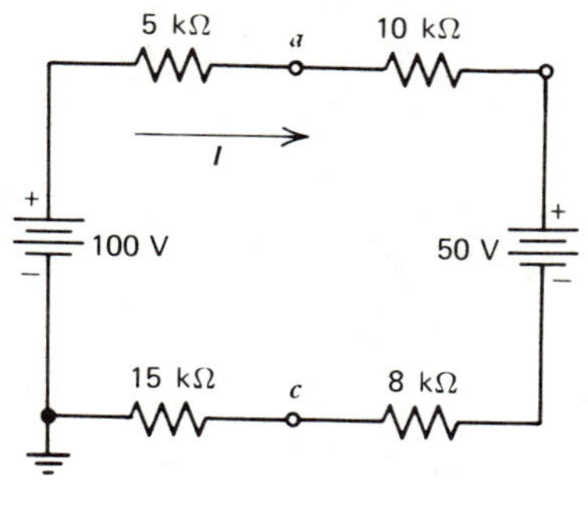

Figure 1-5 Circuit for Ex. 1-10.

Solution

Let I represent the current in the circuit. Then, according to the KVL:

$$V_{5k} + V_{10k} + V_{8k} + V_{15k} = (E_1 + E_2) \qquad (1\text{-}15)$$

$$I \times 5\,\text{k}\Omega + I \times 10\,\text{k}\Omega + I \times 8\,\text{k}\Omega + I \times 15\,\text{k}\Omega = 100 - 50$$

$$I \times 38\,\text{k}\Omega = 50\text{ V}$$

$$I = 1.32\text{ mA}$$

$$V_a = 100 - (5\,\text{k}\Omega \times 1.32\text{ mA})$$

$$= \mathbf{93.4\ volts}$$

$$V_b = 93.4 - (10\,\text{k}\Omega \times 1.32\text{ mA})$$

$$= \mathbf{80.2\ volts}$$

$$V_c = (15\,\text{k}\Omega \times 1.32\text{ mA})$$

$$= \mathbf{19.8\ volts}$$

$$V_{ac} = V_a - V_c$$

$$= 93.4 - 19.8$$

$$= \mathbf{73.6\ volts}$$

Kirchhoff's current law (KCL) states that the algebraic sum of currents at an electrical junction is always equal to zero. This means that at any time, the total current flowing out of a junction must be equal to the total current flowing into the junction. Figure

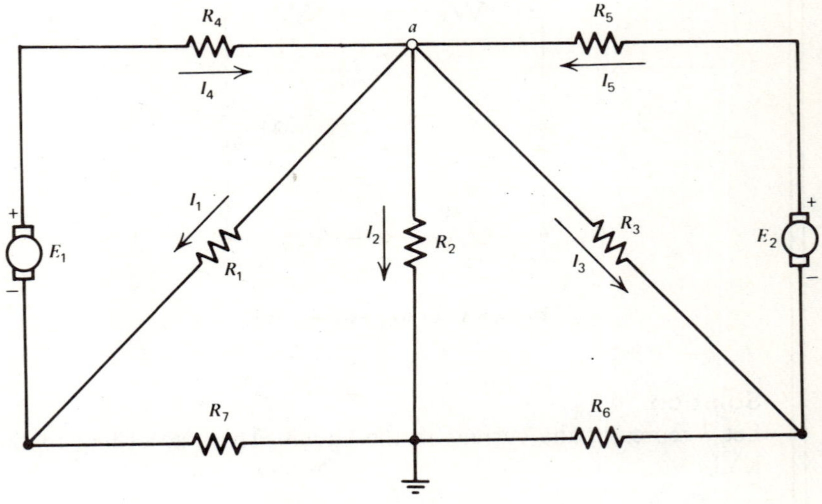

Figure 1-6 Circuit to illustrate KCL.

1-6 shows three currents, I_1, I_2, and I_3, flowing out of the junction a, and two currents, I_4 and I_5, flowing into it.

$$I_1 + I_2 + I_3 = I_4 + I_5 \text{ amperes (A)} \tag{1-16}$$

or

$$I_1 + I_2 + I_3 - (I_4 + I_5) = 0$$

Example 1-11

Figure 1-7 shows a simple electric circuit. Find the junction voltage V_a.

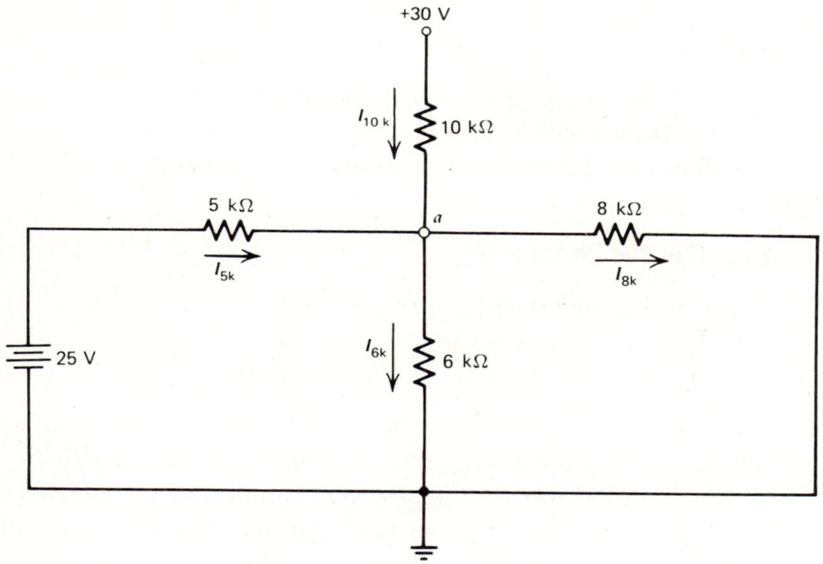

Figure 1-7 Circuit for Ex. 1-11.

Solution

According to KCL:

$$I_{5k} + I_{10k} = I_{6k} + I_{8k} \tag{1-16}$$

Then, by Ohm's law (Eq. 1-1),

$$\frac{25 - V_a}{5\text{ k}\Omega} + \frac{30 - V_a}{10\text{ k}\Omega} = \frac{V_a}{6\text{ k}\Omega} + \frac{V_a}{8\text{ k}\Omega}$$

Solving for V_a yields

$$V_a = \mathbf{13.5\ volts}$$

Note: The algebraic solution for V_a is left as an exercise for the reader.

1-11 Series and Parallel Circuits

Three basic circuit components—resistance, capacitance, and inductance—have been discussed. Note that Eqs. (1-4), (1-9), and (1-13) may be written as:

$$V = I \cdot Z \text{ volts (V)} \qquad (1\text{-}17)$$

where:

V is the voltage drop across the component (V)
I is the current in the component (A)
Z is the ohmic value of the component called *impedance*, in ohms (Ω)

Note that the impedance

$$\begin{aligned} Z &= R \text{ if the component is a resistor } (\Omega) \\ &= -jX_c \text{ if the component is a capacitor } (\Omega) \\ &= jX_L \text{ if the component is an inductor } (\Omega) \end{aligned}$$

In an electric circuit, two or more impedances are said to be connected in series if the same current passes through them. All of the circuits shown in Fig. 1-8 are examples of series-connected impedances. In each case the *same* current passes through all of the impedances in that circuit. Consider the series-connected *RLC* circuit shown in Fig. 1-8d. According to KVL:

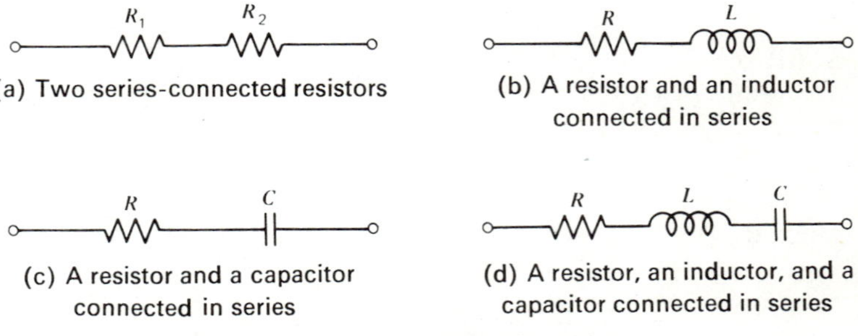

(a) Two series-connected resistors
(b) A resistor and an inductor connected in series
(c) A resistor and a capacitor connected in series
(d) A resistor, an inductor, and a capacitor connected in series

Figure 1-8 Series-connected circuits.

$$
\begin{aligned}
E_i &= V_R + V_L + V_C \\
&= IR + I(jX_L) + I(-jX_c) \quad \text{(using Ohm's law)} \\
&= I(R + jX_L - jX_c) \\
\frac{E_i}{I} &= R + jX_L - jX_c \qquad \text{(using Eq. 1-17)} \\
Z_t &= R + jX_L - jX_c \qquad (1\text{-}18a) \\
&= Z_R + Z_L + Z_c \qquad (1\text{-}18b)
\end{aligned}
$$

where:

- E_i is the applied emf (V)
- V_R is the voltage drop across the resistor (V)
- V_L is the voltage drop across the inductor (V)
- V_c is the voltage drop across the capacitor (V)
- I is the current in the circuit (A)
- jX_L is the impedance of the inductor (Ω)
- $-jX_c$ is the impedance of the capacitor (Ω)
- R is the resistance of the resistor (Ω)
- Z_t is the overall or total impedance of the circuit, as seen by the power source (E_i) (Ω).

Equation (1-17) demonstrates that the total combined impedance of several series-connected impedances is equal to the algebraic sum of their individual impedances.

Example 1-12

Figure 1-9 shows a 15-V power source connected to three series-connected resistors. Calculate the voltage across each resistor.

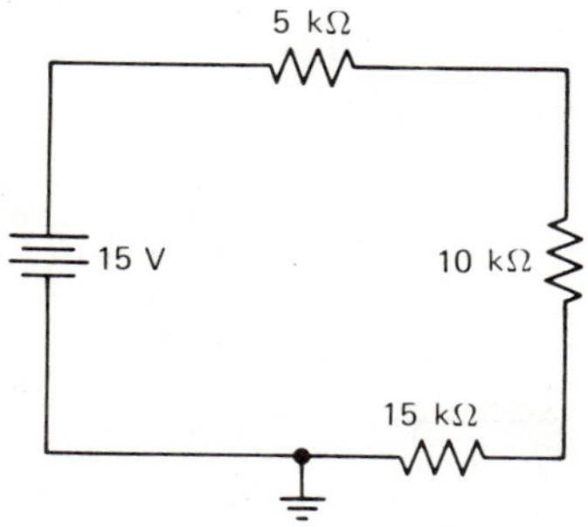

Figure 1-9 Circuit for Ex. 1-12.

Solution

$$R_t = 10\ \text{k}\Omega + 5\ \text{k}\Omega + 15\ \text{k}\Omega \qquad (1\text{-}17)$$
$$= 30\ \text{k}\Omega$$
$$I_t = \frac{V_t}{R_t}$$
$$= \frac{15V}{30\ \text{k}\Omega}$$
$$= 0.5\ \text{mA}$$
$$V_{10k} = 0.5\ \text{mA} \times 10\ \text{k}\Omega$$
$$= \mathbf{5\ V}$$
$$V_{5k} = 0.5\ \text{mA} \times 5\ \text{k}\Omega$$
$$= \mathbf{2.5\ V}$$
$$V_{15k} = 0.5\ \text{mA} \times 15\ \text{k}\Omega$$
$$= \mathbf{7.5\ V}$$

The circuit shown in Fig. 1-10 is an example of parallel-connected impedances. Impedances Z_1 and Z_2 are connected in parallel with each other, and impedances Z_3, Z_4, and Z_5 are also connected in parallel with each other. Note that the same current, I_t, enters and comes out of each parallel combination. Thus it can be concluded that the current that enters a parallel-connected combination of impedances is equal to the one that comes out of it. Also note that

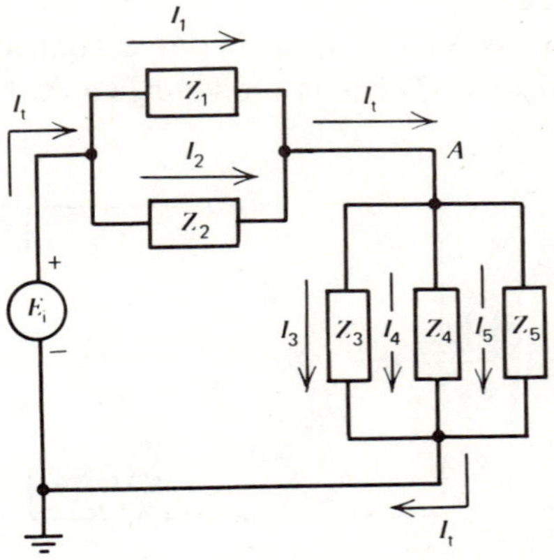

Figure 1-10 Parallel-connected impedances.

parallel-connected impedances are connected together at both ends. Therefore, the voltage drop across each impedance is the same in a set of parallel-connected impedances. Consider junction *A* in Fig. 1-10. In accordance with KCL:

$$I_t = I_3 + I_4 + I_5 \tag{1-16}$$

$$\frac{E_A}{Z_{eq}} = \frac{V_3}{Z_3} + \frac{V_4}{Z_4} + \frac{V_5}{Z_5}$$

But $E_A \equiv V_3 \equiv V_4 \equiv V_5$. Therefore,

$$\frac{1}{Z_{eq}} = \frac{1}{Z_3} + \frac{1}{Z_4} + \frac{1}{Z_5} \text{ siemens (S)} \tag{1-19}$$

where:

I_t is the total current entering the parallel combination (A)
I_3 is the current flowing in Z_3 (A)
I_4 is the current flowing in Z_4 (A)
I_5 is the current flowing in Z_5 (A)
E_A is the potential at point *A* (V)
V_3 is the voltage drop across Z_3 (V)
V_4 is the voltage drop across Z_4 (V)
V_5 is the voltage drop across Z_5 (V)
Z_{eq} is the overall total combined impedance of the parallel combination of Z_3, Z_4, and Z_5 (Ω)

Similarly, the equivalent impedance of the parallel combination of Z_1 and Z_2 is:

$$\frac{1}{Z_{1,2}} = \frac{1}{Z_1} + \frac{1}{Z_2} \text{ siemens (S)}$$

Solving for $Z_{1,2}$ yields:

$$Z_{1,2} = \frac{Z_1 \cdot Z_2}{Z_1 + Z_2} \text{ ohms } (\Omega) \tag{1-20a}$$

Often we encounter two resistances connected in parallel. In that case, the Z's in Eq. (1-20a) are replaced by corresponding R's. Accordingly:

$$R_{1,2} = \frac{R_1 \cdot R_2}{R_1 + R_2} \text{ ohms } (\Omega) \qquad (1\text{-}20b)$$

For the most part, we will be working with **ac** circuits containing only resistances. Calculations for equivalent impedance become considerably simpler because of the 0° phase angle in that case. Example 1-13 below is intended only to indicate the complex nature of calculations when there are capacitances and inductances in an **ac** electric circuit. The reader should acquire a working knowledge of the j operator from elsewhere if he/she finds it difficult to follow Ex. 1-13.

Example 1-13

Consider the circuit shown in Fig. 1-10. Assume $Z_1 = j600$ ohms, $Z_2 = -j500$ ohms, $Z_3 = 680$ ohms, $Z_4 = 566$ ohms, and $Z_5 = 700$ ohms. The input E_i is a 25-V_{rms}, 10-kHz sine wave. Find:

a. Equivalent impedance of Z_1 and Z_2.
b. Equivalent impedance of Z_3, Z_4, and Z_5.
c. Overall total impedance of the circuit.
d. Current drawn from the source by the entire circuit.

Solution

Note from the impedance angles that Z_1 is an inductor, Z_2 is a capacitor, and Z_3, Z_4, and Z_5 are resistors.

a.

$$Z_{eq} = \frac{Z_1 \cdot Z_2}{Z_1 + Z_2} \qquad (1\text{-}20)$$

$$= \frac{j600 \cdot (-j500)}{j600 + (-j500)}$$

$$= \frac{300 \times 10^4}{j100}$$

$$= -j3 \text{ k}\Omega$$

$$= \mathbf{3\,\underline{/-90^\circ}\ k\Omega}$$

b.

$$\frac{1}{Z_{eq}} = \frac{1}{Z_3} + \frac{1}{Z_4} + \frac{1}{Z_5}, \text{ siemens (S)} \qquad (1\text{-}19)$$

$$= \frac{1}{680} + \frac{1}{566} + \frac{1}{700}$$

$$= 46.67 \times 10^{-4} \text{ S}$$

or

$$Z_{eq} = \frac{1}{46.67\ 10^{-4} \text{ S}}$$

$$= \mathbf{214.3\ \angle 0^\circ\ \Omega}$$

c.

$$Z_t = Z_{eq1} + Z_{eq2} \text{ ohms } (\Omega) \qquad (1\text{-}18)$$

$$= 214.3 - j3000$$

$$= \mathbf{3.01\ \angle -85.9^\circ\ k\Omega}$$

d.

$$I = \frac{E_i}{Z_t} \text{ amperes (A)} \qquad (1\text{-}17)$$

$$= \frac{25\ \angle 0^\circ}{3.01\ \angle -85.9^\circ}$$

$$= \mathbf{8.31\ \angle 85.9^\circ\ mA}$$

1-12 Superposition Theorem

Often, circuits are encountered containing *more* than one power source. To analyze and assess the combined effect of *all* of the power sources on a load may become a tedious and cumbersome job. The Superposition theorem simplifies the task in such cases. It states that the current (voltage) in an element in an electrical circuit containing two or more power sources is equal to the algebraic sum of individual currents (or voltages) resulting from each source taken independently (while the other voltage sources are shorted and current sources are open-circuited).

Figure 1-11 shows an electrical circuit containing two power sources, E_1 and E_2. Suppose that we want to find the current in the 10-kΩ resistor. According to the Superposition theorem, the current in it is really made up of two currents, one resulting from E_1 and the other resulting from E_2.

The current from E_1 is calculated by analyzing the circuit when the source voltage E_2 is replaced by its internal resistance. Then, in a similar manner, the current from E_2 is calculated by replacing

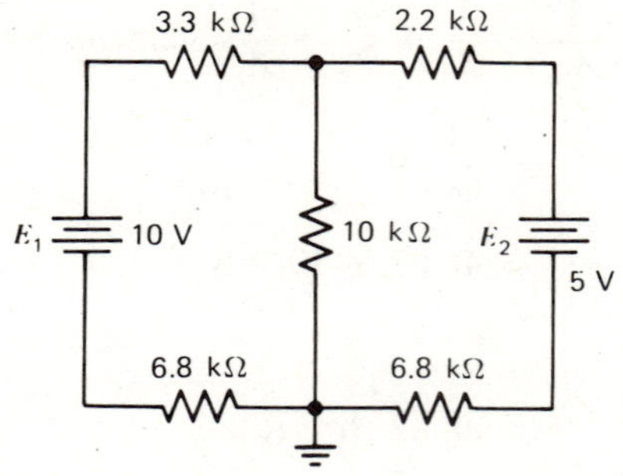

Figure 1-11 Circuit illustrating Superposition theorem.

E_1 by its internal resistance. *The total current is the algebraic sum of these two currents.*

Example 1-14

Refer to the circuit of Fig. 1-11. Find the total current flowing in the 10-kΩ resistor, using Superposition.

Solution

Let us first find I_{10k} as a result of E_1. The equivalent circuit is drawn in Fig. 1-12a. Note that the 5-V source is replaced by a short circuit.

$$I_1 = \frac{10 \text{ V}}{3.3 \text{ k}\Omega + (10 \parallel 9) \text{ k}\Omega + 6.8 \text{ k}\Omega}$$

$$= 0.675 \text{ mA}$$

Thus,

$$I_{10k1} = \frac{0.675 \text{ mA } (10 \parallel 9) \text{ k}\Omega}{10 \text{ k}\Omega}$$

$$= 0.32 \text{ mA}$$

This is the current in the 10-kΩ resistor resulting from the 10-V source.

To find I_{10k2} as a result of E_2, refer to Fig. 1-12b. Note that the 10-V source is replaced by a short circuit.

$$I_2 = \frac{5 \text{ V}}{2.2 \text{ k}\Omega + (10 \parallel 10.1) \text{ k}\Omega + 6.8 \text{ k}\Omega}$$

$$= 0.355 \text{ mA}$$

$$I_{10k2} = \frac{0.355 \text{ mA} \times (10 \parallel 10.1) \text{ k}\Omega}{10 \text{ k}\Omega}$$

$$= 0.18 \text{ mA}$$

This is the current in the 10-kΩ resistor resulting from the 5-V source.

Having found individual currents, then,

$$\text{Total } I_{10k} = I_{10k1} + I_{10k2}$$
$$= (0.32 + 0.18) \text{ mA}$$
$$= 0.50 \text{ mA}$$

This is the actual current in the 10-kΩ resistor resulting from both of the sources.

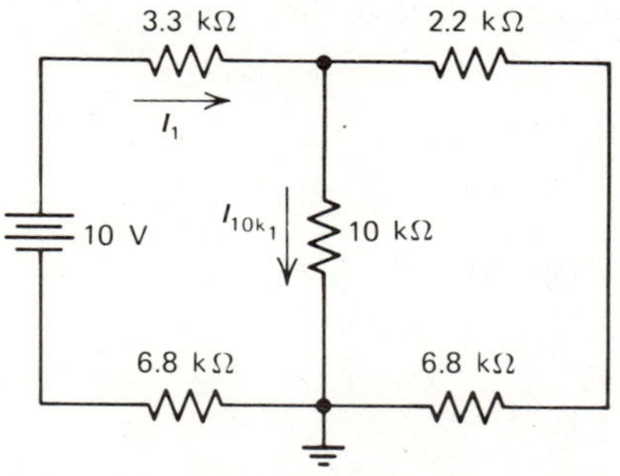

(a) Equivalent circuit as seen by 10 V supply

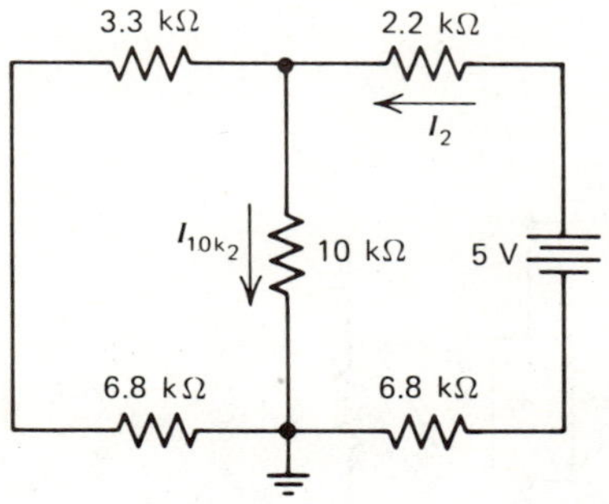

(b) Equivalent circuit as seen by 5 V supply

Figure 1-12 Circuit for Ex. 1-14.

Example 1-15

Refer to the circuit of Fig. 1-13. Find the current flowing in the 200-ohm load.

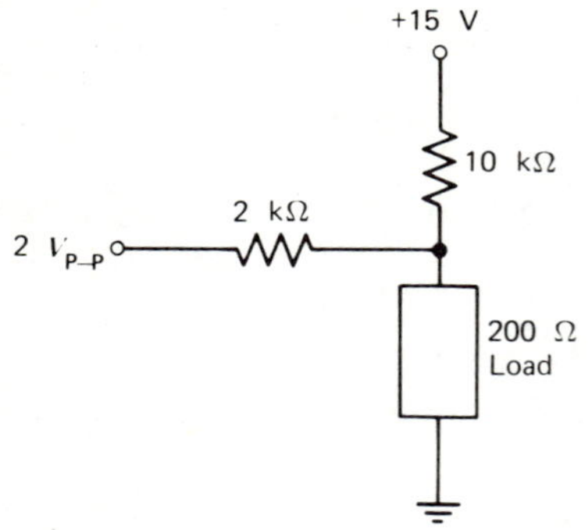

Figure 1-13 Circuit for Ex. 1-15.

Solution

Figure 1-14a shows an equivalent circuit that is used to calculate the **dc** current in the 200-ohm load.

$$I_{\mathbf{dc}} = \frac{15\text{ V}}{10\text{ k}\Omega + (2\text{ k}\Omega \parallel 200)}$$

$$= \frac{15\text{ V}}{10{,}182\ \Omega}\ (2\text{ k}\Omega \parallel 200\ \Omega = 182\ \Omega)$$

$$= 1.47\text{ mA }\mathbf{dc}$$

$$I_{\mathbf{dc}}(200\ \Omega) = \frac{180}{200} \times 1.47\text{ mA} = 1.32\text{ mA}$$

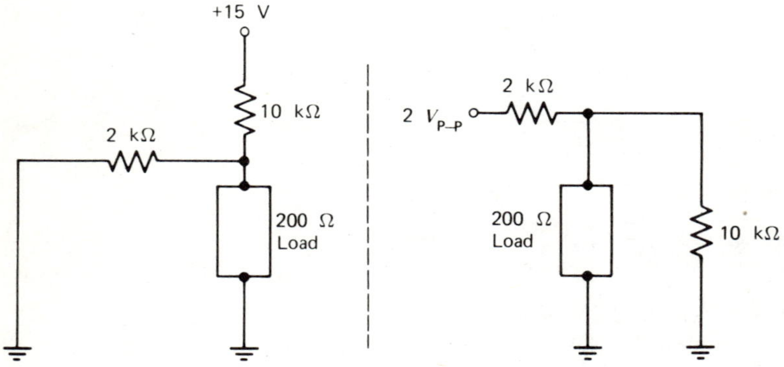

(a) Equivalent circuit used to calculate current as a result of +15 V source only

(b) Equivalent circuit used to calculate current as a result of 2 $V_{p\text{-}p}$ source only

Figure 1-14 Equivalent circuits for using Superposition theorem.

Figure 1-14b is the equivalent circuit that is used to calculate the **ac** current flowing in the 200-ohm load.

$$I_{\mathbf{ac}} = \frac{2\ \mathrm{V_{p\text{-}p}}}{2\ \mathrm{k\Omega} + (10\ \mathrm{k\Omega} \parallel 200)}$$

$$= \frac{2\ \mathrm{V}}{2196\ \Omega} \quad (10\ \mathrm{k} \parallel 200 = 196\ \Omega)$$

$$= 0.91\ \mathrm{mA_{p\text{-}p}}$$

$$I_{\mathbf{ac}}(200\ \Omega) = \frac{196}{200} \times 0.91 = 0.89\ \mathrm{mA_{p\text{-}p}}$$

The total current in the 200-ohm load is:

$$I_{\mathrm{t}} = I_{\mathbf{dc}} + I_{\mathbf{ac}}$$

or

$$= 1.32\ \mathrm{mA}\ (\mathbf{dc}) + 0.89\ \mathrm{mA_{p\text{-}p}}\ (\mathbf{ac})$$

$$= (1.32 \pm 0.445)\ \mathrm{mA}$$

$$= \mathbf{+1.765\ mA}\ \text{to}\ \mathbf{+0.875\ mA\ excursions}$$

1-13 Thévenin's Theorem

Sometimes it is easier to analyze the circuit response of a load to a source if the circuit can be represented by a single power source and an equivalent impedance connected in series with it. Thévenin's theorem does just that. It states that any complex linear electric circuit can be represented by an equivalent circuit consisting of only one voltage source and only one impedance in series with it.

Figure 1-15a shows a simple electric circuit and Fig. 1-15b shows its Thévenin's equivalent circuit. Both of these circuits will react in a similar manner toward the load (R_L) because one is equivalent to the other. The next problem is to find the values of V_{th} and R_{th}.

To find the values of V_{th} and R_{th}, the load is disconnected from the rest of the circuit. Then V_{th} is the open-circuit voltage that appears between the load terminals. Figure 1-16a shows this circuit. The open-circuit voltage between the load terminals (*t* and *h*) is:

$$V_{th} = \left(\frac{R_2}{R_1 + R_2}\right) E_i \text{ volts (V)} \qquad (1\text{-}21)$$

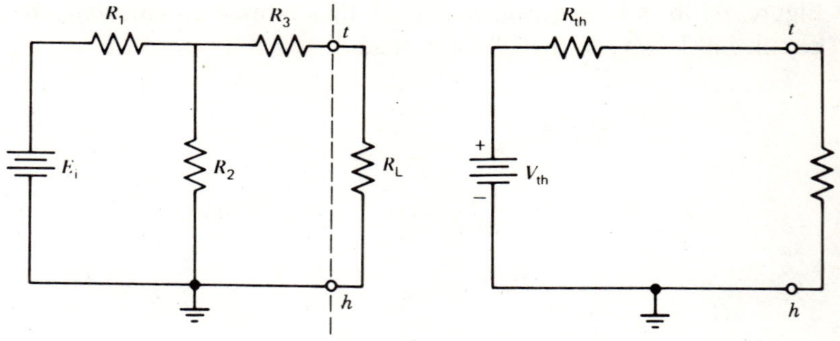

(a) A circuit (b) Thévenin's equivalent

Figure 1-15 Circuit to demonstrate Thévenin's equivalent.

Note that there is no voltage drop across R_3.

A circuit for calculating R_{th} is shown in Fig. 1-16b. R_{th} is the open-circuit impedance that appears between the load terminals when the voltage source is replaced by its internal resistance. In this case,

$$R_{th} = R_3 + (R_1 \parallel R_2) \text{ ohms } (\Omega) \tag{1-22}$$

For the most part, circuits that will be analyzed in this book contain only resistances. Thus, the term R_{th} is used instead of Z_{th}. It is to be noted that Thévenin's theorem applies to **ac** as well as to **dc** circuits.

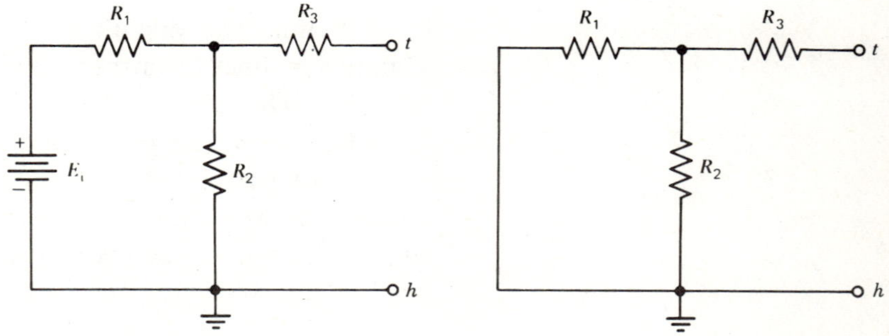

(a) Equivalent circuit for V_{th} (b) Equivalent circuit for R_{th}

Figure 1-16 Equivalent circuits for calculating V_{th} and R_{th}.

All circuits can be analyzed by a variety of other circuit analysis techniques. However, the reader will find that in some cases Thévenin's theorem makes the circuit analysis much easier.

Example 1-16
Find the Thévenin's equivalent for the circuit shown in Fig. 1-17.

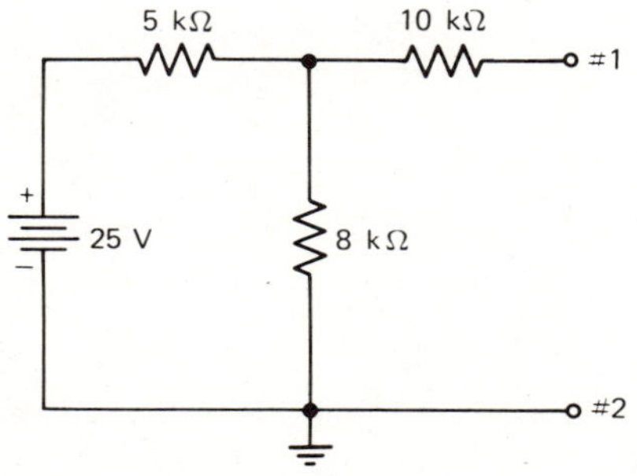

Figure 1-17 Circuit for Ex. 1-16.

Solution
The Thévenin's equivalent circuit is shown in Fig. 1-18a. The open circuit voltage (V_{th}) between terminals #1 and #2 is really the voltage drop across the 8 kΩ resistor. Thus

$$V_{th} = \frac{8\ \text{k}\Omega}{5\ \text{k}\Omega + 8\ \text{k}\Omega} \times 25\ \text{V}$$
$$= \mathbf{15.4\ V}$$

Figure 1-18c shows the circuit for finding R_{th}. Thus

$$R_{th} = 10\ \text{k}\Omega + (5 \parallel 8)\ \text{k}\Omega$$
$$= \mathbf{13.1\ k\Omega}$$

Thus in Fig. 1-18a, $V_{th} = 15.4$ V and $R_{th} = 13.1$ kΩ.

Example 1-17
A load of 5 kΩ is connected between terminals #1 and #2 of Fig. 1-17. Find the current in and the voltage across it.

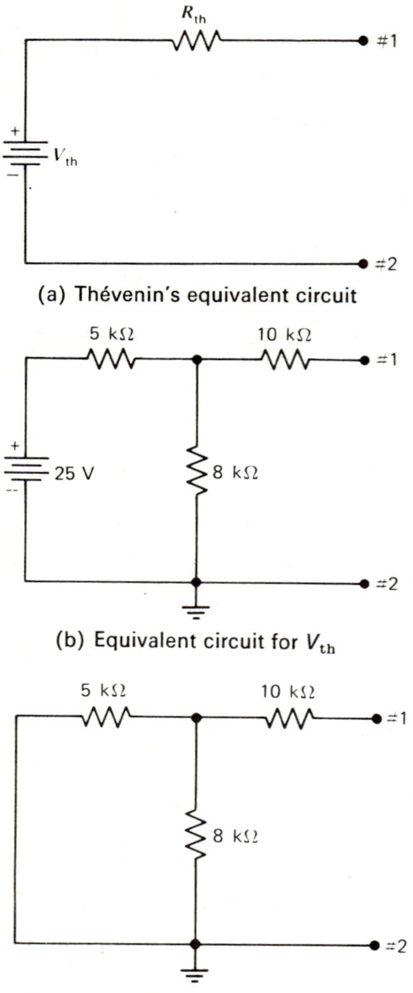

(a) Thévenin's equivalent circuit

(b) Equivalent circuit for V_{th}

(c) Equivalent circuit for R_{th}

Figure 1-18 Circuits for Thévenin's equivalent.

Solution

Using the Thévenin's equivalent circuit of Ex. 1-16, the circuit shown in Fig. 1-19 is drawn. Thus

$$I_{5k} = \frac{15.4\text{ V}}{13.1\text{ k}\Omega + 5\text{ k}\Omega}$$

$$= \mathbf{0.85\ mA}$$

$$V_{5k} = 0.85\text{ mA} \times 5\text{ k}\Omega = \mathbf{4.25\ V}$$

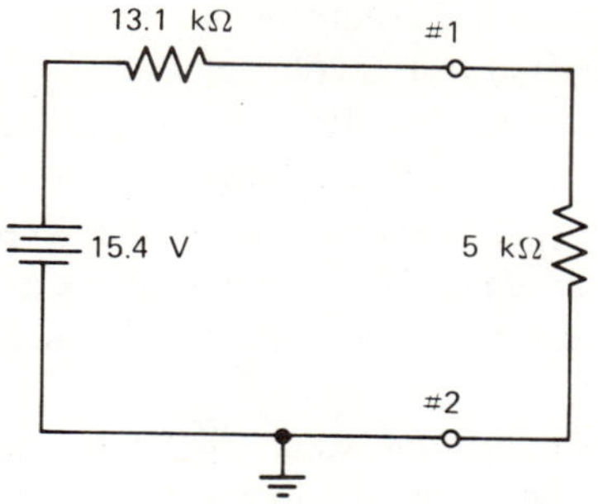

Figure 1-19 Circuit for Ex. 1-17.

The current can also be calculated without using the Thévenin's equivalent circuit, in the following manner. Consider the original total circuit:

$$R_t = 5\text{ k}\Omega + 8\text{ k}\Omega \parallel (10\text{ k}\Omega + 5\text{ k}\Omega)$$
$$= 5\text{ k}\Omega + 8\text{ k}\Omega \parallel 15\text{ k}\Omega$$
$$= 5\text{ k}\Omega + 5.2\text{ k}\Omega$$
$$= 10.2\text{ k}\Omega$$
$$I_t = \frac{25\text{ V}}{10.2\text{ k}\Omega}$$
$$= 2.45\text{ mA}$$

Voltage drop across the $15\text{ k}\Omega \parallel 8\text{ k}\Omega$ combination:

$$V = 2.45\text{ mA} \times 5.2\text{ k}\Omega$$
$$= 12.8\text{ volts}$$

Thus

$$I_{5k} = \frac{12.8\text{ V}}{15\text{ k}\Omega}$$
$$= \mathbf{0.85\text{ mA}}$$

which confirms the results obtained by using the Thévenin equivalent circuit above.

1-14 Summary

There are many laws and theorems used to analyze electronic circuits in general. For analyzing pulse and switching circuits in particular, the following are frequently used:

1. *Ohm's law* states that the current in a component (resistor) is always directly proportional to the voltage across it.
2. *Kirchhoff's voltage law* (KVL) states that around a closed loop (path) in an electric circuit, the algebraic sum of voltage drops across various components is equal to the sum of applied emf's in that loop.
3. *Kirchhoff's current law* (KCL) states that the algebraic sum of currents (entering and leaving) at any electrical junction is always equal to zero.
4. *The Superposition theorem* states that the voltage across (or current through) a component in an electrical circuit containing two or more power sources can be calculated by taking the algebraic sum of the voltages across (or currents through) that component resulting from each individual power source.
5. *Thévenin's theorem* states that a two-terminal electrical circuit can be replaced by an equivalent circuit containing only one power source and only one resistance connected in series with it. The equivalent one-power source (V_{th}) is the open-circuit voltage between the two terminals and the equivalent resistance (R_{th}) is the resistance that appears between the two terminals when the power sources in the original circuit are replaced by their internal resistances.
6. It is also worth noting that:
   a. The above theorems and laws hold true for **dc** as well as for sinusoidal **ac** circuits.
   b. A pure inductor behaves like a short circuit and a capacitor behaves like an open circuit to **dc**.
   c. When the operating voltage is a sinusoidal **ac**,
      (1) The reactance of an inductor $(X_L) = +j2\pi fL = 2\pi fL \angle 90^\circ$.
      (2) The reactance of a capacitor $(X_c) = -j(1/2\pi fC) = (1/2\pi fC) \angle -90^\circ$.
7. Important equations:

$$I = \frac{Q}{t} \tag{1-1}$$

$$i = \frac{dq}{dt} \tag{1-2}$$

$$V = I \cdot R \tag{1-3}$$

$$P = V \cdot I \tag{1-4}$$

$$Q = C \cdot V \tag{1-5}$$

$$i = C\frac{dv}{dt} \tag{1-6}$$

$$v = \frac{1}{C}\int i \cdot dt \tag{1-7}$$

$$v = \frac{I}{C} \cdot t \tag{1-8}$$

$$X_c = \frac{v_c}{I_c} \tag{1-9}$$

$$-jX_c = -j\frac{1}{2\pi fC} \tag{1-10}$$

$$v_L = L\frac{di}{dt} \tag{1-11}$$

$$i_L = \frac{1}{L}\int v\, dt \tag{1-12}$$

$$jX_L = \frac{v_L}{i_L} \tag{1-13}$$

$$X_L = 2\pi fL \tag{1-14}$$

$$V_{R_1} + V_{R_2} + V_{R_3} = E_1 + E_2 \tag{1-15}$$

$$I_1 + I_2 + I_3 = I_4 + I_5 \tag{1-16}$$

$$V = I \cdot Z \tag{1-17}$$

$$Z_t = Z_1 + Z_2 + Z_3 \tag{1-18}$$

$$\frac{1}{Z_{eq}} = \frac{1}{Z_1} + \frac{1}{Z_2} + \frac{1}{Z_3} \tag{1-19}$$

$$R_1 \parallel R_2 = \frac{R_1 \cdot R_2}{R_1 + R_2} \tag{1-20}$$

1-15 Glossary

Capacitance. A property of a component that opposes the change in voltage across it. Its unit of measurement is the farad.

Impedance. The term is applicable only to **ac** circuits. It is the measure of total opposition that a circuit containing *RL*, *RC*, or *RLC* presents to the current (**ac**) that flows through it. Its unit of measurement is the ohm. It has a magnitude as well as phase angle. It is a complex quantity.

Inductance. A property of a component that opposes the change in current through it. Its unit of measurement is the henry.

Reactance. The term is applicable only to **ac** circuits. It is the measure of opposition that a capacitor or an inductor presents to the sinusoidal **ac** current that flows through it. Disregarding the phase angle, it is the **ac** equivalent of resistance in **dc** circuits. There is the capacitive reactance (X_c) and the inductive reactance (X_L). Its unit of measurement is the ohm.

Resistance. It opposes the flow of current and dissipates electrical energy as heat. Its unit of measurement is the ohm.

1-16 Recapitulation

Return to the Objectives and the Self-Evaluation Questions given at the beginning of this chapter and see how well you can answer them. If you cannot answer certain questions satisfactorily, place a check mark (√) next to them and review appropriate parts of the chapter to find the answers. No attempt should be made to solve the Problems (Sec. 1-17), before answering the Self-Evaluation Questions. Ability to solve the Problems should serve as feedback regarding your understanding of the material presented in this chapter. Answers to the Problems are given on p. 513.

1-17 Problems

1-1 Consider the circuit shown in Fig. 1-20. Find the current in the 8-kΩ resistor if $E_i = 10$ V.

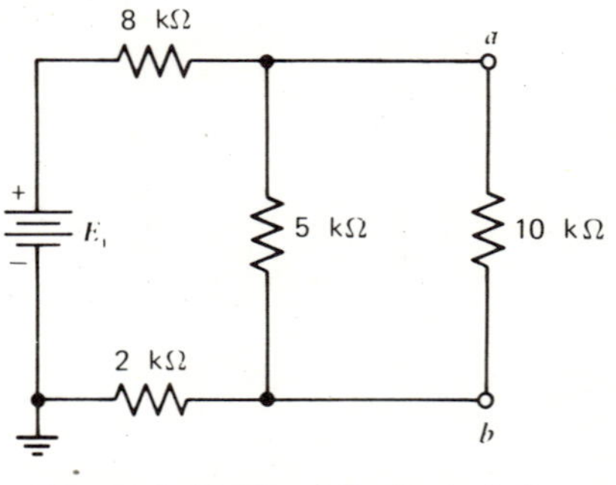

Figure 1-20 Circuit for Prob. 1-1.

1-2 Consider again the circuit shown in Fig. 1-20. Current in the 2-kΩ resistor is 2.5 mA. Find I_{8k}, I_{5k}, and I_{10k}.

1-3 Refer to Prob. 1-2 above. Find E_i, V_{8k}, V_a, V_{ab}.

1-4 Consider the circuit shown in Fig. 1-21. Find I_{5k}, I_{3k}, I_{6k}, and I_{15k}.

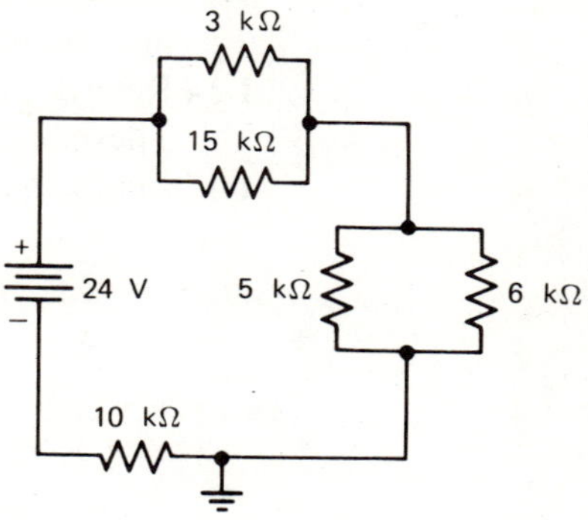

Figure 1-21 Circuit for Prob. 1-4.

1-5 A 0.047-μF capacitor has 400 V **dc** across it. Find the charge on the capacitor.

1-6 A **dc** current of 0.25 mA is charging a 0.1-μF capacitor. Calculate the charge on the capacitor (in volts) five seconds after it has been charging.

1-7 A 120-V, 60-hertz source is connected across a 0.5-μF capacitor. Calculate the current in the capacitor.

1-8 A 10-V, 200-kHz input is applied to a series-connected *RC* circuit where $R = 15$ MΩ and $C = 100$ pF. Find the current in the circuit.

1-9 Refer to Prob. 1-8. It is desired that the voltage drop across the capacitor be $\frac{1}{10}$ of the voltage drop across the resistor. Find the value for another capacitor that when connected in place of the 100 pF capacitor, produces the desired results.

1-10 A 12-V, 1-MHz input is applied to a series-connected *RL* circuit where $R = 50$ kΩ and $L = 10$ mH. Find the reactance of the coil.

1-11 Refer to Prob. 1-10. Find the voltage across the coil.

1-12 Figure 1-22 shows a simple electrical circuit that contains two power

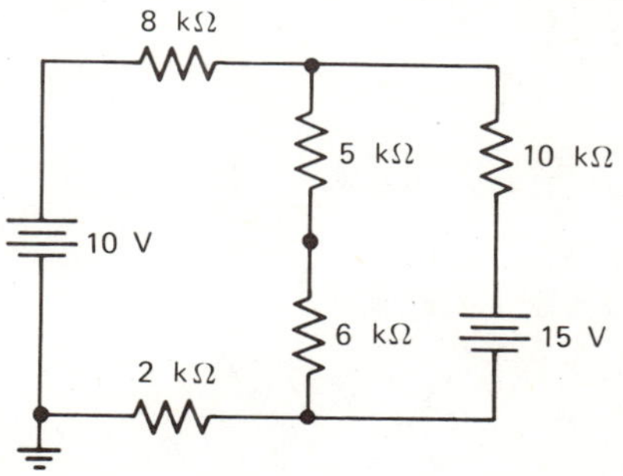

Figure 1-22 Circuit for Prob. 1-12.

sources, 10 V and 15 V. Using Superposition theorem, find the current in the 2-kΩ resistor.

1-13 Refer to the circuit of Fig. 1-22 again. Using Thévenin's theorem, find the voltage drop across the 6-kΩ resistor.

1-14 Refer to question 1-12. Find the current in the 2-kΩ resistor, using Thévenin's theorem.

1-15 Figure 1-23 shows an electrical circuit. Using Thévenin's theorem,

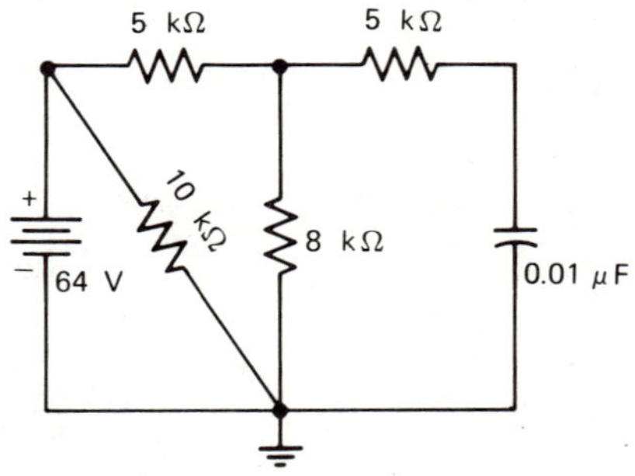

Figure 1-23 Circuit for Prob. 1-15.

reduce it to a simple equivalent series-connected *RC* circuit with a single input voltage source.

1-18 Experiment

Title **Basic Electrical Laws and Circuit Theorems**

Theory Ohm's law relates the current in, the voltage across, and the impedance (resistance) of an electrical component in a circuit. Mathematically stated:

$$V = I \cdot R \tag{1-4}$$

or

$$V = I \cdot Z \tag{1-17}$$

Kirchhoff's voltage law (KVL) states that around a closed loop (path) in an electric circuit, the sum of voltage drops across various components is equal to the algebraic sum of the applied emf's in that loop. Stated in equation form:

$$V_{R_1} + V_{R_2} + V_{R_3} + \cdots + V_{R_n} = E_1 + E_2 + \cdots + E_n \tag{1-15}$$

Kirchhoff's current law (KCL) states that the algebraic sum of currents at an electrical junction is always equal to zero. Stated in equation form:

$$I_1 + I_2 + I_3 \cdots + I_n = 0 \tag{1-16}$$

Thévenin's theorem states that any complex two-terminal electrical circuit can be replaced by an equivalent circuit containing only one voltage source and only one resistance (impedance) connected in series with it. The method used to calculate the equivalent voltage source (V_{th}) and the equivalent series-connected resistance (R_{th}), is explained in Sec. 1-11.

Note that the electrical laws and theorems hold true for **ac** as well as for **dc** circuits. It is important to consider the phase angle when doing **ac** calculations.

$$X_C = 1/2\pi fC \underline{/-90^\circ} \quad (1\text{-}10)$$

$$X_L = 2\pi fL \underline{/90^\circ} \quad (1\text{-}14)$$

Equipment One multimeter having **dc** and **ac** current-measuring capability: Digilin Model 340A, or equal.
One **dc** power supply: ANATEK Model 25-2.00, or equal.
One multifunction generator: EXACT Model 126, or equal.
Resistors: 330 Ω, 470 Ω, 680 Ω, and 1.0 kΩ.
Potentiometer: 1.0 kΩ.
Capacitors: 0.1 μF and 0.01 μF.
Inductor: 30 mH.

Procedure

1. Connect the circuit shown in Fig. E1-1. Measure and record the following currents (I's) and **dc** voltage drops (V's). Record all of the observed data in Table E1-1: I_{330}, I_{1k}, I_{470}, I_{680}, V_{330}, V_{1k}, V_{470}, and V_{680}.

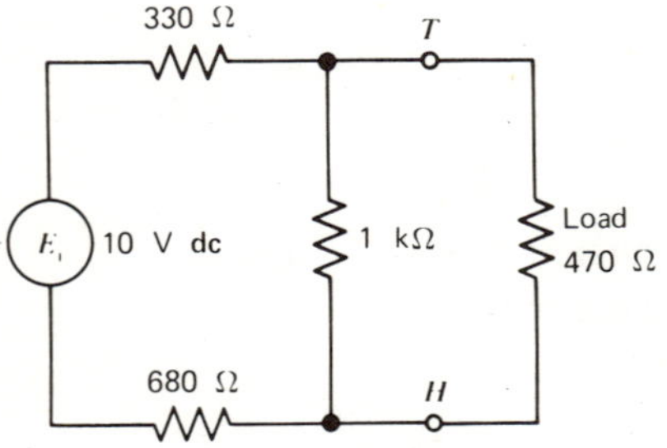

Figure E1-1 A circuit.

2. Change the input to a 10-V_{rms}, 5-kHz sine wave (multifunction generator) and repeat step 1. Use appropriate scales on the multimeter (or an *ac* milliammeter). Record the observed data in Table E1-1.
3. Replace the 470-Ω resistor with a 0.01-μF capacitor and repeat all measurements of step 2. Record the observed data in Table E1-1.
4. Replace the 0.01-μF capacitor with a 0.1-μF capacitor and repeat

step 3 (Table E1-1). Note that now, higher current scales must be used because the load reactance has decreased.

5. Replace the 0.1-μF capacitor (step 4) with a 30-mH coil. Measure and record circuit currents and voltage drops as was done in step 1. Note that here it is I_L and V_L in place of I_{470} and V_{470}. Record your observed results in Table E1-1.
6. Restore the circuit to that of Fig. E1-1. Disconnect the 470-Ω-load resistor. Measure and record the voltage drop across the two output terminals (T and H) to which the load was connected. This is V_{th}. Record it in Table E1-2.
7. Remove the input supply (source) and use a switch to short the two input terminals. Using an ohmmeter, measure and record the resistance between the two output terminals. This is R_{th}. Record it in Table E1-2.
8. Connect the circuit shown in Fig. E1-2. Adjust the **dc** power supply to obtain V_{th} (as measured in step 6). Adjust the potentiometer to obtain R_{th} (as measured in step 7).

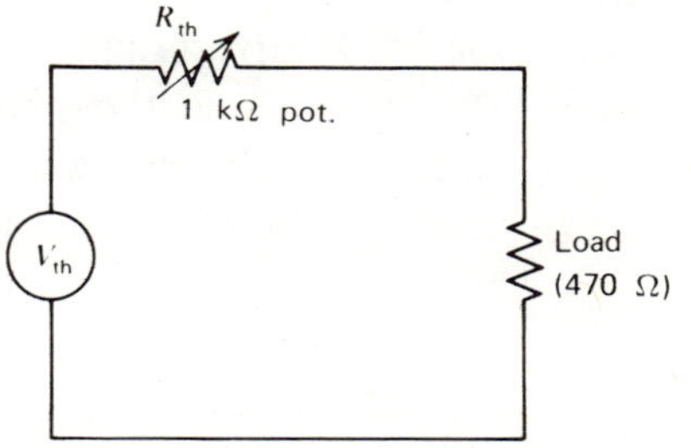

Figure E1-2 Thévenin's equivalent.

9. Measure and record in Table E1-3 the voltage drop appearing across the 470-Ω load.
10. Change the input to a $V_{th(rms)}$, 5-kHz sine wave (multifunction generator). Measure and record (Table E1-3) the voltage drop appearing across the 470-Ω load.
11. Replace the 470-Ω-load resistor with a 0.1-μF capacitor. Measure and record (Table E1-3) the output voltage appearing across the capacitive load.
12. Disconnect and put all test equipment in its proper place.

Calculations

1. Consider the circuit conditions described in step 1 of the Procedure. Using the nominal resistance values and 10 V **dc** as input, calculate various currents and voltage drops in the circuit. Record your theoretically calculated results in Table E1-1.

2. Calculate and record (Table E1-1) various currents and voltage drops as was done in step 1 above, for the circuit conditions described in:
   a. Step 2 of the procedure (Run 2).
   b. Step 3 of the procedure (Run 3).
   c. Step 4 of the procedure (Run 4).
   d. Step 5 of the procedure (Run 5).
3. Using the MEASURED values as the standard values in each Run of Table E1-1, calculate and record (Table E1-1) the percent error between all calculated (theoretical) and measured values.
4. Refer to the circuit of Fig. E1-1. Consider the 470-Ω resistor as load. Calculate and find the Thévenin's equivalent for the rest of the circuit (Fig. E1-2). Enter the calculated values for V_{th} and R_{th} in Table E1-2.
5. Using the measured values as the standard values for V_{th} and R_{th}, calculate the percent error and enter it in Table E1-2.
6. Fill in Table E1-4 from the values recorded in Table E1-1. The Run numbers in Table E1-4 correspond to the Run numbers in Table E1-1. Remember to include the phase angle when calculating current and voltage drops that are not in phase.
7. Complete Table E1-3 from values recorded in Table E1-1.

Tabulation of Results

Table E1-1

Run	Circuit Conditions	Results	I_{330}	I_{1k}	I_{680}	I_{Load}	V_{330}	V_{1k}	V_{680}	V_{Load}
	Calculation step 1	Calculated								
1	Procedure step 1	Measured								
		Percent error								
	Calculation step 2a	Calculated								
2	Procedure step 2	Measured								
		Percent error								
	Calculation step 2b	Calculated								
3	Procedure step 3	Measured								
		Percent error								
	Calculation step 2c	Calculated								
4	Procedure step 4	Measured								
		Percent error								
	Calculation step 2d	Calculated								
5	Procedure step 5	Measured								
		Percent error								

Table E1-2

Results	V_{th}	R_{th}
Calculated		
Measured		
Percent error		

Table E1-3

Circuit Conditions	Results	Voltage Across Load Using Fig. E1-1	Voltage Across Load Using Fig. E1-2
$E_1 = 10\ V_{dc}$	Measured		
Load = 470 Ω	Calculated		
$E_1 = 10\ V_{rms}$	Measured		
Load = 470 Ω	Calculated		
$E_1 = 10\ V_{rms}$	Measured		
Load = 0.1 μF	Calculated		

Table E1-4

Run	Results	KCL		KVL	
		$I_{1k} + I_{load}$	I_{330}	Input Volts	$V_{330} + V_{680} + V_{load}$
1	Calculated			10	
	Measured				
	Percent error				
2	Calculated			10	
	Measured				
	Percent error				
3	Calculated			10	
	Measured				
	Percent error				
4	Calculated			10	
	Measured				
	Percent error				
5	Calculated			10	
	Measured				
	Percent error				

Questions

1. From the calculated percent errors in Tables E1-1, E1-2, and E1-4, describe how your calculated results compare with the experimentally observed ones.
2. In step 4 of the procedure, the load capacitance was increased from 0.01 μF to 0.1 μF. This should have resulted in decreased load

reactance and consequently, a reduction in the voltage drop across it. Do your results (Table E1-1) confirm this? Explain.

3. The tabulation of Table E1-4 is intended to confirm KCL and KVL. Are these laws confirmed? Explain.
4. The tabulation of Table E1-3 is intended to confirm that the Thévenin's theorem does work and that it holds true for **ac** as well as for **dc** circuits. Do the values recorded in Table E1-3 confirm this? Explain.

Discussion

1. Summarize your conclusions from the observed and calculated results of this experiment.
2. Account for serious discrepancies between the observed and calculated results in all tables.
3. Discuss an experimental procedure that might have been used in advance of all voltage and current measurements and that might have reduced the error between the observed and calculated values.

Chapter 2 Common Waveforms

2-1 Objectives

The objectives of this chapter are to:

1. Define and explain the meaning of a *function.*
2. Describe the various functions that are commonly encountered in pulse and switching circuits.
3. Describe and explain the properties of a *cyclical waveform.*
4. Describe the various waveforms commonly encountered in pulse and switching circuits.
5. Compare an *ideal* pulse with a *practical* pulse.
6. Define and explain the rise and fall times (t_r and t_f).
7. Discuss and explain the relationship between the *average* value (**dc**), the *peak* value, and the *peak-to-peak* value of a pulse waveform.
8. Discuss and explain the relationship between the pulse width (t_p), the rest period (t_{rp}), the pulse repetition frequency (PRF), and the pulse repetition time (PRT).
9. Explain the meaning of duty cycle as applied to pulse waveforms.

2-2 Self-Evaluation Questions

Watch for the answers to the following questions as you read the chapter. They should help you to understand the material presented. When you have finished the chapter, return to this section and be sure that you can answer all the questions. Then test your understanding by working out solutions to the Problems at the end of the chapter.

It is strongly recommended that these questions are satisfactorily answered before attempting to solve the Problems (Sec. 2-19).

1. Explain the meaning of a function.
2. Draw graphs to show your understanding of a step function, a ramp function, and an exponential function.
3. Define a cyclical waveform.
4. How is the "period" of a cyclical waveform related to its frequency?
5. Draw graphs to show your understanding of waveforms: sine wave, square wave, pulse wave, sawtooth wave, triangular wave, and trigger pulse (spikes).
6. How does a *practical* pulse waveform differ from an *ideal* pulse waveform?
7. Define the following terms:
   a. Exponential rise.

  b. Exponential fall.
  c. Rise time (t_r).
  d. Fall time (t_f).
  e. Tilt time.
8. Show how to calculate the *average* value of a pulse waveform.
9. Explain the following terms as applied to amplitude of a waveform:
  a. Peak-to-peak value.
  b. Peak value.
  c. Average (dc) value.
10. The following terms are frequently used when describing a pulse waveform. Explain their meanings:
  a. Pulse width (t_p).
  b. Rest period (t_{rp}).
  c. Pulse repetition time (PRT).
  d. Pulse repetition frequency (PRF).
11. Define duty cycle and state it in an Eq. form.

2-3 A Function

When one quantity is dependent upon some other variable quantities, it is said to be a *function* of the various variable quantities. In simple terms, a function is a representation of how any one quantity varies with respect to others.

In electronics engineering work, a function usually means the description of the way a current or voltage varies with respect to time. There are several functions that are of specific interest to pulse and switching work.

2-4 The Step Function

This is the most important function for pulse and switching work. A *step function* makes an instantaneous jump from one steady value to another steady value. Figure 2-1 shows two step functions. A

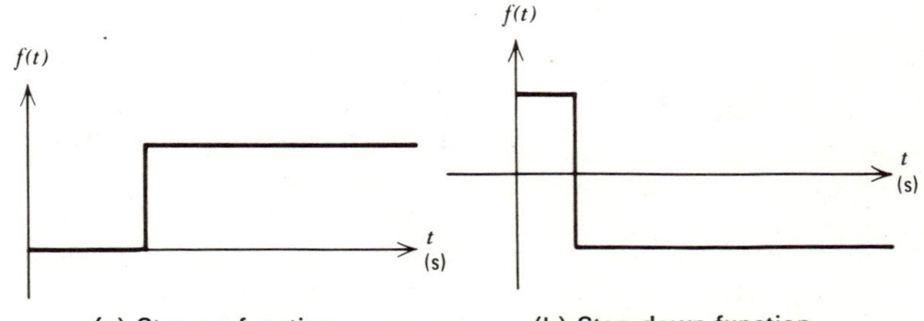

(a) Step-up function (b) Step-down function

Figure 2-1 Step functions.

step function may *increase* instantaneously (Fig. 2-1a) or *decrease* instantaneously (Fig. 2-1b). A *step* merely means an *instantaneous change* in level.

2-5 The Ramp Function

A *ramp* function is one that increases or decreases *linearly* with time. Figure 2-2 shows three such functions. In each case, the *slope* of the function is constant or linear. Note that the **dc** level of the function is immaterial. It is the *shape* or the linear change in function with respect to time that designates it as a ramp.

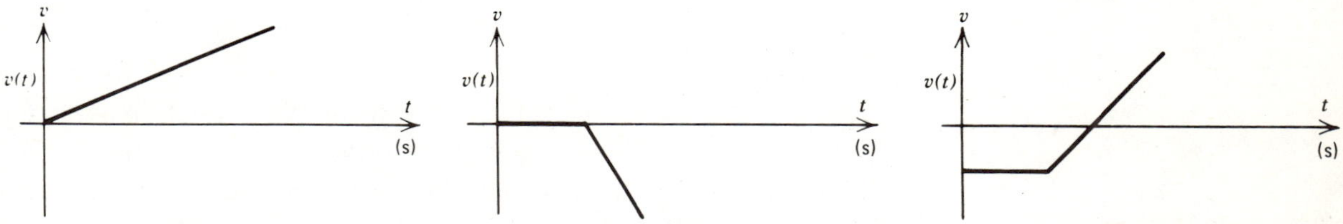

(a) Positive ramp (b) Negative ramp from zero reference (c) Positive ramp with nonzero reference

Figure 2-2 Ramp functions.

2-6 The Exponential Function

An *exponential* function is a function that increases or decreases *exponentially* with time. In pulse and switching circuits, the terms e^x and e^{-x} are encountered frequently. Each is an exponential quantity whose graph is known as an *exponential curve*. Figure 2-3 shows three exponential functions (curves) at various **dc** reference levels. Note again that the **dc** level of the function is not important. It is the shape of the graph (function) that makes it an exponential function.

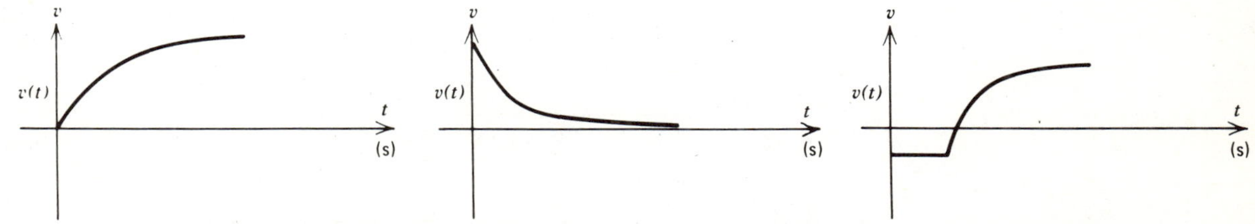

(a) Exponentially increasing function (b) Exponentially decaying function (c) Exponentially increasing function from a nonzero reference

Figure 2-3 Exponential functions.

2-7 Cyclical Waveforms

A waveform or a waveshape refers to the shape of the graph of a function. A *cyclical* waveform is the graph of a function that repeats itself after a certain given period of time. The *repetition rate* of the function (in other words, the number of times that a function or cycle repeats itself per unit time) is called the frequency (f) or the pulse repetition rate (PRR) of the function. It is measured in cycles per second, or simply hertz (Hz). The time (T) for one cycle is called the *time period,* or simply *period.*

The period (T) and the frequency (f) in cycles/second or Hz are related by Eq. (2-1):

$$\text{PRR} = f = \frac{1}{T} \text{ hertz (Hz)} \tag{2-1}$$

Figure 2-4 shows a cyclical waveform of a function $v(t)$. One complete cycle is shown by its time period, T. Every other cycle is exactly the same as this one. The waveform is a positive ramp function from 0 to t_1. From t_1 to t_3, it stays at a constant value, V_1. It is a negative ramp from t_3 to t_5, where it assumes a constant **dc** value of $-V_2$ until it reaches t_7. From t_7 to t_8, the waveform is a positive ramp function. This completes one cycle.

Obviously a cyclical waveform may be made up of more than only one type of function. Its amplitude always varies between two extremes. During each cycle the amplitude goes through a *maximum* value and a *minimum* value. The spread between the maximum value and the minimum value is called the *peak-to-peak* **(p-p)** amplitude of the waveform. The waveform of Fig. 2-4 varies between

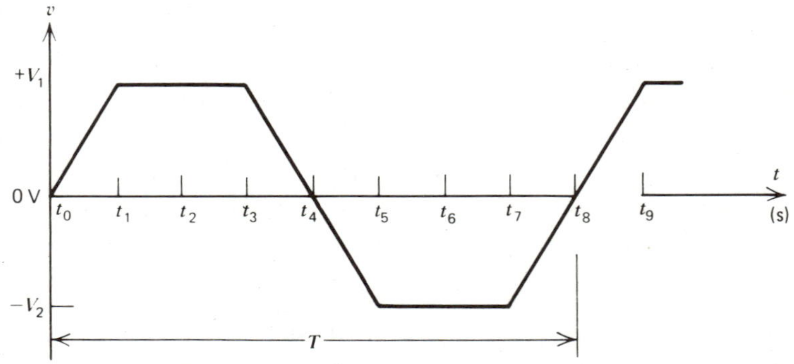

Figure 2-4 Cyclical waveform.

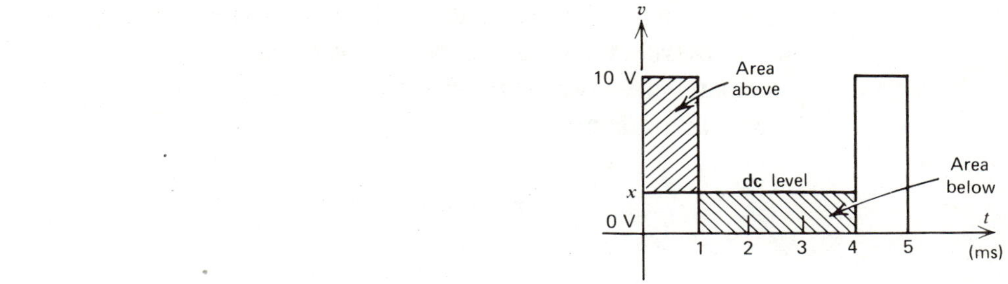

Figure 2-5 Average value of a waveform.

two extremes, $+V_1$ volts and $-V_2$ volts. Peak-to-peak amplitude of this waveform is $V_1 + V_2$. Peak-to-peak amplitude of the waveform shown in Fig. 2-9c is 10 V. It is written as 10 $V_{p\text{-}p}$.

Sometimes the amplitude of a waveform is given in terms of its *peak* value. The *peak value* is one-half the peak-to-peak value. It is abbreviated as V_p. A 10-$V_{p\text{-}p}$ waveform has a peak amplitude value of 5 V. It is written as 5 V_p. The V_p of the waveform shown in Fig. 2-4 is $(V_1 + V_2)/2$.

Average or **dc** value of a cyclical waveform is the value (positive or negative level) where the area below (one cycle of waveform) is equal to the area above. If any cyclical voltage waveform is measured by a **dc** voltmeter, the voltmeter reads the average value of the waveform. A typical pulse waveform is shown in Fig. 2-5. Its **dc** level or average value is shown by a horizontal dotted line. The shaded area above this line must be equal to the shaded area below this line. Mathematically,

$$V_{\mathrm{dc}} = \frac{1}{2\pi}\int_0^{2\pi} v(\theta)\cdot d\theta \text{ volts (V)*} \qquad (2\text{-}2)$$

where:

> V_{dc} is the average value of the function.
> $v(\theta)$ is the instantaneous amplitude of the function at angle θ.

Sometimes, the area above and/or area below cannot be easily calculated. In that case, Eq. (2-2) must be used.

* Calculus is used merely to show the equation in its general form. No calculus is required for the problems or examples.

Note that θ represents an angle measured either in radians (rad) or in degrees (°). One complete cycle consists of 360° or 2π radians. Radians are easily converted into degrees or vice-versa using the conversion factor:

$$2\pi \text{ radians} = 360° \tag{2-3a}$$

$$1 = \frac{360°}{2\pi \text{ radians}}$$

$$= 180/\pi \text{ deg/rad}$$

$$\approx 57.3\ °/\text{rad} \tag{2-3b}$$

Conversely,

$$360° = 2\pi \text{ radians}$$

$$1 = 2\pi/360°$$

$$= \frac{\pi}{180} \text{ deg/rad}$$

$$\approx 17.45 \times 10^{-3} \text{ rad/deg} \tag{2-3c}$$

Note that the conversion factors, Eqs. (2-3b) and (2-3c), represent unity. Multiplying any quantity by an appropriate conversion factor *does not change* that quantity. It merely represents it in different units.

Example 2-1
For the waveform shown in Fig. 2-5, calculate:

a. $V_{p\text{-}p}$.
b. Period (T).
c. Frequency (f).
d. The **dc** value.
e. Area above **dc** level.
f. Area below **dc** level.

Solution
a. $V_{p\text{-}p} = 10$ V
b. $T =$ **4 ms**

c. $f = \dfrac{1}{4\text{ ms}}$

$= \mathbf{250\ Hz}$

d. Assume that the **dc** value is x volts. Applying the area-above and area-below criteria:

$1(10 - x) = 3x$

$x = \mathbf{2.5\ V}$

e. Area above $= (10 - 2.5) \times 1$

$= \mathbf{7.5\ V\ ms}$

f. Area below $= 2.5 \times 3$

$= \mathbf{7.5\ V\ ms}$

Example 2-2

Consider the waveforms shown in Fig. 2-4. Assume $V_1 = +5$ V and $V_2 = -5$ V. Calculate:

a. $V_{p\text{-}p}$.
b. V_p.
c. The **dc** value.
d. Period (T).

Solution

a. $V_{p\text{-}p} = 5 - (-5)$

$= \mathbf{10\ V}$

b. $V_p = 10/2$

$= \mathbf{5\ V}$

c. $V_{dc} = \mathbf{0\ V}$, because the area above is equal to the area below the 0 V line.

d. $T = \mathbf{t_8}$ **seconds**

Example 2-3

Convert:

a. $\pi/4$ radians into degrees.
b. 70° into radians.

Solution

a. $\pi/4 \text{ rad}\left(\frac{180°}{\pi \text{ rad}}\right) = \mathbf{45°}$

b. $70°\left(\frac{\pi \text{ rad}}{180°}\right) = \mathbf{1.22\ rad}$

2-8 Sinusoidal (Sine) Wave

In pulse work, commonly encountered cyclical or periodic waveforms are the *sinusoidal* wave, the *square* wave, the *pulse* wave, the *triangular* wave, and the *sawtooth* wave.

Figure 2-6 shows a typical sine wave, which may be expressed as:

$$v = V_m \sin \theta \tag{2-4}$$

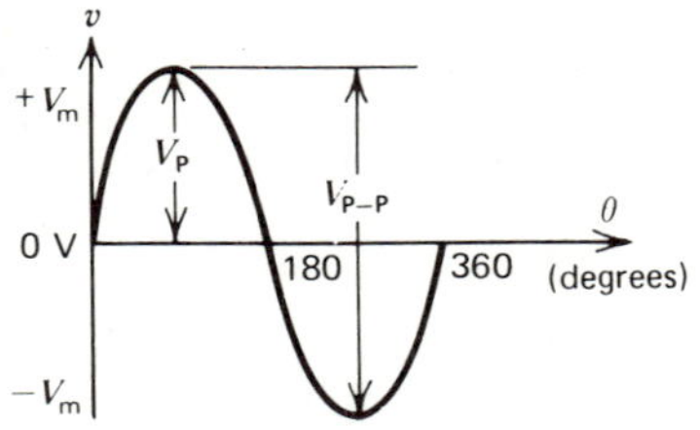

Figure 2-6 Sine wave.

Note that its:

1. Peak-to-peak value is $V_{p\text{-}p} = 2\,V_m$ (2-5)*

2. Peak or maximum value is $V_p = V_m$ (2-6)*

3. rms value is $V_{rms} = 0.707\,V_m.$ (2-7)*

4. dc value (half cycle) is $V_{dc} = 0.637\,V_m.$ (2-8)*

5. Form factor, F_f, is $F_f = \frac{V_{rms}}{V_{dc}};$ (2-9)*

The **dc** value (half cycle) is shown on Fig. 2-7.

* These are listed as reminders only. The reader is asked to consult reference material if difficulty in comprehending them is experienced.

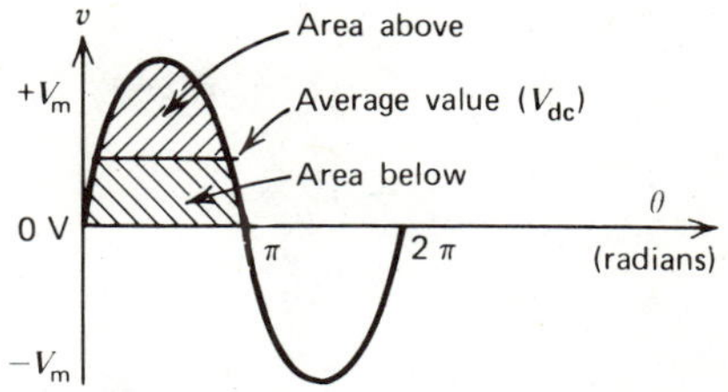

Figure 2-7 Average value of one-half cycle of a sine wave.

Example 2-4

A voltage waveform is given by:

$$v = 10 \sin \theta$$

Calculate its:

a. Peak-to-peak value.
b. Peak value.
c. **rms** value.
d. **dc** value of one-half cycle.
e. Form factor.

Solution

a. $$V_{p\text{-}p} = \mathbf{20\ V}$$

b. $$V_p = \mathbf{10\ V}$$

c. $$V_{rms} = V_m \times 0.707 \qquad (2\text{-}7)$$
$$= 10 \times 0.707$$
$$= \mathbf{7.07\ V}$$

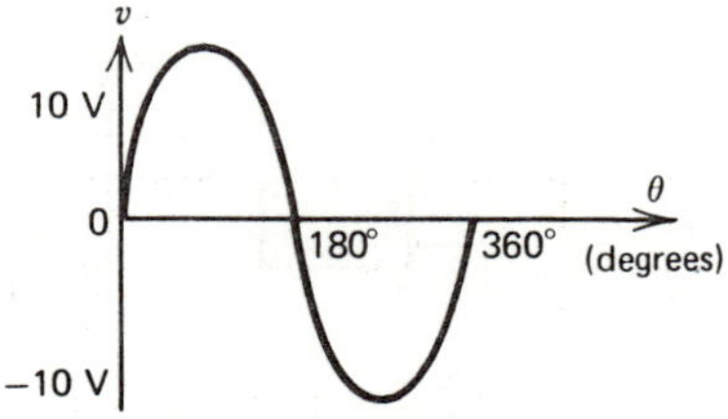

Figure 2-8 Waveform for Ex. 2-4.

d. V_{dc} for one-half cycle

$$= V_m \times 0.637 \qquad (2\text{-}8)$$

$$= 10 \times 0.637$$

$$= \mathbf{6.37\ V}$$

e.

$$F_f = \frac{V_{rms}}{V_{dc}} \qquad (2\text{-}9)$$

$$= \frac{7.07\ V}{6.37\ V}$$

$$= \mathbf{1.11}$$

2-9 Rectangular Wave

Figure 2-9 shows various *rectangular* waveforms. Basically they are similar in shape and differ only in their **dc** or average levels. It is the waveform shape that classifies a waveform. Each cycle of a rectangular wave is made up of two step-functions; one a step *up* and the other a step *down*. Note from Fig. 2-9 that in each case,

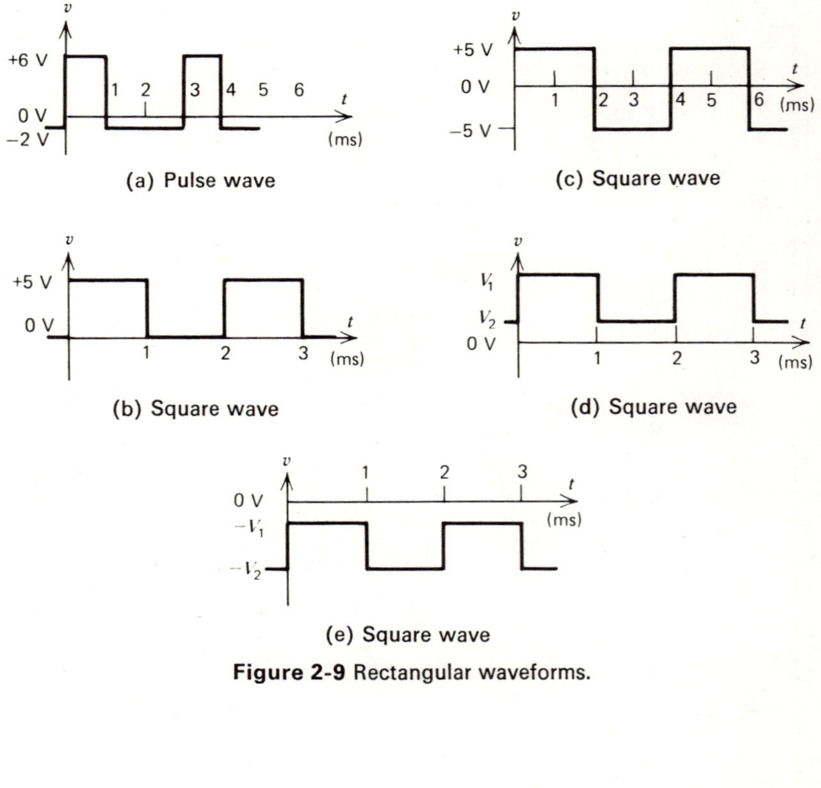

Figure 2-9 Rectangular waveforms.

the amplitude varies between two values and there occurs a sudden and instantaneous jump from one value to another value.

2-10 Square Wave

A type of a rectangular wave that stays at one level for exactly the same length of time as at the other level is called a *square* wave. For any given cycle, the wave given in Fig. 2-9b is at +5 V for 1 ms and is at 0 V also for 1 ms. It is therefore a square wave by definition. The time period (T) for this wave is 2 ms. The waveforms shown in Figs. 2-9c, 2-9d, and 2-9e are all square waves.

Example 2-5

For the square wave shown in Fig. 2-9c, calculate its:

a. Peak-to-peak value.
b. Peak value.
c. Average value.
d. Average value for one-half cycle.

Solution

a. $V_{p\text{-}p} = \mathbf{10\ V}$
b. $V_p = \mathbf{5\ V}$
c. The wave is symmetrical about the 0 axis. Therefore the area above equals the area below. Thus, $V_{dc} = \mathbf{0}$
d. $V_{ave} = V_p = \mathbf{5\ V}$

2-11 Pulse Wave

The *pulse* waveform is another type of a rectangular waveform. It is one of the most important waveforms insofar as the pulse and switching circuits are concerned. Essentially, a pulse wave stays at one (rest) level for a considerably longer period of time than at another (pulse) level. Figure 2-10 shows a typical pulse wave. The

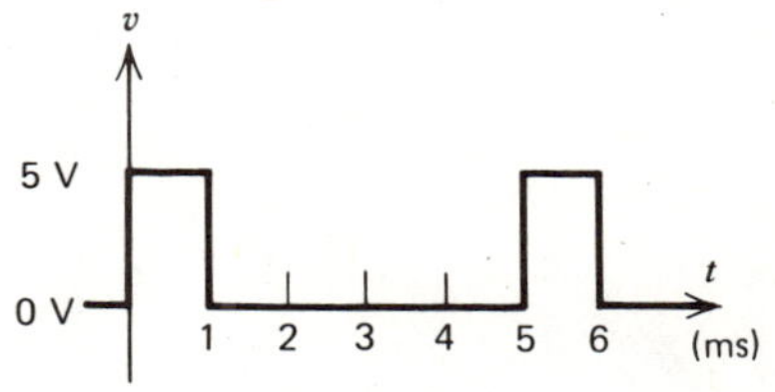

Figure 2-10 Pulse wave.

waveform is at +5 V for only 1 ms and is at 0 V for 4 ms. We could say that there is a voltage (pulse) for 1 ms and no voltage (no pulse) for 4 ms. The time when there is no pulse is called the *rest period* (t_{rp}). The time duration when there is a pulse is known as the *pulse width* (t_p).

Example 2-6
For the pulse wave shown in Fig. 2-9a, find:

a. $V_{p\text{-}p}$.
b. t_{rp}.
c. t_p.

Solution
a. $V_{p\text{-}p} = 6 + 2 =$ **8 V**
b. $t_{rp} =$ **2 ms**
c. $t_p =$ **1 ms**

Example 2-7
For the pulse wave shown in Fig. 2-10, calculate

a. t_{rp}.
b. t_p.

Solution
a. Rest period, $t_{rp} =$ **4 ms**
b. Pulse width, $t_p =$ **1 ms**

Some typical pulse waveforms are shown in Fig. 2-11. Note that during the rest period, a pulse wave does not necessarily have to stay at the 0-V level. It can be at any level. The voltage level at which the waveform stays for a longer period (rest period) is also called the *base line.* The total time for a cycle (period) in a pulse waveform is usually called the pulse repetition time (PRT). It is equivalent to time period T. The pulse repetition frequency (PRF) or the pulse repetition rate (PRR) is the frequency of the pulse waveform (Eq. 2-1).

$$\text{PRT} = t_p + t_{rp} = T \text{ seconds (s)} \quad (2\text{-}10)$$

$$\text{PRR} = \text{PRF} = \frac{1}{\text{PRT}} = \frac{1}{t_{rp} + t_p} = \frac{1}{T}, \text{ pulses/second (pps)} \quad (2\text{-}11)$$

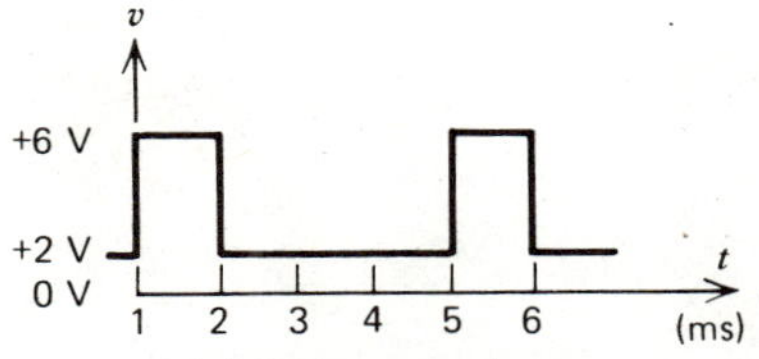

(a) Pulse waveform

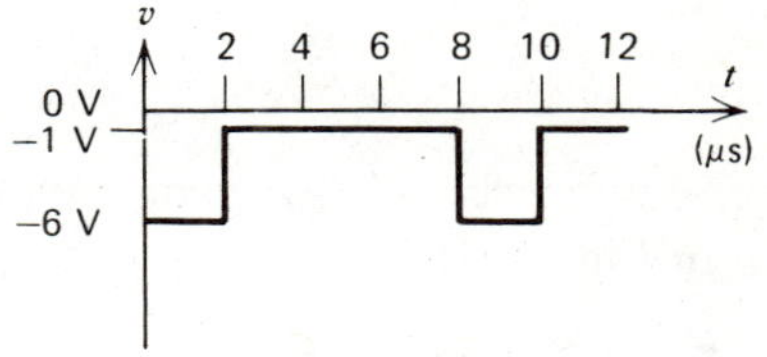

(b) Pulse waveform

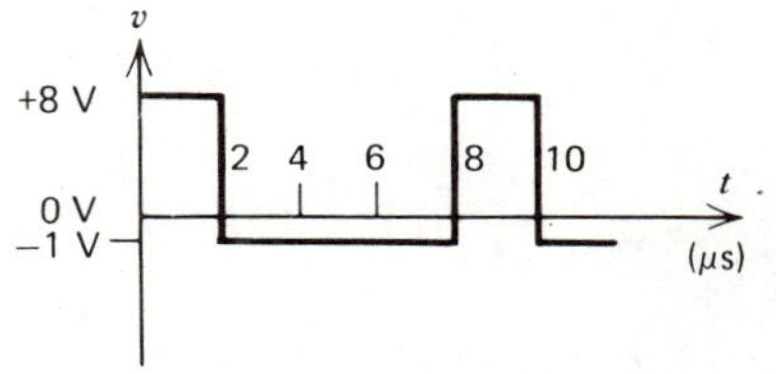

(c) Pulse waveform

Figure 2-11 Pulse waveforms.

Usually, a pulse wave is used to turn (or electronically switch) a circuit ON or OFF. The circuit is ON during t_p and is OFF during t_{rp}. A term known as the *duty cycle* is used to determine the percentage of a cycle that the switch is ON. It is the ratio of t_p to PRT.

$$\text{percent duty cycle} = \frac{t_p}{\text{PRT}} \times 100 \qquad (2\text{-}12)$$

Example 2-8

Refer to the waveform shown in Fig. 2-11a. Determine:

a. t_p.
b. PRT.
c. PRF.

Solution

a. t_p = **1 ms**
b. PRT = **4 ms**

c. $\text{PRF} = \dfrac{1}{\text{PRT}}$

$= \dfrac{1}{4\text{ ms}}$

$=$ **250 pulses per second (PPS)**

Example 2-9

Determine the (a) duty cycle and (b) average (**dc**) value of the pulse waveform shown in Fig. 2-11a.

Solution

a. $\text{Duty cycle} = \dfrac{t_p}{\text{PRT}} \times 100$

$= \frac{1}{4} \times 100$

$=$ **25 percent**

(2-12)

b. The pulse wave of Fig. 2-11a is redrawn in Fig. 2-12. The average value is shown as x. The area above this **dc** line (x) should be equal to the area below it.

$$\text{Area above} = 1(6 - x)$$

$$\text{Area below} = 3(x - 2)$$

$$1(6 - x) = 3(x - 2)$$

$$4x = 12$$

$$x = \mathbf{+3}$$

Thus the **dc** value of the pulse waveform is **+3 volts.**

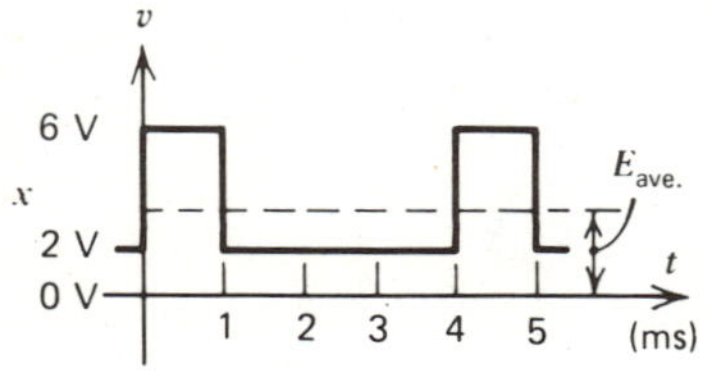

Figure 2-12 Average value of a pulse wave.

Example 2-10

For the waveform shown in Fig. 2-11c, calculate:

a. t_p.
b. t_{rp}.
c. PRT.
d. PRF.
e. Duty cycle.
f. **dc** value.

Solution

a. $t_p = \mathbf{2\ \mu s}$
b. $t_{rp} = \mathbf{6\ \mu s}$
c. $\text{PRT} = 2\ \mu\text{s} + 6\ \mu\text{s}$

$$= \mathbf{8\ \mu s}$$

d. $$\text{PRF} = \frac{1}{8\ \mu\text{s}} \qquad (2\text{-}11)$$

$$= \mathbf{1.25 \times 10^5\ PPS}$$

e. $$\text{Duty cycle} = \frac{2\ \mu\text{s}}{8\ \mu\text{s}} \times 100 \qquad (2\text{-}12)$$

$$= \mathbf{25\ percent}$$

f. Let the **dc** value be x volts. Then,

$$\text{Area above} = \text{Area below}$$

$$(8 - x)2 = 6(x + 1)$$

$$16 - 2x = 6x + 6$$

$$x = \mathbf{+1.25\ volts}$$

2-12 The Ideal Pulse Wave

A pulse waveform is termed as *ideal* if the step function changes from one level to another level *instantaneously* (in zero time). The pulse waveforms shown in Fig. 2-11 are ideal because they rise from one level to another level instantaneously. The pulse wave shown in Fig. 2-10 rises from 0 V to 5 V at time t_0. It is also shown to fall to the 0-V level after 1 ms in zero time and is also ideal. It is the instantaneous transition of current/voltage from one level to another that classifies a waveform as an ideal pulse.

2-13 The Practical Pulse Wave

Practically, it is impossible to generate an *ideal* pulse. A practical pulse is shown in Fig. 2-13. The difference between an ideal pulse and a practical pulse is that the practical pulse takes a certain *finite* time to go from one level to another. The pulse shown in Fig. 2-13 goes from 0 volts to 10 volts in t_0 to t_3 seconds. It comes back to 0 volts from the 10-volt level in t_4 to t_7 seconds. The term *rise time* (t_r) is a very important term in pulse circuits.

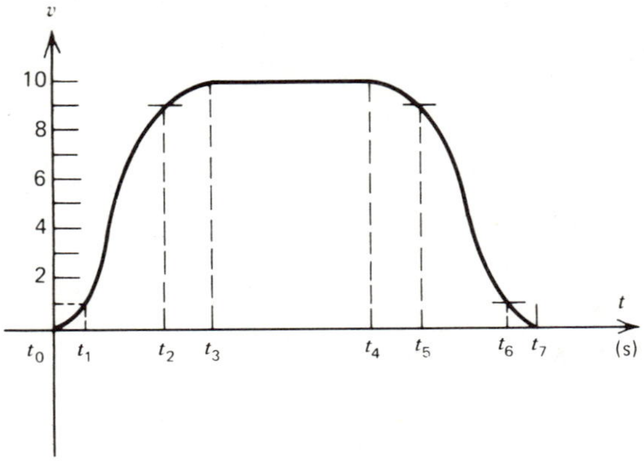

Figure 2-13 t_r and t_f in a practical pulse waveform.

Rise time is the time that a pulse takes to rise from 10 percent of its final value to 90 percent of its final value. The pulse shown in Fig. 2-13 rises to 1 volt (10 percent) at time t_1 and reaches 9 volts (90 percent) at time t_2. The time duration t_1-to-t_2 is the *rise time*.

Another term used in conjunction with the pulse waveforms is the *fall time* (t_f). The decay or fall time is the time required for the trailing edge of the waveform to decay from 90 percent of its peak value to 10 percent of its peak value. In other words, t_f is the time that the pulse takes to fall from 0.9 V_p to 0.1 V_p. In Fig. 2-13, the fall time is the time that the pulse takes to fall from 9 volts to 1 volt (t_5 to t_6).

Both rise time and the fall time determine the *quality* of a pulse waveform. The shorter these "times," the squarer (more ideal) the pulse. Of course, the pulse is ideal if the rise time and the fall time (t_r and t_f) are *both* zero, but this is impossible.

Example 2-11

Find the rise time and the fall time for the pulse wave shown in Fig. 2-14.

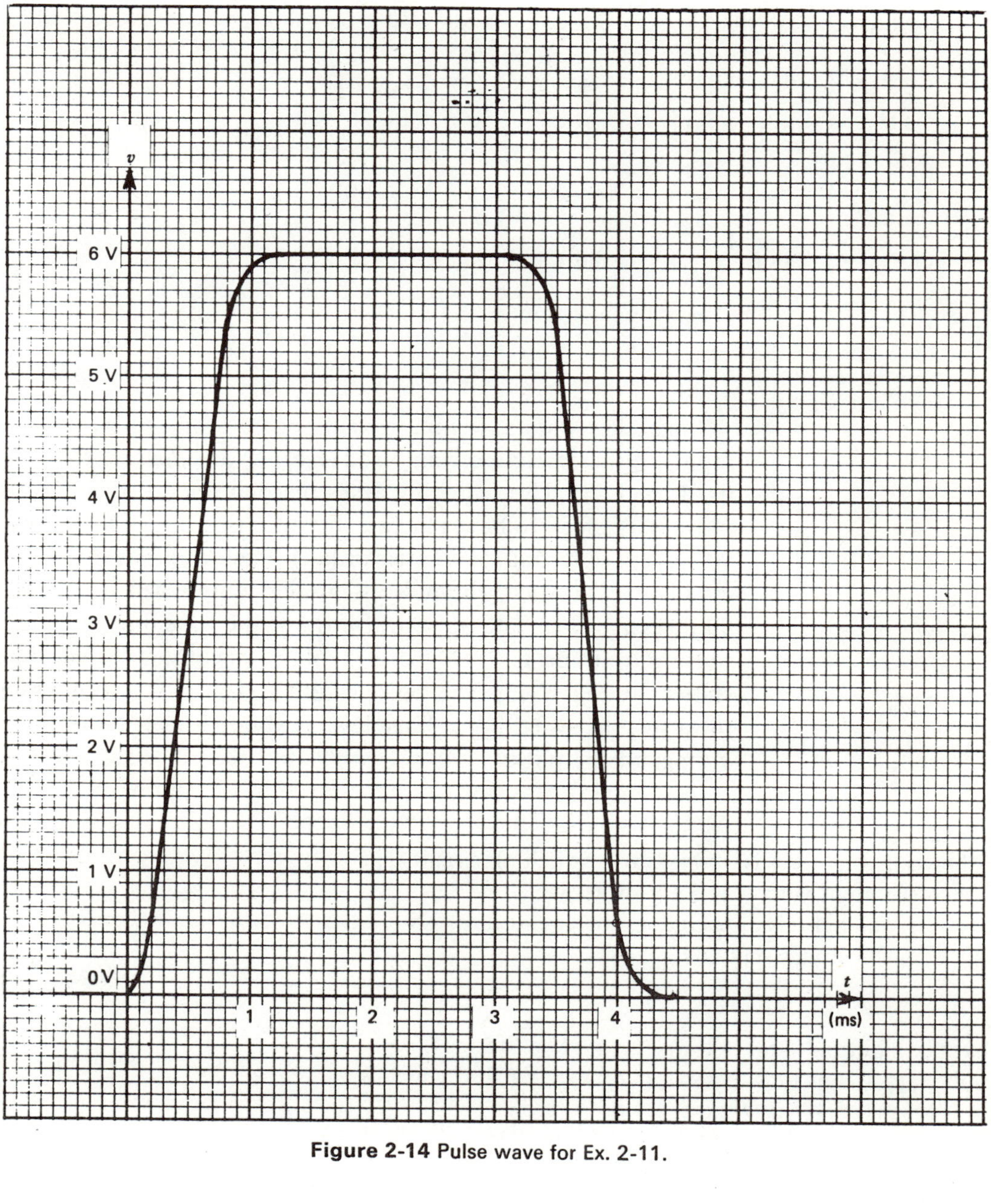

Figure 2-14 Pulse wave for Ex. 2-11.

Solution

10% of 6 volts = 0.6 V

90% of 6 volts = 5.4 V

time to rise to 0.6 V = 0.2 ms (graph)

time to rise to 5.4 V = 0.8 ms (graph)

t_r, rise time = (0.8 − 0.2) ms

= **0.6 ms**

Similarly, the time take to fall from 5.4 V to 0.6 V (fall time):

$$t_f = (4.0 - 3.5) \text{ ms}$$

$$= \mathbf{0.5\ ms}$$

Most electronic switching circuits operate on the basis that the input is either *above* a certain fixed level (A) or *below* another certain fixed level (*B*). The switching takes place at *both* of these levels. A pulse is assumed to have reached the switching level when it reaches 90 percent of its final value (rise or fall). Thus, as far as the switching circuit is concerned, the pulse is present only when it is *above* its 90 percent value. The pulse width (t_p) is the time during which the pulse amplitude is ≧ 90 percent of its maximum value. This is shown in Fig. 2-13. The pulse width (t_p) is $t_5 - t_2$. In practice, however, the switching voltage level is about 50 percent of the peak voltage of the pulse.

Example 2-12

The pulse shown in Fig. 2-14 is applied to a switching circuit that turns ON when the input voltage is 5.4 V or above:

a. How long after the pulse is applied does the circuit turn ON?

b. How long does the circuit stay ON?

Solution

a. Time that the input pulse takes to rise to 5.4 V = **0.8 ms** (graph)

b. The circuit stays ON as long as the input pulse ≧ 5.4 V, which is equal to t_p because 5.4 V is 90 percent of 6 V.

$$t_p = (3.5 - 0.8) \text{ ms}$$

or

$$= \mathbf{2.7\ ms}$$

The time at which one circuit switches with respect to other circuits in an electronic unit is an important consideration in switching circuits. There may be problems if the switching does not take place at a desired time. On the other hand, sometimes it is desirable that one circuit should switch a little later (or earlier) than another circuit. The time that a pulse takes to reach 90 percent of its maximum amplitude level is really a *delay* time (assuming that switching takes place at the 90 percent level). The pulse

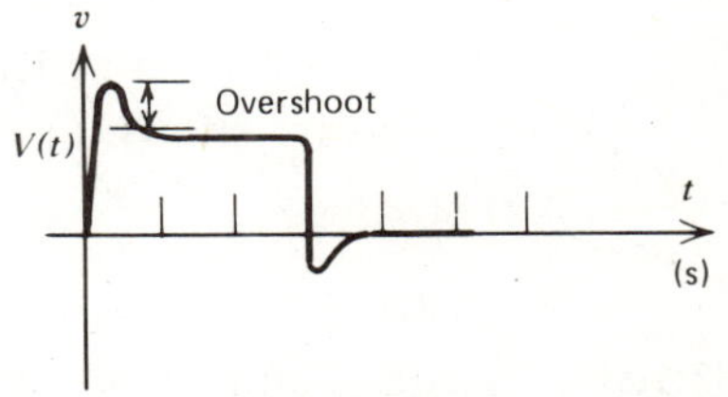

(a) Overshoot

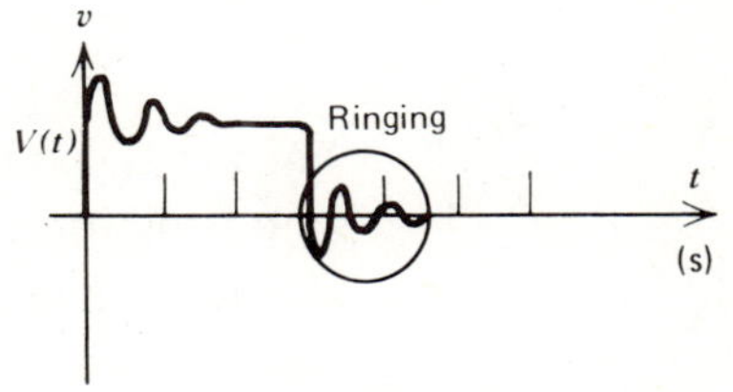

(b) Ringing

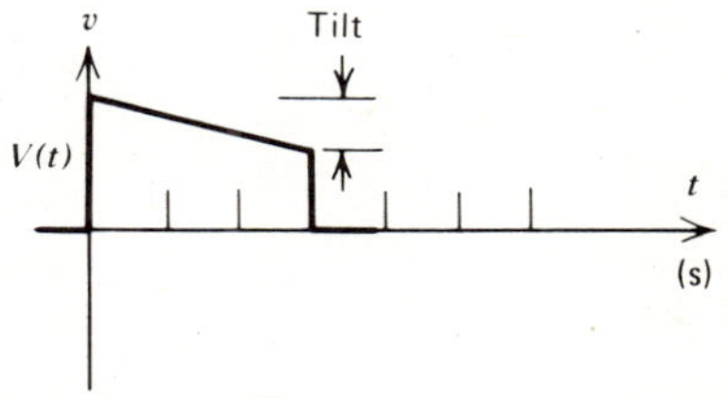

(c) Tilt

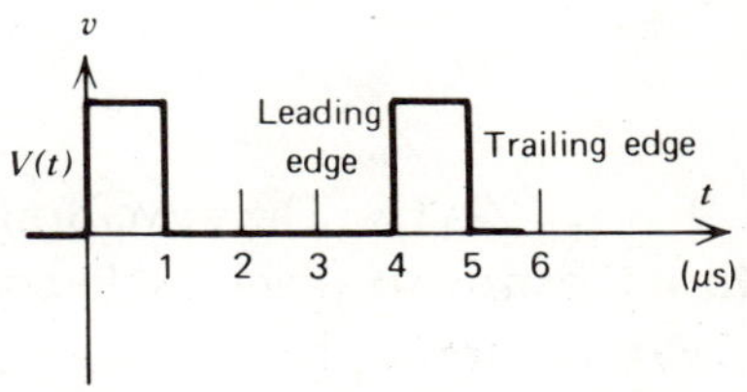

(d) Leading and trailing edge

Figure 2-15 Some characteristics found in practical pulse waveforms.

appears at the input terminals at $t = 0$, but the switching does not take place until it reaches 90 percent of its final value.

In practice, however, circuits are designed such that the switching takes place as the pulse reaches its 20 percent to 50 percent value (range). This reduces the delay time as well as assures switching, even when the amplitude of the input pulse deteriorates a little.

It has been shown that all practical pulse waveforms have both a finite rise time and a finite fall time. Three other characteristics that sometimes occur in pulse waveforms are *overshoot*, shown in Fig. 2-15a; *ringing*, shown in Fig. 2-15b; and *tilt*, shown in Fig. 2-15c.

The so-called *leading edge* and *trailing edge* of a pulse waveform are shown in Fig. 2-15d.

2-14 Sawtooth Wave

Figure 2-16 shows a typical sawtooth waveform. Time for one complete cycle (T) is from t_0 to t_1, that is,

$$T = t_1 - t_0$$

Each cycle contains two functions: a ramp function (t_0 to t_1) and a step function (at t_1). The ramp duration sometimes is also known as the *sweep* time and the step duration as the *flyback* time. These terms emanate from the use of this type of waveform in the horizontal deflection of the electronic beam in the cathode ray tube (CRT) of a cathode ray oscilloscope (CRO).

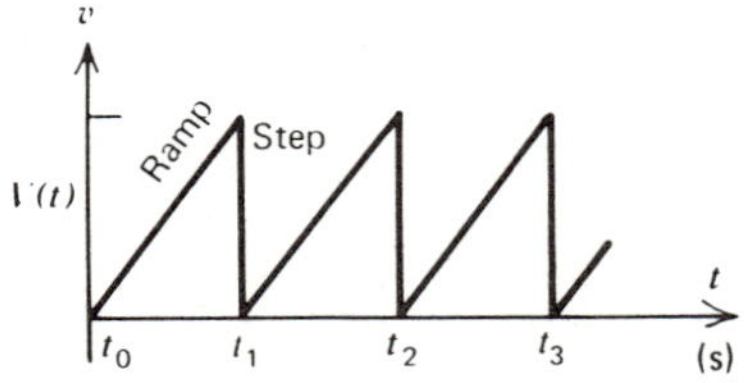

Figure 2-16 Sawtooth wave.

2-15 Other Common Waveforms

The waveform shown in Fig. 2-17a is called the *triangular* waveform. Each cycle consists of two ramp functions. The waveform shown has a negative ramp for one-half of the cycle and a positive ramp for the other half of the cycle.

Figure 2-17b shows a *staircase* waveform. One complete cycle of Fig. 2-17b is shown to consist of three rising functions and one

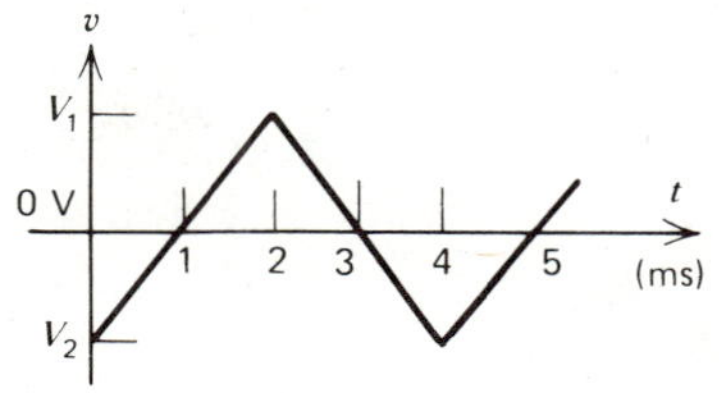

(a) Triangular wave

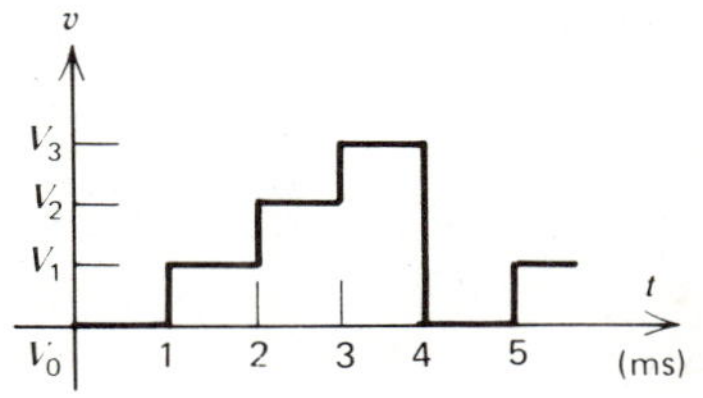

(b) Staircase wave

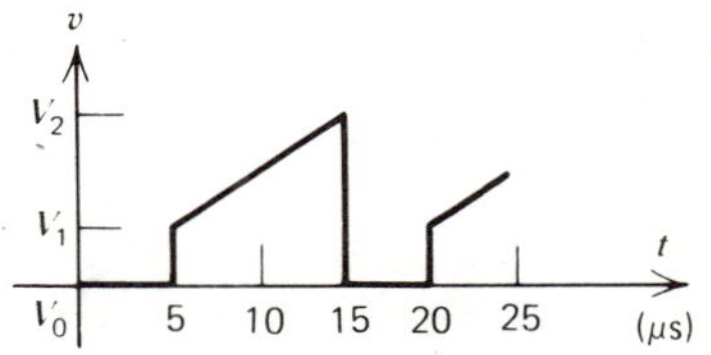

(c) Trapezoidal wave

Figure 2-17 Common waveforms.

descending step. In actual fact, a staircase may consist of any number of step functions. It is the general shape of the waveform that gives it the name *staircase.*

A so-called *trapezoidal* waveform is shown in Fig. 1-17c. Each cycle of a trapezoidal wave consists of two step-functions and a ramp function.

It is worth noting that the amplitude of a waveform has *no* bearing on its classification. The waveform or the waveshape means exactly what it says. It is the shape that classifies the waveform.

2-16 Summary

1. A function describes the way one quantity varies with respect to some other variable quantities.
2. In pulse and switching circuits, we are usually concerned with the way current or voltage varies with respect to time.

3. The step function, the ramp function, and the exponential function are the most important functions in pulse work.
4. One or more types of functions combine to produce various types of cyclical waveforms.
5. The pulse wave, the square wave, and the sawtooth wave are the commonly encountered waveforms in pulse circuits.
6. The pulse wave in particular is the most commonly used waveform in digital logic and switching circuits.
7. Important equations:

$$\text{PRR} = f = \frac{1}{T} \tag{2-1}$$

$$V_{\text{dc}} = \frac{1}{2\pi}\int_0^{2\pi} v(\theta)\cdot d\theta \tag{2-2}$$

$$2\pi \text{ rad} = 360° \tag{2-3}$$

$$v(\theta) = V_m \sin \theta \tag{2-4}$$

$$V = V_{\text{rms}} = 0.707\, V_m \text{ (sine wave)} \tag{2-7}$$

$$\text{PRR} = \text{PRF} = \frac{1}{\text{PRT}} \tag{2-11}$$

$$\text{percent duty cycle} = (t_p/\text{PRT}) \times 100 \tag{2-12}$$

2-17 Glossary

Cyclical waveform. A waveform that repeats itself.

dc value. The average value of a pulsating cyclical waveform.

Duty cycle. The ratio of the average pulse duration to the total average pulse-repetition time. It is also equivalent to the product of average pulse duration and pulse repetition rate (PRR).

Fall time (t_f). The time a pulse takes to fall from 90 percent of its peak amplitude to 10 percent of its peak value.

Function. The representation of the way one quantity varies with respect to others.

Pulse repetition frequency (PRF). The pulse repetition frequency is the number of pulses per second (PPS). It is the reciprocal of the PRT.

Pulse repetition rate (PPR). Same as the PRF.

Pulse repetition time (PRT). The time for one complete cycle of a pulse waveform.

Pulse wave. A rectangular waveform that stays at one level (t_p) for considerably shorter time than at the other level (rest period).

Pulse width (t_p). The time that a pulse is above 90 percent of its maximum value.

Ramp function. A ramp function increases or decreases linearly, usually with respect to time.

Rise time (t_r). The time a pulse takes to rise from 10 percent of its final value to 90 percent of its final value.

Sawtooth wave. A waveform that rises linearly with time and then (almost) instantaneously falls to its starting value at the end of each cycle.

Square wave. A rectangular waveform that stays at one level for one half-cycle and at another level for the other half-cycle.

Step function. A function that changes from one value to another value (almost) instantaneously.

2-18 Recapitulation

Return to the Objectives and Self-Evaluation Questions given at the beginning of this chapter and see how well you can answer them. If you cannot answer certain questions satisfactorily, place a check mark (√) next to them and review appropriate parts of the chapter to find the answers. No attempt should be made to solve the Problems (Sec. 2-19) before answering the Self-Evaluation Questions. Ability to solve the Problems should serve as a test and feedback regarding your understanding of the material presented in this chapter. Answers to Problems are given on p. 513.

2-19 Problems

2-1 A function is given by the equation $f(x) = 20 \cos x$. Draw one cycle of this function.

2-2 One cycle of a voltage wave is given by:

$$V(t) = 10 \text{ V for first half of the cycle}$$
$$V(t) = -10 \text{ V for last half of the cycle.}$$

Draw one cycle of this waveform.

2-3 A square wave varies between +10 V and +2 V. Calculate the average value of this waveform.

2-4 The voltage waveform shown in Fig. 2-18 is measured by a **dc** voltmeter. Find the voltmeter reading.

2-5 Find the average value of the pulse waveform shown in Fig. 2-19.

2-6 Find the duty cycle of the waveform shown in Fig. 2-19.

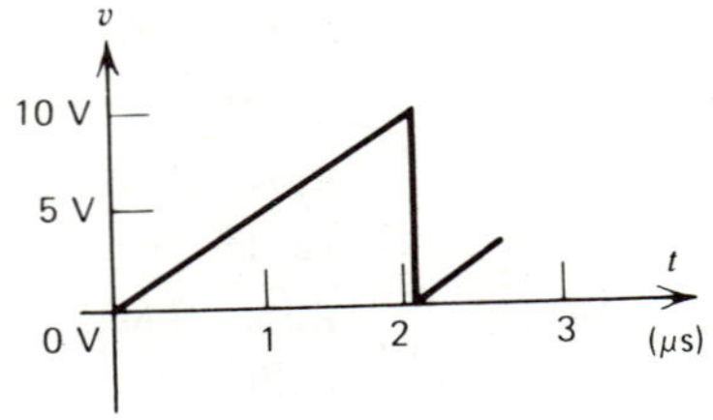

Figure 2-18 Sawtooth wave for Prob. 2-4.

2-7 For the waveform shown in Fig. 2-19, find:
 a. The peak-to-peak value.
 b. The pulse width.
 c. The rest period.
 d. The PRT.
 e. The PRF.

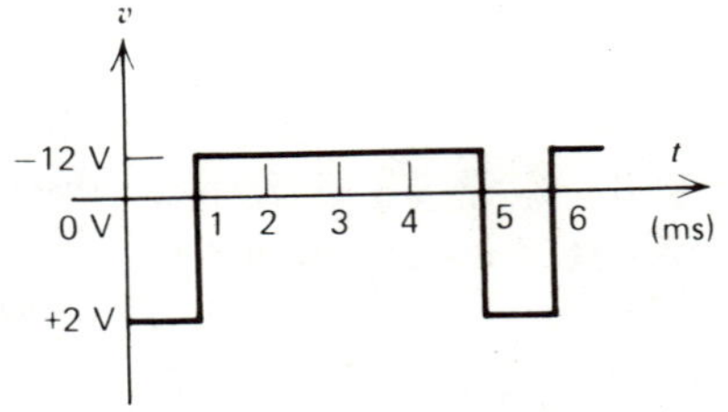

Figure 2-19 Pulse wave for Probs. 2-6 and 2-7.

2-8 For the pulse waveform shown in Fig. 2-20, find:
 a. The rise time.
 b. The fall time.
 c. The pulse width.

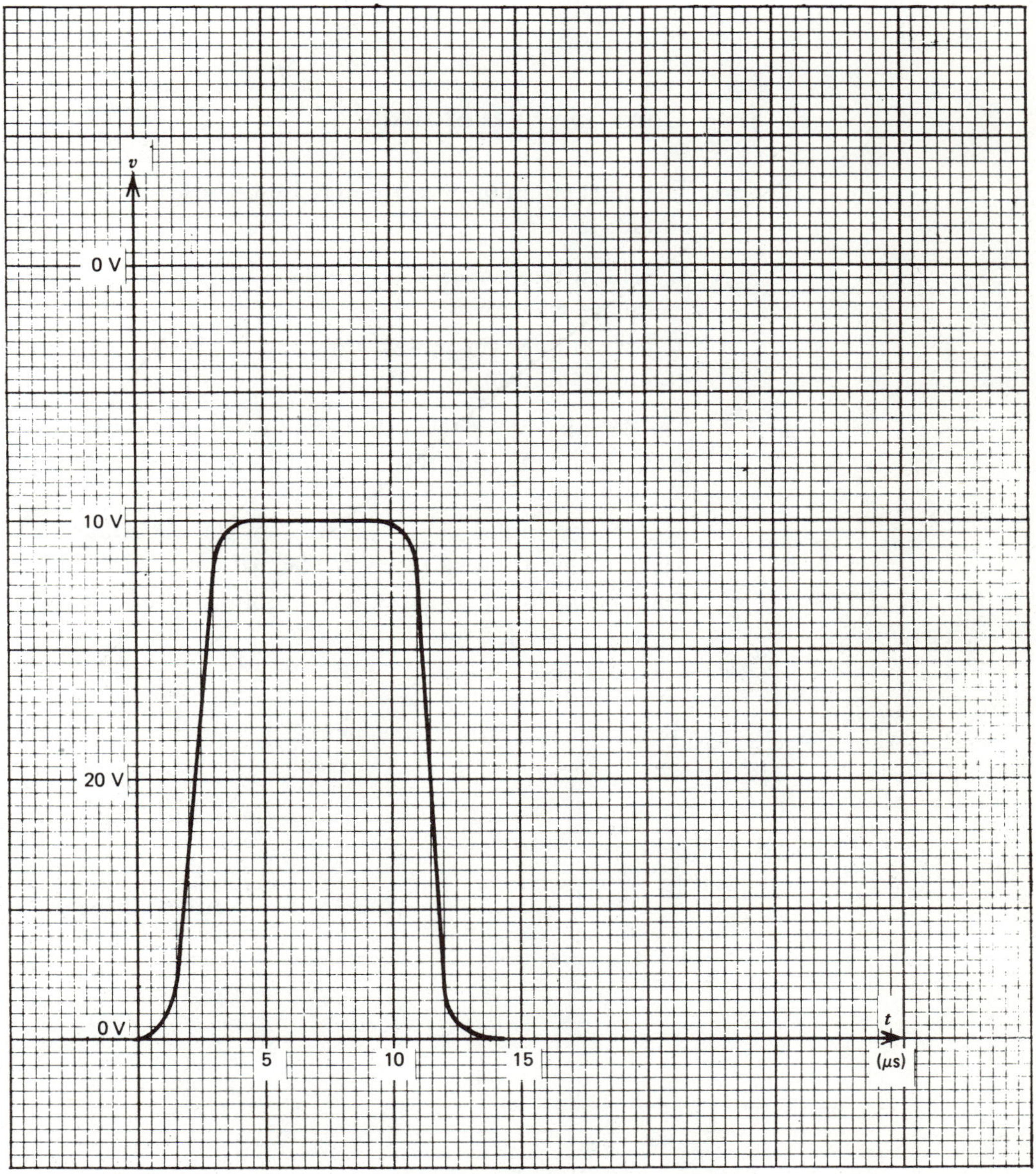

Figure 2-20 Pulse wave for Prob. 1-8.

2-20 Experiment

Title **Common Waveforms**

Theory The most commonly encountered waveforms in pulse and switching circuits are:

1. Pulse waveform.
2. Square waveform.
3. Sawtooth (sweep) waveform.
4. Sine waveform.

The **dc** value of a pulse or a square waveform is calculated using the area-above and the area-below method (Ex. 2-9).

Equipment Multifunction generator: EXACT model 126, or equal.
Multimeter: RCA model WV 98C, or equal.
Dual beam oscilloscope (CRO): Philips model PM3230, or equal.
Resistors: 4.7 kΩ and 10 kΩ.

Procedure

1. Connect the circuit as shown in Fig. E2-1. E_i represents the multifunction generator. Adjust voltage (E_i) to obtain a 6-$V_{p\text{-}p}$, 10-kHz sine wave (using CRO).

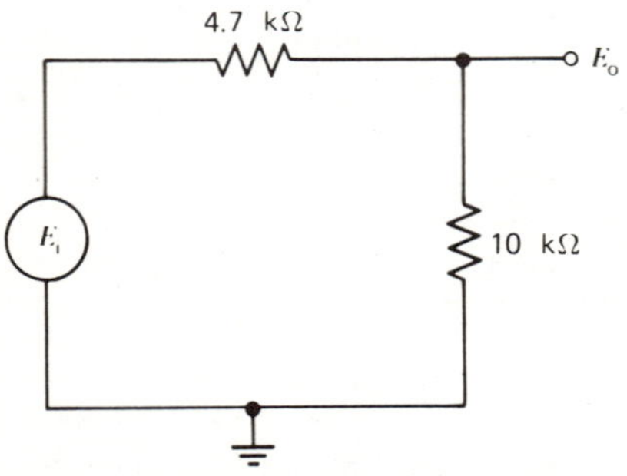

Figure E2-1 Circuit for the experiment.

2. Connect vertical input A of CRO across the input (E_i) and vertical input B across the output (E_o). Adjust the horizontal time scale and the vertical amplitude scale of CRO for each beam to obtain

about two cycles of the waveforms of sufficient amplitude. The two waveforms should be distinctly visible on the CRO screen. Note that ground or the 0 level should be established for each beam before taking any amplitude (voltage) readings.

For this and the experiments that are to follow, always set the **ac-0-dc** mode switch of CRO to the **dc** (direct-coupled) position unless otherwise specified. This is because we are usually concerned with displaying the exact **dc** levels of the waveform in question.

3. Observe (CRO) and draw the input as well as the output waveforms with respect to ground level in Table E2-1. Make sure that you have recorded the peak values (vertical volts per division) and the time scale (horizontal, time per division) accurately for each channel.

Note: In experiments that are to follow, the Procedure step 2 as outlined above will be omitted. The reader is expected to go through this procedure every time waveforms are displayed (to be observed) on CRO screen.

4. Change the input (multifunction generator) to a 6-$V_{p\text{-}p}$, 10-kHz:
a. Square wave; repeat Procedure step 3.
b. Sawtooth wave; repeat Procedure step 3.

5. Adjust the multifunction generator so that E_i is a 6-$V_{p\text{-}p}$, 1-kHz sine wave. Use CRO to monitor the input amplitude. Using the multimeter, measure and record the *rms* values of the input as well as the output (volts) in Table E2-2.

6. Change the input to a 0 V–6 V, 1-kHz square wave. Observe and draw the output waveform in Table E2-3 when the mode switch of CRO is at
a. Ground or 0-level position.
b. dc position.
c. ac position.

7. Repeat step 6 for a 0 V–6 V sawtooth waveform.

8. Disconnect and return all of the equipment and components to their proper places.

Calculations

1. Calculate the period of the square wave described in Procedure step 4a. Record it in Table E2-4.

2. Calculate and record in Table E2-4 the percent error between the measured and the theoretically expected values. Use observed values as the standard values.

3. Consider the circuit conditions described for the sinusoidal waveform in Procedure step 5. Calculate and record in Table

E2-2 the **rms** values of both input and output waveforms from the peak-to-peak values observed.

4. Calculate and record in Table E2-2 the percent error between the observed and the theoretically calculated values. Use the observed values as the standard values.
5. From the **ac/dc** output waveforms drawn in Table E2-3, find the observed average (**dc**) values for the:
   a. Square wave.
   b. Sawtooth wave.
   Record your findings in Table E2-5.
6. Calculate and record in Table E2-5 the **dc** values for:
   a. The output waveform of Procedure step 6.
   b. The output waveform of Procedure step 7.
7. Calculate and record in Table E2-5 the percent error between the observed and the theoretically expected values of V_{dc}. Use the observed values as the standard values.

Tabulation of Results

Table E2-1

	Sine Wave	Square Wave	Sawtooth Wave
Input waveform			
Output waveform			
Time period			

Table E2-2

V_{rms}	Measured	Calculated	Percent error
Input			
Output			

Table E2-3

Waveform	**Square Wave**	**Sawtooth Wave**
Input		
(**dc**) output wave		
(**ac**) output wave		

Table E2-4

	Observed Period	**Calculated Period**	**Percent Error**
Square wave, Procedure step 4a			

Table E2-5

	Measured dc Value	**Calculated dc Value**	**Percent Error**
Square wave, Procedure step 6			
Sawtooth wave, Procedure step 7			

Questions

1. From the calculated percent errors in Tables E2-2, E2-4, and E2-5, describe how your calculated results compare with the experimentally observed ones.
2. The tabulation of Table E2-2 is intended to verify the conversion factor used to convert $V_{p\text{-}p}$ to V_{rms} for a sine wave. Is it confirmed? Explain.
3. The tabulation of Table E2-3 is intended to demonstrate that a CRO may be used to measure the average (**dc**) value of a waveform. Has it been accomplished? Explain.

4. Results recorded in Table E2-4, in a way, compare the accuracy of the multifunction generator and CRO (time scale). Explain how you can check the accuracy of each instrument.

5. The tabulation of Table E2-5 is intended to confirm the area-below and the area-above method of calculating **dc** value of a waveform. Has it been confirmed? Explain.

Discussion

1. Summarize your conclusions from the observed and calculated results of this experiment.

2. Account for serious discrepancies between the observed and calculated results.

Chapter 3 Linear Wave Shaping Circuits

3-1 Objectives

The objectives of this chapter are to:

1. Review the characteristics of a low-pass *RC* (integrator) circuit.
2. Analyze the response of an *RC* low-pass circuit to a step (pulse) input.
3. Discuss the effects of a long, medium, or short time constant on the output waveform of an *RC* low-pass circuit when the input is a pulse wave.
4. Review the characteristics of an *RC* high-pass (differentiator) circuit.
5. Discuss the response of an *RC* high-pass circuit to a pulse waveform.
6. Discuss the effects of a long, medium, or short time constant on the output waveform of an *RC* high-pass circuit when the input is a pulse wave.
7. Review important characteristics of an *RL* circuit.
8. Explain the response of a series-connected *RLC* circuit to a step input.
9. Explain the damping characteristics of an *RLC* circuit.
10. Review the function and operation of a simple attenuator.
11. Explain the function of a compensating capacitor in the attenuator circuit.

3-2 Self-Evaluation Questions

Watch for the answers to the following questions as you read the chapter. They should help you to understand the material presented. When you have finished the chapter, return to this section and be sure that you can answer all the questions. Then test your understanding by working out solutions to the Problems.

It is strongly recommended that these questions be satisfactorily answered before attempting to solve the Problems (Sec. 3-15).

1. What is meant by an *RC low*-pass circuit and why is it also called an *integrator*.
2. A step input is applied to an *RC* low-pass circuit. Draw the circuit current waveform and the voltage waveform that appear across the capacitor.
3. What is meant by the time constant (τ) of an *RC* circuit?
4. A pulse wave is applied to an *RC* low-pass circuit. Give equations that may be used to find the steady-state output voltage waveform.
5. Discuss briefly the effects of a *long*-, *medium*-, or *short*-time-constant

circuit on the output waveform of a *low*-pass *RC* circuit when the input is a pulse wave.

6. What is meant by a *high*-pass *RC* circuit and why is it sometimes called a *differentiator*?
7. A step input is applied to an *RC high*-pass circuit. Find and draw the voltage waveform appearing across the resistor (output).
8. A pulse wave is applied to an *RC high*-pass circuit. Give equations that may be used to find the steady-state output voltage waveform.
9. Discuss and explain the effects of a *long- medium-*, or *short*-time-constant circuit on the output waveform of a *high*-pass *RC* circuit when the input is a pulse wave.
10. Discuss briefly the response of an *RL* circuit to a step input, and explain why *RL* circuits are rarely used for waveshaping.
11. A step input is applied to a series-connected *RLC* circuit. Find an expression for the current in the circuit.
12. Discuss the damping characteristics of an *RLC* circuit.
13. What is the function of an *attenuator*? Explain.
14. Describe the function of a compensating capacitor and show how its value is calculated for a given attenuator circuit.

3-3 *RC* Low-Pass Circuit

An *RC* low-pass circuit is shown in Fig. 3-1a. If a sinusoidal input, e_i, is applied to it, then the output, e_o, is given by:

$$e_o = e_i\left(\frac{-jX_c}{R - jX_c}\right) \text{ volts (V)} \qquad (3\text{-}1)$$

where:

e_o is the output voltage (V)
e_i is the input voltage (V)
X_c is the capacitive reactance of C (Ω)
R is the ohmic value of R (Ω)
j is the imaginary axis of the complex plane

Equation (3-1) shows that e_o is a frequency-dependent quantity, since

$$X_c = \frac{1}{2\pi fC}.$$

(a) *RC* low-pass circuit

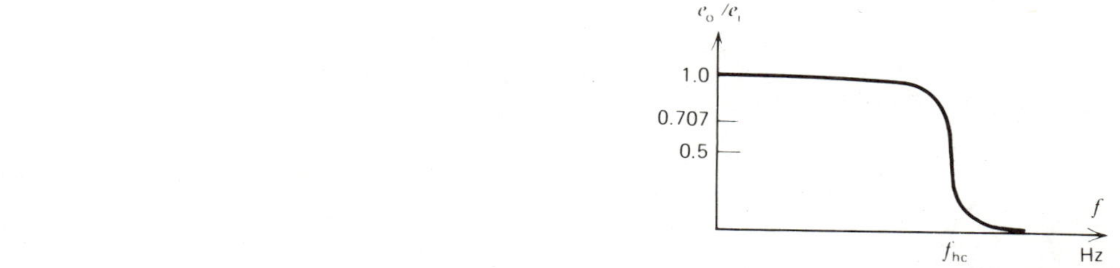

(b) Frequency response of a low-pass circuit

Figure 3-1 A low-pass *RC* circuit and its frequency response.

At *low* frequencies, when $-jX_c \gg R$, Eq. (3-1) reduces to:

$$e_o \cong e_i \text{ (input supply voltage)}$$

At *high* frequencies, when $-jX_c \ll R$, the ratio of the term in the parentheses in Eq. (3-1) is almost zero. Under this condition (high frequency), Eq. (3-1) reduces to:

$$e_o \cong 0$$

Thus, a low-pass circuit behaves like a filter that attenuates signals of high frequencies and does *not affect* signals of *low* frequencies. A typical frequency-response curve of a low-pass filter is shown in Fig. 3-1b. The frequency at which the output (e_o) becomes 70.7 percent of the input (e_i) is called the *cutoff* frequency, f_{hc} or the *half-power point* frequency, and is given by:

$$f_{hc} = \frac{1}{2\pi RC} = \frac{1}{2\pi\tau} \text{ hertz (Hz)} \tag{3-2}$$

where τ (tau), the time constant of the filter, is:

$$\tau = RC \text{ seconds (s)} \tag{3-3}$$

Example 3-1

Refer to the low-pass circuit shown in Fig. 3-1a. Assume that $R = 5\ \text{k}\Omega$ and $C = 0.01\ \mu\text{F}$. Find:

a. The time constant.
b. The cutoff frequency of the filter.

Solution

a.

$$\tau = RC \qquad (3\text{-}3)$$
$$= 5\ \text{k}\Omega \times 0.01\ \mu\text{F}$$
$$= \mathbf{50\ \mu s}$$

b.

$$f_{hc} = \frac{1}{2\pi RC} \qquad (3\text{-}2)$$
$$= \frac{1}{2\pi \times 5\ \text{k}\Omega \times 0.01\ \mu\text{F}}$$
$$= \mathbf{3.18\ kHz}$$

3-4 Step Input to an *RC* Low-Pass

Figure 3-2a shows a **dc** input, E_i, applied to an *RC* low-pass circuit through a switch. Closing the switch is similar to applying a step input to the circuit.

Let us analyze the manner in which the capacitor charges when a step input is applied.

$$E_i = v_R + v_C$$
$$= iR + v_C$$

but

$$i = C\frac{dv_c}{dt} \qquad (1\text{-}6)$$

$$\therefore E_i = RC\frac{dv_c}{dt} + v_C$$

Solving this differential equation* yields the instantaneous voltage across the capacitor, or,

$$v_C = E_i(1 - e^{-(t/RC)}) \text{ volts (V)} \qquad (3\text{-}4a)$$

* Calculus is used to show the derivation of Eq. (3-4) only. No example or problem requires knowledge of calculus.

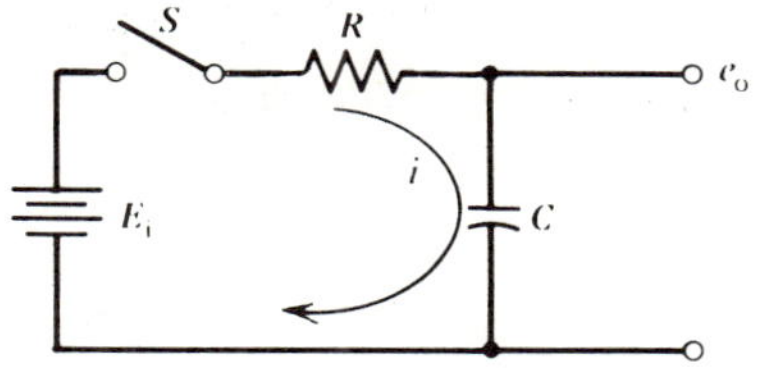

(a) An *RC* low-pass circuit

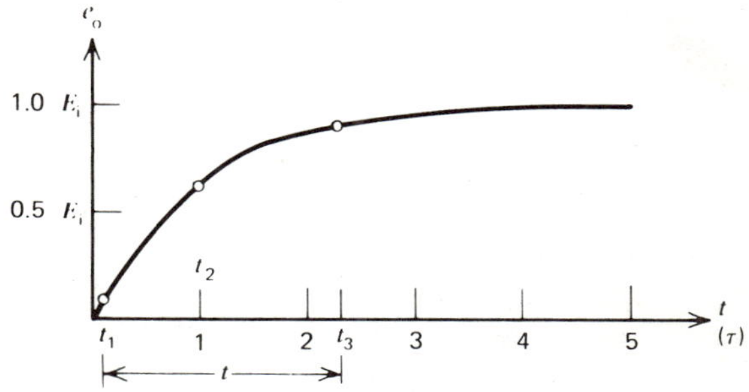

(b) Transient response

Figure 3-2 An *RC* low-pass circuit and its transient response.

The capacitive *charge* curve is shown in Fig. 3-2b.

The instantaneous voltage (v_C) of Eq. (3-4a) can be written in its other forms as follows:

$$v_C = E_i - E_i e^{-(t/RC)} \tag{3-4b}$$

or

$$= E_i - E_i e^{-(t/\tau)} \tag{3-4c}$$

or

$$= I_0 R - I_0 R e^{-(t/\tau)} \tag{3-4d}$$

or

$$= I_0 R[1 - e^{-(t/\tau)}]$$

where:

$I_0 = \dfrac{E_i}{R}$, the initial (maximum) current in amperes (A)

$$\tau = RC \text{ in seconds (s)} \tag{3-3}$$

Let us identify some important characteristics of the capacitor charge curve shown in Fig. 3-2b. Equation (3-3) may be solved to find that:

1. At time $t_0 = 0$, $v_C = 0$ V (initially).
2. At time $t_1 = 0.1\tau$, $v_C = 0.1E_i$ (10% of E_i).
3. At time $t_2 = \tau$, $v_C = 0.632E_i$ (63.2% of E_i).
4. At time $t_3 = 2.3\tau$, $v_C = 0.9E_i$ (90% of E_i).
5. At time $t_4 = 5\tau$, $v_C = 0.993E_i \approx E_i$.
6. The slope of the capacitor charge curve is calculated by differentiating Eq. (3-4). Its initial slope at $t = 0$ works out to be:

$$\frac{dv_{C_0}}{dt} = \frac{E_i}{\tau}, \frac{\text{volts}}{\text{seconds}} \text{ (v/s)} \tag{3-5}$$

Note that the rise time (t_r) of the capacitor charge curve is the time to go from 10 percent E_i to 90 percent E_i, as shown in Fig. 3-2b. Using the above values:

$$\begin{aligned} t_r &= t_3 - t_1 \\ &= 2.3\tau - 0.1\tau \\ &= 2.2\tau \text{ seconds (s)} \end{aligned} \tag{3-6}$$

Example 3-2

Consider the RC low-pass circuit shown in Fig. 3-2a. Assume that $E_i = 10$ V, $R = 10$ kΩ, and $C = 0.047\ \mu$F. Find:

a. The time constant (τ) of the circuit.
b. The initial slope of the capacitor charge curve.
c. Time for the capacitor to *fully* charge.
d. The initial current in the circuit.
e. The voltage across the capacitor 1 ms after the switch has been closed.
f. How long after the switch has been closed that the capacitor voltage reaches 8 volts.
g. The rise time (t_r).

Solution

a.

$$\begin{aligned} \tau &= RC \qquad (3\text{-}3) \\ &= 10\text{ k}\Omega \times 0.047\ \mu\text{F} \\ &= 0.47 \times 10^{-3} \\ &= \mathbf{0.47\ ms} \end{aligned}$$

b. Initial slope $= \dfrac{E_i}{\tau}$ (3-5)

$$= \frac{10 \text{ V}}{0.47 \times 10^{-3} \text{ s}}$$

$$= 21.3 \times 10^3 \text{ volts/second}$$

$$= \mathbf{2.3 \text{ kV/s}}$$

c. Total time, $t_t = 5\tau$

$$= 5 \times 0.47 \text{ ms}$$

$$= \mathbf{2.35 \text{ ms}}$$

d. $I_0 = \dfrac{E_i}{R}$

$$= \frac{10 \text{ V}}{10 \text{ k}\Omega}$$

$$= 1 \times 10^{-3} \text{ A} = \mathbf{1.0 \text{ mA}}$$

e. $v_c = E_i - E_f e^{-(t/\tau)}$ (3-4c)

$$v = 10 - 10e^{-(1.0 \text{ ms}/0.47 \text{ ms})}$$

$$= 10 - 10 \times 0.119$$

$$= 10 - 1.19$$

$$= \mathbf{8.81 \text{ volts}}$$

f. $v_c = E_i - E_i e^{-(t/\tau)}$ (3-4c)

$$8 = 10 - 10e^{-(t/\tau)}$$

$$e^{-(t/\tau)} = \frac{10 - 8}{10}$$

$$= 0.2$$

$$\frac{t}{\tau} = 1.61$$

Since $\dfrac{t}{\tau} = 1.61$, then solving for t yields:

$$t = 1.61 \times \tau = 1.61 \times 0.47 \text{ ms}$$

$$= \mathbf{0.757 \text{ ms}}$$

g.

$$t_r = 2.2\tau$$
$$= 2.2 \times 0.47 \text{ ms}$$
$$= \mathbf{1.03\ ms}$$

Example 3-3

Consider the circuit shown in Fig. 3-3a.

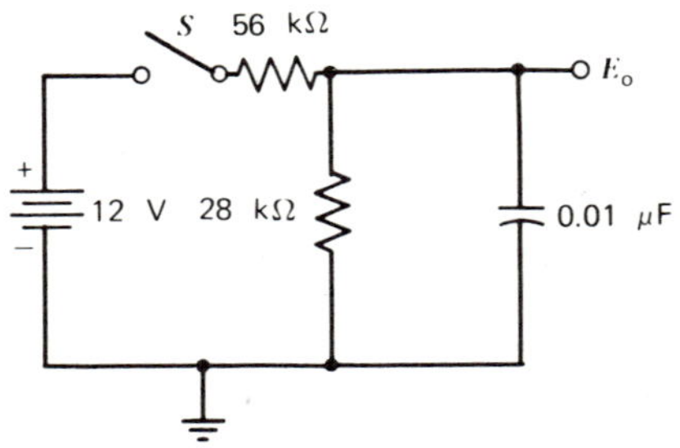

(a) Circuit for Ex. 3-3

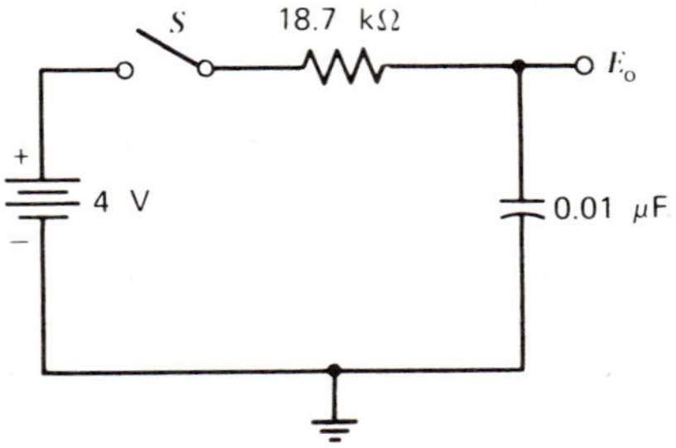

(b) Thévenin's equivalent

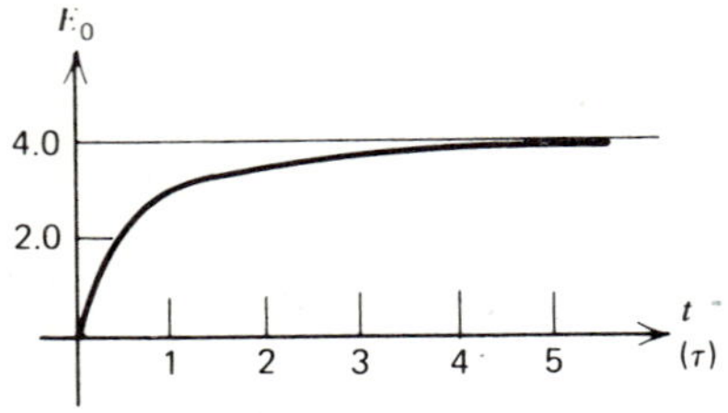

(c) The capacitor charge curve

Table 3–1

t	v_c (volts)
1 τ	2.53
2 τ	3.45
3 τ	3.80
4 τ	3.93
5 τ	3.97

(d) Table for the capacitor charge curve

Figure 3-3 Circuits and waveforms for Ex. 3-3.

a. Find charge on the capacitor 0.2 ms after the switch is closed.
b. Sketch the capacitor charge curve by finding values for the capacitor voltage in 1τ increments.

Solution

The circuit of Fig. 3-3a can be reduced to its Thévenin equivalent circuit (Fig. 3-3b) in the following manner:

$$E_{th} = \frac{28\ k\Omega}{(56 + 28)\ k\Omega} \times 12\ V$$

$$= 4\ \text{volts}$$

$$R_{th} = \frac{56 \times 28}{56 + 28}\ k\Omega$$

$$= 18.67\ k\Omega$$

a.

$$\tau = R_{th} \times C$$

$$\tau = 18.67\ k\Omega \times 0.01\ \mu F = 0.187\ ms$$

$$v_c = E_{th} - E_{th}e^{-(t/\tau)}$$

or

$$v_c = 4 - 4e^{-(0.2\,ms/0.187\,ms)}$$

$$= 4 - 4 \times 0.343$$

$$= \mathbf{2.63\ volts}$$

b. Calculate values for v_c to fill in Table 3-1 (Fig. 3-3d) in the following manner:

$$v_c\tau_1 = 4[1 - e^{-(1\tau/\tau)}]$$

$$= 4(1 - 0.368)$$

$$= \mathbf{2.53\ volts}$$

$$v_c\tau_2 = 4[1 - e^{-(2\tau/\tau)}]$$

$$= 4(1 - 0.135)$$

$$= \mathbf{3.45\ V}$$

and so on in increments of 1τ for values of t (Fig. 3-3d). The capacitor charge curve is drawn in Fig. 3-3c.

Suppose that the capacitor in Fig. 3-2a has some *initial* charge *before* the switch is closed or the step function is applied. This situation occurs frequently in pulse work. Let us analyze the manner in which the capacitor charges if such is the case.

An analysis of Eq. (3-4) shows that the capacitor voltage is trying to change in value by an amount equal to the initial voltage across the resistor. When t is significantly large as compared to τ, the change in the capacitor voltage is equal to the initial voltage across the resistor (I_0R). Writing Eq. (3-4d) again:

$$v_c = I_0R - I_0Re^{-(t/\tau)} \qquad \text{(3-4d)}$$

where:

$I_0 = E_i/R$, when there is *no* initial charge on the capacitor, and
$I_0 = (E_i \pm V_{ci})/R$, when there *is* an initial charge (V_{ci}) on the capacitor.

The "$\pm$" sign above is used to indicate that the value of the initial current (I_0) depends upon the polarity of the initial charge on the capacitor with respect to the applied voltage. Substituting this value of initial current in Eq. (3-4d) gives:

$$\Delta v_c = \left[\frac{(E_i \pm V_{ci})}{R} \times R\right] - \left[\frac{(E_i \pm V_{ci})}{R} \times R\right]e^{-(t/\tau)}$$

$$(v_c \pm V_{ci}) = (E_i \pm V_{ci}) - (E_i \pm V_{ci})e^{-(t/\tau)}$$

$$v_c = E_i - (E_i \pm V_{ci})e^{-(t/\tau)} \text{ volts (V)} \qquad \text{(3-7)}$$

where all terms have been defined earlier.

Example 3-4

Consider the circuit described in Ex. 3-1. Find the voltage that will appear across the capacitor (E_o), 0.8 ms after the switch has been closed if:

a. The initial charge on the capacitor is +4 V.
b. The initial charge on the capacitor is −2 V.

Solution

a.

$$v_c = E_i - (E_i - V_{ci})e^{-(t/RC)} \qquad \text{(3-7)}$$

$$= 10 - (10 - 4)e^{-(0.8\,\text{ms}/0.47\,\text{ms})}$$

(Note the sign of V_{ci})

$$= 10 - 6e^{-1.7}$$

$$= 10 - 6 \times 0.182$$

$$= \mathbf{8.91\ V}$$

b.

$$v_c = E_i - (E_i - V_{ci})e^{-(t/\tau)} \quad (3\text{-}7)$$
$$= 10 - (10 + 2)e^{-1.7}$$

(Note the sign of V_{ci})

$$= 10 - 12 \times 0.18$$
$$= \mathbf{7.81\ V}$$

Example 3-5

Refer to the circuit of Fig. 3-4a. Assume that the initial charge on the capacitor is 0 V (switch at position #2 for a time longer than 5τ). The switch is toggled between positions #1 and #2 and is kept at each position for 6 μs. Draw the voltage waveform that appears across the capacitor for two cycles of switching.

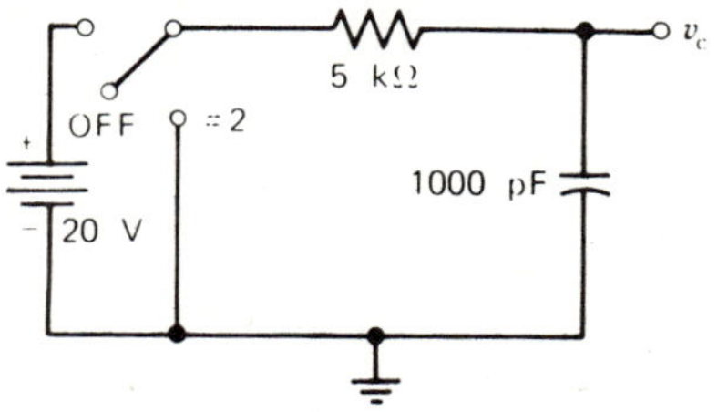

(a) Circuit for Ex. 3-5

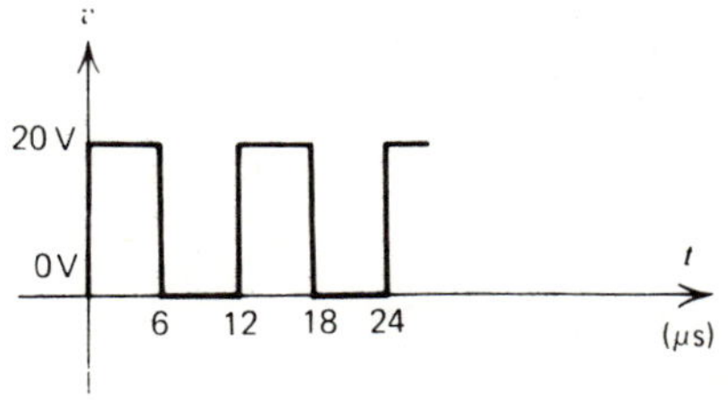

(b) Input waveform

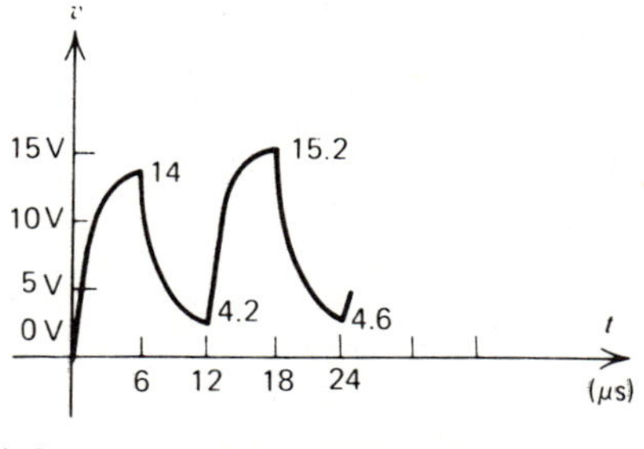

(c) Output waveform (across the capacitor)

Figure 3-4 Circuit and input/output waveforms for Ex. 3-5.

Solution

The series-connected *RC* circuit "sees" 20 V for 6 μs (position 1) and 0 V for 6 μs (position 2). Then the cycle repeats itself. Two complete cycles of the resultant input waveform applied to the *RC* circuit are shown in Fig. 3-4b.

$$\tau = 5\ \text{k}\Omega \times 0.0001\ \text{pF}$$
$$= 5\ \mu\text{s}$$
$$e^{-(t/\tau)} = e^{-(6\mu\text{s}/5\mu\text{s})}$$
$$= e^{-1.2}$$
$$= 0.3$$

At

$$t = 0$$
$$v_c = 0$$

1. From $t = 0$ to $t = 6\ \mu$s,

$$E_i = 20\ \text{V}$$

Thus at $t = 6\ \mu$s:

$$v_c = 20 - (20 + 0) \times 0.3 \qquad \text{(Using Eq. 3-7)}$$
$$= 20 - 20 \times 0.3$$
$$= 14\ \text{V}$$

2. From $t = 6\ \mu$s to $t = 12\ \mu$s,

$$E_i = 0\ \text{V}$$

Thus at $t = 12\ \mu$s

$$v_c = 0 - (0 - 14)e^{-(t/\tau)} \qquad \text{(Using Eq. 3-7)}$$
$$= 14 \times 0.3$$
$$= 4.2\ \text{V}$$

3. From $t = 12\ \mu$s to $t = 18\ \mu$s,

$$E_i = 20\ \text{V}$$
$$V_{c1} = 4.2\ \text{V}$$

Thus at $t = 18\ \mu s$,

$$v_c = 20 - (20 - 4) \times 0.3 \qquad \text{(Using Eq. 3-7)}$$
$$= 20 - 4.8$$
$$= \mathbf{15.2\ V}$$

4. From $t = 18\ \mu s$ to $t = 24\ \mu s$,

$$E_1 = 0\ \text{V}$$
$$V_{c1} = 15.2\ \text{V}$$

Thus at $t = 24\ \mu s$:

$$V_c = 15.2 \times 0.3$$
$$= \mathbf{4.6\ V}$$

This completes the two cycles. The resulting waveform is drawn in Fig. 3-4c.

Example 3-6
Refer to Ex. 3-5. After how many cycles does the capacitor charge/discharge voltage reach the two steady-state levels? Calculate these two voltage levels.

Solution
Two cycles have been completed. The capacitor voltage at the first half of the third cycle:

$$v_c = E_1 - (E_1 + V_{c1})e^{-(t/\tau)} \qquad (3\text{-}7)$$
$$\therefore v_3 = 20 - (20 - 4.6)0.3$$
$$= 20 - 4.6$$
$$= \mathbf{15.4\ V}$$

The capacitor voltage at the end of the third cycle:

$$v_c = 15.4 \times 0.3$$
$$= \mathbf{4.62\ V}$$

This value is practically the same as the one obtained for the end of the second cycle. Thus, the capacitor voltage will vary from approximately 15.4 V to 4.6 V from then on. It takes *three* cycles to reach the steady-state voltage levels of 15.4 V and 4.62 V.

3-5 Pulse Wave Input to an RC Low-Pass

As noted in Sec. 2-9, one cycle of a *square* wave is made up of two step-functions. The switching arrangement of Ex. 3-5 demonstrates this and generates a square wave input to the circuit. It could be made to generate a *pulse* wave by keeping the switch at *one* position considerably longer than the other. Thus, one way to analyze the response of an *RC* low-pass circuit to a pulse wave input is the step-by-step method of Ex. 3-5. We are sometimes interested only in the *steady-state* or the settled conditions. Depending on circuit values, the method used in Ex. 3-5 could be a long and tedious one. The steady-state voltage levels can be calculated *directly* by solving the *steady-state* charge and discharge equations.

Let us solve Ex. 3-6 for the steady-state output voltage levels, using a direct equation method. The input wave (Fig. 3-4b) and the approximate output steady-state waveform (Fig. 3-4c) are redrawn in Fig. 3-5.

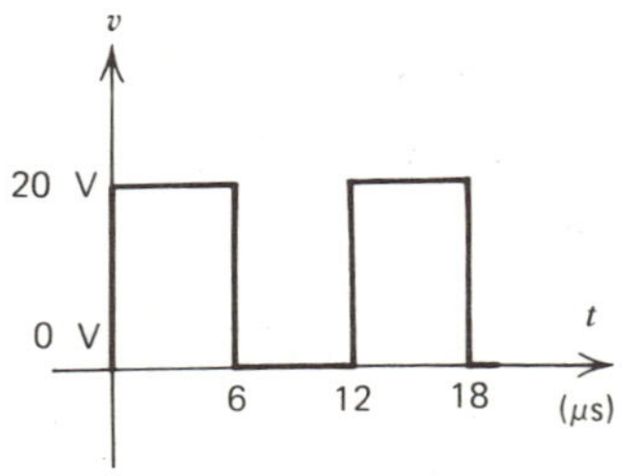

(a) Input waveform

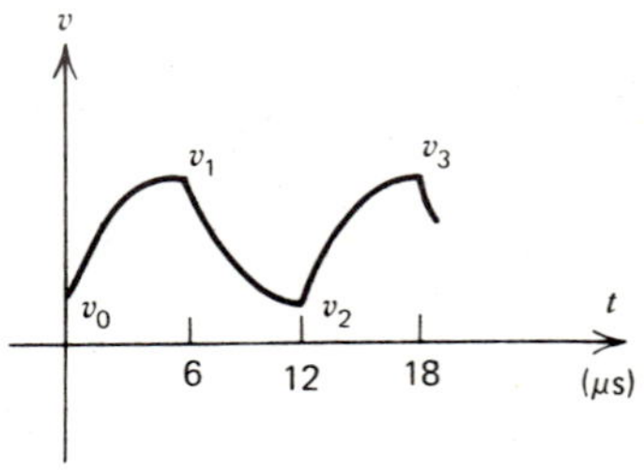

(b) Output waveform

Figure 3-5 Input and output waveforms for Ex. 3-6.

Let us assume that the waveform shown in Fig. 3-5b has reached its steady state. Then,

$$v_0 = v_2, \quad \text{the steady-state lower level} \tag{3-8}$$

and

$$v_1 = v_3, \quad \text{the steady-state upper level} \tag{3-9}$$

For the first half of a cycle, the waveform goes from v_0 to v_1 in 6 μs (Fig. 3-5b). Writing this voltage rise in equation form (Eq. 3-7) yields:

$$v_1 = E_i - (E_i - v_0)e^{-(t/\tau)}$$

$$\therefore v_1 = 20 - (20 - v_0)e^{-(6\mu s/5\mu s)}$$

$$= 20 - (20 - v_0)0.3$$

$$= 14 + 0.3v_0 \tag{3-10}$$

Similarly, the equation for the second half of a cycle (v_1 to v_2) is:

$$v_2 = E_i - (E_i - v_1)e^{-(t/\tau)} \tag{3-11}$$

$$\therefore v_2 = 0 - (0 - v_1)e^{-(6\mu s/5\mu s)}$$

$$= 0.3\, v_1$$

or

$$v_0 = 0.3\, v_1$$

Substituting this value of v_0 into Eq. (3-10) yields:

$$v_1 = 14 + 0.3 \times 0.3\, v_1$$

$$0.91\, v_1 = 14$$

$$v_1 = \frac{14}{0.91}$$

$$= \mathbf{15.4\ volts}$$

Substituting this value in Eq. (3-12) yields:

$$v_2 = 0.3 \times 15.4$$

$$= \mathbf{4.62\ volts}$$

This result agrees with the result obtained in Ex. 3-6, but the calculations are more direct and much simpler. Calculations for a pulse wave are similar, except that the time for the pulse width is usually different from the rest period, as shown in Ex. 3-7.

Example 3-7

A pulse wave shown in Fig. 3-6a is applied to an *RC* low-pass circuit whose $\tau = 50\ \mu s$.

a. Sketch an approximate steady-state output waveform.

b. Calculate the upper and the lower steady-state levels.

Solution

a. The approximate steady-state output waveform is shown in Fig. 3-6b.

b. Writing the charge equation from v_0 to v_1:

$$\begin{aligned} v_1 &= 5 - (5 + v_0)e^{-(25\ \mu s/50\ \mu s)} \\ &= 5 - (5 + v_0)e^{-0.5} \\ &= 5 - (5 + v_0)0.607 \\ &= 1.965 - 0.607v_0 \\ v_1 + 0.607v_0 &= 1.965 \qquad (3\text{-}13) \end{aligned}$$

Similarly, the equation that applies during the rest period is

$$v_2 = -5 - (-5 + v_1)e^{-(75\ \mu s/50\ \mu s)}$$

or

$$\begin{aligned} v_0 &= -5 - (-5 + v_1)e^{-1.5} \\ &= -5 - (-5 + v_1)0.223 \\ &= -3.885 + 0.223v_1 \\ 0 &= 0.223v_1 - v_0 - 3.885 \qquad (3\text{-}14) \end{aligned}$$

Solving Eqs. (3-13) and (3-14) in the following manner, yields:

$$\begin{aligned} 0.223v_1 - v_0 &= 3.885 \\ 0.223v_1 + 0.135v_0 &= 0.438 \qquad (\text{Eq. 3-13} \times 0.223) \\ -1.135v_0 &= 3.447 \\ v_0 &= -\left(\frac{3.447}{1.135}\right) \\ &= -3.04 \text{ volts} \end{aligned}$$

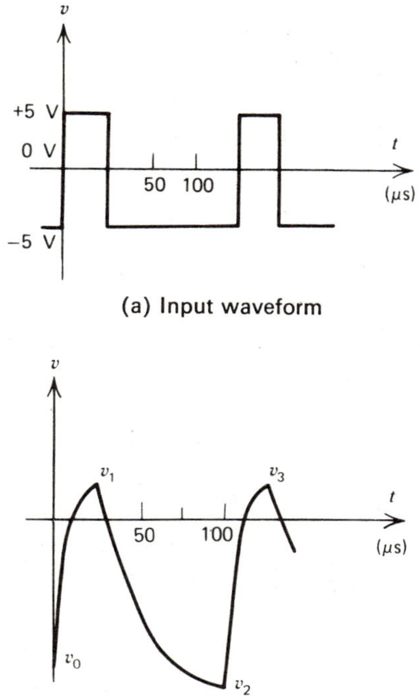

Figure 3-6 Input and output waveforms for Ex. 3-7.

From Eq. (3-14):

$$\begin{aligned} v_1 &= 1.965 - 0.607v_o \\ &= 1.965 - 0.607 \times 3.04 \\ &= 1.965 - 1.845 \\ &= 0.12 \text{ volts} \end{aligned}$$

Thus the upper level = **0.12 volts.**
And the lower level = **−3.03 volts.**

An examination of Eq. (3-7) reveals that the charge on the capacitor (v_c) is basically determined by the ratio t/τ when E_i and V_{ci} are constant. For a pulse waveform, τ of the low-pass circuit may be short, long, or medium compared to t_p, the pulse duration (not the period) of the input pulse wave.

The output waveforms that result during each of these three conditions are shown in Fig. 3-7.

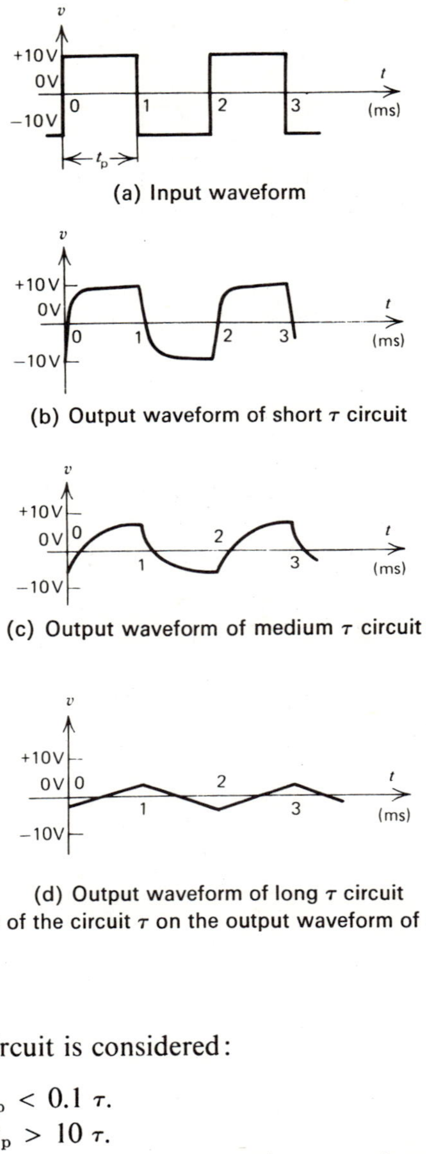

Figure 3-7 Effects of the circuit τ on the output waveform of an *RC* low-pass circuit.

The τ of a circuit is considered:

a. *Long* when $t_p < 0.1\ \tau$.
b. *Short* when $t_p > 10\ \tau$.
c. *Medium* when τ lies between the abovementioned two extremes, that is, $10\ \tau > t_p > 0.1\tau$.

where:

t_p is the pulse duration or width or capacitor charge (or discharge) time allowed by the *input* waveform (s).
τ is the circuit time constant (s).

Note that when circuit τ is *very long* compared to the pulse width, t_p, the step input is converted to a ramp output (Fig. 3-7d). The low-pass circuit under these conditions behaves like and is sometimes called an *integrator*, that is, the output is the integral of the input waveform.

3-6 RC High-Pass Circuit

An *RC* high-pass circuit is shown in Fig. 3-8a. If a sinusoidal input, e_i, is applied to it, then the output, e_o, is given by:

$$e_o = \left(\frac{R}{R - jX_c}\right) e_i \text{ volts (V)} \tag{3-15}$$

where:

e_o is the output voltage, volts (V)
e_i is the input voltage, volts (V)
X_c is the capacitive reactance of C (Ω)
R is the resistance (Ω)
j is the imaginary axis of the complex plane

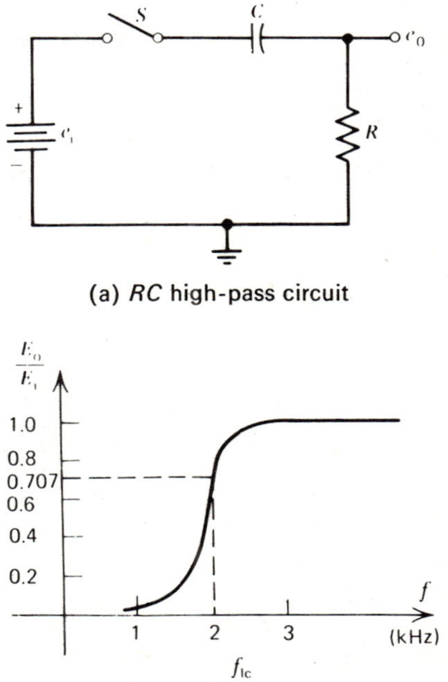

(a) *RC* high-pass circuit

(b) Frequency-response curve

Figure 3-8 Frequency response of a high-pass *RC* circuit.

Equation (3-15) shows that e_o is a frequency-dependent quantity ($X_c = 1/2\pi fc$). At low frequencies, $-jX_c \gg R$. Under this condition, the ratio of the parenthetical expression in Eq. 3-15 becomes almost zero, resulting in:

$$e_o \cong 0$$

At high frequencies, $-jX_c \ll R$. Under this condition, the ratio of the parenthetical expression in Eq. 3-15 is unity, resulting in $e_o \cong e_i$.

Thus we see that a *high*-pass filter *attenuates* signals of *low* frequencies and does not affect the high-frequency signals. A typical frequency-response curve of a high-pass filter is shown in Fig. 3-8b. The frequency at which the output becomes 70.7 percent of the input is called the lower-cutoff frequency, f_{1c}. It is given by the expression

$$f_{1c} = \frac{1}{2\pi} \times \frac{1}{RC} = \frac{1}{2\pi} \times \frac{1}{\tau} \text{ hertz (Hz)} \qquad (3\text{-}16)$$

where all terms have been defined earlier. Note that Eq. (3-16) is identical to Eq. (3-2).

Example 3-8

Consider the high-pass circuit shown in Fig. 3-8a. Assume $R = 25\ \text{k}\Omega$ and $C = 0.1\ \mu\text{F}$. Find f_{1c} of the filter.

Solution

$$f_{1c} = \frac{1}{2\pi} \times \frac{1}{RC} \qquad (3\text{-}16)$$

$$= \left(\frac{1}{2 \times 3.14}\right) \times \left(\frac{1}{25\ \text{k}\Omega \times 0.1\ \mu\text{F}}\right)$$

$$= \mathbf{63.7\ Hz}$$

3-7 Pulse Wave Input to an *RC* High Pass

The *RC* high-pass circuit (Fig. 3-8a) differs from the *RC* low-pass circuit (Fig. 3-1a) in that the components *R* and *C* are interchanged. In the high-pass circuit, the output is taken across the resistor rather than the capacitor. An algebraic equation for the output

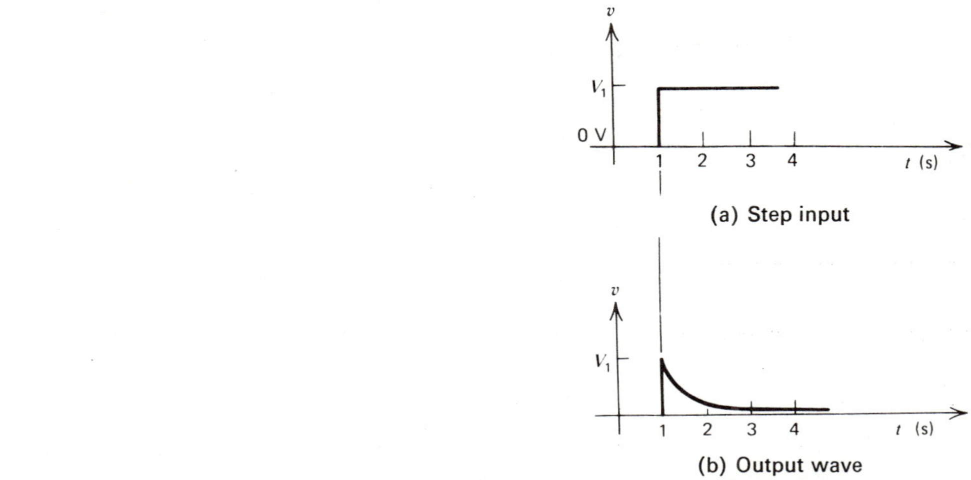

Figure 3-9 Step input to an *RC* high-pass circuit.

voltage can be derived using Eq. (3-7) when the input is a step function as shown in Fig. 3-9a.

$$\begin{aligned} v_R &= E_i - v_C \\ &= E_i - [E_i - (E_i \pm V_{ci})]e^{-(t/\tau)} \\ &= E_i - E_i + (E_i \pm V_{ci})e^{-(t/\tau)} \\ &= (E_i \pm V_{ci})e^{-(t/\tau)} \text{ volts (V)} \end{aligned} \tag{3-17}$$

When the initial charge on the capacitor V_{ci} is zero, Eq. (3-17) simplifies to:

$$v_R = E_o = E_i e^{-(t/\tau)} \text{ volts (V)} \tag{3-18}$$

Figure 3-9b shows the response of an *RC* high-pass circuit to a step input. Note the exponential decay of the output as given by Eq. (3-18).

Example 3-9

The input (E_i) applied to the circuit of Fig. 3-10 is a step function of 15 V **dc**. Find:

a. The output voltage (e_o) 0.1 ms after the switch has been closed.

b. The voltage across the capacitor 0.1 ms after the switch has been closed.

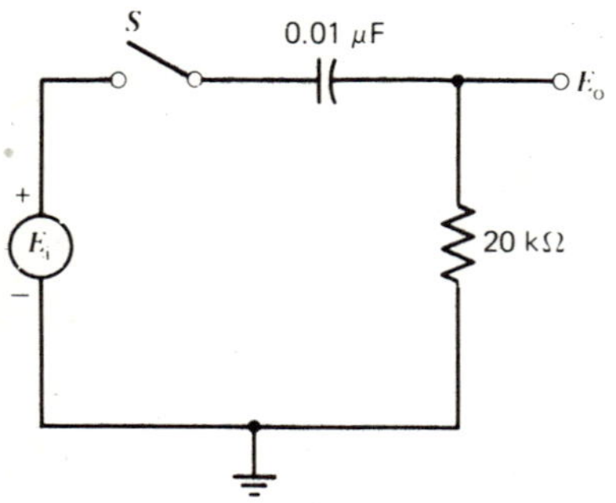

Figure 3-10 An *RC* high-pass circuit for Ex. 3-9.

Solution

a.

$$\tau = RC \tag{3-3}$$

$$= 20 \text{ k}\Omega \times 0.01\ \mu\text{F}$$

$$= 0.2 \text{ ms}$$

$$e_o = E_i e^{-(t/\tau)}$$

$$= 15e^{-(0.1 \text{ ms}/0.2 \text{ ms})}$$

$$= 15e^{-0.5}$$

$$= 15 \times 0.607$$

$$= \mathbf{9.1 \ volts}$$

b.

$$v_c = E_i - v_R$$

$$= 15 - 9.1$$

$$= \mathbf{5.9 \ V}$$

Example 3-10

Input to the circuit of Fig. 3-10 is a square wave shown in Fig. 3-11a. Draw the first three cycles of the output voltage waveform after the switch has been closed.

Solution

Immediately after the switch is closed, the entire 25 V appears across the resistor. Thus e_o at t_o is 25 V.

From t_o to t_1, e_o decays in accordance with Eq. (3-18).

$$t_p = 0.25 \text{ ms}$$

$$\tau = 0.2 \text{ ms}$$

$$e^{-t/\tau} = e^{-(0.25\text{ ms}/0.2\text{ ms})}$$
$$= e^{-1.25}$$
$$= 0.2865$$
$$V(t_1) = E_i e^{-(t/\tau)}$$
$$= 25 \times 0.2865$$
$$= \mathbf{7.16\ V}$$

Right at $t = t_1$,

$$v_R = 7.16\text{ V}$$
$$v_c = 25 - 7.16$$
$$= 17.84\text{ V}$$

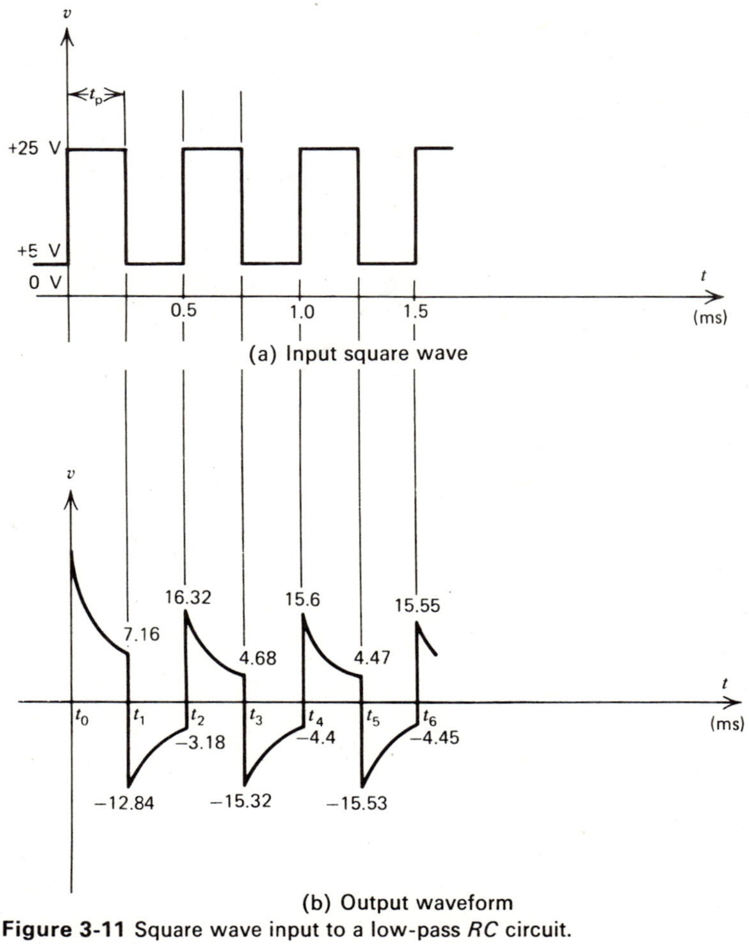

Figure 3-11 Square wave input to a low-pass *RC* circuit.

Immediately after t_1,

$$E_i = +5\text{ V}$$
$$v_{ci} = 17.84\text{ V}$$
$$e_o = +5 - 17.84$$
$$= -\mathbf{12.84\text{ V}}$$

From t_1 to t_2, the input stays at +5 V. V_{ci} = 17.84 V. Applying Eq. (3-18) gives:

$$v(t_2) = (5 - 17.84) \times 0.2865$$
$$= -12.84 \times 0.2865$$
$$= -\mathbf{3.68\text{ V}}$$

Immediately after t_2:

$$E_i = 25\text{ V}$$
$$V_{c_2} = 5 + 3.68$$
$$= 8.68\text{ V}$$
$$e_o = \mathbf{16.32\text{ V}}$$

From t_2 to t_3, the input stays at 25 V. V_{ci} = 8.68 V. Applying Eq. (3-17) gives:

$$v_{(t_3)} = (25 - 8.68) \times 0.2865$$
$$= \mathbf{4.68\text{ V}}$$
$$V_{c3} = 25 - 4.68$$
$$= 20.32\text{ V}$$

Immediately after t_3:

$$E_i = 5\text{ V}$$
$$e_o = (5 - 20.32)$$
$$= -\mathbf{15.32}$$

From t_3 to t_4:

$$v_{(t_4)} = -15.32 \times 0.2865$$
$$= -\mathbf{4.4\text{ V}}$$
$$V_{c_4} = 5 + 4.4$$
$$= 9.4\text{ V}$$

Immediately after t_4:

$$E_1 = 25$$
$$v_c = 9.4$$
$$\therefore e_o = 25 - 9.4$$
$$= \mathbf{15.6\ V}$$

From t_4 to t_5:

$$v_{(t_5)} = 15.6 \times 0.2865$$
$$= \mathbf{4.47\ V}$$
$$v_{c_5} = 25 - 4.47$$
$$= 20.53\ \text{V}$$

Immediately after t_5:

$$E_1 = 5\ \text{V}$$
$$e_o = 5 - 20.53$$
$$= \mathbf{-15.53\ V}$$

From t_5 to t_6:

$$v_{t_6} = 15.53 \times 0.2865$$
$$= \mathbf{-4.45\ V}$$

Three cycles of the output waveform of Ex. 3-10 are drawn in Fig. 3-11b. The waveform has settled down to its steady-state conditions. It is important to note the following points:

1. The instantaneous transitions in the output waveform that occur at the time of the instantaneous transitions in the input voltage are equal to the *peak-to-peak* value of the input wave.
2. One basic property of a capacitor is that it does not allow **dc** to go through it. This is shown to be true by the output waveform of Fig. 3-11b. The input is really a pulsating voltage. It could be thought of as being made up of a 20-$V_{p\text{-}p}$, 2-kHz square wave (**ac**) and a 15 V **dc**. The steady-state output waveform contains only **ac**. As a matter of fact, the steady-state output voltage waveform will be exactly the same if the input was simply a 20-$V_{p\text{-}p}$, 2-kHz (**ac**) square wave. This is confirmed by the results of Ex. 3-11.

3. The output voltage waveform is symmetrical around the zero axis.

Example 3-11
Input to the circuit of Fig. 3-10 is a 20-$V_{p\text{-}p}$, 2-kHz square wave. Find and draw the steady-state output voltage waveform.

Solution
The input and the output waveforms are drawn in Fig. 3-12. Equation (3-18) applies to the exponential decay.

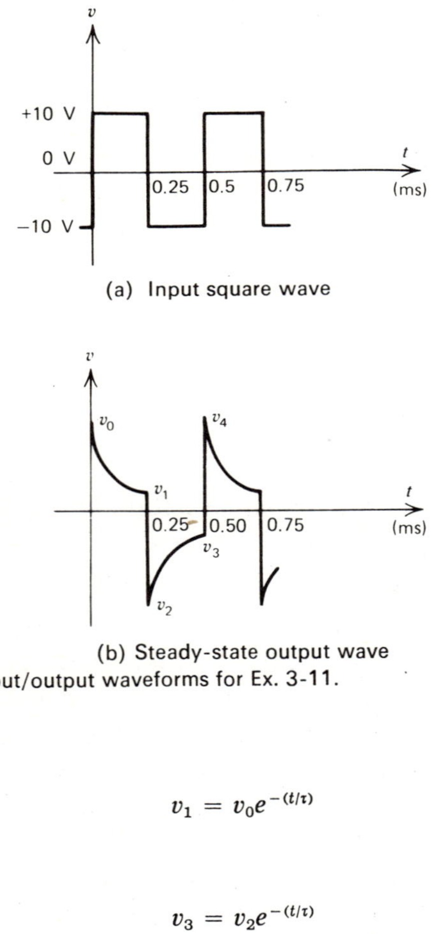

Figure 3-12 Input/output waveforms for Ex. 3-11.

Accordingly:

$$v_1 = v_0 e^{-(t/\tau)} \qquad (3\text{-}19)$$

and

$$v_3 = v_2 e^{-(t/\tau)} \qquad (3\text{-}20)$$

also,

$$v_1 - v_2 = 20 \qquad (3\text{-}21)$$

and

$$v_4 - v_3 = 20 \qquad (3\text{-}22)$$

$$v_o - v_3 = 20 \qquad (v_o = v_4)$$

As in Ex. 3-10:

$$e^{-(t/\tau)} = 0.2865$$

Solve the above four Equations for v_1, v_2, v_3, and v_o.

$$v_1 - v_2 = 20$$

$$0.2865\, v_o - v_2 = 20 \qquad (3\text{-}23)$$

also

$$v_4 - v_3 = 20$$

$$v_o - 0.2865\, v_2 = 20 \qquad (3\text{-}24)$$

Solving Eqs. (3-24) and (3-25):

$$0.2865\, v_o - v_2 = 20$$

$$0.2865\, v_o - 0.082\, v_2 = 5.7$$

$$0.918\, v_2 = 14.3$$

$$v_2 = \mathbf{-15.6\ V}$$

$$v_3 = \mathbf{-4.4\ V}$$

$$v_1 - v_2 = 20 \qquad (3\text{-}21)$$

$$v_1 = 20 + v_2$$

$$v_1 = 20 + (-15.6)$$

$$= \mathbf{4.4\ V}$$

Similarly,

$$v_o = 20 + (-4.4)$$

$$= \mathbf{15.6\ V}$$

These directly computed voltage levels for the steady-state output waveform are practically the same as were found in Ex. 3-10.

Example 3-12
A pulse waveform of the type shown in Fig. 3-13a is applied to an RC high-pass circuit whose time-constant τ is 10 μs. Find and draw the steady-state output waveform.

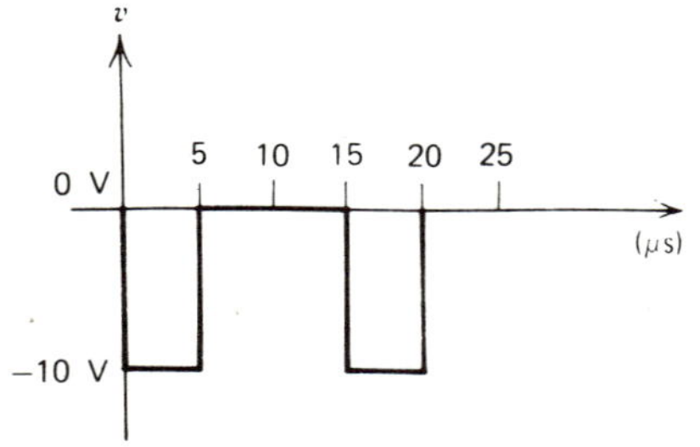

(a) Input pulse wave

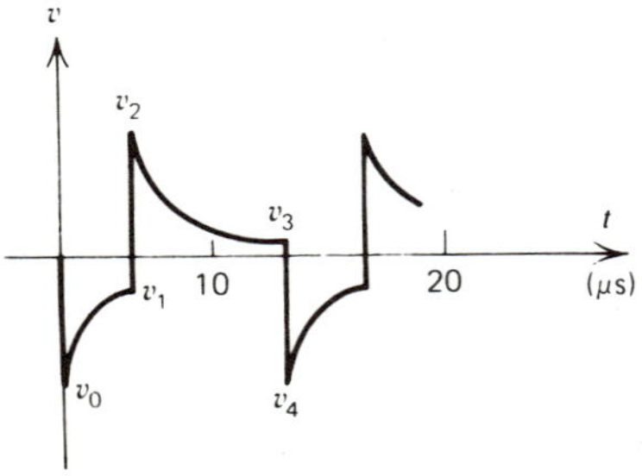

(b) Steady-state output wave

Figure 3-13 Input/output waveforms for Ex. 3-12.

Solution
An approximate output waveform is shown in Fig. 3-13b. The four voltage levels that are to be computed are labeled as v_0, v_1, v_2, and v_3.

Using Eq. (3-18) gives:

$$v_3 = v_2 e^{-(10\,\mu s/10\,\mu s)}$$
$$= v_2 e^{-1} = 0.368\ v_2$$

and

$$v_1 = v_0 e^{-(5\,\mu s/10\,\mu s)}$$
$$= v_0 e^{-0.5}$$
$$= 0.607\ v_0$$

also,

$$v_2 - v_1 = 10 \text{ and } v_3 - v_0 = 10$$

or

$$v_2 - 0.607\, v_0 = 10$$

and

$$0.368\, v_2 - v_0 = 10$$

Solving these two equations gives:

$$0.368\, v_2 - v_0 = 10$$

$$0.368\, v_2 - 0.223\, v_0 = 3.68$$

$$-0.777\, v_0 = 6.32$$

$$v_0 = \mathbf{-8.13\ V}$$

$$v_1 = -8.13 \times 0.607$$

$$= \mathbf{-4.94\ V}$$

$$v_2 = 10 - 4.94$$

$$= \mathbf{5.06\ V}$$

$$v_3 = 10 - 8.13$$

$$= \mathbf{1.87\ V}$$

The shape of the output waveform of an *RC* high-pass circuit depends upon the value of the circuit τ (compared to t_p). τ may be long, short, or medium for a given set of circuit conditions. The output waveforms that result under these three situations for an *RC* high-pass are shown in Fig. 3-14. The output waveform shown in Fig. 3-14b (long τ) is very similar to the input waveform. However, usually a slight tilt exists in it. A common way to describe the extent of tilt present in a waveform is to describe its *fractional tilt* F_t.

$$F_t = \frac{t_p}{t_t} \tag{3-25}$$

where:

t_p is the pulse width
t_t is the tilt time to reach 0 V (Fig. 3-14b)

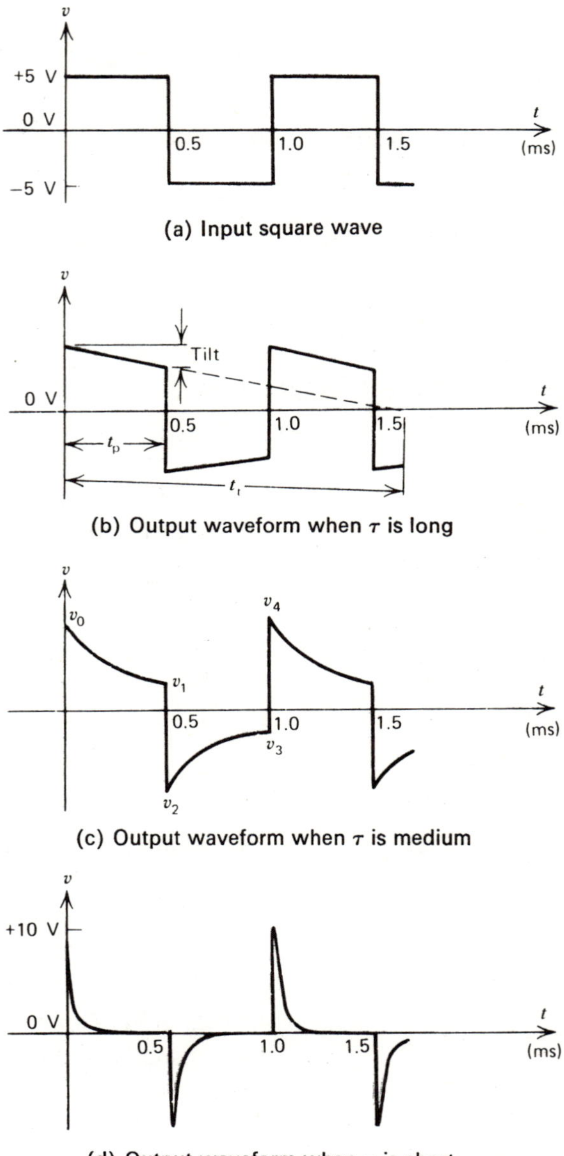

Figure 3-14 Effects of τ on the output waveform of an *RC* low-pass circuit.

Note that the *smaller* the ratio F_t (fractional tilt), *the lower* the tilt. The output waveform under the short τ condition is shown in Fig. 3-14d. It basically consists of positive and negative spikes. It is really a *differentiated* square wave. The *RC* high-pass circuit under these conditions (short τ) is also known as a *differentiator*. This type of waveform is sometimes used to trigger devices in other electronic circuits. Consequently, it is also referred to as the trigger pulse. It is also possible to obtain either positive or negative triggers. We simply clip off either the negative or the positive pulses from the differentiated wave (Fig. 3-14d) using clippers (Sec. 4-6).

3-8 Step Input to an *RL* Circuit

Simple *RL* circuits are seldom used in switching circuits, especially in integrated circuits (IC). However there are times when an inductor is still used, as in the case of current sweep generation (magnetic deflection) and in tunnel diode monostable/astable circuits. Also, the inherent inductance of wiring especially noticeable in high-frequency circuits must sometimes be considered. Let us briefly review the response of an *RL* circuit to a step input. Consider the circuit shown in Fig. 3-15a.

$$E_i = v_L + v_R$$

$$= L\frac{di}{dt} + iR$$

Solution of the above differential equation is:

$$i = \frac{E_i}{R}(1 - e^{-(t/\tau)})$$

where

$$\tau = \frac{L}{R}$$

$$v_R = iR = E_i(1 - e^{-(t/\tau)}) = E_i(1 - e^{-(tR/L)}) \qquad (3\text{-}26)$$

and

$$v_L = E_i - v_R$$

$$= E_i - [E_i(1 - e^{-(t/\tau)})]$$

$$= E_i e^{-(t/\tau)} = E_i e^{-(tR/L)} \qquad (3\text{-}27)$$

Compare Eqs. (3-26) and (3-27) with Eqs. (3-4) and (3-18). Note that the exponential rise- or fall-controlling quantity (τ) is equal to

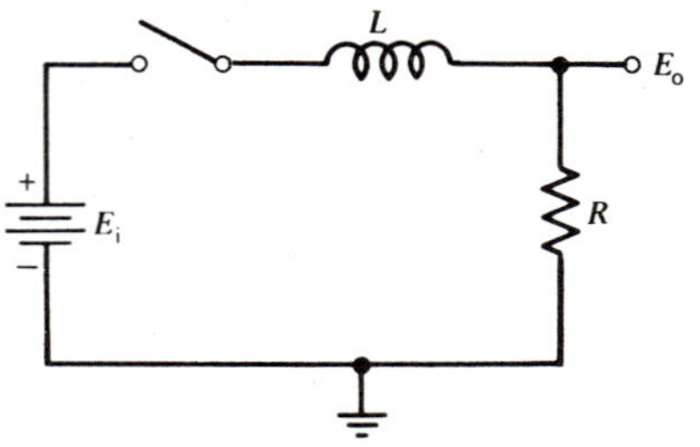

(a) Series-connected *RL* circuit

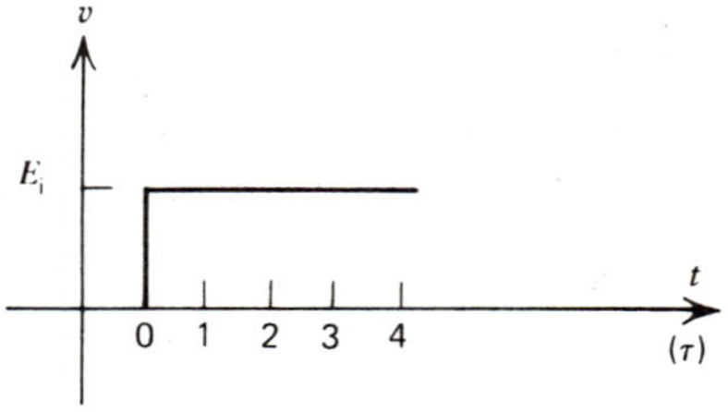

(b) Step input

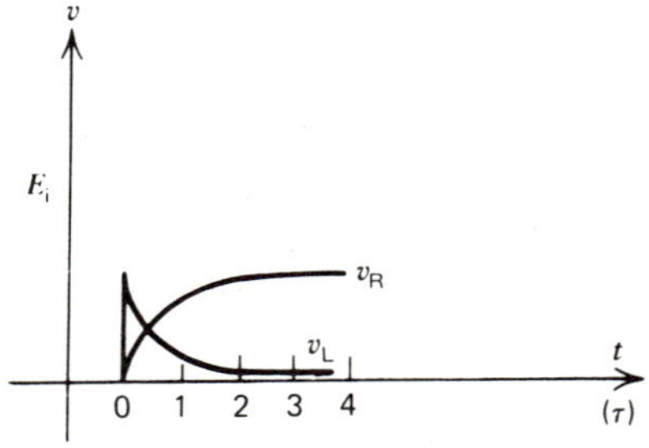

(c) Voltages across *L* and *R*

Figure 3-15 Response of an *RL* circuit to a step input.

$R \times C$ in an *RC* circuit and is equal to L/R in an *LR* circuit. It is further noted that the equation for v_R in an *RC* circuit is similar to equation for v_L in *RL* circuit. The same is true for v_c in an *RC* circuit and v_R in an *RL* circuit. Thus the response of circuit shown in Fig. 3-16a is similar to the response of circuit shown in Fig. 3-16b. Also, the response of the circuit shown in Fig. 3-16c is similar to the response of the circuit shown in Fig. 3-16d.

Example 3-13

Consider the *RL* circuit shown in Fig. 3-15a. Assume that $R = 100\ \Omega$, $L = 20$ mH, and $E_i = 25$ V. Find the voltage across the inductor 0.3 ms after the switch has been closed.

Solution

$$\tau = L/R$$
$$= 20 \text{ mH}/100\ \Omega$$
$$= 0.2 \text{ ms}$$
$$v_L = E_i e^{-(t/\tau)}$$
$$= 25e^{-(0.3 \text{ ms}/0.2 \text{ ms})} \qquad (3\text{-}27)$$
$$= \mathbf{5.58\ V}$$

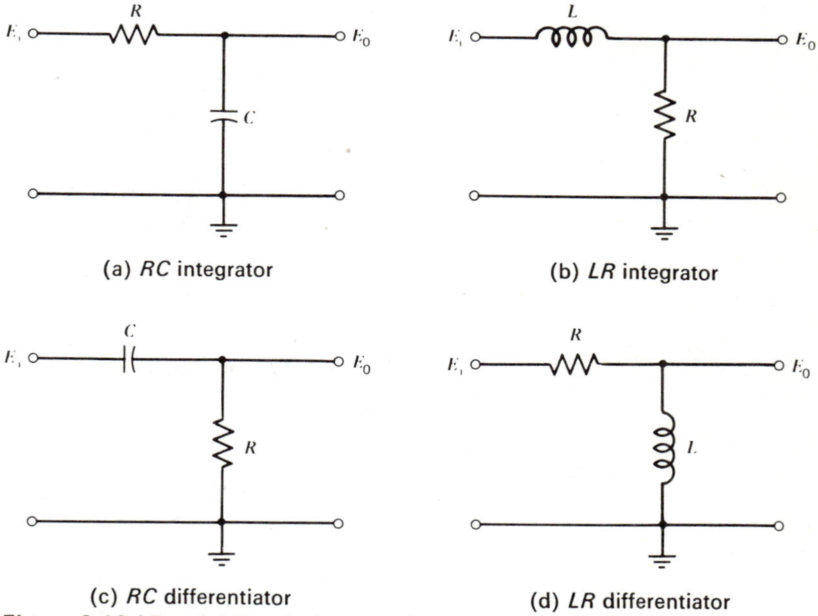

Figure 3-16 *LR* and *RC* equivalent circuits.

3-9 Step Input to an *RLC* Circuit

A series-connected *RLC* circuit is shown in Fig. 3-17a. Its analysis is a little more complex compared with the *RC* or *RL* circuits previously discussed. If the input is a step function, we see that:

1. Initially, there is no current at the instant the switch is closed. Current must rise in a finite time, because the inductor (*L*) opposes any instantaneous change in current.
2. The current eventually ceases to flow, however, when the capacitor (*C*) is fully charged.

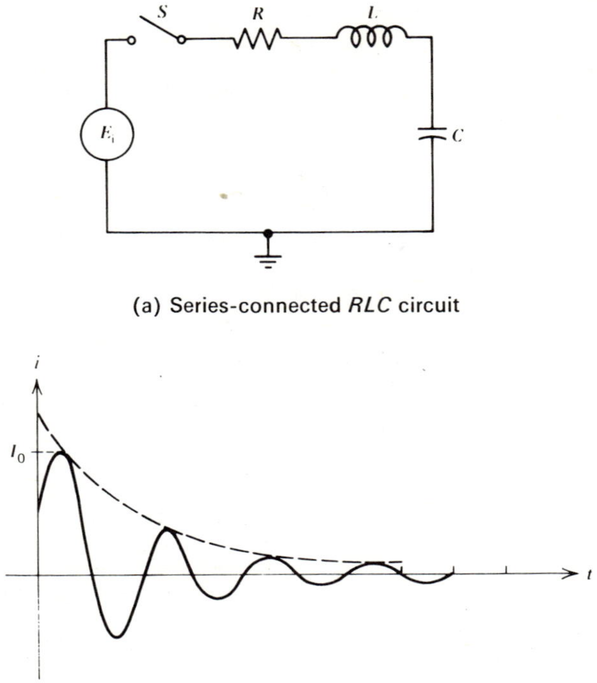

(a) Series-connected *RLC* circuit

(b) Current in the underdamped circuit

Figure 3-17 Step input to a series-connected *RLC* circuit.

An equation can be written using Kirchhoff's voltage law:

$$e_i = v_R + v_L + v_c$$

$$= iR + L\frac{di}{dt} + \frac{1}{C}\int i\,dt \qquad (3\text{-}28)$$

Equation (3-28) is a complex differential equation. Its solution is beyond the scope of this text, especially when E_i is not a step function. When the input *is* a step function, that is, when E_i is a pulse, Eq. (3-28) may be solved for i to yield:

$$i = \frac{2E_i}{\sqrt{(4L/C - R^2)}} \cdot e^{-(Rt/2L)} \sin \frac{\sqrt{(4L/C) - R^2}}{2L} \cdot t \text{ amperes (A)} \qquad (3\text{-}29)$$

where:

i is the instantaneous current in the circuit, amperes (A)
E_i is the magnitude of the step input voltage, volts (V)
t is the time (seconds) elapsed after the step input is applied
R is the resistance in ohms (Ω)
C is the capacitance in farads (F)
L is the inductance in henrys (H)

Equation (3-29) is the equation for the circuit current (i) shown in Fig. 3-17b. It is partially of the form $i = I_0 e^{-x} \sin wt$. Analysis of Eq. (3-29) indicates that:

a. The maximum value of the current (I_0) is given by the term:

$$I_0 = \frac{2E_i}{\sqrt{(4L/C) - R^2}} \text{ amperes (A)} \tag{3-30}$$

b. The current decays exponentially with time constant L/R. The exponential decay of the maximum value (shown as dashed line in Fig. 3-17b) is considered to be practically *complete* when:

$$\frac{Rt}{2L} = 5$$

or

$$t = \frac{10L}{R} \text{ seconds (s)} \tag{3-31}$$

$$= 10\,\tau$$

c. The current is sinusoidal in nature and the frequency is:

$$w = \frac{\sqrt{(4L/C) - R^2}}{2L}$$

$$f = \frac{1}{2\pi} \cdot \frac{\sqrt{(4L/C) - R^2}}{2L} \text{ hertz (Hz)} \tag{3-32}$$

To summarize **a.**, **b.**, and **c.** above, we may conclude that the current rises to a maximum value I_0 and then decays exponentially to zero while oscillating sinusoidally, as shown in Fig. 3-17b.

3-10 Damping Characteristics of an RLC Circuit

In Eq. (3-29), the quantity under the square root sign, that is, $[(4L/C) - R^2]$, is important. Values for maximum current I_0 in Eq. (3-30) and the frequency of oscillations in Eq. (3-32) are controlled by it. Depending on whether $(4L/C)$ is greater than, equal to, or less than R^2, this quantity may be any one of three forms: a positive real number, a zero, or an imaginary number. Let us examine these three possible situations individually:

1. When $(4L/C) - R^2 > 0$, the current in the circuit reaches its maximum value (I_0), oscillates sinusoidally, and ultimately decays exponentially to zero (Fig. 3-17b). Exponentially decaying sinusoidal oscillations classify this circuit as *underdamped.* An underdamped circuit or system oscillates around its final (stable) value before settling down. An underdamped circuit is also known as a ringing circuit. The ringing behavior of the current in an underdamped circuit is again shown in Fig. 3-18a.
2. When $(4L/C) - R^2 = 0$, the maximum current value has no meaning. The frequency of oscillations is zero because $R^2 = (4L/C)$ in Eq. (3-32). The *RLC* circuit under these conditions is said to be *critically damped.* Its behavior is shown in Fig. 3-18b. It does *not* oscillate but does decay from maximum to zero. The current reaches its final zero value in the shortest time possible without oscillating.
3. When $(4L/C) - R^2 < 0$, that is, a *negative* number, its square root is an *imaginary* number because $R^2 > (4L/C)$. This means that I_0 (in Eq. 3-30) has an *imaginary* value and the frequency of oscillations is also imaginary. In simple language, the current never reaches its maximum value and under these conditions, does not oscillate. The circuit is said to be *overdamped.* An overdamped circuit (system) does not oscillate but in fact "crawls" to the final state slowly. It may never reach the final state for a given time period. This situation is shown by the current waveform of Fig. 3-18c.

Sometimes the damping conditions are expressed in terms of the quality factor Q of the circuit. Q for an *RLC* circuit is given by:

$$Q = \frac{wL}{R} = \frac{1}{R}\sqrt{\frac{L}{C}} \qquad (3\text{-}33)$$

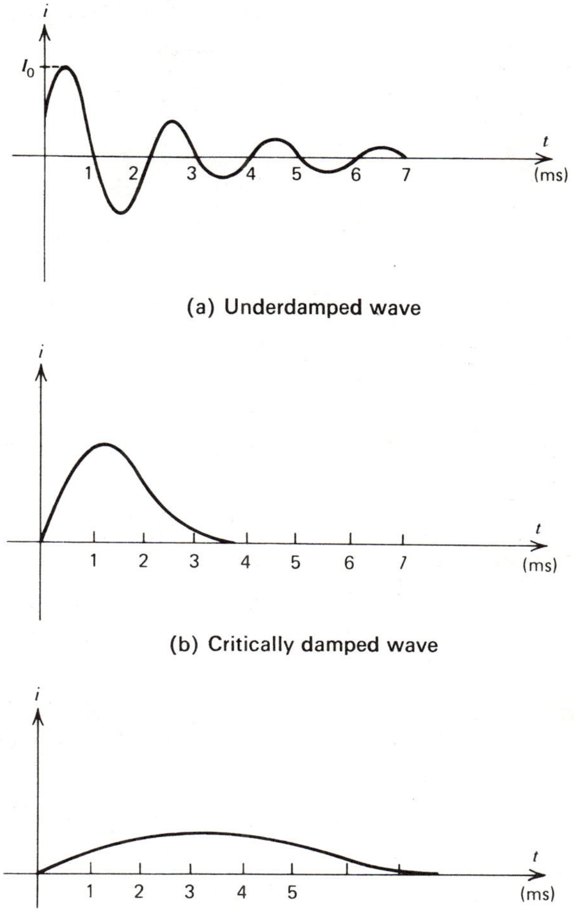

(a) Underdamped wave

(b) Critically damped wave

(c) Overdamped wave

Figure 3-18 Effect of damping on current in the circuit.

Then,

$$f = \frac{1}{2\pi} \cdot \frac{\sqrt{(4L/C) - R^2}}{2L} \tag{3-32}$$

or

$$= \left(\frac{1}{2\pi\sqrt{LC}}\right) \cdot \sqrt{1 - \frac{1}{4Q^2}} \tag{3-34}$$

or

$$= f_0\sqrt{1 - \left(\frac{1}{4Q^2}\right)} \tag{3-35}$$

For a critically damped circuit:

$$(4L/C) = R^2$$
$$R = 2\sqrt{L/C}$$

Substituting this in Eq. (3-33 above,)

$$Q = \left(\frac{1}{2\sqrt{L/C}}\right) \cdot \sqrt{\frac{L}{C}}$$
$$= 0.5$$

Thus, for a critically damped circuit, $Q = 0.5$. It can be shown that if Q is less than 0.5, the circuit is *overdamped.* The circuit is *underdamped* if its Q is greater than 0.5.

To further illustrate the meaning of damping characteristics of a system, consider a **dc** moving-coil voltmeter used to measure voltage at a point in a circuit where the voltage is known to be 10 V. Right after the voltmeter is connected, if:

a. The voltmeter needle quickly jumps to 10 V and then oscillates around this value for some time before coming to rest at the 10-V reading, then the meter movement mechanism is *underdamped.*
b. The voltmeter needle goes fairly quickly to the 10-V reading but does *not* oscillate, then the meter movement mechanism is *critically damped.*
c. The voltmeter needle *slowly* crawls toward the 10-V reading (it may not reach it) and does *not* oscillate, then the meter movement mechanism is *overdamped.*

Most pulse systems are designed so that they are *critically damped.* The pulse circuit should switch as soon as the pulse is applied. An *overdamped* circuit has a *pulse-delaying* effect. A critically damped circuit performs its function in the *shortest* time possible without oscillations. Oscillations resulting from an *underdamped* circuit are undesirable because they may result in false switching, as well as pulse delaying.

Example 3-14

A 10-$V_{p\text{-}p}$, 10-kHz square wave is applied to a series-connected *RLC* in which R = 100 ohms, C = 200 pF, and L = 20 μH.

a. Find the degree of damping.

b. Find the frequency of oscillations if underdamped, using Eq. (3-32) and Eq. (3-35).
c. Draw the voltage waveform that appears across R, L, and C.
d. Find the number of cycles (N) of oscillations before they completely decay out.

Solution

a. $$\left(\frac{4L}{C}\right) - R^2 = \left(\frac{4 \times 20\ \mu\text{H}}{200\ \text{pF}}\right) - (100)^2$$

$$= 40 \times 10^4 - 10^4$$

$$= 39 \times 10^4,$$ which is a *positive real* number.

∴ the circuit is *underdamped.*

b. Frequency of oscillations, using Eq. (3-32),

$$f = \frac{1}{2\pi} \times \frac{\sqrt{(4L/C) - R^2}}{2L} \tag{3-32}$$

$$= \frac{1}{2 \times 3.14} \times \frac{\sqrt{39 \times 10^4}}{2 \times 20 \times 10^{-6}}$$

$$= \mathbf{2.485\ MHz}$$

Frequency of oscillations, using Eq. (3-35),

$$Q = \frac{1}{R}\sqrt{\frac{L}{C}} \tag{3-33}$$

$$Q = \left(\frac{1}{100}\right)\sqrt{\frac{20 \times 10^{-6}}{200 \times 10^{-12}}} = 3.16$$

$$f_0 = \frac{1}{2\pi} \times \frac{1}{\sqrt{LC}}$$

$$= \frac{1}{2 \times 3.14} \times \frac{1}{\sqrt{20 \times 10^{-6} \times 200 \times 10^{-12}}} = 2.52 \times 10^6\ \text{Hz}$$

$$f = f_0\sqrt{1 - \frac{1}{4Q^2}} \tag{3-35}$$

$$= (2.52 \times 10^6)\sqrt{1 - \frac{1}{4 \times (3.16)^2}}$$

$$= \mathbf{2.49\ MHz}$$

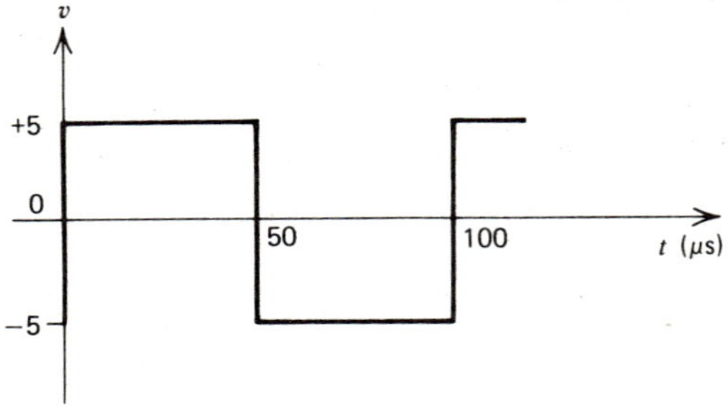

(a) Input square wave

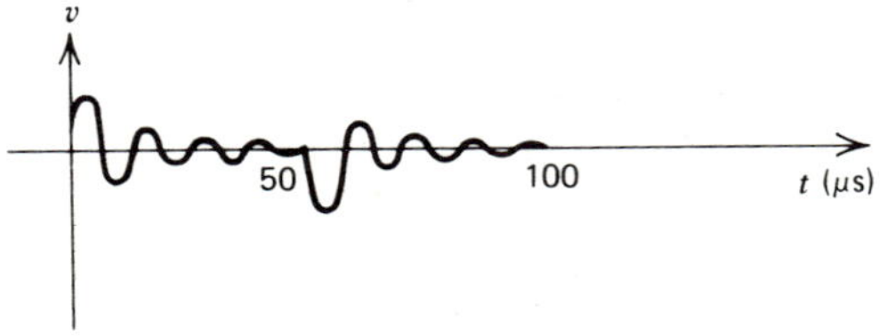

(b) Voltage waveform across *R*

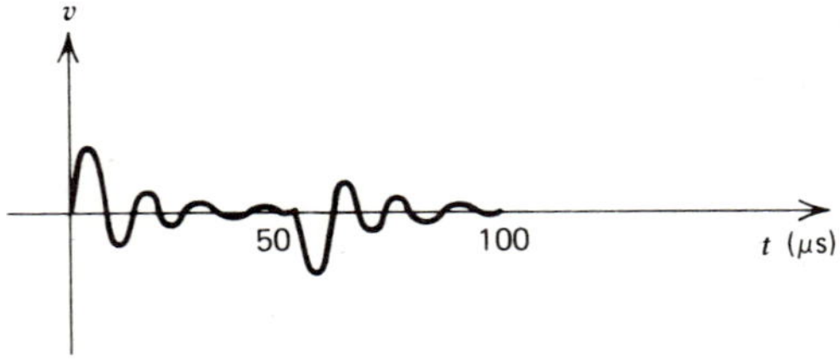

(c) Voltage waveform across *L*

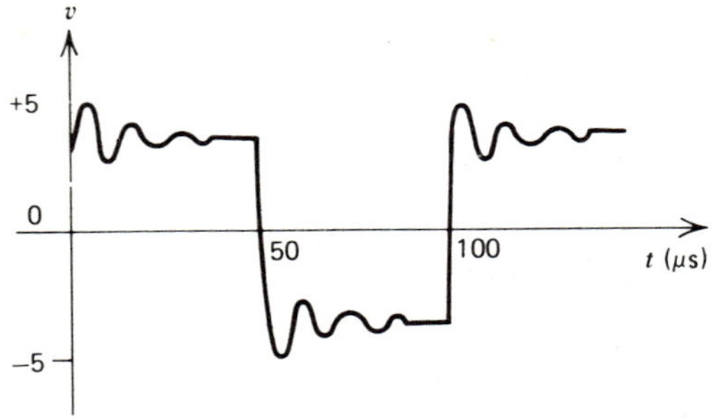

(d) Voltage waveform across *C*

Figure 3-19 Waveforms for Ex. 3-14.

c. Voltage waveforms that will appear across R, L, and C are drawn in Fig. 3-19.

$$\text{Time for one cycle} = \frac{1}{f} = \frac{1}{2.49 \times 10^6}$$

$$= \mathbf{0.402\ \mu s}$$

d. Decay time,

$$t = \frac{10L}{R} \tag{3-31}$$

$$= \frac{10 \times 20 \times 10^{-6}}{100} = \mathbf{2\ \mu s}$$

Number of oscillations,

$$N = \frac{\text{decay time}}{\text{time period}}$$

$$= \frac{2\ \mu s}{0.402\ \mu s}$$

$$\cong \mathbf{5}$$

3-11 Attenuators

An attenuator is a circuit or a device used to lower the amplitude of an electrical signal. Basically, it is a voltage-divider circuit. The output voltage from it is a fraction of the input voltage.

A simple attenuator circuit is shown in Fig. 3-20. The output voltage appearing at E_o is given by:

$$E_o = E_i \frac{R_2}{R_1 + R_2}$$

$$E_o/E_i = \frac{R_2}{R_1 + R_2} \tag{3-36}$$

The expression E_i/E_o is known as the *attenuation factor*. When a load, R_L, is connected at the output of this circuit, the attenuation factor becomes:

$$E_o/E_i = \frac{R_2 \parallel R_L}{R_1 + R_2 \parallel R_L} \tag{3-37}$$

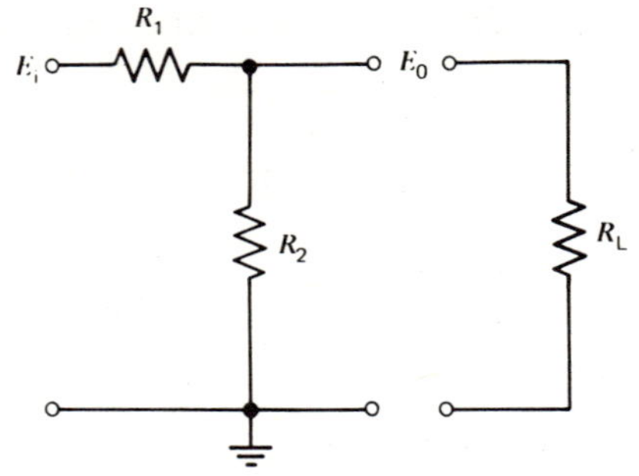

Figure 3-20 Attenuator circuit.

Equation (3-37) reduces to Eq. (3-36) when $R_L \gg R_2$.

The attenuator circuit considered so far (Fig. 3-20) contains only resistances. Therefore it does not introduce any distortion into the electrical signal to be attenuated. However, when one or more than one of the components are either inductors or capacitors, then as has already been discussed, the waveform may get distorted. Let us examine the effect of a capacitive load on the attenuator circuit.

Assume that a capacitor, C, is connected as load to the attenuator circuit, as shown in Fig. 3-21a. The Thévenin equivalent of this circuit is shown in Fig. 3-21b. Note that

$$V_{th} = E_i \frac{R_2}{R_1 + R_2}$$

and

$$R_{th} = \frac{R_1 \cdot R_2}{R_1 + R_2}$$

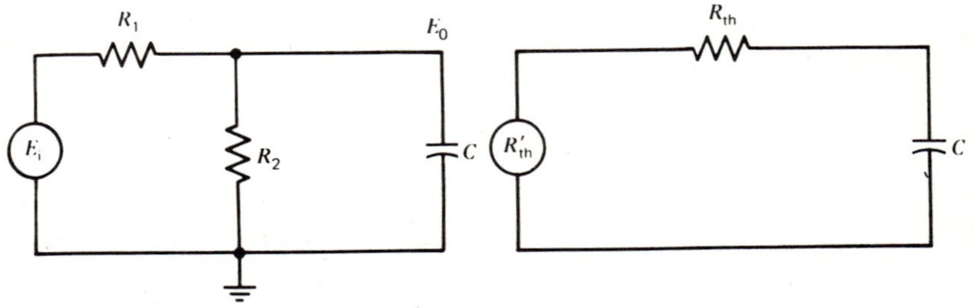

(a) Capacitively loaded attenuator (b) Thévenin's equivalent

Figure 3-21 Effects of a capacitive load on the attenuated waveform.

Note that Fig. 3-21b is an *RC* **low-pass** circuit whose time constant is $R_{th} \times C$. A pulse wave input to this circuit will get distorted (integrated). The extent of the distortion depends upon the time constant of the circuit as compared with the pulse width of the input. The smaller the pulse width, the greater the distortion.

This distortion situation is remedied by connecting a so-called *compensation capacitor*. To compensate the attenuator circuit of Fig. 3-21a, the compensation capacitor is connected across R_1 as shown in Fig. 3-22a. To follow the compensation process and the rationale behind it, note that all cyclical waveforms consist of three distinct parts:

1. The **dc** or average value.
2. The fundamental sinusoidal signal, whose frequency is the same as the cyclical waveform in question.
3. Sinusoidal *harmonics* whose frequencies are whole number *multiples* of the *fundamental*, each having a certain specific amplitude and relative phase.

Thus we see that all cyclical waveforms really consist of a **dc** value and a number of harmonically related sinusoidal waveforms.

When a sine wave is applied to the *uncompensated* attenuator circuit of Fig. 3-21a, the attenuation factor is given by:

$$E_o/E_i = \frac{R_2 \parallel X_c}{R_1 + R_2 \parallel X_c} \tag{3-38}$$

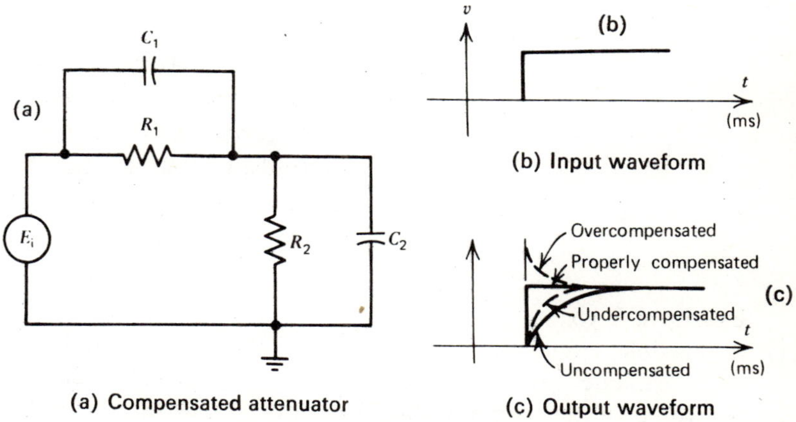

(a) Compensated attenuator (c) Output waveform

Figure 3-22 Compensating an attenuator.

The attenuation factor as given by Eq. (3-38), is a frequency-dependent quantity. Signals (sinusoidal) of *different* frequencies are attenuated by *different* amounts. Thus, the constituent signals of a nonsinusoidal waveform are attenuated by different amounts when applied to an *uncompensated* attenuator, which results in the distorted output.

When a compensation capacitor is connected as shown in Fig. 3-22a, the equation for the amplification factor becomes;

$$E_o/E_i = \frac{Z_2}{Z_1 + Z_2} \tag{3-39}$$

where:

$$Z_1 = R_1 \parallel X_{c1}$$

and

$$Z_2 = R_2 \parallel X_{c2}$$

If time constant $R_1 \times C_1$ is made equal to time constant $R_2 \times C_2$, then Eq. (3-39) reduces to Eq. (3-36), which is independent of the signal frequency. The attenuator that has $R_1 \times C_1 = R_2 \times C_2$ is known as the *properly compensated* attenuator. In a properly compensated attenuator, the input signal is attenuated by the same amount regardless of whether it is a **dc**, sinusoidal, or anything else in nature. Thus, for proper compensation:

$$R_1C_1 = R_2C_2 \tag{3-40}$$

Note that it is critical to have $C_1R_1 = C_2R_2$ for proper compensation. The circuit is considered to be *undercompensated* if $C_1R_1 > C_2R_2$, *overcompensated* if $C_1R_1 < C_2R_2$. The circuit response to a step (pulse) input is shown in Fig. 3-22c for the three possible circuit conditions, namely, undercompensated, overcompensated, and properly compensated.

Example 3-15

A possible circuit for an **ac/dc**, "10 down" attenuator probe of a cathode ray oscilloscope is shown in Fig. 3-23a. Find the value of the compensation capacitor C_1, when C_2 of CRO is 50 pF.

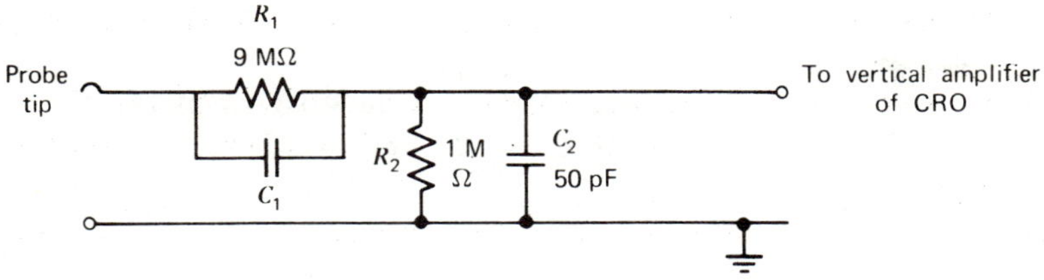

Figure 3-23 Circuit for Ex. 3-15.

Solution

For proper compensation,

$$C_1R_1 = C_2R_2 \tag{3-40}$$

$$C_1 = \frac{C_2R_2}{R_1}$$

$$= \frac{50 \text{ pF} \times 1 \text{ M}\Omega}{9 \text{ M}\Omega}$$

$$= \mathbf{5.556 \text{ pF}}$$

3-12 Summary

1. When a step input is applied to a series-connected *RC* circuit, the current that flows through it decays exponentially from its initial value of *E*/*R* to zero. The capacitor voltage rises exponentially toward the input voltage, while the voltage across the resistance decays exponentially to zero. The following capacitor-charge equation is worth memorizing.

$$v_c = E_i - (E_i \pm V_{ci})e^{-(t/RC)} \tag{3-7}$$

where:

v_c is the charge on the capacitor, volts (V)
E_i is the input voltage, volts (V)
V_{ci} is the initial charge on the capacitor, volts (V)
RC is the time constant (τ) of the circuit, seconds (s)

2. The time constant of a series-connected *LR* circuit is *L*/*R*, and that of a series-connected *RC* circuit is the product $R \times C$.
3. A series-connected *LR* or *RC* *low-pass* circuit is sometimes also called an integrator. Similarly, a series-connected *RL* or *RC* *high-pass* circuit is sometimes also called the differentiator circuit.
4. Whenever possible, an *RC* circuit is used in place of an *RL* circuit,

because an inductor is usually bulky and expensive as compared to a capacitor.

5. An RLC circuit is underdamped (oscillations) when $(4L/C) > R^2$, critically damped when $(4L/C) = R^2$, and overdamped when $(4L/C) < R^2$.
6. Attenuators should be properly compensated to prevent distortion.
7. Important equations:

$$\tau = RC \tag{3-3}$$

$$v_C = E_i[1 - e^{-(t/\tau)}] \tag{3-4}$$

$$t_r = 2.2\tau \tag{3-6}$$

$$v_C = E_i - (E_i \pm V_{ci})e^{-(t/\tau)} \tag{3-7}$$

$$v_R = (E_i \pm V_{ci})e^{-(t/\tau)} \tag{3-17}$$

$$v_R = E_i e^{-(t/\tau)} \tag{3-18}$$

$$F_t = \frac{t_p}{t_t} \tag{3-25}$$

$$i = \frac{2E_i}{\sqrt{(4L/C) - R^2}} \cdot e^{-(Rt/2L)} \cdot \sin \frac{\sqrt{(4L/C) - R^2}}{2L} \cdot t \tag{3-29}$$

$$f = \frac{1}{2\pi} \cdot \frac{\sqrt{(4L/C) - R^2}}{2L} \tag{3-32}$$

$$C_1R_1 = C_2R_2 \tag{3-40}$$

3-13 Glossary

Critically damped. A system (circuit) that goes to its final steady-state position in the shortest possible time without oscillating whenever a step input has been applied.

Differentiator. A circuit whose output is proportional to the derivative of input.

High-pass. A high-pass circuit allows the high-frequency sinusoidal signals to pass through practically unattenuated and attenuates the low-frequency signals.

Integrator. A circuit whose output is proportional to the integral of input.

Low-pass. A low-pass circuit allows the low-frequency sinusoidal signals to pass through practically unattenuated and attenuates the high-frequency signals.

Overdamped. An overdamped system reaches the final steady-state position slowly and without oscillating after a step input has been applied.

Steady-state waveform. A waveform that does not change with time is called the steady state or the settled waveform.

Time constant (τ). The time constant of a circuit is the time that a voltage (or current) takes to reach 63 percent of its final charge value after the charge has been initiated in an electrical circuit.

Underdamped. A system (circuit) is said to be underdamped when a step input to it results in exponentially decaying sinusoidal oscillations.

3-14 Recapitulation

Return to the Objectives and Self-Evaluation Questions given at the beginning of this chapter and see how well you can answer them. If you cannot answer certain questions satisfactorily, place a check mark (✓) next to them and review appropriate parts of the chapter to find the answers. No attempt should be made to solve the Problems (Sec. 3-15) before answering the self-evaluation questions. Ability to solve the Problems should serve as a test and feedback regarding your understanding of the material presented in this chapter. Answers to Problems are given on p. 514.

3-15 Problems

3-1 A 0.01-μF capacitor and a 10-kΩ resistor are connected in low-pass filter configuration. Find the cut-off frequency (f_{hc}) for this circuit.

3-2 A 22-kΩ resistor and a capacitor are used in a *low-pass* filter circuit. The upper cut-off frequency of this circuit should be 20 kHz. Find the value of the capacitor.

3-3 A 20-$V_{p\text{-}p}$, 100-kHz sine wave is applied to an integrator circuit where $C = 0.01$ μF. Find an appropriate value of R such that the output is 14.1 $V_{p\text{-}p}$.

3-4 An input of 10 V **dc** is applied to the circuit of Prob. 3-1 through a switch. Find the output voltage (capacitor) 0.15 ms after the switch is closed.

3-5 Find the rise time of the output waveform of Prob. 3-4.

3-6 Draw the capacitor-charge curve for Prob. 3-4. Show calculations for:

a. Initial slope.

b. Total charge time.

c. Time taken to charge to 63 percent of the final value.

3-7 Figure 3-24 shows a voltage-divider circuit feeding a capacitive load. Find and draw the voltage waveform that appears across the capacitor after the switch is closed. Calculate its:
a. Initial slope.
b. Total charge time.
c. Rise time.
d. Time taken to charge to 63 percent of the final value.

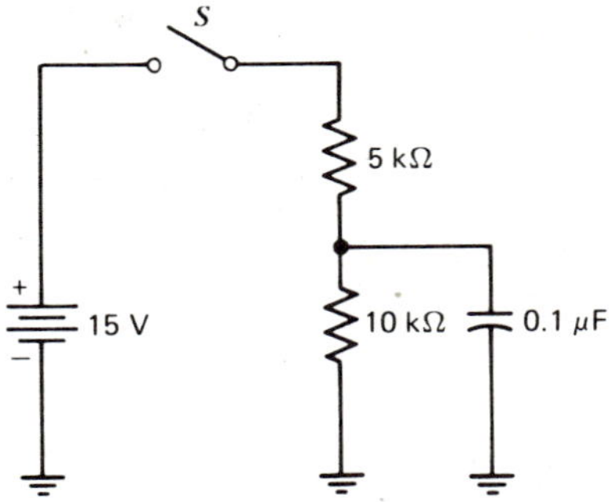

Figure 3-24 Circuit for Prob. 3-7.

3-8 An 18-$V_{p\text{-}p}$, 2-kHz square wave is applied to the circuit of Fig. 3-24, in place of the 15-V **dc** source. Sketch the first two cycles of the voltage waveform that will appear across the capacitor after the switch has been closed.

3-9 Carry on the step-by-step calculations of Prob. 3-8 until the output voltage waveform reaches steady-state conditions.

3-10 Refer to the conditions outlined in Prob. 3-8. Find and sketch the steady-state capacitor voltage waveform using the *direct equations* method.

3-11 A 68-kΩ resistor and a 100-pF capacitor are connected to form a low-pass filter. A 5-$V_{p\text{-}p}$, 10-kHz square wave input is applied to it. Find and sketch the steady-state output waveform. Is it a long, medium, or short τ circuit?

3-12 A pulse wave of Fig. 3-25 is applied to the circuit of Prob. 3-11. Find and sketch the steady-state output waveform.

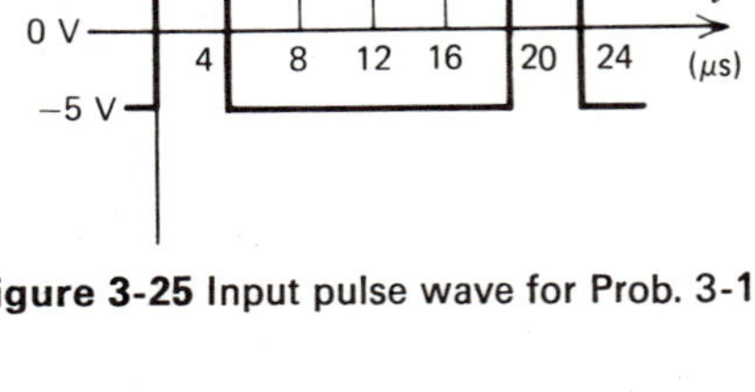

Figure 3-25 Input pulse wave for Prob. 3-12.

3-13 The frequency of the input voltage in Prob. 3-11 is increased to 100 kHz. Find and sketch the steady-state output voltage waveform.

3-14 Refer to the circuit of Fig. 3-26. Find the output voltage level 5 ms after the switch has been closed.

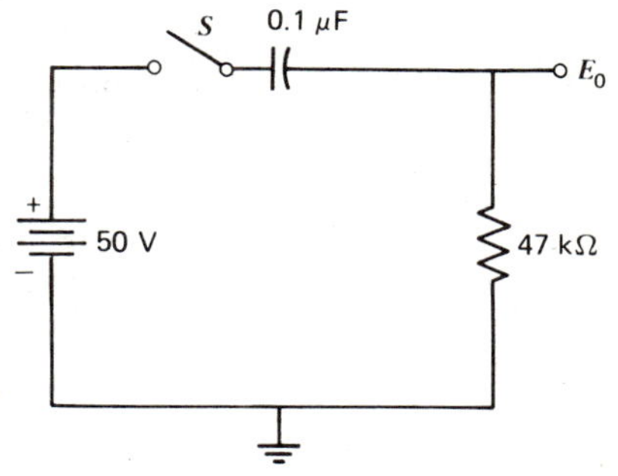

Figure 3-26 Circuit for Prob. 3-14.

3-15 Refer to the circuit of Fig. 3-26. The 50-V **dc** supply is replaced by a 50-$V_{p\text{-}p}$, 100-Hz square wave input. Find voltage levels and draw the first three cycles of the output voltage waveform, using the step-by-step method.

3-16 Refer to Prob. 3-15. Find and sketch the steady-state output voltage waveform using the *direct equations* method.

3-17 Refer to Prob. 3-15. The 47-kΩ resistor is replaced by a 5 kΩ resistor. Find and sketch the steady-state output voltage waveform.

3-18 Figure 3-27 shows a coupling circuit for a transistor amplifier. The input pulse waveform is as shown in Fig. 3-25. Find and sketch the steady-state voltage waveform that appears at the input terminals of the amplifier.

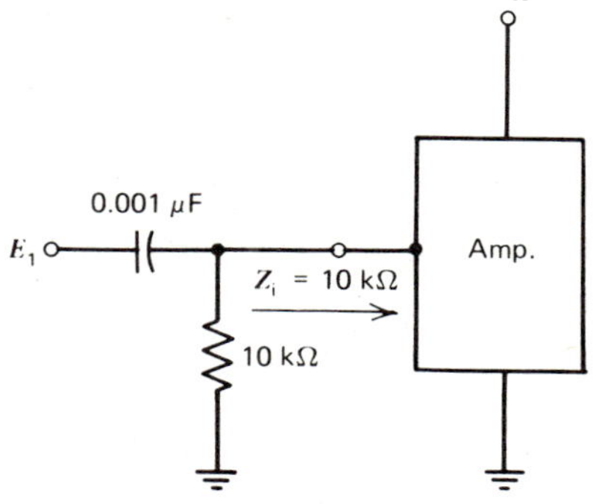

Figure 3-27 Circuit for Prob. 3-18.

3-19 Figure 3-26 shows an *RC* differentiator circuit. Find and draw an equivalent *RL* circuit.

3-20 Repeat Prob. 3-14. The 0.1 μF-capacitor is now replaced by a 200 mH coil and the 47 kΩ-resistor is replaced by a 47 Ω-resistor.

3-21 A series-connected *RLC* circuit is shown in Fig. 3-28.

a. Is it an over-, under-, or critically damped circuit?

b. Find the frequency of oscillation if it is an underdamped circuit.

c. Find the value of resistance that will make it a critically damped circuit.

3-22 Refer to the circuit of Fig. 3-28. Draw the current waveform that exists in the circuit after the switch is closed.

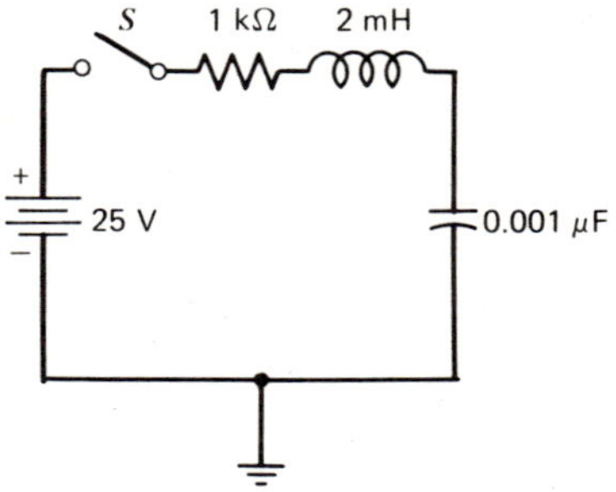

Figure 3-28 Circuit for Prob. 3-22.

3-23 The 25-V **dc** supply in the circuit of Fig. 3-28 is replaced by 20-$V_{p\text{-}p}$, 25-kHz square wave. Sketch the current waveform that appears in the circuit.

3-24 Refer to Prob. 3-22. Sketch the voltage waveform that appears across the capacitor.

3-25 Refer to Prob. 3-22. Sketch the voltage waveform that appears across the capacitor if it is made a *critically damped* circuit.

3-26 The 25-V **dc** power source in the circuit of Fig. 3-28 is replaced by a pulse wave input of Fig. 3-25 and the 0.001-μF capacitor is replaced by a 100-pF capacitor. Sketch the voltage waveform that appears across the resistor.

3-27 Refer to the circuit shown in Fig. 3-22a. Assume that $R_1 = 10$ kΩ, $R_2 = 2$ kΩ, and $C_2 = 100$ pF. Calculate the proper value of the compensating capacitor C_1.

3-28 Refer to the CRO probe circuit shown in Fig. 3-23. Assume that $R_1 = 38$ MΩ, $R_2 = 2$ MΩ and $C_2 = 200$ pF. Find the attenuation factor of this probe and the proper value of compensating capacitor C_1.

3-16 Experiment

Part A

Title RC **Low-Pass and High-Pass Circuits**

Theory When a step voltage (input) is applied to a series-connected *RC* circuit, the current in the circuit starts from its maximum initial value (E_i/R) and then exponentially decays as C charges. As C charges, the voltage across it rises exponentially and the voltage across R decays exponentially. The equation for the capacitor charge curve is:

$$v_C = E_i - (E_i \pm V_{ci})e^{-(t/\tau)} \text{ volts (V)} \tag{3-7}$$

The instantaneous voltage across R is given by:

$$v_R = (E_i \pm V_{ci})e^{-(t/\tau)} \text{ volts (V)} \tag{3-18}$$

where:

$$\tau = RC \text{ seconds (s)} \tag{3-3}$$

An *RC* circuit connected in *low-pass* configuration (Fig. 3-1a) is sometimes also called an *integrator*. The output is taken across the capacitor. See Ex. 3-6 for the *integrated* square wave analysis.

An *RC* circuit connected in *high-pass* configuration (Fig. 3-10a) is sometimes also called a *differentiator*. The output is taken across the resistor. See Ex. 3-11 for the *differentiated* square wave analysis.

Equipment Multifunction generator: EXACT model 126, or equal.
Multimeter: RCA model WV 98C, or equal.
Dual beam oscilloscope: PHILIPS model 3230, or equal.
Capacitors: 0.1 μF, 0.01 μF, and 0.001 μF.
Resistors: 15 kΩ, 3.3 kΩ, and 100 kΩ.

Procedure

A. ***RC* low-pass circuit**

1. Connect the *RC* low-pass circuit shown in Fig. E3-1.

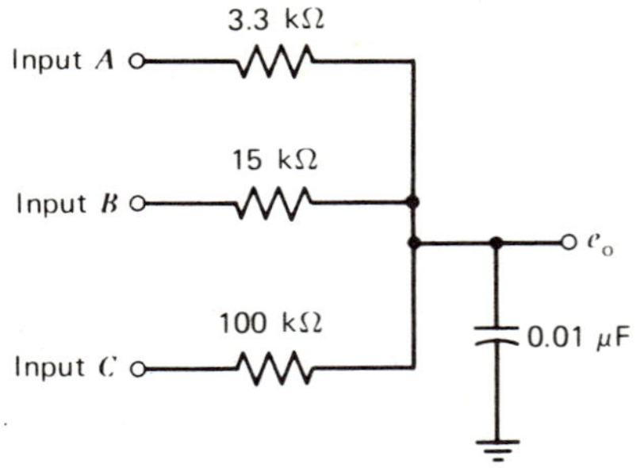

Figure E3-1 *RC* low-pass circuits.

2. Apply a 5-$V_{p\text{-}p}$, 2.5-kHz square wave at the input point *A*. Observe (CRO) and record the input as well as the output waveforms in in Table E3-1.
3. Change the input to point *B* and repeat Procedure step 2.
4. Change the input to point *C* and repeat Procedure step 2.

B. ***RC* high-pass circuit**
5. Connect the *RC* high-pass circuit shown in Fig. E3-2.

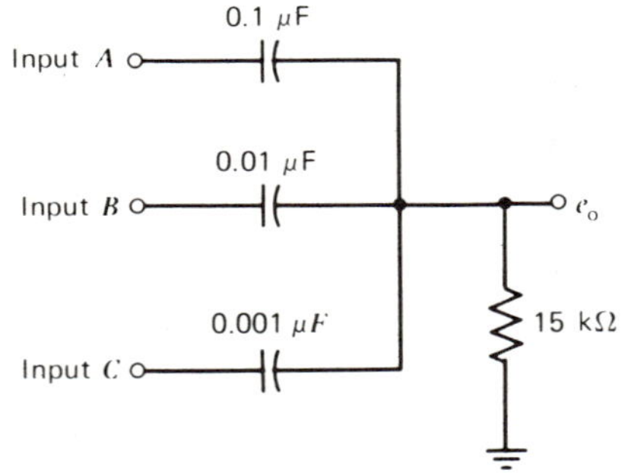

Figure E3-2 *RC* high-pass circuits.

6. Apply a 5-$V_{p\text{-}p}$, 2.5-kHz square wave at input point *A*. Observe (CRO) and record the input as well as the output waveforms in Table E3-2.
7. Change the input to point *B* and repeat Procedure step 6.
8. Change the input to point *C* and repeat Procedure step 6.
9. Disconnect and return all of the test equipment to its proper place.

Calculations

1. Consider the *RC* low-pass circuit used in Procedure steps 2, 3, and 4, respectively. Calculate τ and steady-state output waveforms (Ex. 3-7) in each case. Enter your results in Table E3-1. Assess whether τ is long, medium, or short in each case. It should be recorded in Table E3-1 in the following manner:

$$\tau = k \text{ ms (long)}$$

2. Consider the *RC* high-pass circuits used in Procedure steps 6, 7, and 8. Calculate τ and the steady-state output voltage waveforms (Ex. 3-11) in each case. Record your results in table E3-2. Assess whether τ is long, medium, or short in each case.

Tabulation of Results

Table E3-1

Waveform	Procedure step 2 $R = 3.3\ \text{k}\Omega$	Procedure step 3 $R = 15\ \text{k}\Omega$	Procedure step 4 $R = 100\ \text{k}\Omega$
Input			
Output (Measured)			
Output (Calculated)			
τ	$\tau =$	$\tau =$	$\tau =$

Table E3-2

Waveform	Procedure step 6 $C = 0.1\ \mu\text{F}$	Procedure step 7 $C = 0.01\ \mu\text{F}$	Procedure step 8 $C = 0.001\ \mu\text{F}$
Input			
Output (Observed)			
Output (Theoretical)			
τ	$\tau =$	$\tau =$	$\tau =$

Questions

1. Examine the observed and theoretical waveforms in Tables E3-1 and E3-2. Describe how your observed waveforms compare with the theoretically expected ones in each case.
2. For the circuit conditions described in Tables E3-1 and E3-2, you were asked to calculate τ and indicate whether it is long, short, or medium. Explain how you determined whether the τ was long, short, or medium.
3. Explain the results (wave shapes) observed in Procedure steps 3 and 7. Why did the square wave get distorted?

Discussion

1. Summarize your conclusions as far as the observed and expected output waveforms recorded in Tables E3-1 and E3-2 are concerned.
2. Discuss the possibility of using an *RC* low-pass circuit to convert a square wave to a triangular wave.
3. Discuss the possibility of using an *RC* high-pass circuit to convert a square wave to a series of positive and negative spikes.

Part B

Title

A. Damping Characteristics of an RLC Circuit
B. Compensating an Attenuator

Theory

When a step input is applied to a series-connected *RLC* circuit, the resulting circuit current *i*, is given by:

$$i = \frac{2E}{\sqrt{4L/C - R^2}} \cdot e^{-(R/2L)\cdot t} \cdot \sin \frac{\sqrt{(4L/C) - R^2}}{2L} \cdot t \qquad (3\text{-}29)$$

The circuit is said to be:

a. underdamped if $(4L/C) > R^2$ (oscillations)
b. overdamped if $(4L/C) < R^2$ (no oscillations)
c. critically damped if $(4L/C) = R^2$ (only one peak)

In an underdamped circuit, the frequency of oscillations is given by:

$$f = \left(\frac{1}{2\pi}\right)\left(\frac{\sqrt{(4L/C) - R^2}}{2L}\right) \qquad (3\text{-}32)$$

A step input when applied to a capacitively loaded attenuator (voltage divider) may get distorted. The extent of distortion depends upon the circuit τ. To improve this undesirable situation, a *compensating* capacitor usually is added to the circuit, as shown in Ex. 3-15.

Equipment Dual beam CRO: PHILIPS model 3230, or equal.
Multifunction generator: EXACT model 126, or equal.
Multimeter: RCA model WV 98C, or equal.
Capacitors: 0.01 μF and 0.001 μF.
Inductor: 30 mH.
Resistors: 1 kΩ and 10 kΩ.
Potentiometer: 10 kΩ.

Procedure

A. ***RLC* circuit**

1. Connect the *RLC* circuit shown in Fig. E3-3.

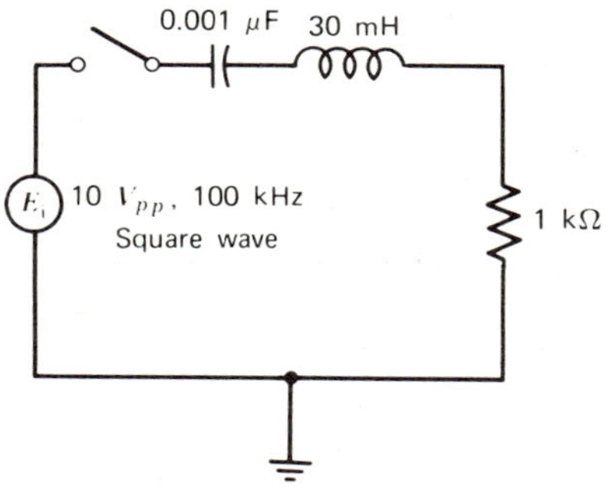

Figure E3-3 An *RLC* circuit.

2. Connect input *A* of CRO across E_i and input *B* across the resistor. In this way, observe the input voltage waveform and the current waveform that exist in the circuit. Draw these two waveforms in Table E3-3.

3. The current waveform recorded in Procedure step 2 above should have indicated it to be an underdamped circuit. Note and record the frequency of oscillations (ringing) of the circuit current in Table E3-4.

4. Connect a 10-kΩ potentiometer in series with the 1-kΩ resistor. Vary the potentiometer back and forth and observe its effects on the circuit current waveform. You should find that the circuit becomes overdamped (no ringing) as the circuit resistance is increased.

5. Adjust the 10-kΩ potentiometer such that the output waveform becomes critically damped (only one sharp peak). Measure and record in Table E3-4 the total resistance of the two series-connected resistors for this setting.

6. Disconnect the circuit. Measure and record the **dc** resistance of the coil R_L = ________.

B. Attenuator

7. Connect the capacitively loaded, uncompensated attenuator shown in Fig. E3-4.

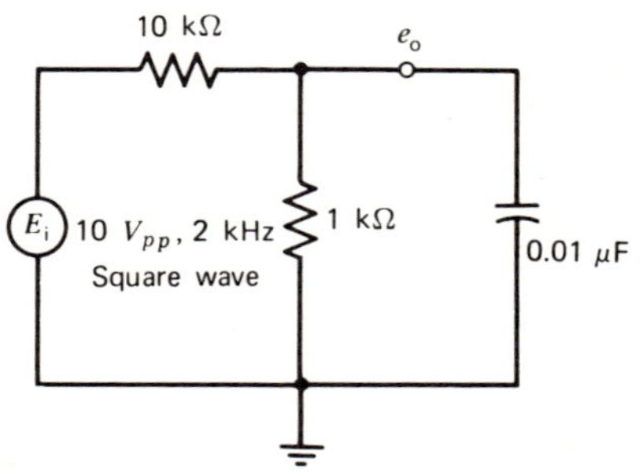

Figure E3-4 Uncompensated attenuator.

8. Observe, note, and draw the input as well as the output waveforms in Table E3-5.

9. Connect a 0.001-μF capacitor (compensating) across the 10-kΩ resistor. Observe, note, and draw in Table E3-5 the output waveform of this compensated attenuator.

10. Disconnect and return all the test equipment to its proper place.

Calculations

1. Consider the *RLC* circuit shown in Fig. E3-3. Find by calculations whether it is an underdamped, overdamped, or critically damped circuit.

2. In step 1 above, you should have concluded it to be an underdamped circuit. Calculate and enter in Table E3-4 its frequency of oscillations.

3. Calculate a value of the circuit resistance that will critically damp the circuit. Enter your results in Table E3-4.

4. Calculate percent errors for Table E3-4. Treat the theoretical value as the standard value. Remember that **dc** resistance of the coil forms a part of the total circuit resistance. Include it in your calculations.

Tabulation of Results

Table E3-3

Waveforms	
Input	
Circuit current	

Table E3-4

	Frequency of Oscillations	**Critically Damping Resistance**
Observed		
Theoretical		
Percent error		

Table E3-5

Waveforms	
Input	
Uncompensated output	
Compensated output	

Questions

1. Examine the current waveform drawn in Table E3-3. Is it an overdamped, critically damped, or underdamped circuit?
2. Examine Table E3-4. Compare the observed values with the theoretically expected ones.
3. Waveforms in Table E3-5 are intended to show the attenuator compensation technique. Has it been accomplished?
4. Check by calculations (Eq. 3-40) whether the circuit of Procedure step 9 is properly compensated.
5. Explain the results observed in Procedure step 5.

Discussion

1. Summarize your conclusions from the observed and calculated results of this experiment.
2. Account for serious discrepancies between the observed and the theoretically expected values.
3. List some instances where knowledge gained in this experiment may be utilized gainfully.

Chapter 4 The Diode Switch

4-1 Objectives

The objectives of this chapter are to:

1. Review the characteristics of a semiconductor diode.
2. Identify the properties of a semiconductor diode that affects its switching characteristics.
3. Characterize the (so-called) ideal diode.
4. Demonstrate the use of a diode as a switch.
5. Demonstrate the use of a diode switch in limiter circuits.
6. Demonstrate the use of a diode switch in clampers.
7. Demonstrate the use of a diode switch in comparators.
8. Identify circuit elements and conditions that control the diode switching time.

4-2 Self-Evaluation Questions

Watch for the answers to the following questions as you read the chapter. They should help you to understand the material presented. When you have finished the chapter, return to this section and be sure that you can answer all the questions. Then test your understanding by working out solutions to the Problems.

It is strongly recommended that these questions be satisfactorily answered before attempting to solve the Problems (Sec. 4-18).

1. Draw typical *characteristic curves* of a semiconductor diode. Identify and explain:
   a. *Knee* voltage.
   b. *Peak inverse* voltage (PIV).
   c. *Forward resistance* region.
   d. *Reverse resistance* region.
   e. *Forward bias* region.
   f. *Reverse bias* region.
2. Explain what is meant by *junction capacitance* of a semiconductor diode.
3. List some desirable characteristics of a *switching* diode.
4. List the characteristics of an *ideal* diode.
5. Give one example showing the use of a semiconductor diode as a switch.
6. List typical *specifications* for a switching diode.
7. Define a *limiter* or a *clipper* circuit.

8. Give one example showing the use of a diode in a *parallel-limiter* circuit.
9. Give one example showing the use of a diode in a *series-limiter* circuit.
10. A sine wave is to be converted into a square wave. Show how it is done using a *dual clipper*.
11. Explain the purpose and function of a *clamper* circuit.
12. Clampers are also sometimes called the **dc** restorers. Explain why.
13. Give one example showing the use of a diode in a *comparator circuit*.
14. Identify and explain conditions that influence the *switching time* of a diode.

4-3 A Practical Diode

The standard *symbol* for a semiconductor diode is shown in Fig. 4-1a. The *characteristic curve*, sometimes called the volt-ampere curve, is shown in Fig. 4-1b. Figures 4-1c and 4-1e show conditions when the diode is *forward-biased* and when it is *reverse-biased*, respectively. The *forward* current is symbolized as I_f and the reverse current as I_r. I_f flows in the direction of the diode arrow. I_r flows

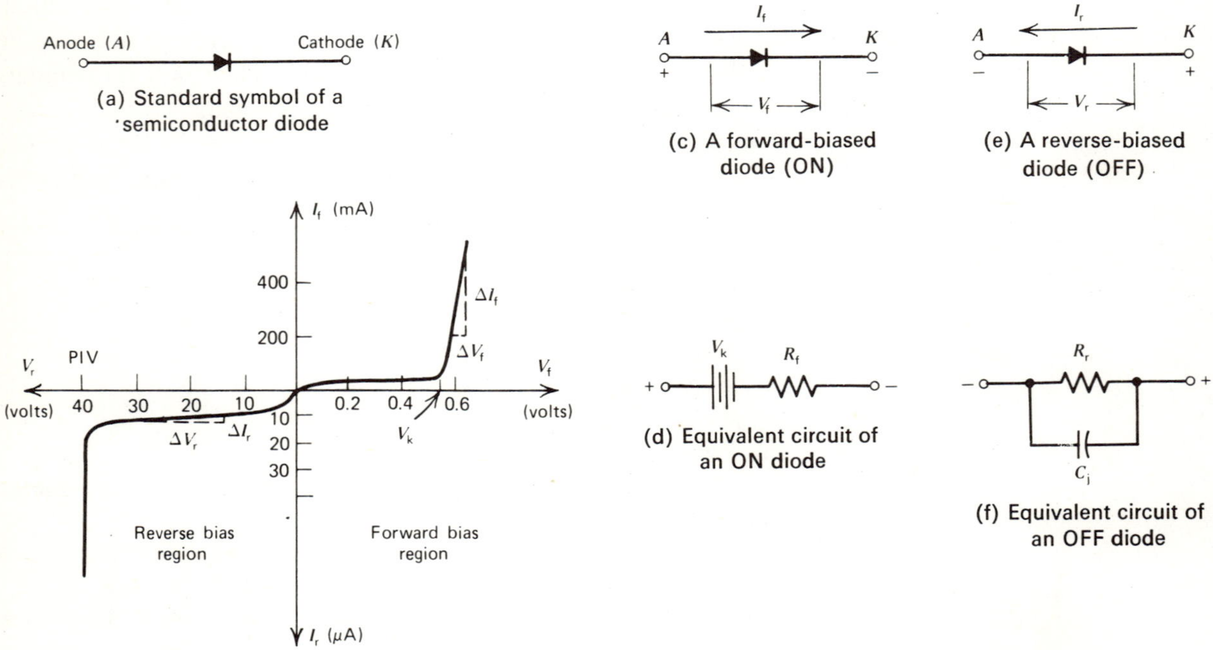

(a) Standard symbol of a semiconductor diode

(b) Characteristic curve of a practical diode

(c) A forward-biased diode (ON)

(d) Equivalent circuit of an ON diode

(e) A reverse-biased diode (OFF)

(f) Equivalent circuit of an OFF diode

Figure 4-1 Characteristics of a practical diode.

in the opposite direction of the diode arrow. I_r is also sometimes called the *reverse leakage* current.

Consider the characteristic curve shown in Fig. 4-1b. It shows two distinct regions of diode operation, namely, the *forward-bias* region and the *reverse-bias* region. Let us examine each of these two regions, one at a time.

4-3.1 Forward-Biased Diode

In the forward region, the equivalent resistance of a diode is known as its *forward* resistance (R_f). The diode is termed as being ON only when it is *conducting* in the forward direction. In fact, an ideal diode *only* conducts in the forward direction.

The forward resistance, R_f, is the inverse of the slope of the characteristic curve in the forward-bias region. Note from Fig. 4-1b that this slope of the curve ($1/R_f$) varies with the forward-bias voltage (V_f), especially when V_f is less than the *knee* voltage (V_K). The slope of the curve is almost linear after V_K. The value of R_f is higher below V_K and decreases as V_f increases. Above V_K, the value of R_f is low and is almost constant.

The value of R_f that the manufacturer gives corresponds to the equivalent forward resistance of the diode when $V_f > V_K$. R_f, at any given time, is

$$R_f = \frac{\Delta V_f}{\Delta I_f} \text{ ohms } (\Omega) \tag{4-1a}$$

Example 4-1

The following data is obtained from the forward characteristic of a silicon diode: $V_f = 0.80$ V at a current of 0.6 mA and $V_f = 0.81$ V at a current of 6.0 mA. Calculate the forward resistance of the diode, R_f.

Solution

$$R_f = \frac{\Delta V_f}{\Delta I_f} \qquad \text{(4-1a)}$$

$$= \frac{(0.81 - 0.80)\text{ V}}{(6.0 - 0.6)\text{ mA}}$$

$$= \frac{10\text{ mV}}{5.4\text{ mA}}$$

$$= \mathbf{1.852\ \Omega}$$

It follows from Eq. (4-1a) that at any forward voltage V_f (greater than V_K), R_f is given by:

$$R_f = \frac{V_f - V_K}{I_f - I_K} \text{ ohms } (\Omega) \qquad (4\text{-}1b)$$

where:

V_f is the forward voltage across the diode, volts (V)
I_f is the forward current in the diode, amperes (A)
V_K is the knee voltage, volts (V)
I_K is the current in the diode when $V_f = V_K$, amperes (A)

Usually $I_f \gg I_K$. Thus Eq. (4-1b) may be reduced to:

$$R_f = \frac{V_f - V_K}{I_f} \text{ ohms } (\Omega) \qquad (4\text{-}1c)$$

or,

$$V_f = I_f R_f + V_K \text{ volts (V)} \qquad (4\text{-}2a)$$

Example 4-2
From the value of R_f obtained in Ex. 4-1, calculate the forward voltage drop across the diode when carrying a current of 2 mA, using a value of 0.7 V for the knee voltage, V_K.

Solution

$$V_f = I_f R_f + V_K \qquad (4\text{-}2a)$$

$$= (2 \text{ mA})(1.852\ \Omega) + 700 \text{ mV}$$

$$= 3.7 \text{ mV} + 700 \text{ mV} = 703.7 \text{ mV} = \mathbf{0.7037\ V}$$

In accordance with Eq. (4-2a), the equivalent circuit of a forward-biased diode is drawn in Fig. 4-1d. The practical values of R_f range from less than one ohm (above V_K) to several thousand ohms, particularly below V_K. When operated above V_K, the diode is ON and is equivalent to a closed switch.

The term $I_f R_f$ in Eq. (4-2a) is usually much smaller than V_K, as shown in Ex. 4-2 above. Consequently, Eq. (4-2a) may be further reduced to:

$$V_f = V_K \text{ volts (V)} \tag{4-2b}$$

The typical value of V_K for a germanium diode is 0.3 V and for a silicon diode is 0.7 V. These values are used almost universally in the literature of diodes.

4-3.2 Reverse-Biased Diode The equivalent resistance of a *reverse-biased* diode is known as its reverse resistance (R_r). The diode is termed as being OFF whenever it is operated in the reverse-bias region. A small reverse current (I_r), actually exists in a practical diode when it is OFF. I_r is zero in an *ideal* diode (open switch). When reverse-biased, an ideal diode can be considered an open switch.

R_r is the inverse of the slope of the characteristic curve when the diode is OFF.

$$R_r = \frac{\Delta V_r}{\Delta I_r} \tag{4-3a}$$

Note from Fig. 4-1b that R_r is also a dynamic quantity. The manufacturer merely gives its average value. For a given V_r, it may be approximated as:

$$R_r \cong \frac{V_r}{I_r} \text{ ohms } (\Omega) \tag{4-3b}$$

where:

V_r is the reverse voltage across the diode (V)
I_r is the reverse current in the diode (A)

In accordance with Eq. (4-3b), the equivalent circuit of an OFF diode is drawn in Fig. 4-1f.

Example 4-3

When operating in the reverse mode of its characteristic, a diode exhibits a reverse current of 2 μA when a reverse voltage of 100 V is applied. Calculate the reverse resistance of the diode.

Solution

$$R_r = V_r/I_r \tag{4-3b}$$
$$= 100 \text{ V}/2\ \mu\text{A} = \mathbf{50\ M\Omega}$$

4-3.3 Junction Capacitance The symbol C_j (Fig. 4-1f) stands for the equivalent junction capacitance of an OFF diode. It is also dynamic in nature and its value partially depends upon the magnitude of V_r. Its existence gives rise to distortion in the switching waveform, especially in high-speed switching circuits. Manufacturers of switching diodes have now managed to reduce C_j to negligible values.

4-3.4 Peak Inverse Voltage (PIV) The *peak inverse voltage* (PIV) is the maximum allowable reverse-bias voltage for a given diode. It should not be exceeded for *safe* diode operation.

4-3.5 Diode Power Rating In addition to PIV, C_j, R_f, R_r, and V_k, the power rating of a diode is an important characteristic that must always be considered. The manufacturer usually lists either the maximum power rating (P_{max}) or the maximum allowable forward current (I_{max}) of the diode. Knowing one, the other can be easily calculated using Eq. (1-4c):

$$P = I^2 \cdot R \tag{1-4c}$$

$$P_{max} = I^2_{max} \cdot R_f = I_{f(max)} \times V_f \tag{4-4}$$

where:

P_{max} is the power rating of the diode, watts (W)
I_{max} is the maximum allowable forward current, amperes (A)
R_f is the forward resistance of the diode, ohms (Ω)

Example 4-4

A diode reaches its maximum power rating of 2 W when operating in the forward mode at the forward voltage of 900 mV. Calculate:

a. The maximum allowable forward current, $I_{f(max)}$
b. The forward diode resistance, R_f

Solution

a.

$$I_{f(max)} = P_{max}/V_f \tag{4-4}$$

$$= 2 \text{ W}/0.9 \text{ V} = \mathbf{2.2 \text{ A}}$$

b.

$$R_f = P_{max}/I^2_{max} \tag{4-4}$$

$$= 2/(2.2)^2 = \mathbf{0.405\ \Omega}$$

4-4 An Ideal Diode

As indicated earlier, every *practical* diode has some small forward-resistance, high reverse-resistance, finite peak inverse-voltage ratings, finite power-rating, finite knee-voltage, and finite junction-capacitance.

The desirable qualities of a practical diode are that it should have very low forward resistance, very high reverse resistance, very high peak inverse voltage, sufficiently high power rating, very low knee voltage, and very low junction capacitance.

An *ideal* diode is a diode that has the best possible (*ideal*) characteristics:

a. Its forward resistance is *zero*. This means that when forward-biased, an ideal diode acts like a *short* circuit (or closed switch).
b. Its reverse resistance is *infinite*. This means that when an ideal diode is reverse-biased, it acts like an *open circuit* (or open switch). There is no reverse or leakage current.
c. Its peak inverse voltage is *infinite*. Any magnitude of reverse-bias voltage cannot destroy it.
d. Its power rating is *infinite*. An infinite amount of forward current cannot destroy it. Any amount of forward current is permissible.
e. Its knee voltage (V_K), is 0 V. In the forward-biased configuration, voltage across the diode is zero volts.
f. Its junction capacitance is *zero*.

Of course, *no* practical semiconductor diode behaves like an ideal diode. But it is often easier to analyze circuit behavior if the diodes used are first considered ideal.

4-5 The Diode as a Voltage-Controlled Switch

It is customary to call a diode ON when forward-biased and *conducting*. A diode is said to be OFF when reverse-biased and *not conducting*. When ON, an ideal diode acts like a *closed* switch (short circuit) and when OFF, it acts like an *open* switch (open circuit). This equivalency situation is shown in Fig. 4-2. A practical diode approximates an ideal diode. It has very low forward resistance (R_f) and extremely high reverse resistance (R_r).

When analyzing circuits containing diodes, the procedure is as follows:

1. Determine whether the diode is ON or OFF.
2. Replace the diode by its appropriate equivalent circuit.

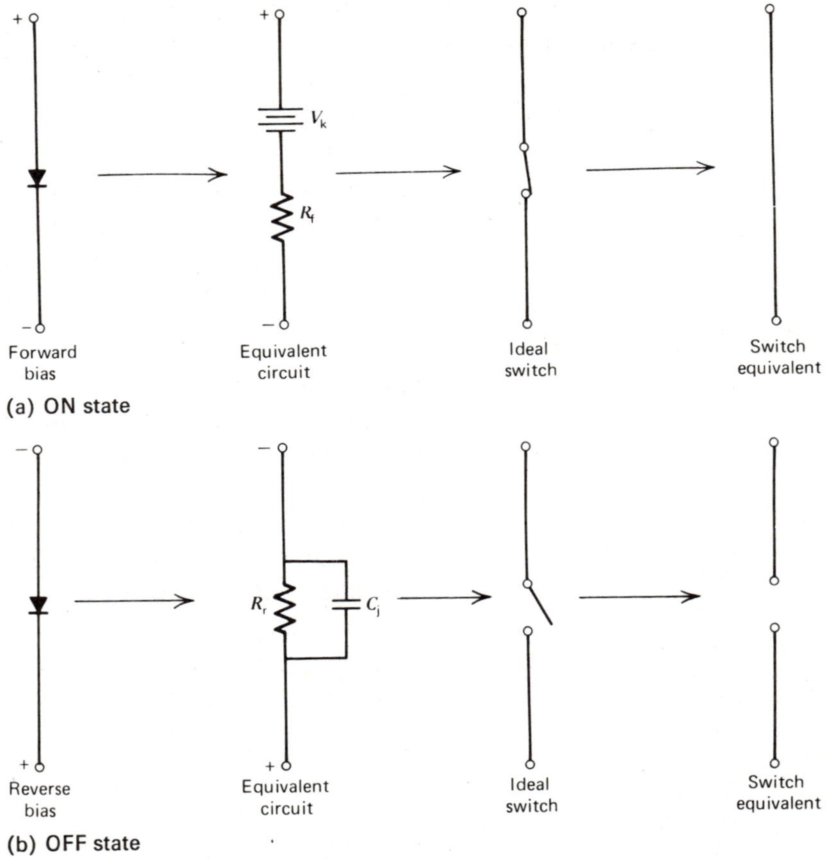

Figure 4-2 ON/OFF states of a diode switch.

3. If the input is a dynamic one (changing), as in the case of Fig. 4-3d, first locate and identify the times when the diode switches from ON to OFF and then from OFF to ON.
4. Draw equivalent circuits for each of the two states (ON or OFF) and analyze each individually.

Example 4-5

Consider the diode switching circuit shown in Fig. 4-3a. Assume the diode D as ideal. Find the output voltage (E_o) when:

a. $E_i = +12$ V.

b. $E_i = -12$ V.

Solution

a. When $E_i = +12$ V, D is forward-biased and acts as a closed switch.

$$\therefore E_o = E_i$$
$$= +12 \text{ V}$$

b. When $E_i = -12$ V, D is reverse-biased and acts as an open switch.

$$\therefore E_o = 0 \text{ V}$$

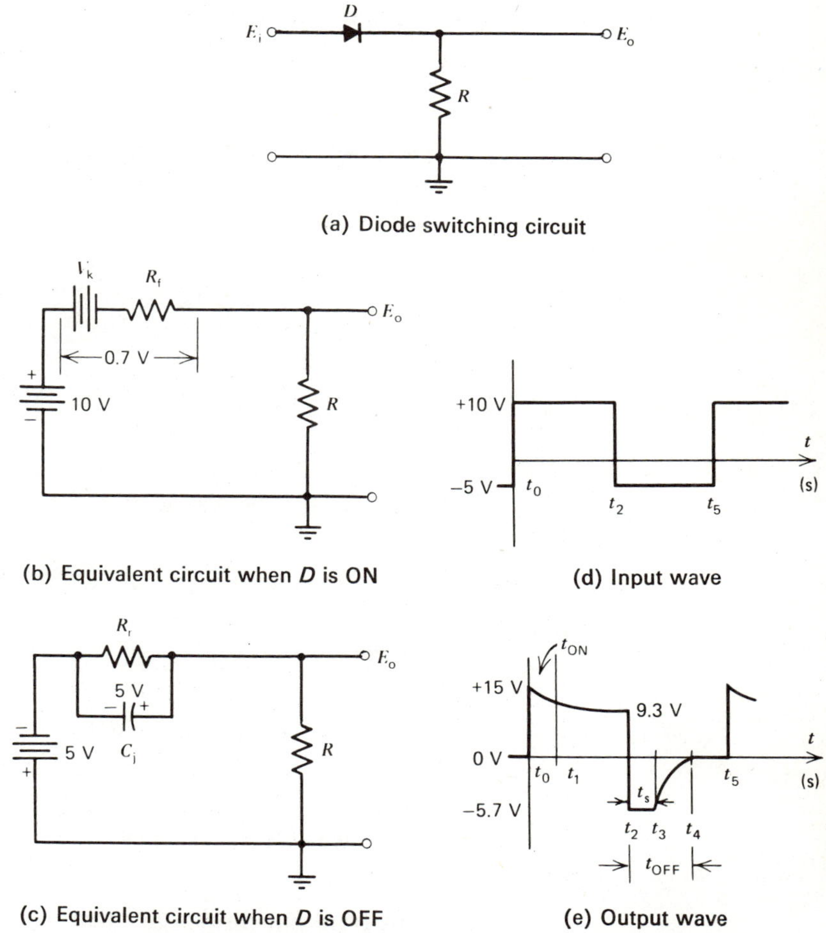

Figure 4-3 Transients in diode switching.

4-6 Transient Characteristics of Switching Diodes

Practical diodes take a *finite* time to turn *fully* ON *after* the turn-ON forward bias has been applied. This time delay between the time that the diode turns fully ON and the time that the forward bias was applied to it is known as its *turn-ON* time (t_{ON}).

If a diode is ON and is turned OFF by applying to it a step reverse-bias, it will also take a certain finite time to turn OFF. The time delay between the time when the step reverse-bias is applied and the time when the diode actually gets turned OFF is known as its *turn-OFF* time (t_{OFF}).

For most practical purposes, t_{ON} and t_{OFF} can be ignored because usually they are quite small compared with the applied pulse width (t_p). In high-speed switching circuits, where t_p itself is small, t_{ON} and t_{OFF} become significant and cannot be ignored.

Consider the circuit shown in Fig. 4-3a. Assume that the input is an ideal pulse wave, as shown in Fig. 4-3d. The diode is reverse-biased by -5 V just prior to time t_0. The diode junction has -5V across it, as shown in Fig. 4-3c. Instantaneously, at t_{0+}, after the input switches from -5 V to $+10$ V, the voltage at E_o becomes $+15$ V (Fig. 4-3a). Then, of course, the small junction capacitance (C_j) discharges from -5 V to 0 V and V_D, voltage across the diode, becomes 0.7 V, as shown in Fig. 4-3b. During this time, therefore, the output voltage goes from $+15$ V to 9.3 V and stays there as long as the input is at $+10$ V (see Figs. 4-3b and 4-3e).

The junction capacitance (C_j) discharges through resistor R. The discharge time constant τ_d is RC_j. In accordance with the (exponential) general charge/discharge equation, voltage v_C for this particular instance may be written as:

$$v_{Cj} = V_1 - (V_1 + V_{DO})e^{-(t/RC_j)}$$

where:

V_1 is the new input voltage applied to the circuit to turn the diode ON, volts (V)

V_{DO} is the initial voltage across the diode (when the diode was OFF), volts (V)

This transient-discharge equation applies only until C_j reaches 0 V. After that, the steady-state conditions exist.

4-6.1 Turn-on Time (t_{ON})

The time that C_j takes to discharge to 0 V and V_D to become 0.7 V is the *turn-ON time*. It is shown as t_0 to t_1 in Fig. 4-3e. In practical circuits, a small turn-ON time is desired. In other words, the sooner C_j discharges to 0 V, the better. This is accomplished by discharging C_j faster. One way to do this is by making the discharge time-constant (RC_j) small in the above equation. Thus the smaller the C_j and/or R, the shorter the turn-ON time. Another way to reduce the turn-ON time is to make C_j discharge from a small negative toward a much larger positive voltage. This makes it reach 0 V faster. C_j discharges very quickly to 0 V if it starts discharging from a small negative toward a large positive voltage. This is true if the input pulse swings between a *low negative* voltage and a *large positive* voltage.

4-6.2 Turn-off Time (t_{OFF})

When the input is −5 V, the equivalent circuit of Fig. 4-3c applies. The output is 0 V during the negative part of the input if D is an ideal diode. Practically, though, the output instantly falls to −5.7 V as the input changes from +10 V to −5 V. It stays at −5.7 V for a certain length of time called the *storage time* (t_s) or the recovery time, and then exponentially decays to 0 V as C_j charges to −5 V. The time that the output takes to go from −5.7 V to 0 V is the *turn-OFF time* (t_{OFF}). Turn-OFF time for the waveform shown in Fig. 4-3e is from t_2 to t_4.

4-6.3 Storage Time

When a diode is ON, *minority* charge-carriers accumulate at the junction. The extent of the accumulation depends on the current flowing through the diode. Reducing the current lowers the accumulated charge-carriers at the junction.

The storage time (t_s) is the time taken to sweep accumulated minority carriers *away* from the junction. This time is shown in Fig. 4-3e as t_s (t_2 to t_3). It really is a time delay as far as the switching action is concerned, because the diode only begins to turn OFF after t_s has elapsed (Fig. 4-3e). The reverse-bias voltage across the diode sweeps the minority carriers away. The magnitude of the reverse-bias voltage determines the time (t_s) that minority carriers take to be swept away. Thus, a low positive input ensures that only a small quantity of minority carriers accumulate at the junction during the ON time. A high reverse-bias sweeps away the minority carriers from the junction faster during the turn-OFF time. Thus, an input pulse that swings between a small positive voltage and a large negative voltage results in a reduced storage time and an improved output waveform.

However, this is exactly opposite to the input amplitude conditions required to reduce the turn-ON time (t_{ON}). Clearly, the solution lies not in controlling the input waveform, but rather in obtaining diodes having extremely low junction capacitance (C_j). Such diodes are called *switching diodes*.

The exponential decay time shown as t_3 to t_4 in Fig. 4-3e is the time that C_j takes to charge to the input voltage level. Again, this is controlled by the time constant RC_j. Reducing the time constant tends to make the charge time shorter. Thus, the time constant RC_j should really be made as short as possible to reduce the t_{ON} and t_{OFF} times. This is done by using the lowest possible value of R and diodes having the lowest possible value of C_j.

Note that the above discussion only took into consideration switching transients introduced by the diode capacitance, C_j, while the diode switched its states.

Example 4-6

Consider the circuit shown in Fig. 4-3a. Let $R = 56\ \text{k}\Omega$ and D be a silicon diode whose $R_f = 200\ \Omega$, $R_r = 50\ \text{M}\Omega$, and $C_j = 5\ \text{pF}$. A 20-$V_{p\text{-}p}$, 100-kHz square wave input gives rise to a storage time of 50 μs. Calculate the turn-ON and the turn-OFF times.

Solution

When $E_i = +10$ V, the circuit of Fig. 4-4a applies. Note that C_j had an initial charge of 10 V when the input was -10 V. Writing the discharge equation gives:

$$0 = 10 - (10 + 10)e^{-(t_{ON}/\tau)} \qquad \text{(Using Eq. -37)}$$

or,

$$e^{-(t_{ON}/\tau)} = \frac{10}{20}$$

$$e^{-(t_{ON}/\tau)} = 0.5$$

then,

$$t_{ON}/\tau = 0.693$$

C_j charges and discharges through a 56-kΩ resistor.

$$\therefore \tau = RC_j$$

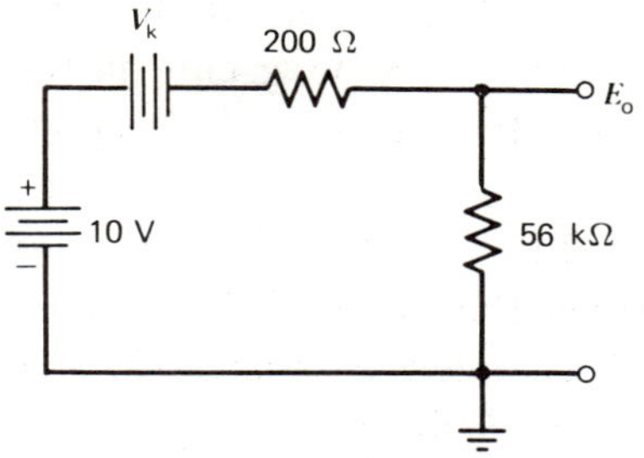

(a) Equivalent circuit when *D* is forward-biased

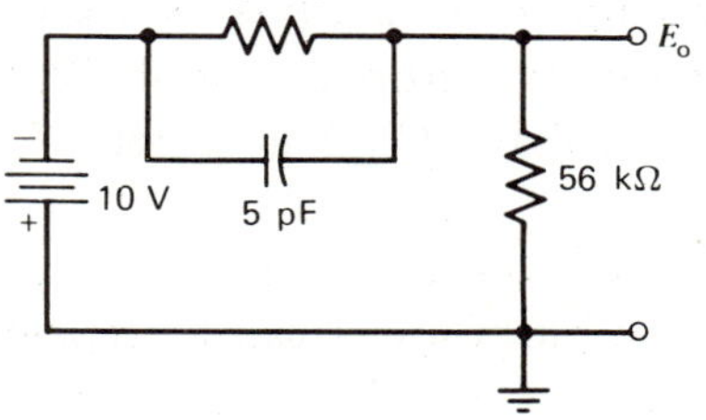

(b) Equivalent circuit when *D* is reverse-biased

Figure 4-4 Equivalent circuits for Ex. 4-6.

or,

$$= (56 \times 10^3) \times (5 \times 10^{-12})$$

or,

$$= 0.28\ \mu s$$

$$\therefore\ t_{ON} = 0.693 \times (0.28 \times 10^{-6})$$

or,

$$= \mathbf{0.194\ \mu s}$$

When $E_i = -10$ V, the circuit of Fig. 4-4b applies.

$$t_{OFF} = t_s + 5\tau \qquad (t_s = \text{storage time} = 50\ \mu s)$$
$$= 50\ \mu s + 5 \times 0.28\ \mu s$$
$$= \mathbf{51.4\ \mu s}$$

Note that neither R_f nor R_r of the diode is involved in the above calculations!

4-7 Series Diode Limiter

A *limiter* circuit is a *waveshaping* circuit. It either limits the amplitude of a waveform to a certain value or cuts off the positive or negative half of a waveform. It is also called a *clipper* circuit. Figure 4-5 shows the *series diode limiter* circuit. The diode (D) is in series with load R_L. Resistor R is known as the *bleeder* resistor. The value of R is at least 10 times the value of R_L, so that the bleeder does not load the circuit or change the value of R_L materially. The parallel combination of R and R_L is thus approximately R_L. The basic purpose of the bleeder R is to provide a discharge path for any residual capacitive voltage after the load has been disconnected.

In Fig. 4-5, when the input voltage E_i is positive, D is forward-biased. Depending upon whether it is a silicon or germanium diode, the approximate voltage drop across it (V_D) is either 0.7 V or 0.3 V. The output voltage (V_{R_L}) is simply $E_i - V_D$, or approximately E_i.

During the negative half of the input cycle, the diode is reverse-biased. For analysis purposes, it can be replaced by its equivalent resistance (R_r). The output voltage (E_o) in this OFF configuration (using the voltage-divider rule) is given by:

$$E_o = -E_i\left(\frac{R_L}{R_L + R_r}\right) \text{ volts (V)} \qquad (4\text{-}5a)$$

where:

E_i is any value of the input reverse-bias voltage. But usually $R_L \ll R_r$, and whenever this is the case,

$$\therefore E_o \cong -E_i\left(\frac{R_L}{R_r}\right) \approx 0, \text{ since } R_r \gg R_L \qquad (4\text{-}5b)$$

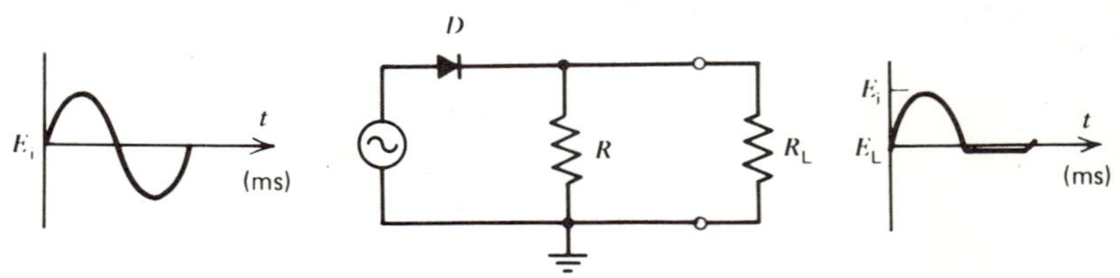

(a) Input voltage (b) Series diode limiter circuit (c) Output voltage

Figure 4-5 Series diode limiter.

Example 4-7

Refer to the circuit of Fig. 4-5b. Assume D as ideal and $R_L = 50\ \text{k}\Omega$. E_i is a 10-V, 10-kHz sine wave. Analyze operation and draw the output voltage waveform.

Solution

1. D is ON as long as the input is positive and above zero volts. As an *ideal* diode, it acts like a *closed* switch. The equivalent circuit that applies under these conditions is drawn in Fig. 4-6a. The

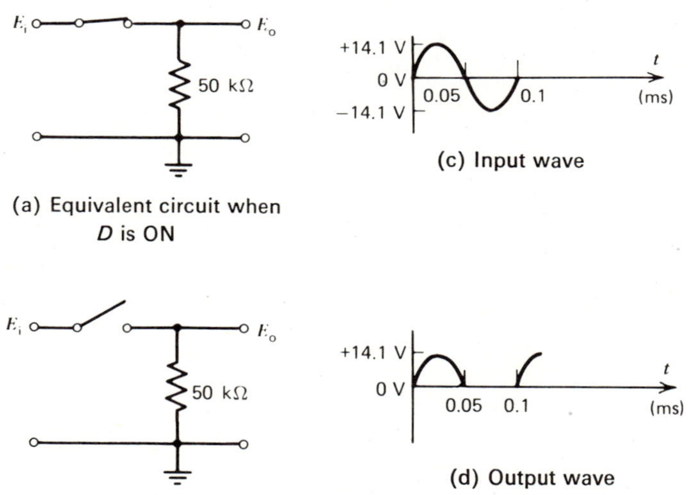

(a) Equivalent circuit when D is ON

(c) Input wave

(b) Equivalent circuit when D is OFF

(d) Output wave

Figure 4-6 Equivalent circuits and waveforms for Ex. 4-7.

output voltage is the same as the input voltage. Its *maximum* amplitude is:

$$V_m = 10\sqrt{2}$$
$$= \mathbf{14.1\ V}$$

2. When the input is negative, the diode D is *reverse*-biased (OFF). As an *ideal* diode, it acts like an *open* switch, as shown in Fig. 4-6b. Thus, under these conditions, the output voltage waveform remains at the 0-V level.

One complete cycle of the idealized output voltage waveform is drawn in Fig. 4-6d.

Example 4-8

Now repeat Ex. 4-7 for a *practical* silicon diode whose $V_K = 0.7$ V and whose $R_r = 10$ MΩ. Find and sketch the voltage waveform that appears across the 50-kΩ load.

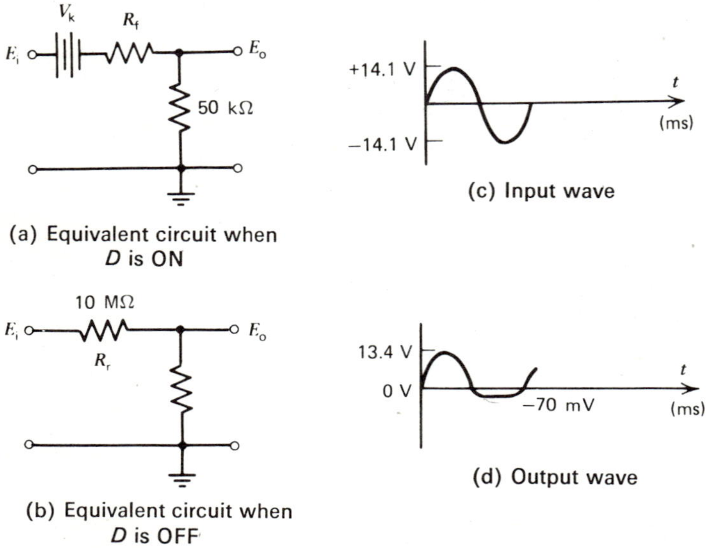

Figure 4-7 Equivalent circuits and waveforms for Ex. 4-8.

Solution

1. D is ON as long as the input is above zero volts, that is, during the positive half-cycle. During this period, the equivalent circuit of Fig. 4-7a applies. Therefore, under these conditions:

$$E_o(t) = E_i(t) - 0.7 \text{ V}$$

$$+E_{o(\text{max})} = 14.1 - 0.7$$

$$= \mathbf{13.4 \text{ V}}$$

2. When the input is negative, the diode is reverse-biased (OFF). The equivalent circuit of Fig. 4-7b applies.

$$E_o = -E_i\left(\frac{R_L}{R_r + R_L}\right) \text{ volts (V)} \qquad \text{(4-5a)}$$

$$E_o = -E_i\left(\frac{50 \text{ k}\Omega}{10 \text{ M}\Omega + 50 \text{ k}\Omega}\right)$$

$$E_{o(\text{max})} \cong \frac{-14.1}{10.05 \text{ M}\Omega} \times 50 \text{ k}\Omega$$

$$\cong \mathbf{-0.07 \text{ V}}$$

The output waveform is drawn in Fig. 4-7d. Note that the output waveform using an *ideal* diode in Ex. 4-7 is a good approximation of the practical diode. Thus, for a *quick estimation* of the circuit behavior, diodes should first be assumed as ideal. Usually this assumption does *not* introduce too much error and simplifies the calculations. Then, if more precision is desired, the exact values can be entered into the appropriate equations, as shown in the above examples.

Example 4-9

Consider the circuit conditions given in Ex. 4-7. Calculate:

a. Minimum PIV.
b. Maximum current rating (I_{max}).
c. Maximum power rating of the diode.

Solution

a. The maximum reverse-bias voltage that appears across the diode is 14.1 V. The diode must be able to withstand this voltage. Thus, its *minimum* PIV ratings should be **14.1 V or higher**, depending on commercial availability.

b. Maximum current in the circuit flows when the input is at its positive peak. Thus,

$$I_{max} = \frac{14.1 - 0.7}{50\ \text{k}\Omega}$$

$$= \mathbf{0.268\ mA}$$

The current rating of the diode (I_{max}) should be at least 0.268 mA (or higher) depending on commercial availability.

$$P_{max} = V_K \cdot I_D(\text{max})$$

$$= 0.7\ \text{V} \times 0.268\ \text{mA}$$

$$= \mathbf{0.188\ mW}$$

4-8 Parallel Diode Limiter

The circuit shown in Fig. 4-8b is called the *parallel diode limiter* circuit. Here diode D is connected in *parallel* with load R_L. Resistor R is now the *current-limiting* resistor. Its function is to keep the diode current below maximum-rated current (I_{max}) when the diode is forward-biased.

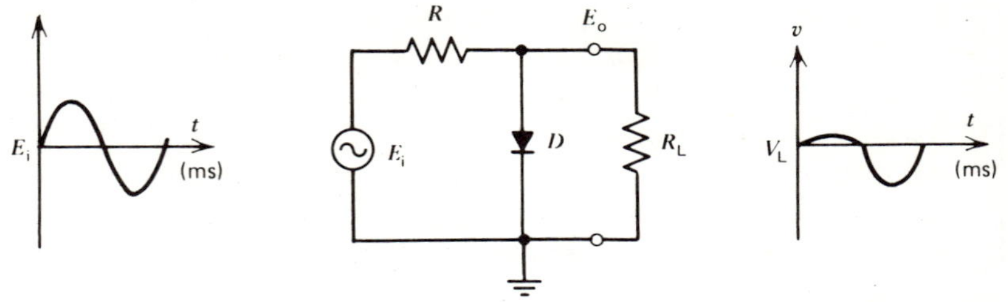

(a) Input voltage (b) Parallel diode limiter circuit (c) Output voltage

Figure 4-8 Parallel diode limiter.

Thus,

$$R \geqq \frac{E_i - V_D}{I_{max}} \approx \frac{E_i}{I_{max}} \tag{4-6}$$

where:

> E_i is the maximum value of forward input voltage (V)
> V_D is the voltage drop across the forward-biased diode (V)
> I_{max} is the maximum current rating of the diode (A)

As long as the input (E_i) is *positive*, diode D is *forward*-biased and the diode's low forward resistance (R_f) is in parallel with R_L. The equivalent resistance of this parallel combination is essentially equal to the low value of R_f. Thus, the low forward-voltage drop across the diode is now really the output voltage, or in other words, $E_o = V_L = V_D$ (Eq. 4-2c). The output voltage is approximately either 0.7 V or 0.3 V, depending upon whether D is silicon or germanium, respectively.

During the *negative* half-cycle of input, D is *reverse*-biased. For analysis purposes, D is replaced by its very high reverse resistance (R_r). The equivalent resistance of the parallel combination of R_L and R_r is essentially R_L. (R_L is usually quite small as compared with R_r.) The output voltage is given by:

$$E_o = -E_i\left[\frac{R_L \parallel R_r}{R + (R_L \parallel R_r)}\right] \tag{4-7a}$$

But because $R_L \ll R_r$, E_o reduces to:

$$E_o \cong -E_i\left(\frac{R_L}{R + R_L}\right) \text{ volts (V)} \qquad (4\text{-}7b)$$

where:

E_i is any value of the reverse input voltage (Note that the polarity of the input voltage is negative in this case simply because D is OFF when E_i is negative.)
R is the current-limiting resistance (Ω)
R_L is the load resistance (Ω)
R_r is the reverse resistance of the diode (Ω)

Example 4-10

The input to the parallel limiter circuit of Fig. 4-8b is a 20-$V_{p\text{-}p}$, 10-kHz square wave. D is a germanium diode ($V_K = 0.3$ V) whose $R_f = 20\ \Omega$, $R_r = 2\ \text{M}\Omega$, and $P = 100$ mW. The load resistor R_L is 100 kΩ. Calculate:

a. Current rating of the diode (I_{max}).
b. A reasonable value for the current limiting resistor (R).
c. PIV required for the diode.
d. The voltage waveform appearing across the 100 kΩ load.

Solution

a.

$$I_{max} = \sqrt{\frac{P_{max}}{R_f}} \qquad (4\text{-}4)$$

$$I_{max} = (P_{max}/R)^{1/2}$$

$$I_{max} = \sqrt{\frac{100 \times 10^{-3}}{20}}$$

$$= \mathbf{70.7\ mA}$$

b.

$$R = \frac{E_i - V_D}{I_{max}} \qquad (4\text{-}6)$$

$$R = \frac{(10 - 0.3)\text{ V}}{70.7\text{ mA}}$$

$$= \mathbf{137\ \Omega}$$

c. The maximum reverse-bias voltage across the diode is -10 V. Thus, PIV $=$ **10 V**.

d. Positive half-cycle of the input ($E_i = +10$ V) keeps the diode ON. For a germanium diode, $V_D = 0.3$ V, which also is the voltage across the load. (Voltage across R_f is negligible.) Thus, $E_o = 0.3$ V for the positive half cycle (see Fig. 4-9a).

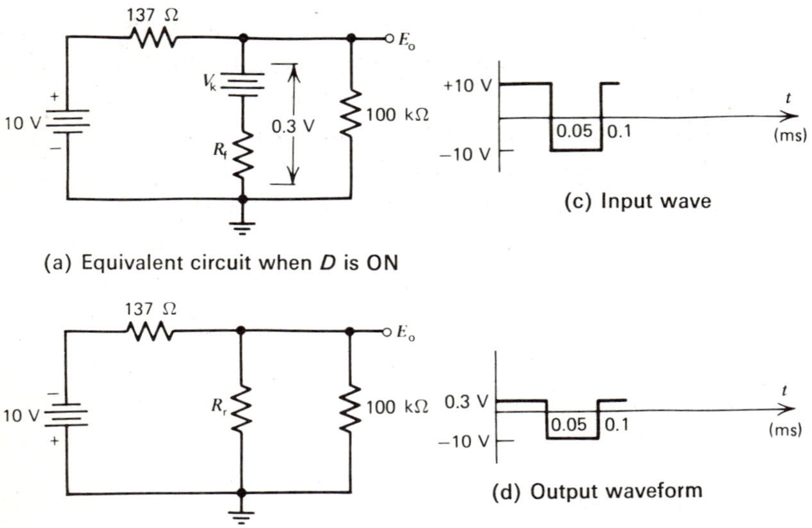

(a) Equivalent circuit when *D* is ON

(b) Equivalent circuit when *D* is OFF

(c) Input wave

(d) Output waveform

Figure 4-9 Equivalent circuits and waveforms for Ex. 4-10.

Equivalent circuit for the negative half-cycle is drawn in Fig. 4-9b. Note that *D* is reverse-biased during this time.

$$E_o = \left[\frac{R_r \parallel R_L}{R + (R_r \parallel R_L)}\right] E_i \text{ volts (V)} \tag{4-7a}$$

$$E_o = -10\left[\frac{2\text{ M} \parallel 100\text{ k}}{137 + (2\text{ M} \parallel 100\text{ k})}\right] \text{ volts (V)}$$

or,

$$= -10\left(\frac{95.24\text{ k}}{137 + 95.24\text{ k}}\right)$$

or,

$$= \mathbf{-9.99\ V \approx -10\ V}$$

Thus, the output voltage (E_o) varies between $+0.3$ V and -10 V, as shown in Fig. 4-9d.

4-9 Biased Clipper

As mentioned earlier, limiters are also sometimes called *clippers*. These are used to clip off a part of the amplitude of a waveform. A biased-clipper circuit is used to clip the amplitude of a waveform at a level other than 0 V. Figure 4-10b shows a parallel-diode limiter circuit with a biasing voltage V_B in series with D.

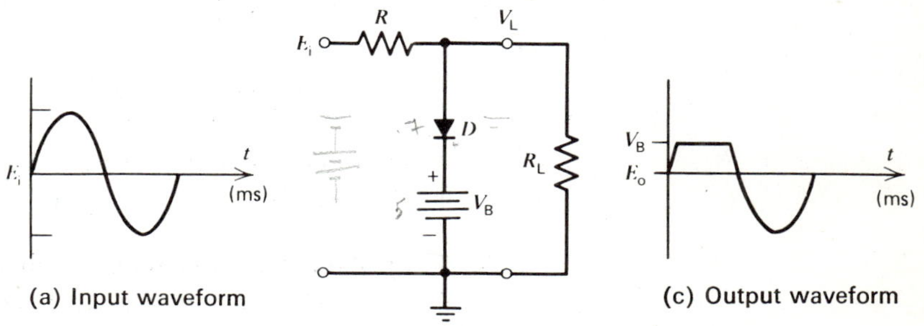

(a) Input waveform

(b) Biased-clipper circuit

(c) Output waveform

Figure 4-10 Biased clipper.

Diode D is forward-biased as long as E_i is more positive than V_B. Under this condition (forward bias),

$$V_L = V_B - V_D \text{ volts (V)} \tag{4-8a}$$

where:

V_L is the voltage across the load (V)
V_B is the **dc** biasing voltage (V)
V_D is the forward voltage across the diode (V)

It has been pointed out earlier that for most practical purposes, $V_D \approx V_K$. Thus, Eq. (4-8a) becomes:

$$V_L = V_B - V_K \text{ volts (V)} \tag{4-8b}$$

For an ideal diode $V_K = 0$. In that case,

$$V_L = V_B \text{ volts (V)} \tag{4-8c}$$

When E_i is less positive than V_B, D becomes reverse-biased (OFF). The equivalent circuit shown in Fig. 4-11a applies under these conditions. Thévenizing the circuit between points A and B

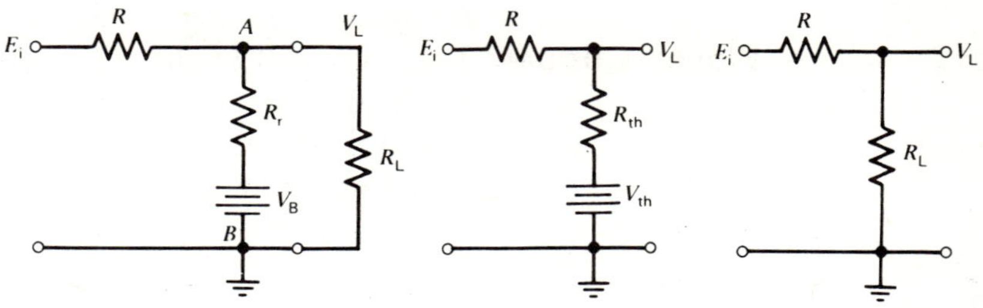

(a) Complete equivalent circuit (b) Simplified equivalent circuit (c) Simplified (further) equivalent circuit

Figure 4-11 Equivalent circuit of the biased clipper (Fig. 4-10) when *D* is reverse-biased.

(Fig. 4-11a) results in an equivalent circuit shown in Fig. 4-11b. V_{th} and R_{th} are found as follows:

$$R_{th} = \frac{R_r \cdot R_L}{R_r + R_L}$$

Since normally,

$$R_r \gg R_L$$
$$R_{th} = R_L$$

and

$$V_{th} = V_B \frac{R_L}{R_r + R_L}$$

but

$$R_r \gg R_L$$
$$V_{th} \cong 0 \text{ V (negligible compared to } E_i)$$

This simple situation is reflected in the circuit shown in Fig. 4-11c. Now calculating V_L reduces simply to:

$$V_L = E_i\left(\frac{R_L}{R_L + R}\right) \text{ volts (V)} \tag{4-9}$$

where:

V_L is the voltage across the load, volts (V)
E_i is any input voltage below $+V_B$, volts (V)
R_L is the load resistor, ohms (Ω)
R is the current-limiting resistor, ohms (Ω)

Usually $R_L \gg R$, then Eq. (4-9) reduces to:

$$V_L \simeq E_i \text{ volts (V)} \tag{4-10}$$

The voltage across R_L in Fig. 4-10b is given by Eq. (4-8) when diode is ON and by Eq. (4-10) when diode is OFF. The resultant output waveform is shown in Fig. 4-10c. Note that this diode-clipper clips off that portion of the input voltage which exceeds V_B. Other variations of this circuit may have either the diode reversed and/or V_B reversed, depending upon the application requirements.

Example 4-11

The input to the circuit of Fig. 4-12b is as shown in Fig. 4-12a. Find and sketch the voltage waveform appearing at E_o. Assume D as ideal.

Solution

1. Determine the time during which the diode is ON and the time during which it is OFF.
2. Replace the diode with its *ideal* equivalent, that is, a closed switch when ON and an open switch when OFF.

Whenever $E_i \geqq +2$ V, the diode is reverse-biased and OFF. It acts like an *open* switch. The output voltage E_o under these conditions is the *same* as the input voltage.

The diode is ON when $E_i \leqq +2$ V. It acts like a *short* circuit, placing the 2 V battery across E_o. The output voltage E_o is simply 2 V **dc** for all input values less than +2 V.

The resultant output waveform can now be drawn for the sinusoidal input of Fig. 4-12a. It is shown in Fig. 4-12c.

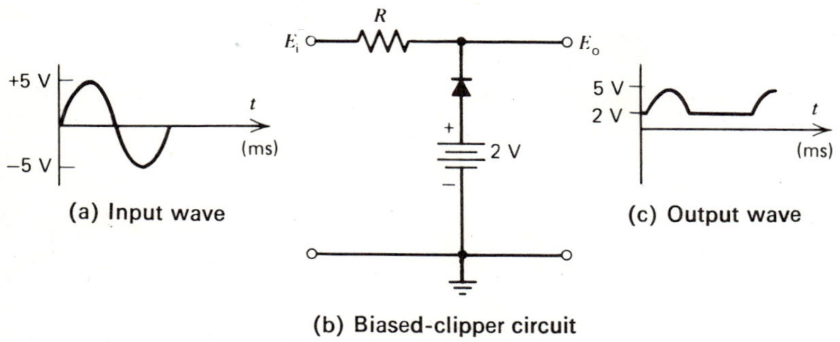

Figure 4-12 Biased-clipper circuit and waveforms for Ex. 4-11.

4-10 Dual Diode Clipper

Figure 4-13c shows a *dual clipper* circuit. It essentially is a biased clipper circuit having *two* biased clipping diodes. The $D_1 V_{B_1}$ combination clips off the input at V_{B_1}. The $D_2 V_{B_2}$ combination clips off the input at V_{B_2}. The input and output waveforms are shown in Figs. 4-13a and 4-13b, respectively.

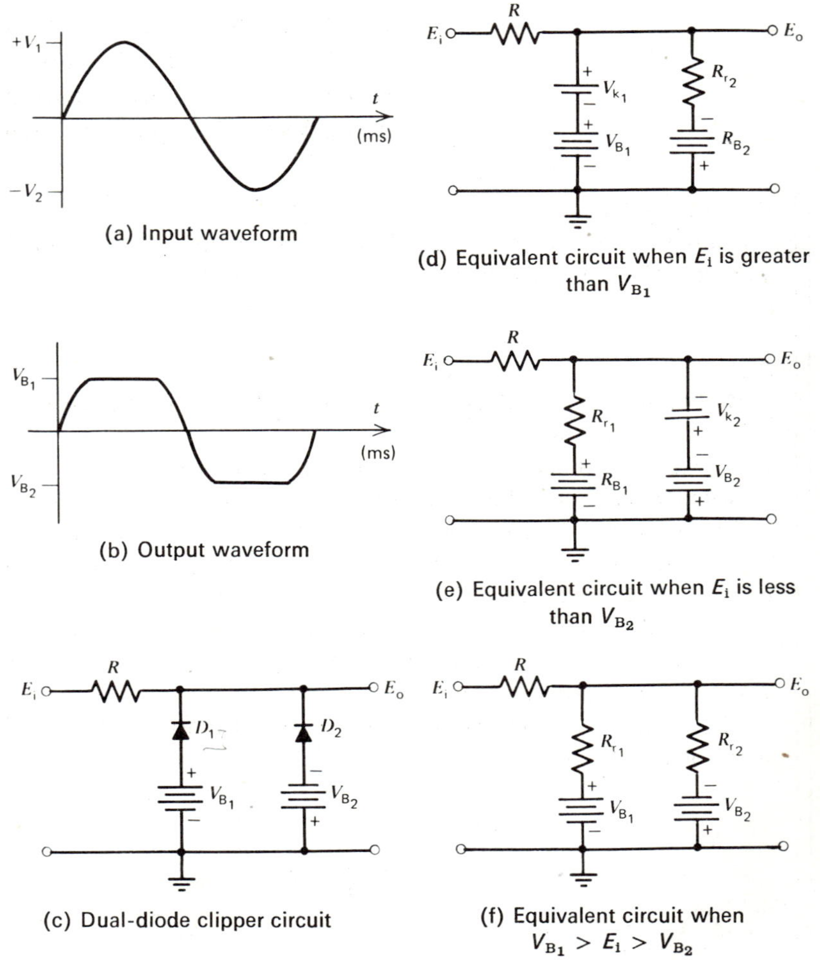

(a) Input waveform
(b) Output waveform
(c) Dual-diode clipper circuit
(d) Equivalent circuit when E_i is greater than V_{B_1}
(e) Equivalent circuit when E_i is less than V_{B_2}
(f) Equivalent circuit when $V_{B_1} > E_i > V_{B_2}$

Figure 4-13 Dual-diode clipper.

When $E_i \geqq V_{B_1}$, D_1 is forward-biased and D_2 is reverse-biased. The equivalent circuit under this condition is shown in Fig. 4-13d. The output voltage level is

$$E_o = V_{B_1} + V_{K_1} \text{ volts (V)} \tag{4-11}$$

where:

V_{B_1} is the bias voltage at D_1
V_{K_1} is the knee voltage for D_1

When $E_i \leqq V_{B_2}$, D_1 is reverse-biased and D_2 is forward-biased. This condition is shown in Fig. 4-13e. The output voltage level here is

$$E_o = -(V_{B_2} + V_{K_2})$$

When $V_{B1} < E_i < V_{B_2}$, *both* of the diodes are reverse-biased (OFF). The equivalent circuit under this condition is shown in Fig. 4-13f.

To simplify calculations for E_o, first find the Thévenin equivalent circuit for the V_{B_1} R_{r_1} R_{r_2} V_{B_2} combination between points E_o and ground. Accordingly:

$$V_{th} = V_{B_1} - (V_{B_1} + V_{B_2})\left(\frac{R_{r_1}}{R_{r_1} + R_{r_2}}\right) \tag{4-12}$$

$$R_{th} = R_{r_1} \parallel R_{r_2}$$

$$= \frac{R_{r_1} \cdot R_{r_2}}{R_{r_1} + R_{r_2}} \tag{4-13}$$

Now E_o can be easily calculated:

$$E_o = E_i - V_R$$

$$E_o = E_i - (E_i \pm V_{th})\left(\frac{R}{R + R_{th}}\right) \text{ volts (V)} \tag{4-14a}$$

where all terms have been defined above.

The $\pm$ polarity of V_{th} in Eq. (4-14a) means that V_{th} is added to E_i when its polarity is in series with (aiding) E_i. V_{th} is subtracted from E_i when its polarity opposes E_i. In most practical circuits, $V_{th} \cong 0$ V. If so, then Eq. (4-14a) reduces to:

$$E_o \cong E_i - \left(\frac{R}{R + R_{th}}\right)E_i \tag{4-14b}$$

Again since usually $R_{th} \gg R$, term $RE_i/(R + R_{th})$ is insignificant compared to E_i and can be neglected. E_o becomes:

$$E_o \cong E_i \tag{4-14c}$$

Example 4-12

Input to the dual clipper circuit shown in Fig. 4-14a is a 200-$V_{p\text{-}p}$ 10-kHz sine wave (Fig. 4-14c). Assume both diodes as ideal.

a. Find and sketch the output waveform at E_o.

b. Find PIV rating and I_{max} rating of the two silicon diodes.

Solution

a. Diode D_2 switches ON when the voltage at E_o becomes greater than +10 V. It is OFF as long as the voltage at E_o is below +10 V.

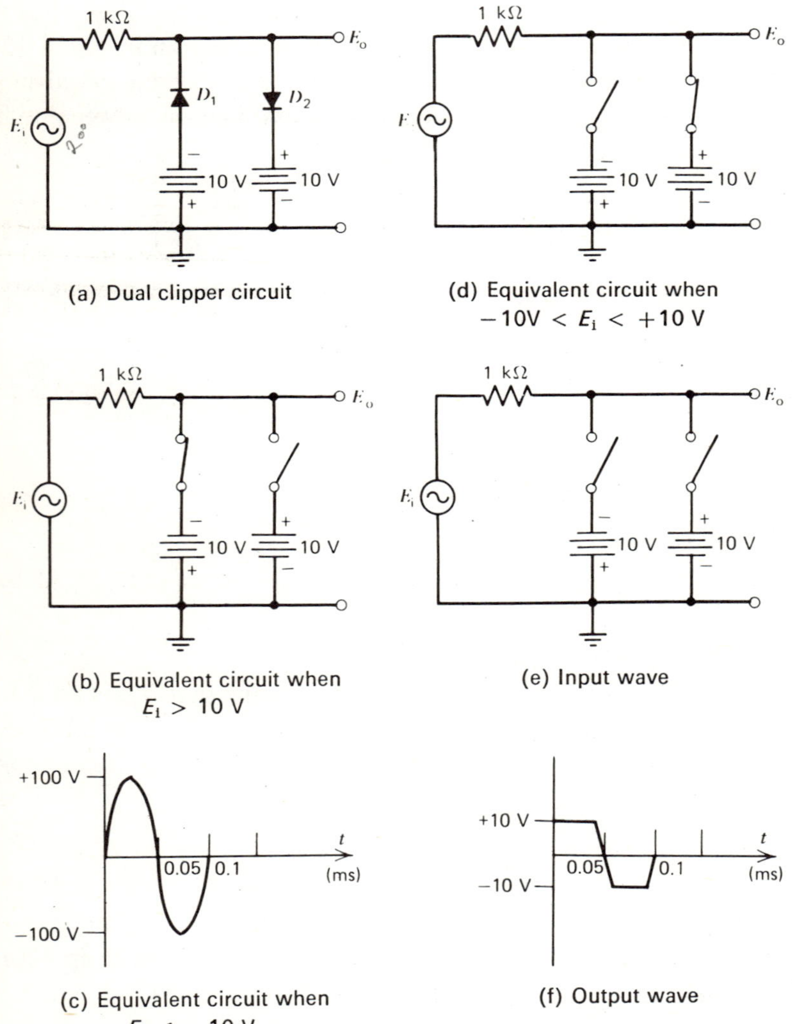

(a) Dual clipper circuit

(d) Equivalent circuit when $-10\text{V} < E_i < +10$ V

(b) Equivalent circuit when $E_i > 10$ V

(e) Input wave

(c) Equivalent circuit when $E_i < -10$ V

(f) Output wave

Figure 4-14 Dual-clipper circuit and waveforms for Ex. 4-12.

Similarly, D_1 turns ON if the voltage at E_o becomes lower than -10 V. It stays OFF as long as the E_o voltage is above -10 V. The equivalent circuit when $E_i > +10$ V is shown in Fig. 4-14b. The equivalent circuit shown in Fig. 4-14c applies when $E_i < -10$ V. When $-10\text{ V} < E_i < +10$ V, the equivalent circuit of Fig. 4-14d applies with both diodes OFF. As a result of the reasons given above, the output waveform at E_o will be similar to the one shown in Fig. 4-14f. Note that with *both* diodes OFF, the *output follows input*.

b. Maximum reverse bias occurs when the input is at its peak point. At the negative peak, the reverse bias across D_2 is 110 V. Similarly, at the positive peak, the reverse bias across D_1 is 110 V. Thus, the PIV ratings for *both* diodes is **110 V.**

Note once again that the maximum forward current through diodes occurs when the input is at its peak values. Maximum current through D_2 flows when the input is 100 V. At that time, current I_{max} is (assume $V_K = O$)

$$I_{max}(D)_2 = \frac{(100 - 10)\text{ V}}{1\text{ k}\Omega} \qquad \text{(Fig. 4-14b)}$$

$$= \mathbf{90\ mA}$$

Similarly, I_{max} for D_1 is

$$I_{max}(D)_1 = \frac{(100 - 10\text{ V})}{1\text{ k}\Omega} \qquad \text{(Fig. 4-14c)}$$

$$= \mathbf{90\ mA}$$

4-11 Differentiator-Clipper Combination

Sometimes only positive (or only negative) *trigger spikes* are required. To produce such spikes, a *differentiator* and *clipper* combination may be employed. It was pointed out in Sec. 3-7 that when an *RC* differentiator circuit has a *short* time-constant (τ), it produces sharp negative and positive spikes (alternately) when a rectangular waveform is applied to it (see Fig. 3-14d).

After the positive and negative spikes have been produced by a differentiator, a clipper circuit is used to clip off either the positive or the negative spikes. A variation of a dual clipper may be employed to clip off the negative spikes and part of the positive spikes, or vice versa.

Figure 4-15b shows a differentiator-clipper circuit. The *RC* combination constitutes a short-time-constant differentiator circuit.

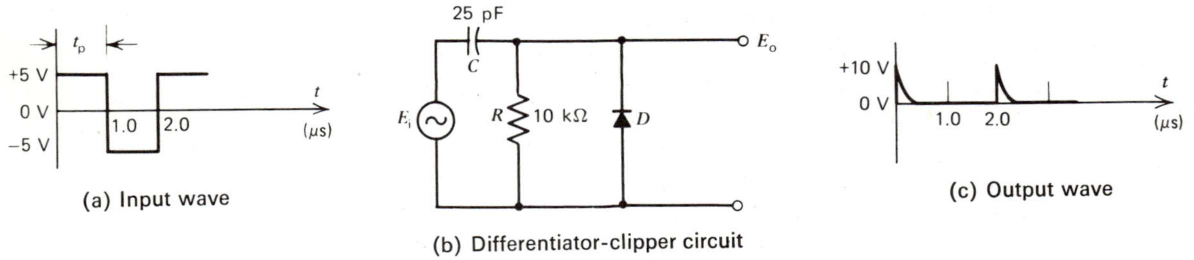

Figure 4-15 Differentiator-clipper circuit and input-output waveforms.

The capacitor fully charges and discharges during each half of the input cycle. Diode D is the clipper. It shorts the negative spikes to ground, leaving only the positive spikes at E_o.

Example 4-13
A differentiator-clipper circuit is shown in Fig. 4-15b. The input voltage is a 10-$V_{p\text{-}p}$, 50-kHz square wave. Assuming the D as an ideal diode, sketch the output voltage waveform.

Solution

$$t_p = \frac{1}{2} \times \frac{1}{f}$$

$$= \frac{1}{2} \times \frac{1}{50\ \text{kHz}}$$

$$= 10\ \mu\text{s},$$

considering only the RC differentiator circuit.

The circuit τ is:

$$\tau = RC$$

$$= 10\ \text{k}\Omega \times 25\ \mu\text{F}$$

$$= 0.25\ \mu\text{s}$$

and

$$5\tau = 5(0.25) = 1.25\ \mu\text{s}$$

Because $t_p > 5\tau$, Fig. 4-15 is a short-time-constant differentiator circuit. The output of a short τ circuit consists of sharp positive and negative spikes.

But D is OFF (open-circuit) during the positive spike and ON (short-circuit) during the negative spike. Therefore, only the positive spikes are seen at the output (E_o), because the negative spikes are shorted to ground through ON D. The output is sketched in Fig. 4-15c.

Example 4-14

Repeat Ex. 4-13, but this time consider D to be a silicon diode whose $R_f = 25$ and $R_r = 10$ M.

Solution

When $E_i = +5$ V, the circuit shown in Fig. 4-16a applies:

$$\begin{aligned}\tau_1 &= RC\\ &= 25 \times 10^{-12} \times (10\ \text{k}\Omega \parallel 10\ \text{M}\Omega)\\ &= 25 \times 10^{-12} \times (10\ \text{k}\Omega)\\ &= 0.25\ \mu\text{s}\\ 5\tau_1 &= 5 \times 0.25\ \mu\text{s} = 1.25\ \mu\text{s}\\ t_p &= 2\mu\end{aligned}$$

so $t_p > 5\tau_1$ and the circuit acts as a differentiator to positive pulses. Therefore, E_o quickly decays to zero and the capacitor charges to $+5$ V.

When $E_i = -5$ V, the circuit shown in Fig. 4-16b applies.

$$\begin{aligned}\tau_2 &= 25 \times 10^{-12} \times (10\ \text{k}\Omega \parallel 25\ \text{k}\Omega)\\ &= 25 \times 10^{-12} \times (25)\\ &= 0.625\ \mu\text{s}\end{aligned}$$

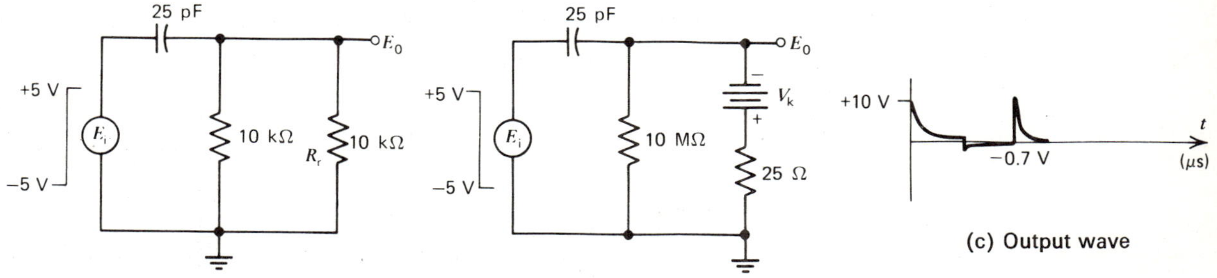

(a) Equivalent circuit when D is OFF (b) Equivalent circuit when D is ON

Figure 4-16 Equivalent circuits and the output waveform for Ex. 4-13.

$$5\tau_2 = (5 \times 0.625)\ \mu s$$
$$= 3.125\ \mu s \text{ (when } t_p = 2.5\ \mu s\text{)},$$

so again, $t_p > 5\tau_2$ and the circuit also is a differentiator to negative pulses.

When -5 V is applied, the capacitor charges to -4.3 V but the circuit current decays very quickly to zero. The output voltage (E_o) decays to 0.7 V (V_K of the silicon diode). The resultant output waveform is shown in Fig. 4-16c. Note the differences between Fig. Fig. 4-16c for the practical diode and Fig. 4-15c for the ideal diode. These differences are significant in some pulse circuitry.

4-12 Diode Clamp

A clamper circuit clamps an **ac** voltage to a desired **dc** reference level (usually 0 V). Either limit of the **ac** voltage may be clamped to a **dc** level while its waveshape remains unchanged. Assume that a diode clamp is to be used to convert a 10-$V_{p\text{-}p}$(± 5 V), 2-kHz square wave (**ac**) to a 0 V–10 V, 2-kHz pulsating square wave. Another way of saying the same thing is that the negative limit (-5 V) of a 10-$V_{p\text{-}p}$, 2-kHz square wave is to be clamped at 0 V. This is shown in Fig. 4-17. Figure 4-18 shows the diode clamp

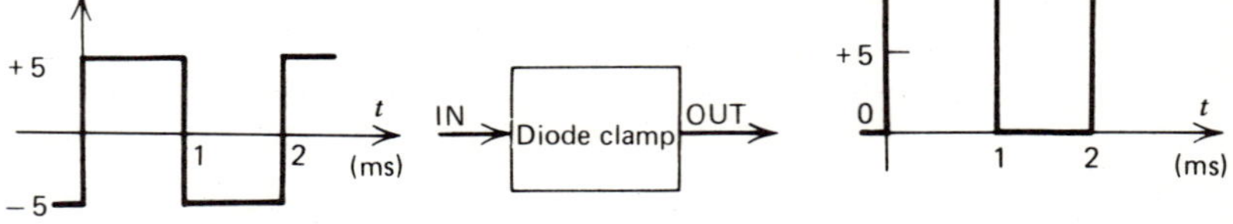

Figure 4-17 Negative peak of the input is clamped at 0 V.

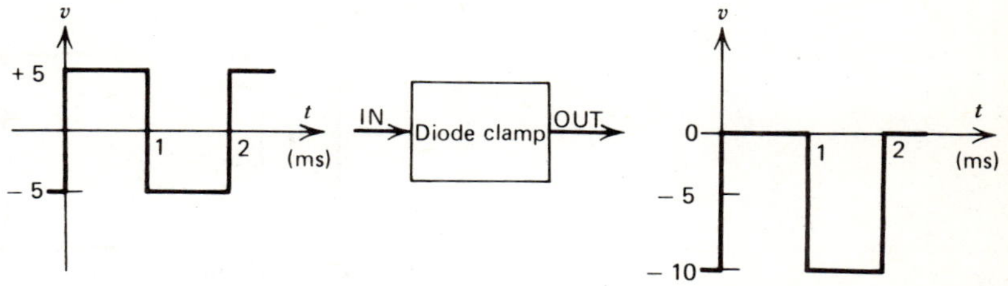

Figure 4-18 Positive peak of the input is clamped at 0 V.

being used to clamp the positive (+5 V) limit of a 10-$V_{p\text{-}p}$, 2-kHz square wave (**ac**) to 0 V.

A basic clamper circuit is shown in Fig. 4-19b. When E_i is at +5 V, D is ON. The equivalent circuit under this condition is shown in Fig. 4-19d. C gets fully charged fairly quickly because the circuit time-constant R_fC is very short. Here,

$$E_o = V_K \text{ volts (V)} \tag{4-15}$$

and

$$V_c = E_i - V_K$$

where:

V_c is the voltage across the capacitor, volts (V)
V_K is the knee voltage for the diode, volts (V)

When E_i becomes −5 V, D is OFF. The equivalent circuit under this condition is shown in Fig. 4-19e. C does not have enough time to discharge significantly because the discharging time-constant

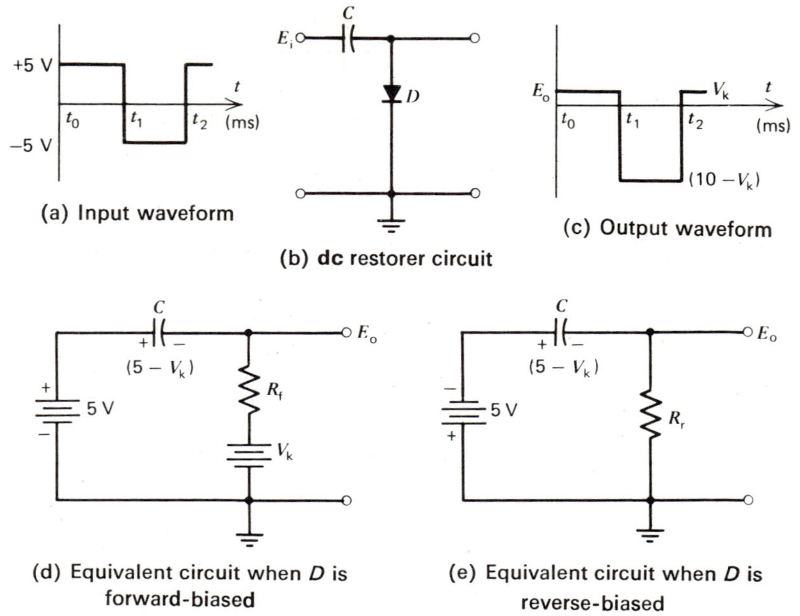

Figure 4-19 dc restorer circuit and input/output waveforms.

R_rC is usually very long compared to the time that -5 V stays at the input (t_p). During this time, -5 V in series with V_c appears at E_o. Thus,

$$\begin{aligned} E_o &= -(5 + V_c) \\ &= -(5 + (5 - V_K)) \\ &= -(10 - V_K) \end{aligned} \tag{4-16}$$

For the rest of the time, E_o varies between V_K and $-(10 - V_K)$ as the input swings between $+5$ V and -5 V. C stays charged to $(5 - V_K)$ volts. Note that:

1. The output voltage has been clamped to V_K.
2. The peak-to-peak amplitude (10 V) of the output is the *same* as the input.
3. The waveform at the output is the same as at the input.

$V_K = 0$ if D is an ideal diode. Then Eqs. (4-15) and (4-16) yield

$$E_o = 0 \text{ V, when } D \text{ is ON}$$

and

$$E_o = -10 \text{ V, when } D \text{ is OFF}$$

Note that a 10-$V_{p\text{-}p}$ **ac** square wave has been converted to a pulsating (0 V to −10 V) square wave. A 0 V to −10 V pulsating square wave consists of two components:

a. ±5 V **ac** wave (input)
b. −5 V **dc** (inserted by the **dc** restorer circuit)

Thus, we see that a −5 V **dc** component *has been inserted* into the original input wave by the clamper circuit. It is for this reason that a clamper is also sometimes called a **dc** *inserter* or a **dc** *restorer*.

If the diode direction in Fig. 4-19b is reversed, the output voltage will swing between 0 V and +10 V, permitting only the positive (rather than negative) excursions of voltage.

If the amplitude of the input voltage decreases for several cycles, the capacitor does not have time to discharge to lowered amplitude. D_1 stays open (OFF) all the time. Thus, the resultant output voltage may be clamped at some undesired negative voltage rather than at zero.

A more *practical* **dc** restorer (clamper) circuit is shown in Fig. 4-20. The addition of a resistor R allows the output to adjust, if the amplitude of the input waveform changes. It does this by directly providing a capacitor discharge path. The output voltage now always remains clamped at zero. Only its amplitude will change in accordance with the change in the input amplitude.

The value of R is usually chosen to be equal to the geometric mean of R_f and R_r of the diode, or:

$$R = \sqrt{R_f \times R_r} \text{ ohms } (\Omega) \tag{4-17}$$

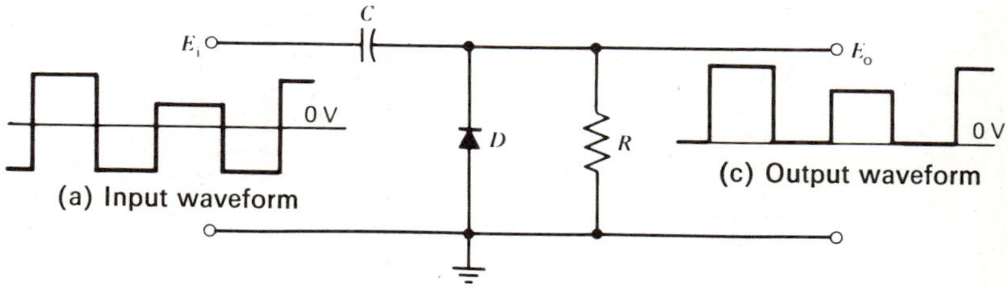

(b) Clamper circuit

Figure 4-20 Practical **dc** restorer (clamper) circuit.

Note that the practical clamper circuit of Fig. 4-20 is similar to the differentiator-clipper combination circuit of Fig. 4-15b. The difference is in the circuit time-constant. In Fig. 4-20, the capacitor, once charged, practically never discharges (long τ), whereas in Fig. 4-15b, the capacitor charges and discharges during each half-cycle of the input waveform (short τ).

Example 4-15

Refer to the circuit of Fig. 4-21a. The input is a 10-$V_{p\text{-}p}$, 5-kHz square wave. C is a 0.01-μF capacitor. The diode D has R_f of 20 Ω and R_r of 50 MΩ. Assume $V_{D(ON)} = 0$.

a. Calculate the forward time-constant (τ_f) and compare it to input waveform (t_p).

b. Calculate the reverse time-constant (τ_r) and compare it to input waveform (t_p).

c. Draw the output waveform.

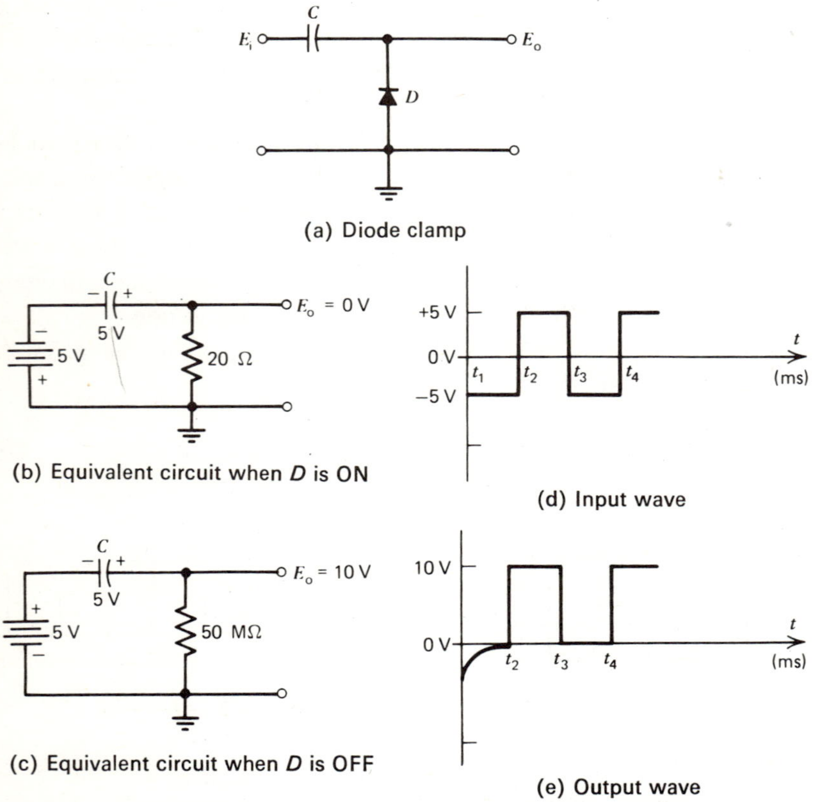

Figure 4-21 Simple diode clamp and input/output waveforms for Ex. 4-15.

Solution

a. $f = 5\text{ kHz} = 1/T$, where T is the period of the input waveform

$$t_p = \frac{T}{2} = \frac{1}{2} \cdot \frac{1}{5\text{ kHz}}$$
$$= 0.1\text{ ms}$$

The forward time constant (Fig. 4-21b) is

$$\tau_f = RC$$
$$= 20 \times 0.01\ \mu\text{F}$$
$$= \mathbf{0.2\ \mu s}$$

τ_f is very small compared to t_p, which gives the capacitor more than enough time to charge and discharge.

b. The reverse-circuit time constant (Fig. 4-21c) is

$$\tau_r = RC$$
$$= 50\text{ M}\Omega \times 0.01\ \mu\text{F}$$
$$= \mathbf{0.5\ s}$$

τ_r is very long compared to t_p, which means that the capacitor does not have enough time to significantly charge and discharge.

The capacitor quickly charges to 5 V during the very first negative half of the input cycle. After that it just stays fully charged during positive and negative halves of all of the following cycles. Thus, the output, as shown in Fig. 4-21e, varies between 0 V and +10 V and is clamped to a 0 volt level.

Example 4-16

The circuit of Ex. 4-15 is to be a practical clamper circuit, similar to the one shown in Fig. 4-20b. Find a reasonable value for R.

Solution

$$R = \sqrt{R_f \times R_r} \qquad (4\text{-}17)$$
$$= \sqrt{20 \times (50 \times 10^6)}$$
$$= \mathbf{31.6\ k\Omega}$$

4-13 Biased Clamp

Sometimes it is necessary to clamp a voltage waveform to a reference level other than 0 V. A *biased clamper* is used for this purpose. It clamps the waveform to the **dc**-biasing voltage (V_B).

A typical biased clamp is shown in Fig. 4-22b. When E_i, shown in Fig. 4-22a, is at $-V_1$ volts, diode D is forward-biased. Capacitor C charges through R_f and R. The equivalent circuit for this condition is drawn in Fig. 4-22d. The charging time constant (τ_f) is:

$$\tau_f = C \cdot (R \parallel R_f)$$

Since R is considerably larger than R_f, the parallel combination of R and R_f is essentially R_f.

$$\therefore \tau_f \cong R_f C \text{ seconds (s)} \tag{4-18}$$

R_f, the forward resistance of the diode, is usually very small, which makes τ_f very small. Consequently, C charges quickly to

$$V_c = V_1 + V_B + V_k \cong V_1 + V_B \text{ volts (V)} \tag{4-19}$$

where:

V_c is the voltage across C when fully charged
V_1 is the peak input voltage swing (negative)
V_B is the biasing **dc** voltage

The output voltage (E_o) is

$$E_{o_1} \cong V_B \tag{4-20}$$

When E_i is at the $+V_2$ level, D is reverse-biased. The equivalent circuit for this condition (D OFF) is shown in Fig. 4-22e. The discharging time-constant τ_r is

$$\tau_r = C \cdot (R_r \parallel R)$$

Since R is considerably smaller than R_r, the reverse resistance of the diode, the parallel combination of R and R_r is essentially R.

$$\therefore \tau_r = RC \text{ seconds (s)} \tag{4-21}$$

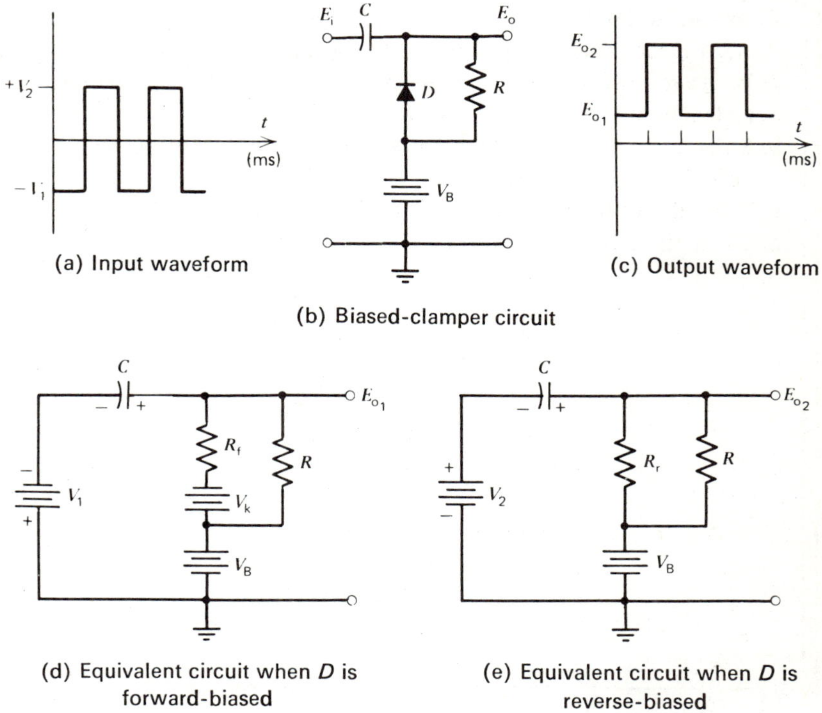

Figure 4-22 Biased clamper and input/output waveforms.

τ_r is made long in comparison to the pulse width for the time during which D is OFF. This is done to assure that C does *not* discharge significantly during this time. The output voltage stays at

$$E_{o_2} = V_2 + V_c \text{ volts (V)} \tag{4-22}$$

where:

V_2 is the peak positive voltage
V_c is the voltage across the capacitor as given by Eq. (4-19)

This completes one cycle. From then on, the output just swings between E_{o_1} and E_{o_2}.

R is calculated using Eq. (4-17). It serves the same purpose as it did for a **dc** restorer. In fact, the biased clamper circuit is a specialized version of the **dc** restorer. It differs only in that it clamps the output to the applied biasing voltage.

Example 4-16

A negative biased clamper is shown in Fig. 4-23a. The input is a pulse waveform shown in Fig. 4-23b ($R_r = 20\ \text{M}\Omega$, $R_f = 50\ \Omega$).

a. Calculate a reasonable value for R.

b. Calculate τ_f and τ_r and compare them to the input waveform period.

c. Compute output voltage levels under condition of **(b)**.

d. Draw the output waveform appearing at E_o.

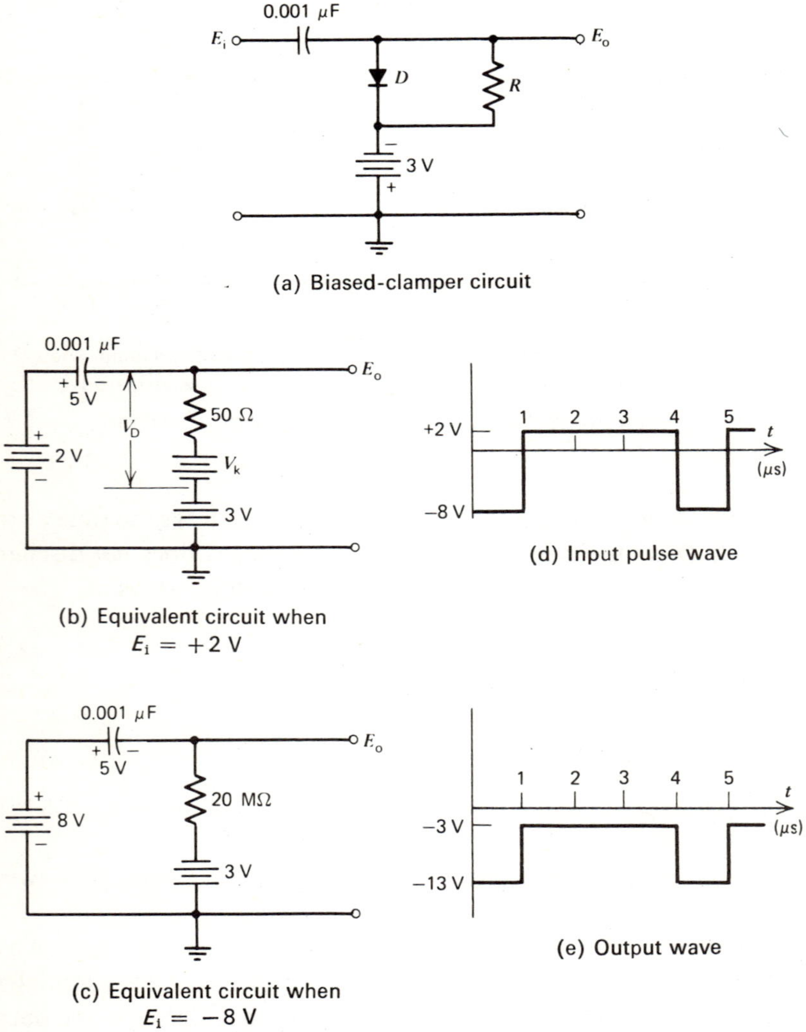

Figure 4-23 Biased clamper and input/output waveforms.

Solution

a. $$R = \sqrt{R_f \times R_r} \quad (4\text{-}17)$$
$$= \sqrt{(50) \times (20 \times 10^6)}$$
$$= \mathbf{31.6\ k\Omega}$$

b. $$\tau_f = R_f C \quad (4\text{-}18)$$
$$= 50 \times 0.001\ \mu F$$
$$= \mathbf{50\ ns}, \text{ which is very small compared to } 3\ \mu s$$
$$\tau_r = RC \quad (4\text{-}21)$$
$$\tau_r = 31.6 \times 10^3 \times 0.001\ \mu F$$
$$= \mathbf{31.6\ \mu s}, \text{ which is long compared to } 1\ \mu s$$

As shown in Fig. 4-23d, when $E_i = +2$ V, τ_f is very small. The capacitor quickly charges to +5 V, as shown in Fig. 4-23b.

$$E_o = E_i + V_c \quad (4\text{-}22)$$
$$= -8 + (-5)$$
$$= \mathbf{-13\ V}$$

c. The output waveform excursions, therefore, vary from −3 V to −13 V, as shown in Fig. 4-23e. The output waveform has been clamped to −3 V because of the −3 V biasing battery!

4-14 Diode Comparator

The circuit shown in Fig. 4-24a is known as a diode *comparator* circuit. Essentially, it is also a biased limiter circuit. As the name implies, it compares an input signal voltage (E_i) with a bias voltage

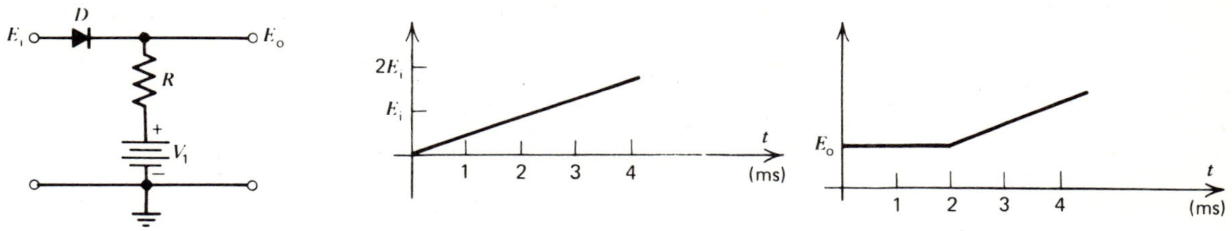

(a) Comparator circuit (b) Input wave (c) Output wave

Figure 4-24 The diode comparator.

(V_1). As long as E_i is less than V_1, the diode (D) is OFF. Under this condition, the voltage signal at E_o is simply equal to V_1. Whenever E_i is higher than V_1, D is ON. Voltage at E_o is then practically equal to E_i. Diode D used in this configuration is known as a *pickoff diode.* Note that the output of a comparator circuit becomes equal to the input signal whenever it *exceeds* the reference voltage (V_1), as shown in Fig. 4-24c. The comparator circuit is used to identify the specific *time* whenever the input signal becomes greater than the reference voltage. In Fig. 4-24c, this occurs at 2 ms.

4-15 Summary

1. A semiconductor diode is said to be ON when conducting and OFF when *not* conducting. It acts as a *voltage-controlled switch* (VCS). The ON condition is similar to a *closed* switch and the OFF condition to an *open* switch.
2. An *ideal* diode has *no* forward resistance, an *infinite* reverse resistance, *no* junction capacitance, *no* knee voltage, an *infinite* peak inverse voltage, and an *infinite* power rating. It switches from ON to OFF (or vice versa) immediately, that is, as soon as the input pulse voltage reaches the switching level.
3. A *practical* diode has a *small* forward resistance, a *finite* (high) reverse resistance, a *small* junction capacitance, a *small* knee voltage, a finite (high) peak inverse voltage, and a finite power rating. It switches from ON to OFF (or vice versa) in a finite (delay) time.
4. The diode switch is used in limiter circuits, clamper circuits, comparator circuits, and so on.
5. The diode turns ON when forward-biased and it turns OFF when reverse-biased. When it is ON, the voltage drop across it is approximately 0.7 V or 0.3 V, depending on whether it is a silicon or germanium diode, respectively. Because of its high reverse resistance (R_r), the voltage drop across it is high when it is OFF.
6. The transient characteristics of a diode are important in high-speed switching circuits. Charge and discharge times of the diode junction capacitance and its series resistance (RC_j) determine the turn-ON and turn-OFF times of the diode itself. The storage time of a diode is a function of the input pulse amplitude.
7. Desirable features of a switching diode are:
   - **a.** Low junction-capacitance (C_j).
   - **b.** Low forward-resistance (R_f).
   - **c.** High reverse-resistance (R_r).
   - **d.** Low knee-voltage (V_K).

8. Important equations:

$$R_f = \frac{\Delta V_f}{\Delta I_f} \tag{4-1}$$

$$V_f = I_f R_f + V_K \tag{4-2}$$

$$R_r = \frac{\Delta V_r}{\Delta I_r} \tag{4-4}$$

$$R \geqq \frac{E_1 - V_D}{I_{max}} \tag{4-6}$$

$$R = \sqrt{R_f \times R_r} \tag{4-17}$$

4-16 Glossary

Clamper. An electronic circuit that clamps an input signal to a certain reference voltage level.

Clipper. Same as **Limiter,** defined below.

Comparator. An electronic circuit that sends out a signal if a certain test signal goes above (or below) a set reference voltage.

Forward resistance. The equivalent resistance of a diode when it is ON.

Knee voltage. The minimum forward-bias voltage necessary to turn the diode ON.

Limiter. An electronic circuit that clips off part of the input waveform.

OFF condition. A diode is said to be OFF when it does **not** conduct as a result of *reverse* bias applied across it.

ON condition. A diode is said to be ON when it conducts as a result of *forward* bias applied across it.

Reverse resistance. The equivalent resistance of a diode when it is OFF.

Storage time. The time period taken by minority carriers to be swept away from the junction.

4-17 Recapitulation

Return to the Objectives and Self-Evaluation Questions given at the beginning of this chapter and see how well you can answer them. If you cannot answer certain questions satisfactorily, place a check mark (√) next to them and review appropriate parts of the chapter to find the answers. No attempt should be made to solve the Problems (Sec. 4-18) *before* the Self-Evaluation Questions are answered. Ability to solve the Problems should serve as a test and feedback regarding your understanding of the material presented in this chapter. Answers to all Problems are given on p. 518.

4-18 Problems

4-1 Figure 4-25 shows typical semiconductor diode graphical characteristics. Find:

a. PIV.
b. V_k.
c. R_f.
d. R_r.

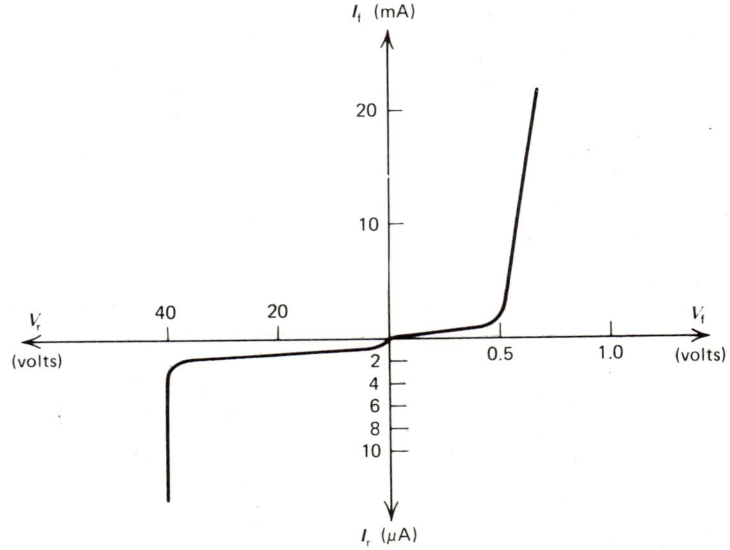

Figure 4-25 Diode characteristics for Prob. 4-1.

4-2 Refer to the circuit of Fig. 4-26. Assume $R_f = 100\ \Omega$ and $R_r = 10\ \text{M}\Omega$ for the silicon diode D. Calculate the diode current.

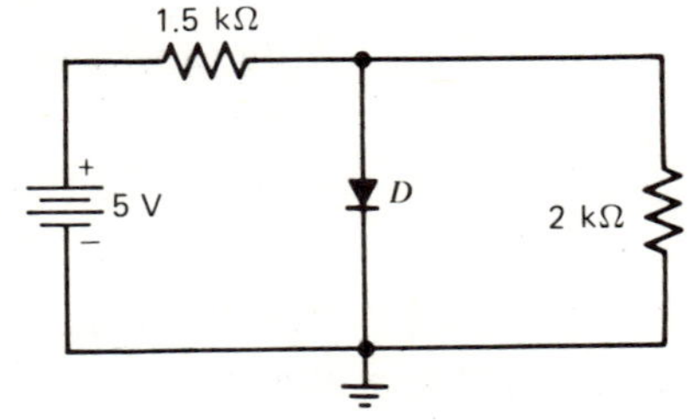

Figure 4-26 Circuit for Probs. 4-2, 4-3, and 4-4.

4-3 What happens to the diode current (Fig. 4-26) if:

a. The Si diode is replaced by a germanium diode?
b. A 2-kΩ resistor is replaced by a 10-kΩ resistor?
c. A 1.5-kΩ resistance is increased to 5 kΩ?

d. The diode burns out and becomes a short circuit?

e. The diode is reversed?

4-4 Diode D in Fig. 4-26 is reversed. Calculate:

a. The diode current.

b. The voltage drop across the diode.

4-5 For the silicon diode used in the circuit of Fig. 4-27, $R_f = 5\ \Omega$ and $R_r = 5\ \text{M}\Omega$. Suggest a reasonable value for R.

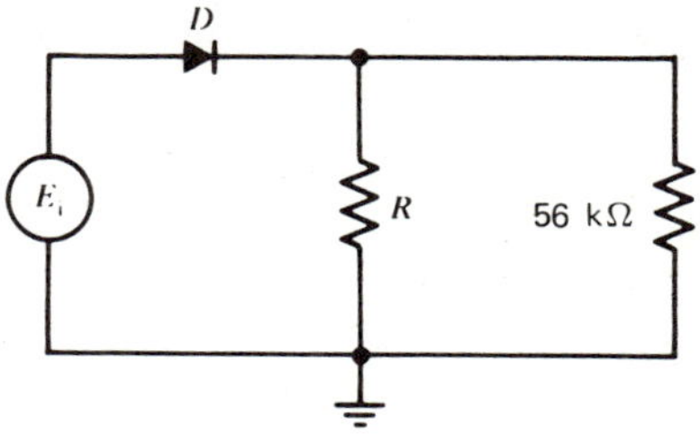

Figure 4-27 Circuit for Probs. 4-5, 4-6, 4-7, 4-8, and 4-9.

4-6 Refer to Prob. 4-5. Calculate the diode current when:

a. $E_i = +10$ V.

b. $E_i = -10$ V.

4-7 Input to the series-limiter circuit of Fig. 4-27 is a 10-$V_{p\text{-}p}$, 5-kHz square wave. Assume that D is an ideal diode and $R = 2\ \text{M}\Omega$. Find and draw the voltage waveform that appears across the 56-kΩ load resistor.

4-8 Repeat Prob. 4-7 with input as a 10-$V_{p\text{-}p}$, 5-kHz sine wave.

4-9 Refer to Prob. 4-5. The input is a 30-$V_{p\text{-}p}$, 10-kHz sine wave. Draw the voltage waveform that appears across the 56-kΩ load resistor.

4-10 Consider the parallel diode limiter circuit shown in Fig. 4-8b. $R = 47\ \Omega$, $R_L = 2\ \text{k}\Omega$, and $E_i =$ 10-$V_{p\text{-}p}$, 10-kHz sine wave. Assume $R_f = 7.5\ \Omega$ and $R_r = 2\ \text{M}\Omega$ for the silicon diode D. Sketch the output waveform.

4-11 Refer to Prob. 4-10. Find diode PIV and wattage rating for safe operation.

4-12 The pulse wave shown in Fig. 4-28a is applied to the biased limiter

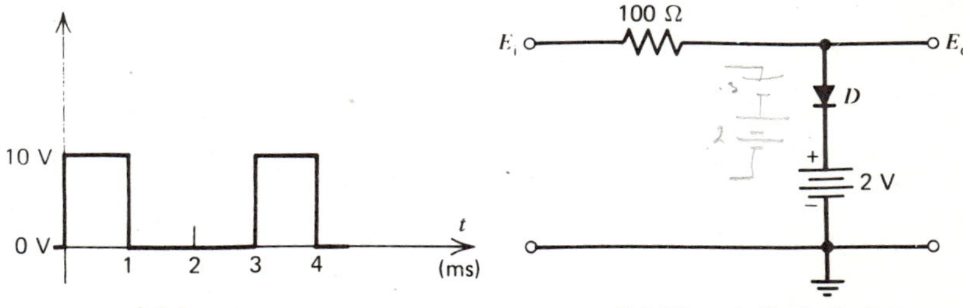

(a) Input wave

(b) Biased-diode limiter

Figure 4-28 Circuit for Probs. 4-12, 4-13, and 4-14.

circuit of Fig. 4-28b. Assume D as ideal. Find and sketch the voltage waveform that appears at E_o.

4-13 Repeat Prob. 4-12 with D a germanium diode whose $R_f = 10\ \Omega$ and $R_r = 2.0\ \mathrm{M}\Omega$.

4-14 Repeat Prob. 4-13 with a 50-kΩ load connected across the output terminals.

4-15 A 50-V, 100-kHz sine wave is to be converted into a 10-$V_{p\text{-}p}$, 100-kHz square wave. Design an appropriate dual clipper circuit. Give PIV and I_{max} ratings for the silicon diodes used.

4-16 Consider the clamper circuit of Fig. 4-29. Find the output voltage waveform when $R = 15\ \mathrm{k}\Omega$, $C = 0.1\ \mu\mathrm{F}$, and $D =$ ideal. The input is a 24-$V_{p\text{-}p}$, 4-kHz square wave.

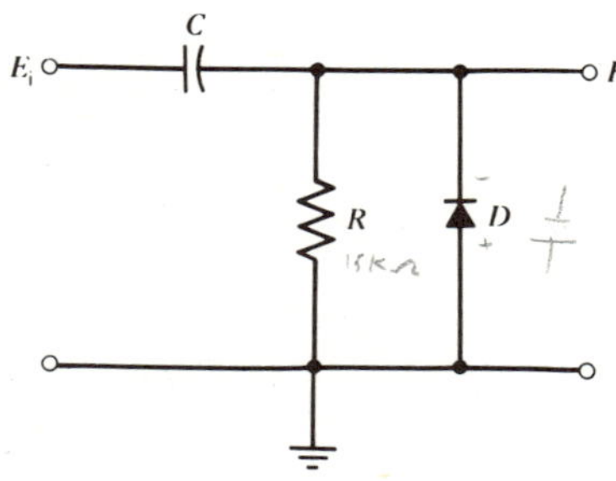

Figure 4-29 Circuit for Probs. 4-16, 4-17, and 4-18.

4-17 Repeat Prob. 4-16 with D as a silicon diode whose $R_f = 7.5\ \Omega$ and $R_r = 50\ \mathrm{M}\Omega$.

4-18 Repeat Prob. 4-17 with D reversed.

4-19 Design a simple diode comparator circuit that yields a signal if the input goes above 6 V **dc**; otherwise it stays at +6 V **dc**.

4-20 Consider the comparator circuit shown in Fig. 4-30. Draw the voltage waveform that appears at E_o.

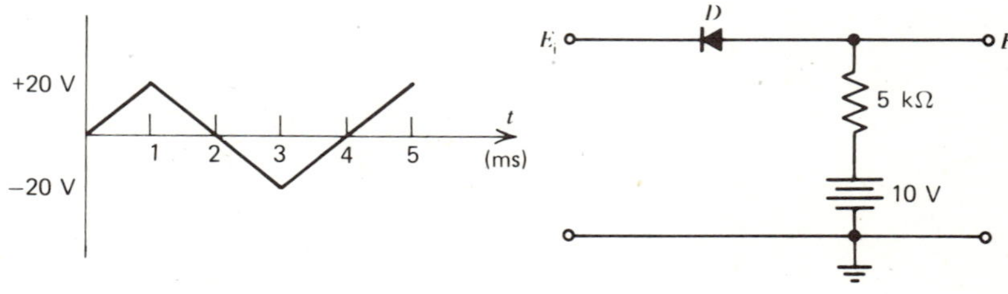

Figure 4-30 Circuit for Prob. 4-20.

4-21 A 25-V, 10-kHz sine wave is available. Using a block diagram, show how can this be converted to the pulse wave of Fig. 4-31.

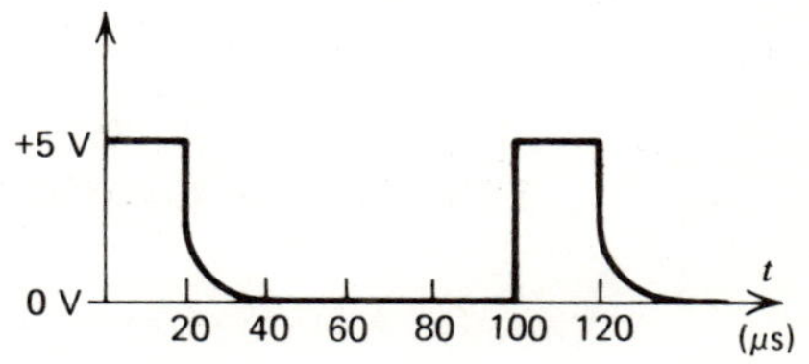

Figure 4-31 Waveform for Prob. 4-21.

4-22 Consider the circuit shown in Fig. 4-3a. *D* is a silicon diode whose $C_j = 2$ pF, $R_f = 80\ \Omega$, and $R_r = 78\ \text{M}\Omega$. A 25-$V_{p\text{-}p}$ square-wave input results in a storage time of 10 ns. *R* is a 1-kΩ resistor. Find:

a. The turn-ON time (t_{ON}).

b. The turn-OFF time (t_{OFF}).

4-19 Experiment

Title **Diode Limiters and Clampers**

Theory A diode limiter circuit is used to clip off a part of a waveform. The diode itself behaves like an open switch when OFF (*reverse-biased*) and a closed switch when ON (*forward-biased*). The diode used to clip off a part of a waveform is connected in series with the load in a series limiter and is connected in parallel with the load in a parallel diode limiter. A series diode limiter is shown in Fig. E4-1 and a parallel diode limiter in Fig. E4-2. R_L is the load.

A diode clamp or a **dc** restorer clamps the input waveform to the 0-V level or to a given **dc** level by inserting a **dc** component into the initial **ac** waveform. A diode clamp is shown in Fig. E4-3. The capacitor charges (fully) during the very first time the diode is turned ON. After that, it always stays charged, because the circuit time-constant is very long when the diode is OFF.

Equipment Multifunction generator: EXACT model 126, or equal.
Multimeter: RCA model WV 89C, or equal.
Dual beam oscilloscope: PHILIPS model 3230, or equal.
Resistors: 470 Ω, 10 kΩ, and 100 kΩ.
Capacitor: 0.01 F 25 V or higher.
Diode: 2N4148.

Procedure

1. Connect the series limiter circuit shown in Fig. E4-1. Connect

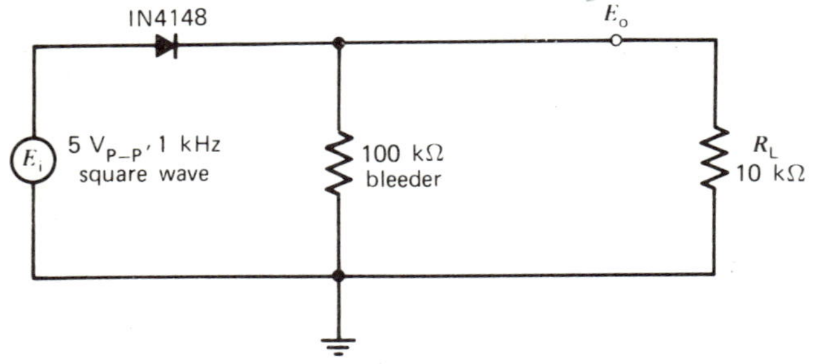

Figure E4-1 Series diode limiter.

one beam of CRO across E_i and the other across E_o. Energize the circuit. Establish ground reference level on the CRO screen.

2. Observe and carefully draw the input (E_i) and the output (E_o) waveforms in Table E4-1.
3. From the output waveform observed above, find and record the value of V_K of the diode. V_K = ——————
4. Reverse the direction of the switching diode. Repeat Procedure step 2. Record your results in Table E4-1.
5. Change the input (multifunction generator) to a 10-$V_{p\text{-}p}$, 1-kHz sine wave. Observe and draw the input and the output waveforms in Table E4-1.
6. Connect the parallel diode-limiter circuit shown in Fig. E4-2.

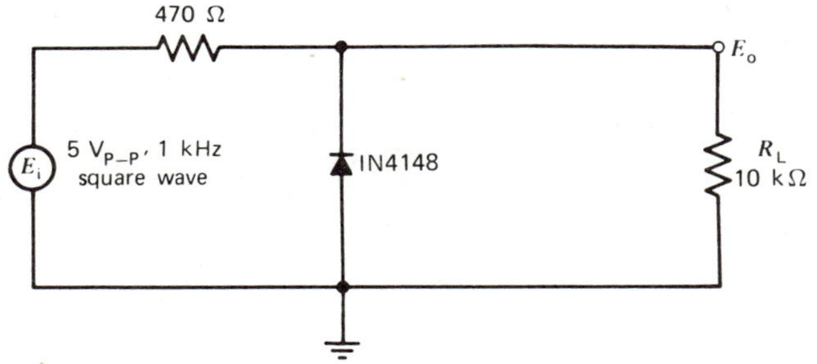

Figure E4-2 Parallel diode limiter.

7. Observe and draw the input and the output waveforms in Table E4-2.
8. Reverse the direction of the diode. Repeat Procedure step 7.
9. Connect the clamper (or dc restorer) circuit shown in Fig. E4-3.

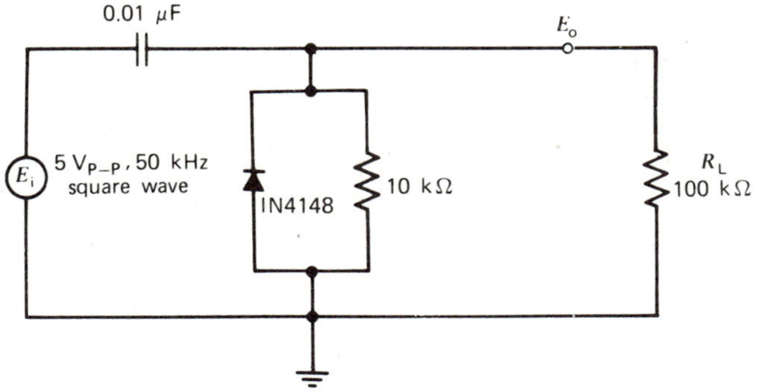

Figure E4-3 Diode clamp or dc restorer.

10. Observe and draw the input and the output waveforms in Table E4-3.
11. Reverse the direction of the diode and repeat Procedure step 10.
12. Disconnect and return the test equipment to its proper place.

Calculations 1. Find (calculate) the theoretically expected output waveforms and draw them in Tables E4-1, E4-2, and E4-3. In each case, calculate and show the theoretical peak amplitude during each half-cycle. Record the theoretical output waveforms in appropriate Tables. Assume $R_f = 5\Omega$, $R_r = 1\ M\Omega$, and $V_D(\text{ON}) = 0.7$ V for the diode.

Tabulation of Results Table E4-1
Series Diode Limiter

Procedure Step	Input Waveform	Output Waveform, Experimental	Output Waveform, Theoretical
2			
4			
5			

Table E4-2
Parallel Diode Limiter

Procedure Step	Input Waveform	Output Waveform, Experimental	Output Waveform, Theoretical
7			
8			

Table E4-3
Diode Clamp

Procedure Step	Input Waveform	Output Waveform, Experimental	Output Waveform, Theoretical
10			
11			

Questions

1. Examine the experimentally observed and the theoretically expected output waveforms in Tables E4-1, E4-2, and E4-3. Describe how your observed waveforms compare with the theoretically expected ones in each case.
2. In Procedure step 3, the experimental value of the knee voltage (V_k) of the diode was recorded. Considering it, determine whether the 1N4148 is a silicon or a germanium diode. Give reasons for your answer.
3. Consider the clamper circuit shown in Fig. E4-3. What will happen to the output waveform if the discharge time-constant of the circuit becomes approximately equal to the discharging pulse width? Does this aspect of the circuit configuration place any limitations on the pulse repetition frequency (PRF)?

Discussion

1. Summarize your conclusions insofar as the observed and the expected output waveforms are concerned. Examine Tables E4-1, E4-2, and E4-3 for this purpose.
2. A series limiter circuit is essentially a half-wave rectifier. Explain.
3. Discuss the process by which a clamper circuit inserts the **dc** component into an **ac** input.

Chapter 5 The Transistor Switch

5-1 Objectives

The objectives of this chapter are to:

1. Review the characteristics of a bipolar transistor.
2. Identify the characteristics of a so-called ideal transistor.
3. Show that the ON-OFF operation of a transistor is equivalent to the ON-OFF operation of an ordinary toggle switch.
4. Show how to determine whether the transistor is ON, OFF, or actively conducting.
5. Explain and discuss the transient characteristics of a transistor switch.
6. Explain the function of a speed-up or commutating capacitor used to improve the switching time of a transistor switch.
7. Explain the operation of a practical transistor inverter.
8. Review the characteristics of a field effect transistor (FET).
9. Compare the transistor switch with a FET switch.
10. Show that the ON-OFF operation of a MOSFET is equivalent to the ON-OFF states of an ordinary toggle switch.
11. Explain the operation of a COS/MOS inverter.

5-2 Self-Evaluation Questions

Watch for the answers to the following questions as you read the chapter. They should help you to understand the material presented. When you have finished the chapter, return to this section and be sure that you can answer all the questions. Then test your understanding by working out solutions to the problems (Sec. 5-15).

It is strongly recommended that these questions be satisfactorily answered before attempting to solve the Problems.

1. Show the standard symbol and the conventional direction of current in:
   a. A *pnp* transistor.
   b. A *npn* transistor.
2. Explain briefly the relationship that exists between the collector current (I_C), the base current (I_B), and the emitter current (I_E) in an actively conducting transistor.
3. Identify and list the important characteristics (ratings) of a typical transistor that are usually given by the manufacturer.
4. Discuss the significance of the characteristic curves of a transistor

and identify the three regions of operation, namely, the *cutoff* region, the *active* region, and the *saturation* region.

5. Describe the characteristics of a so-called ideal transistor.
6. Show that a transistor operating in the *saturated* region is equivalent to an *ON switch* and in the *cutoff* region is equivalent to an *OFF switch*.
7. How would you determine whether a transistor switch is ON or OFF using a simple **dc** voltmeter?
8. Design a simple transistor inverter circuit for the two logic levels of +5 V and +0.3 V.
9. Describe and identify the circuit elements that limit the maximum switching speed of a transistor.
10. Describe the conditions and a method of improving the switching speed of a transistor switch (speed-up capacitor).
11. Discuss briefly the characteristics of a field effect transistor (FET).
12. Compare a bipolar transistor switch with a FET switch.
13. What is the difference between:
    a. A *p*-channel MOSFET and an *n*-channel MOSFET?
    b. A depletion-type MOSFET and an enhancement-type MOSFET?
14. Show that an ON MOSFET is equivalent to a closed switch and an OFF MOSFET is equivalent to an open switch.
15. Explain the operation of a basic COS/MOS inverter.

5-3 Review of the Characteristics of the Bipolar Transistor

The standard symbol for a bipolar *npn* transistor is shown in Fig. 5-1a. The symbol for a *pnp* bipolar semiconductor transistor is shown in Fig. 5-1b. Note that the base-emitter arrow symbol in each case is from *p*-type to *n*-type material. The three leads are labeled as *collector* (*C*), *base* (*B*), and *emitter* (*E*).

The current in a bipolar (junction) transistor is a result of the interaction between the majority charge-carriers and the minority charge-carriers. It is called a bipolar device because the current in it consists of two types of charge carriers. In a *pnp* transistor, the majority carriers are holes and the minority carriers are electrons. In the *npn* transistor, the majority carriers are electrons and the minority carriers are holes. The bipolar junction transistor (BJT) is usually called simply a transistor.

The transistor is a *current-controlled* device. There exist definite relationships between the three currents I_C, I_B, and I_E when the transistor is operated in the active region. These are:

$$I_C + I_B = I_E \tag{5-1}$$

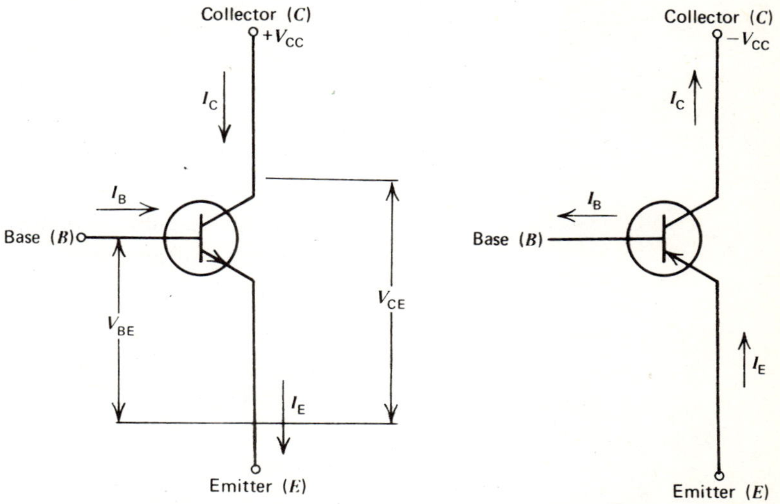

(a) Symbol for an *npn* transistor

(b) Symbol for a *pnp* transistor

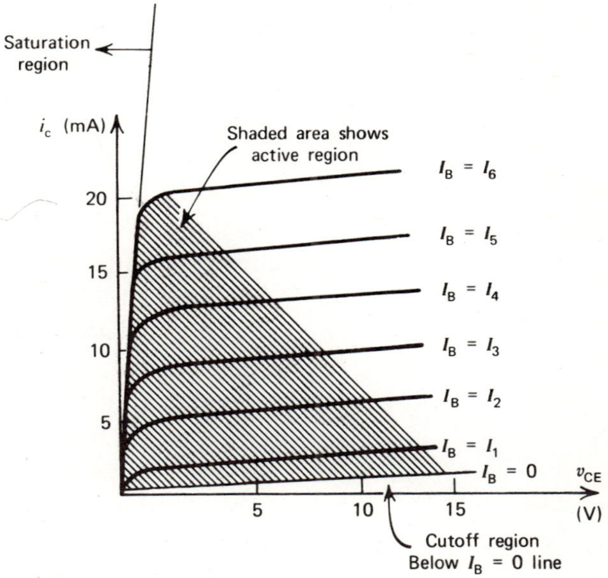

(c) Characteristic curves of a typical *npn* transistor

Figure 5-1 Symbols and characteristic curves of bipolar transistors.

$$\frac{I_C}{I_B} = \beta \text{ (Beta)} \tag{5-2}$$

$$\frac{I_C}{I_E} = \alpha \text{ (Alpha)} \tag{5-3}$$

where:

I_C is the collector current, amperes (A)
I_B is the base current, amperes (A)
I_E is the emitter current, amperes (A)
β is the forward current ratio, commonly called the current gain and symbolized as h_{fe}
α is the forward current ratio of collector current to emitter current

The current ratios β and α are considered relatively constant for a given transistor and hold true only in the active region of operation and within normal range of temperature. Note that the conventional direction of current in a transistor is in the direction of the base-emitter symbolic arrow.

The characteristic curves of a typical *npn* transistor are shown in Fig. 5-1c. Three regions of operation are also shown.

1. When the transistor is operated in the *saturation* region, a maximum collector current exists in the transistor. Any further increase in the base current does *not* result in a corresponding increase in the collector current. Under these conditions, the collector-emitter voltage (V_{CE}) and the base-emitter (V_{BE}) are quite small while both *B-E* and *C-B* junctions are forward-biased.
2. When the transistor is operated in the cutoff region, there exists a very small base current known as the *leakage* current. The leakage current, being comparatively small, is usually neglected. Under the cutoff condition, the transistor junction capacitances play important parts in transistor high-speed switching circuits.
3. The shaded area of operation in Fig. 5-1c is called the active region. In this region, small variations in base current cause correspondingly larger variations in collector current.
4. The important characteristics of a transistor that the manufacturer lists are:

   I_C(max), which is the maximum allowable collector current, amperes (A)

h_{fe} or β, which is the forward current ratio
V_{CE}(ON), which is the collect-emitter voltage (**dc**) when the transistor is saturated, volts (V)
V_{BE}(ON), which is the base-emitter voltage (**dc**) when the transistor is saturated, volts (V)
V_{CC}(max), which is the maximum allowable supply voltage, volts (V)
Q_T, which is the charge storage capacity of the transistor's base-emitter junction in coulombs (C)

5. The two semiconductor materials used to manufacture transistors are *silicon* (Si) and *germanium* (Ge). For a standard Si transistor, it is usually safe to assume that $V_{CE}(\text{ON}) \simeq 0.3$ V and $V_{BE}(\text{ON}) \simeq 0.7$ V, and for a Ge transistor, $V_{CE}(\text{ON}) \simeq 0.1$ V and $V_{BE}(\text{ON}) \simeq 0.3$ V.

5-4 An Ideal Transistor

An *ideal* transistor has the following (best possible) characteristics:

a. Unlimited power or I_C(max) rating.
b. Unlimited supply voltage [V_{CC}(max)] rating.
c. β is constant in the active region.
d. There is no leakage current.
e. $V_{CE}(\text{ON}) = V_{BE}(\text{ON}) = 0$ V.
f. All junction capacitances are zero (nonexistent).

5-5 The Transistor Switch

In pulse circuits, the transistor is used as a *switch*. At any one time, it is either ON or OFF. The input pulse accomplishes this ON-OFF operation either by *saturating* the transistor (ON) or by *cutting it off* (OFF). A simple transistor switch is shown in Fig. 5-2a. When the input is positive, as shown in Fig. 5-2b, and its amplitude is enough to saturate the transistor, the transistor is said to be ON. Under this, the ON condition, the output is equal to V_{CE}(ON) of the transistor. If the transistor is assumed to be ideal, then V_{CE}(ON), or in this case the output E_o, is equal to 0 V. Equivalent circuits under this, the ON condition, are shown in Figs. 5-2c and 5-2d. When the input pulse *reverse-biases* the base-emitter junction (BEJ) of the transistor as shown in Fig. 5-2e, the transistor is turned OFF. Under this, the OFF condition, there is no base or collector current. The transistor acts like an open switch, as shown

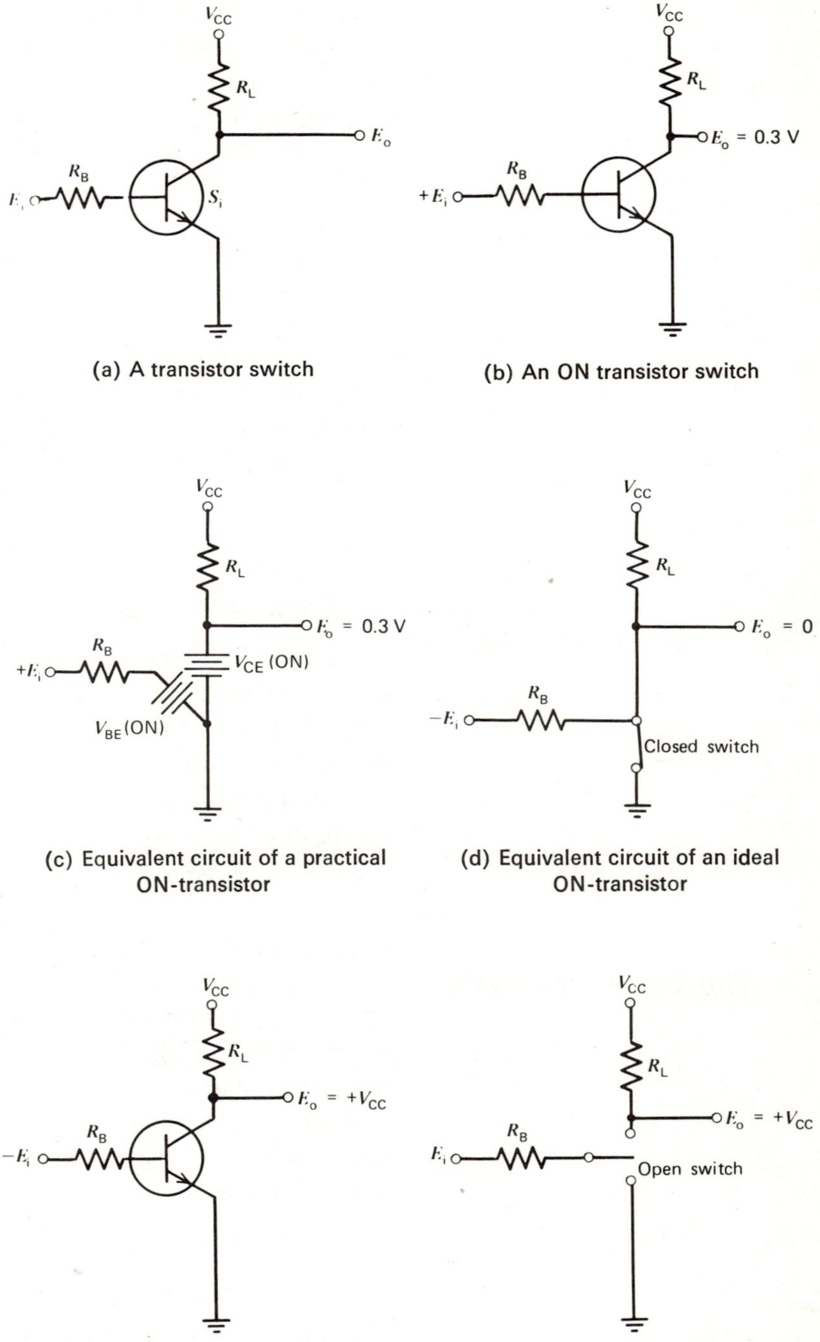

(a) A transistor switch

(b) An ON transistor switch

(c) Equivalent circuit of a practical ON-transistor

(d) Equivalent circuit of an ideal ON-transistor

(e) An OFF transistor switch

(f) Equivalent circuit of an OFF transistor

Figure 5-2 ON/OFF equivalent circuits of a transistor switch.

in Fig. 5-2f, and is termed as being OFF. The output voltage V_{CE} is the same as the supply voltage V_{CC}.

There are two types of transistor switches: the *saturated* type and the *nonsaturated* type. The transistor switch discussed above (Fig. 5-2) is the *saturated* type. When the transistor is being used as a saturated-type switch, it is either cut off (OFF) or saturated (ON), at any given time. It is the most commonly used transistor switch (in switching circuits today), because its design is simple and reliable. The circuits are designed such that small variations in the input signal's amplitude does not affect the switching process. The biggest drawback in a saturated-type switch is that its *turn-ON time* (t_{ON}) and the *turn-OFF time* (t_{OFF}) limit the maximum switching speed (Sec. 5-6) of the transistor.

The *nonsaturating* type of a transistor switch is used when fast switching speed is required. It is operated only in the active region (Fig. 5-1c), where the *delay* time (t_d) and the *storage* time (t_s) are eliminated (Sec. 5-6). Thus the t_{ON} time and the t_{OFF} time are greatly reduced. The biggest disadvantage is that any variation in the input amplitude level is reflected in the output waveform. Our discussion is exclusively devoted to the saturated-type transistor switch, because it is the most commonly used switch, despite its longer switching time.

A *practical* saturating-type transistor switch is designed such that in its ON state, the base current is 150 percent to 200 percent of the minimum value required for saturation. Let us analyze the transistor switch of Fig. 5-2a.

$$I_C(\text{ON}) = \frac{V_{CC} - V_{CE}(\text{ON})}{R_L} \text{ amperes (A)} \tag{5-4}$$

and

$$I_B(\text{ON}) \geqq \frac{I_C(\text{ON})}{\beta} \text{ (usually driven 200 percent into saturation to assure the ON condition)}$$

$$= \frac{V_{R_B}}{R_B} \text{ amperes (A)} \tag{5-5}$$

where:

$I_C(\text{ON})$ is the saturation collector current, amperes (A)
V_{CC} is the supply voltage, volts (V)

V_{CE}(ON) is the collector-emitter junction voltage when the transistor is ON, volts (V)
I_B(ON) is the base current when the transistor is ON, amperes (A)
V_{R_B} is the voltage drop across R_B, volts (V)

It is customary in analyzing transistor switching circuits to assume that the transistors used are *ideal.* This assumption simplifies calculations. For more accurate results, the transistor characteristics given by the manufacturer are used.

5-6 Transients in a Practical Transistor Switch

So far it has been assumed that a transistor switches ON as soon as the input pulse reaches the switching level. In practical transistors, this is *not* the case. There always is a finite *turn-ON time* (t_{ON}) and a finite *turn-OFF time* (t_{OFF}), just as was the case for diodes. An ideal input pulse cannot turn the transistor ON or OFF instantly. The t_{ON} and the t_{OFF} times become significant in high-speed switching circuits (in low-speed switching circuits, t_{ON} and t_{OFF} times are usually insignificant and can be ignored).

Figure 5-3a shows a simple transistor switch. Junction capacitances of the transistor are shown in dotted lines. Figure 5-3b shows an equivalent circuit model of the switch. The junction capacitances (C_{BC} and C_{BE}) charge/discharge during each half-cycle of the input and they take a finite time to do so. This partly controls the t_{ON} and the t_{OFF} times as follows:

1. Just before time t_0 (t_{-0}), input is at $-V_2$ and the transistor is OFF. The output (E_o) is $+V_{CC}$. The junction capacitances charge to values shown in Fig. 5-3b.
2. At time t_{+0} (just after t_0), the input voltage becomes $+V_1$. The charge on C_{BE} starts discharging from $-V_2$ toward 0 V and then charging toward $+V_1$. As it reaches 0 V, the transistor starts conducting. Note that the switch remains OFF until C_{BE} fully discharges to 0 V, even though the input had already switched from $-V_1$ to $+V_2$. This *delay time* (t_d) is shown in Fig. 5-3d.
3. Then, of course, the transistor turns fully ON as V_{BE} becomes equal to V_{BE}(ON). The output goes from V_{CC} to V_{CE}(ON) as the V_{BE} goes from 0 V to V_{BE}(ON). The time associated with this voltage fall is shown as t_1 in Fig. 5-3d.

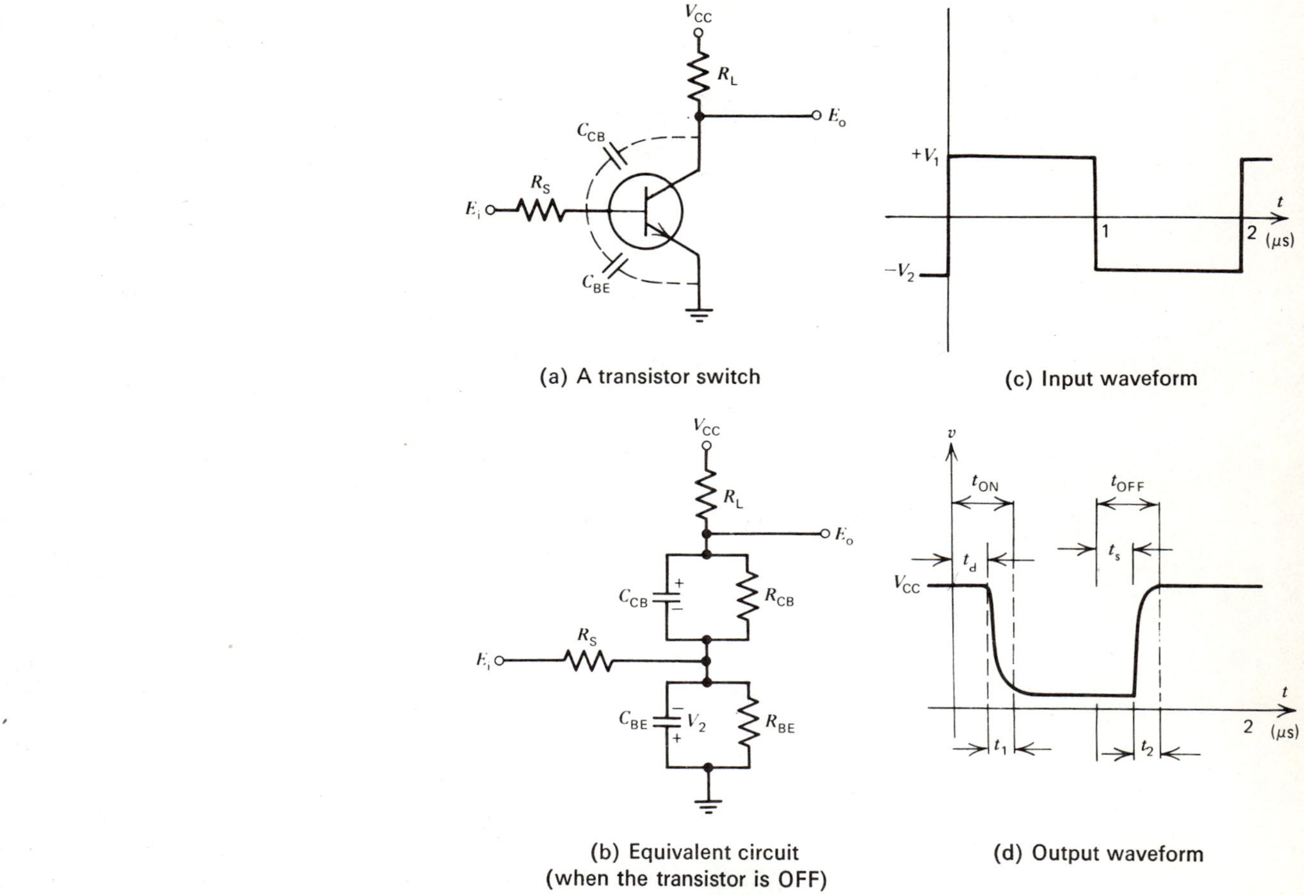

Figure 5-3 Transients in a transistor switch.

4. The total turn-ON time is the sum of t_d and t_1.

$$t_{ON} = t_d + t_1 \text{ (Fig. 5-3d)} \tag{5-6}$$

where:

t_d is the delay time, seconds (s)
t_1 is the time taken by the BEJ voltage to go from 0 V to V_{BE}(ON) seconds (s).

The t_d is directly proportional to the negative amplitude of the input pulse. The smaller the negative amplitude, the smaller will be t_d. Thus, to keep t_{ON} to a minimum, the negative bias that keeps the transistor OFF should be kept to a minimum. The t_{ON} can also be reduced by increasing the positive amplitude of the input pulse. Then the C_{BE} will be charging toward a higher voltage and will reach V_{BE}(ON) sooner.

Let us now consider conditions as the transistor is turned OFF. Just before the input becomes $-V_2$, the transistor is saturated. As a result, a quantity of minority carriers accumulates at the base region. The extent of the quantity of the minority carriers that accumulates at the base region depends upon the extent of transistor saturation. The deeper the transistor is operated in the saturation region, the greater the accumulation. As the input pulse switches from $+V_1$ to $-V_2$, the following happens:

1. The minority carriers are swept away from the base in time t_s, called the *storage time*. The transistor stays OFF during t_s, even though the input has switched to $-V_2$.

2. After t_s has elapsed, BEJ becomes reverse-biased and the output voltage rises exponentially toward V_{cc} as the C_{BE} charges to $-V_2$ This exponential rise in the output voltage is shown in Fig. 5-3d. The total turn-OFF time (t_{OFF}) is:

$$t_{OFF} = t_s + t_2 \text{ (Fig. 5-3d)} \qquad (5\text{-}7)$$

where:

t_s is the storage time, seconds (s)
t_2 is the time taken by the output voltage to rise (exponentially) to V_{CC} from V_{CE}(ON), seconds (s)

The t_s is a function of the extent to which the transistor was saturated and the amplitude of the reverse bias ($-V_2$) applied at the BEJ. The greater the reverse bias, the sooner the minority carriers are swept away. In other words, t_s is directly proportional to the extent of minority-carriers accumulation, and is inversely proportional to the magnitude of the reverse bias applied at BEJ. Thus, to improve or lower t_{OFF}, the transistor should not be operated too deep into saturation and there should exist a maximum allowable reverse bias at the BEJ.

The charge time t_2 is also reduced if C_{BE} is made to charge toward a high negative voltage. The higher the reverse bias across BEJ, the lower the t_2.

Note that the input amplitude conditions that reduce t_{ON}, increase t_{OFF}, and vice versa. Obviously, the answer to reducing the t_{ON} and t_{OFF} lie elsewhere. One way to reduce the charge/discharge time of C_{BE} is to use switching transistors with low C_{BE} values.

Sometimes, a so-called *speed-up* (also known as *switching* or *commutating*) capacitor (C_s) is used to improve the t_{ON} and t_{OFF} times. In such applications, the C_s is connected across the base resistor as shown in Fig. 5-4b. The C_s in series with the BEJ resistance forms a differentiator circuit. As a result, the actual waveform that the base "sees" is drawn in Fig. 5-4b. Note the effect of the differentiator action on the input waveform. Essentially, for a short period of the pulse time, the input amplitude appears much larger than it really is. As has been explained earlier, this reduces the t_{ON} and t_{OFF} times.

To obtain the sharp spikes during the rise and fall times of the input pulse, the time constant of the differentiator must be kept small. Accordingly,

$$t_p \geqq 5\tau$$

$$\geqq 5R_BC_s\ (R_B \parallel R_{BE} \cong R_B)$$

or

$$C_s \leqq \frac{t_p}{5R_B} \tag{5-8}$$

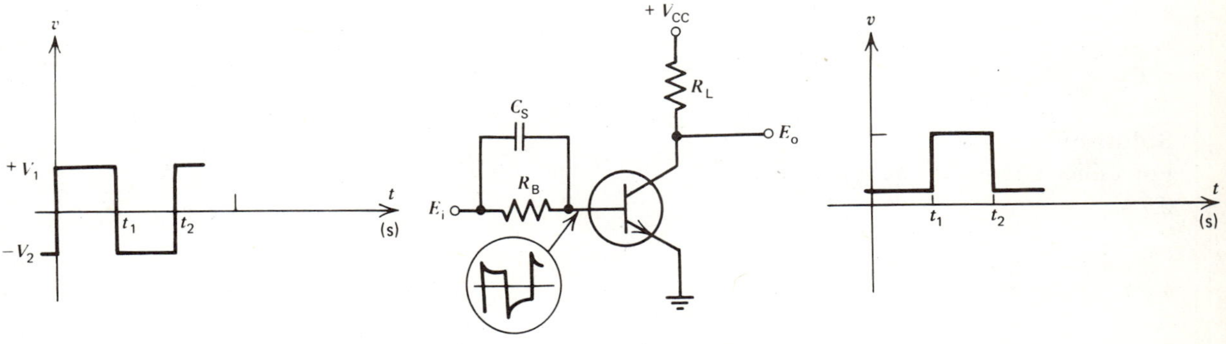

(a) Input

(b) Transistor switch showing the speed-up capacitor

(c) Output

Figure 5-4 Use of a speed-up capacitor (c_s) to improve the switching time.

where:

C_s is the speed-up capacitor, farads (F)
t_p is the input pulse width, seconds (s)
R_B is the base resistor, ohms (Ω)

C_s must also be able to store the charge that accumulates at BEJ. The charge storage capacity (Q_T) of a transistor is given by the manufacturer. Also, the capacitor C_s should have a voltage rating that is about double the peak-to-peak voltage value of the input pulse. Thus,

$$C_s \geqq \frac{Q_T}{2E_{p\text{-}p}} \tag{5-9}$$

where:

Q_T is the charge storage capacity of the transistor, coulombs (C)
$E_{p\text{-}p}$ is the peak-to-peak amplitude of the input voltage, volts (V)

From the above discussion it is clear that the range of C_s is governed by Eqs. (5-8) and (5-9).

Example 5-1

Consider the transistor switch shown in Fig. 5-4b. Assume that $V_{CC} = 12$ V, $E_i = 6\text{-}V_{p\text{-}p}$, 10-kHz square wave; and for the Si transistor, assume Beta = 150, $I_C(\text{max}) = 50$ mA, and $Q_T = 400$ pC. Find:

a. R_L.
b. R_B.
c. C_s.

Solution

For a silicon transistor, we assume $V_{CE}(\text{ON}) = 0.3$ V and $V_{BE}(\text{ON}) = 0.7$ V.

a.
$$R_L = \frac{V_{CC} - V_{CE}(\text{ON})}{I_C(\text{ON})} \tag{5-4}$$

$$= \frac{(12 - 0.3)\text{ V}}{50\text{ mA}}$$

$$= \mathbf{234\ \Omega}$$

b. $$R_B = \frac{V_{R_B}}{I_B(\text{ON})} \qquad (5\text{-}5)$$

$$= \frac{(3 - 0.7)\ \text{V}}{50\ \text{mA} \times (2)} \cdot 150 \quad \text{(overdrive 200 percent)}$$

$$= \mathbf{3.45\ k\Omega}$$

c. $$C_s \leqq \frac{t_p}{5R_B} \qquad (5\text{-}8)$$

$$\leqq \frac{50\ \mu\text{s}}{5 \times 3.45\ \text{k}\Omega}$$

$$\leqq 0.0029\ \mu\text{F}$$

also,

$$C_s \geqq \frac{Q_T}{2E_{p\text{-}p}} \qquad (5\text{-}9)$$

$$\geqq \frac{400\ \text{pC}}{2 \times 6\ \text{V}}$$

$$\geqq 33.33\ \text{pF}$$

Thus,

$$\mathbf{0.0029\ \mu F > C_s > 33.33\ pF.}$$

Note that in practical circuits, $I_C(\text{ON})$ is made considerably smaller than the rated $I_C(\text{max})$.

5-7 The Transistor Inverter

An *inverter* circuit is used extensively in digital logic circuits. Its symbol is shown in Fig. 5-5. It is one of the three basic logic gates and its function is to *invert* the *logic state* of the input.

A voltage level is assigned to each logic state. For example, V_1 may represent the TRUE (1) logic state and V_2 the FALSE (0) logic state. Then, an input of V_1 to an inverter will result in an output of V_2. Also, an input of V_2 volts to an inverter will result in an

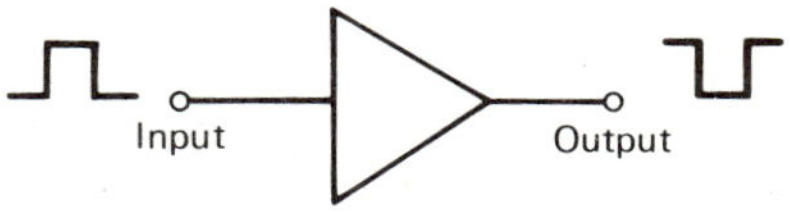

Figure 5-5 Inverter.

output of V_1 volts. An inverter simply changes V_1 to V_2 and vice versa.

Figure 5-6a shows a transistor inverter circuit. This circuit is simply a transistor switch. One input level turns the transistor ON and the other input level turns it OFF. The output has two levels, $V_{CE}(ON)$ and V_{CC}. The circuit is designed such that $E_o = V_{CC}$ when $E_i = V_{CE}(ON)$ and $E_o = V_{CE}(ON)$ when $E_i = V_{CC}$.

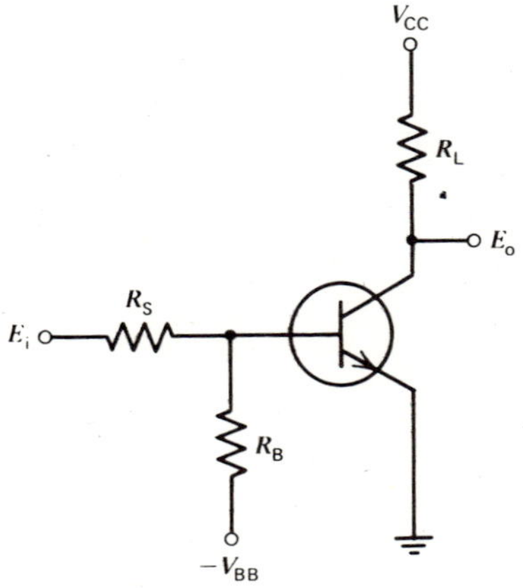

(a) Inverter circuit

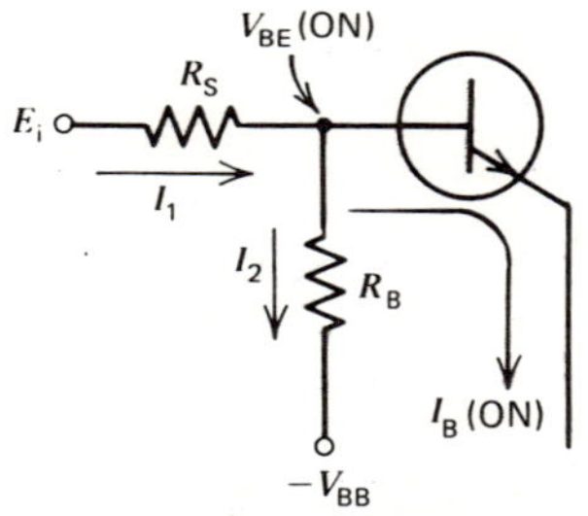

(b) Partial circuit when transistor is ON

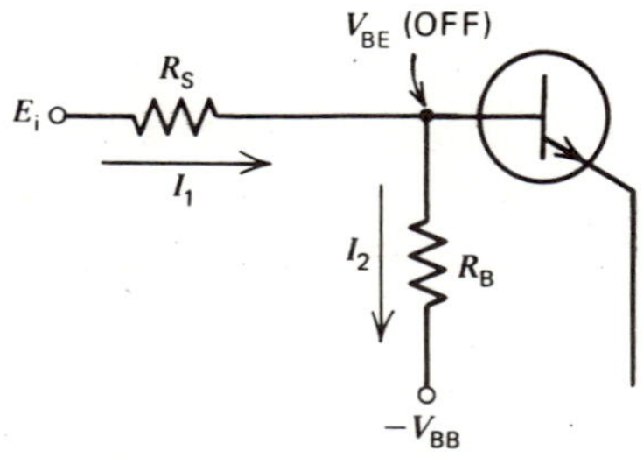

(c) Partial circuit when transistor is OFF

Figure 5-6 Transistor inverter circuit.

Consider the case when $E_i = V_{CC}$. The transistor needs to be ON so that the $E_o = V_{CE}(\text{ON})$. The partial circuit shown in Fig. 5-6b satisfies this condition if:

$$I_1 - I_2 = I_B(\text{ON})$$

$$\frac{V_{CC} - V_{BE}(\text{ON})}{R_S} - \frac{V_{BB} + V_{BE}(\text{ON})}{R_B} = I_B(\text{ON}) \qquad (5\text{-}10)$$

Note again that in practical inverter circuits, $I_B(\text{ON})$ is made approximately 200 percent of the $I_{B_{min}}(\text{ON})$.

When $E_i = V_{CE}(\text{ON})$, the transistor should be OFF. For this condition to be satisfied, the BEJ must be reverse-biased. The partial circuit shown in Fig. 5-6c accomplishes this if:

$$I_1 - I_2 = 0$$

$$\left[\frac{V_{CE}(\text{ON}) + V_{BE}(\text{OFF})}{R_S}\right] - \left[\frac{V_{BB} - V_{BE}(\text{OFF})}{R_B}\right] = 0 \qquad (5\text{-}11)$$

$V_{BE}(\text{OFF})$ is the reverse bias across BEJ when the transistor is OFF. Its numerical value should be below the PIV rating of the BEJ diode of the transistor. A reverse bias of 0.5 V to 1.5 V is considered reasonable.

Note that the inverter circuit shown in Fig. 5-6a differs from a transistor switch (Fig. 5-3a) only in that R_B and V_{BB} have been added. Addition of R_B and V_{BB} provides the reverse bias needed to turn the transistor OFF when the input is at its *LOW* level.

A *pnp* transistor is used if the two logic levels are $-V_{CC}$ and $-V_{CE}(\text{ON})$. The most commonly used logic levels in integrated circuits (IC) today are 0.4. V (LOW) and 5.0 V (HIGH); 0.4 V is the $V_{CE}(\text{ON})$ and 5 V is V_{CC}.

Example 5-2

Design an inverter circuit using a silicon *npn* transistor whose $h_{fe} = 50$, $I_C(\text{max}) = 25$ mA. $V_{BE}(\text{OFF})$ is to be -1 V. A ± 10 V **dc** power supply is available. The two logic levels are 10 V and 0.3 V.

Solution

The inverter circuit is shown in Fig. 5-6a.

$$V_{CC} = +10 \text{ V}$$

$$V_{BB} = -10 \text{ V}$$

$$R_L = \frac{V_{CC} - V_{CE}(\text{ON})}{I_C(\text{ON})} \qquad (5\text{-}4)$$

$$= \frac{(10 - 0.3)\text{ V}}{25 \text{ mA}}$$

$$= 388\ \Omega$$

$$I_B(\text{ON}) = \frac{I_C(\text{ON})}{\beta} \qquad (5\text{-}2)$$

$$= \frac{25 \text{ mA}}{50}$$

$$= 500\ \mu\text{A}$$

To assure complete saturation when the input is HIGH (10 V), overdrive the transistor by a factor of 2 (200 percent). Thus,

$$I_B(\text{ON}) = 500\ \mu\text{A} \times 2$$

$$= 1.0 \text{ mA}$$

Applying Eq. (5-10) for the ON condition (Fig. 5-6b):

$$\left[\frac{V_{CC} - V_{BE}(\text{ON})}{R_S}\right] - \left[\frac{V_{BB} + V_{CE}(\text{ON})}{R_B}\right] = I_B(\text{ON}) \qquad (5\text{-}10)$$

$$\left(\frac{10 - 0.7}{R_S}\right) - \left(\frac{10 + 0.7}{R_B}\right) = 1.0 \text{ mA}$$

$$\left(\frac{9.3}{R_S}\right) - \left(\frac{10.7}{R_B}\right) = 1.0 \text{ mA} \qquad (5\text{-}12)$$

Applying the OFF conditions as shown in Fig. 5-6c:

$$\left[\frac{V_{CE}(\text{ON}) + V_{BE}(\text{OFF})}{R_S}\right] - \left[\frac{V_{BB} - V_{BE}(\text{OFF})}{R_B}\right] = 0 \qquad (5\text{-}11)$$

Assume that $V_{BE}(\text{OFF}) = -1.0$ V. Then,

$$\left(\frac{0.3 + 1.0}{R_S}\right) - \left(\frac{10 - 1.0}{R_B}\right) = 0$$

$$\left(\frac{1.3}{R_S}\right) - \left(\frac{9.0}{R_B}\right) = 0$$

$$R_B = \frac{9}{1.3} R_S$$

$$= 6.92 R_S \qquad (5\text{-}13)$$

Solving Eqs. (5-12) and (5-13) yields:

$$\left(\frac{9.3}{R_S}\right) - \left(\frac{10.7}{6.92 R_S}\right) = 1.0 \text{ mA}$$

$$\left(\frac{9.3}{R_S}\right) - \left(\frac{1.55}{R_S}\right) = 1.0 \text{ mA}$$

$$R_S = \frac{9.3 - 1.55}{1.0 \text{ mA}}$$

$$= 7.75 \text{ k}\Omega$$

$$R_B = 6.92 \times R_S \qquad (5\text{-}13)$$

$$= 6.92 \times 7.75 \text{ k}\Omega$$

$$= 53.63 \text{ k}\Omega$$

Example 5-3

The inverter circuit shown in Fig. 5-6a contains the following component values: $R_L = 1\ \text{k}\Omega$, $R_B = 50\ \text{k}\Omega$, $R_S = 150\ \text{k}\Omega$, $V_{CC} = +12$ V and $V_{BB} = -8$ V. The transistor has $h_{fe} = 100$. $I_C(\text{max}) = 20$ mA. $V_{CE}(\text{ON}) = 0.2$ V and $V_{BE}(\text{ON}) = 0.5$ V. The two logic levels are 0.2 V and +12 V. Check by calculations whether the circuit works satisfactorily as an inverter.

Solution

What is required here is to check whether $I_C(\text{ON})$ is less than the rated $I_C(\text{max})$ and that the transistor is ON when $E_i = 12$ V and is OFF when $E_i = 0.2$ V.

$$I_C(\text{ON}) = \frac{V_{CC} - V_{CE}(\text{ON})}{R_L} \qquad (5\text{-}4)$$

$$= \frac{(12 - 0.2)\text{ V}}{1.0 \text{ k}\Omega}$$

$$= 11.8 \text{ mA}$$

Thus, $I_C(\text{ON})$ is below the rated $I_C(\text{max})$.

During the ON condition (Fig. 5-6b):

$$I_B(\text{ON}) = \left[\frac{V_{CC} - V_{BE}(\text{ON})}{R_B}\right] - \left[\frac{V_{BB} + V_{BE}(\text{ON})}{R_S}\right] \quad (5\text{-}10)$$

$$= \left[\frac{(12 - 0.5)\text{ V}}{50\text{ k}\Omega}\right] - \left[\frac{(8 + 0.5)\text{ V}}{150\text{ k}\Omega}\right]$$

$$= (230 - 57)\ \mu\text{A}$$

$$= 173\ \mu\text{A}.$$

$$I_{B_{min}}(\text{ON}) = \frac{I_C(\text{ON})}{h_{fe}} \quad (5\text{-}2)$$

$$= \frac{11.8\text{ mA}}{100}$$

$$= 118\ \mu\text{A}$$

Thus, there is more than enough base current to turn the transistor ON, since only 118 μA are needed for saturation.

During the OFF condition (Fig. 5-6c), the actual BEJ voltage is:

$$V_{BE}(\text{OFF}) = 0.2 - \left[\frac{(8 + 0.2)\text{ V}}{(150 + 50)\text{ k}\Omega}\right] \cdot 50\text{ k}\Omega$$

$$= (0.2 - 2.05)\text{ V}$$

$$= -1.85\text{ V}$$

Thus, BEJ is reverse-biased and the transistor is OFF. All of the conditions for safe and proper operation have been satisfied. Therefore the circuit works satisfactorily as an inverter.

5-8 Review of FET Characteristics

The field effect transistor (FET) is a *unipolar* device. The current in its conduction channel consists of only one type of charge carriers (majority only). The charge carriers in a *p*-channel FET are holes and in an *n*-channel FET are electrons. The current in the conduction channel is controlled by the charge potential at the gate terminal. Actually, the gate-to-source potential (V_{GS}) controls resistance of the conduction channel that exists between the source (S) terminal and the drain terminal (D) of an FET.

There are *two* types of FETs: one, the *junction field effect transistor* (JFET); and the other, the *metal oxide semiconductor field effect transistor* (MOSFET). The two differ in their physical makeup configuration and to some extent, in their electrical behavior. The MOSFET is used almost exclusively in switching circuits and for this reason we will limit our discussion to a MOSFET switch.

An FET differs from a bipolar transistor in that:

1. The current in the conduction channel of an FET consists of majority carriers only, whereas the current in a bipolar transistor results from interaction between the majority carriers and the minority carriers.
2. As a result of the absence of minority carriers in an FET switch, the storage time (in an FET switch) is nonexistent.
3. Again, as result of the absence of minority carriers, the thermal runaway phenomenon in an FET is eliminated. In fact, the resistance of the conduction channel increases with a rise in temperature.
4. The input impedance of an FET is generally very high, while that of a junction transistor is low.
5. The current in the conduction channel of an FET does not cross any *p-n* junction, which results in a considerable reduction in the internally generated noise.
6. An FET uses comparatively less power, is easier to fabricate, and occupies a very small area on an integrated circuit (IC) chip than does a bipolar transistor.
7. But the FET has a higher input capacitance than a bipolar transistor. This makes the FET a slower switching device. However, the industry has now been able to reduce the input capacitance of an FET by improving the fabricating techniques and has come up with a so-called *D-type* MOSFET. The switching time of a D-type MOSFET compares favorably with the bipolar switches (TTL).
8. An FET must be handled with extreme care at all times. Sometimes the capacitive potentials present in the environment are enough to damage it. All of the FET leads are usually shorted together during transportation and the supply voltages turned off when making any changes in the circuit where an FET is being used.

The physical-construction configuration and the standard symbol of a MOSFET are shown in Fig. 5-7a. Note that the *gate* (*G*) is insulated from the conduction channel by a thin layer of silicon oxide. Thus, there is *no ohmic connection* between the gate

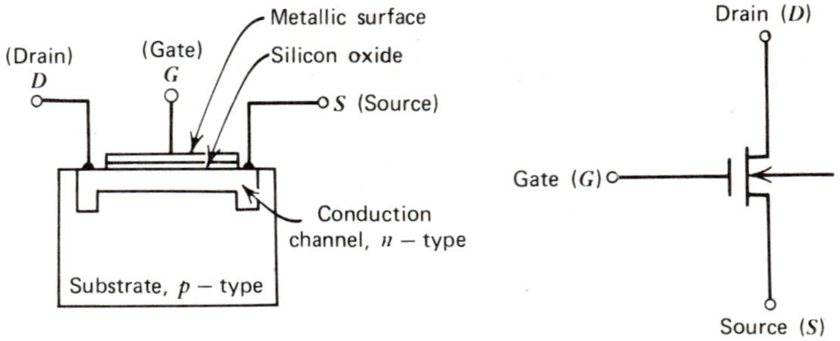

(a) Physical construction

(b) Symbol

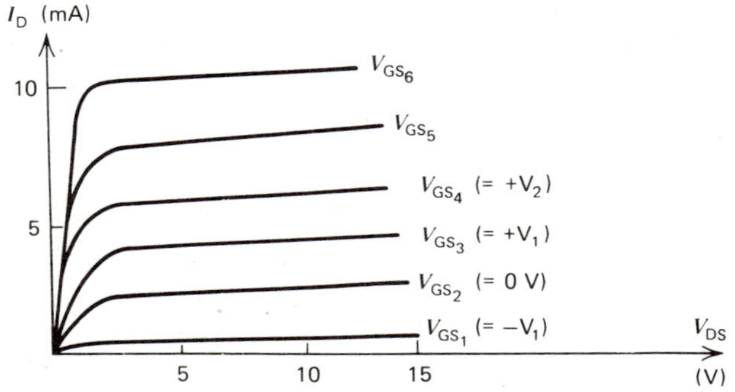

(c) I_D – V_{DS} characteristic curves

Figure 5-7 A depletion-type MOSFET.

and the conduction channel. The symbol shown in Fig. 5-7b is descriptive of this fact. An ohmic connection exists between the *drain* (*D*) and the *source* (*S*) via the conduction channel. The arrow symbol drawn from the substrate toward the channel indicates that it is an *n*-channel MOSFET. This arrow symbol is drawn in the reverse direction for a *p*-channel MOSFET.

The characteristic curves for a depletion-type MOSFET are shown in Fig. 5-7c. Note that a certain minimum negative potential (cutoff) is required at the gate to deplete the conduction channel (no more charge carriers). This is why it is called a *depletion*-type MOSFET (Fig. 5-7c).

In most switching applications, an *enhancement*-type MOSFET is used. Figure 5-8a shows the physical-construction configuration of an enhancement-type *n*-channel MOSFET. Note that the conduction channel between the source and the drain does not exist under normal circumstances (when $V_{GS} = 0$ V). As V_{GS} is increased

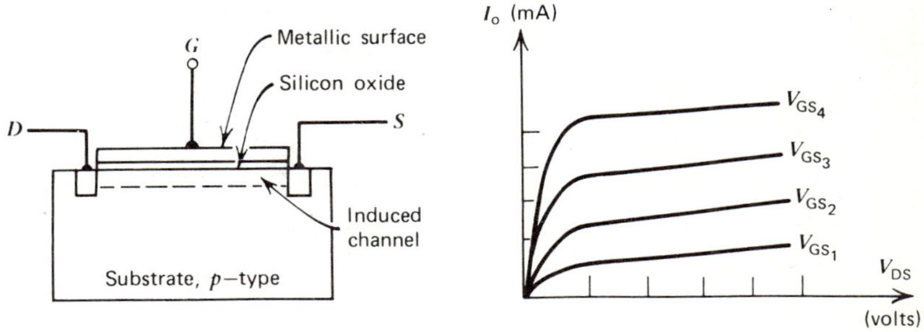

(a) Physical construction (b) $I_D - V_{DS}$ characteristic curves

Figure 5-8 An enhancement-type MOSFET.

(positively), electrons from the *p*-type substrate are attracted toward the gate, and thus an *n*-type channel is formed between the source and the drain. The width of this induced channel depends upon the magnitude of the charge potential at the gate (V_{GS}). In effect, a charge voltage applied at the gate reduces the *ohmic resistance* between the source and the drain by inducing a conduction channel in the substrate. The potential difference between the drain and the source terminals (V_{DS}) has *no control* over the resistance of the conduction channel.

Typical characteristic curves for an *n*-channel enhancement-type MOSFET are shown in Fig. 5-8b. Among other things, the manufacturer usually lists the following parameters of a MOSFET:

1. Channel type (*p* or *n*).
2. V_{DS}, the maximum allowable drain-to-source voltage.
3. I_D, the maximum allowable drain current.
4. BV_{GSS}, the maximum allowable gate-to-source voltage.
5. V_{th}, the threshold voltage required to turn ON the MOSFET.
6. *P*, the power rating of the MOSFET.

Note that only the *n*-channel MOSFETs were discussed above. The *p*-channel MOSFETs are equally important devices. The *p*-channel MOSFETs are manufactured by interchanging *p* and *n* semiconductor materials from that of an *n*-channel MOSFET (Fig. 5-8a).

5-9 The MOSFET Switch

A MOSFET behaves like a *closed* switch when ON and an *open* switch when OFF. This equivalency situation is shown in Fig. 5-9.

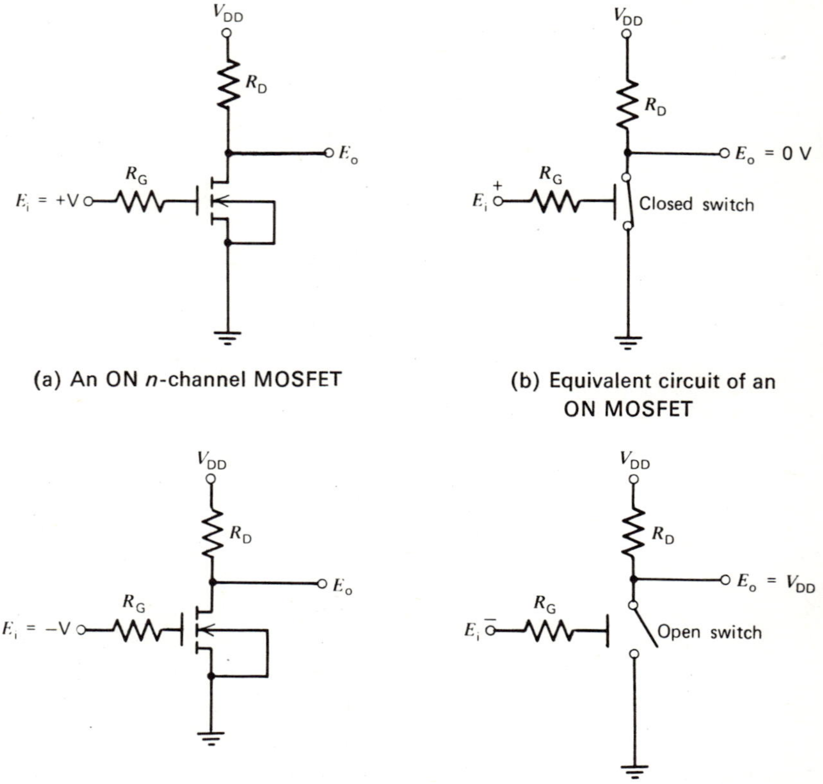

(a) An ON *n*-channel MOSFET

(b) Equivalent circuit of an ON MOSFET

(c) An OFF *n*-channel MOSFET

(d) Equivalent circuit of an OFF MOSFET

Figure 5-9 ON/OFF equivalent circuits of an *n*-channel MOSFET.

In effect, the gate voltage controls the resistance of the conduction channel. The resistance of the conduction channel is *very high* (open switch) when $V_{GS} \leqq$ cutoff voltage, and is *very low* (closed switch) when $V_{GS} \geqq$ turn-on voltage (V_{th}).

Example 5-4

A MOSFET switch is shown in Fig. 5-10b. The input E_i is a 0 V–10 V, 10-kHz square wave. The MOSFET shown is an enhancement-type *n* channel rated as: P = 310 mW, V_{DS} = 15 V, BV_{GSS} = 15 V, and I_D = 6 mA.

Find whether the circuit conditions are within the MOSFET ratings, and sketch the output voltage waveform.

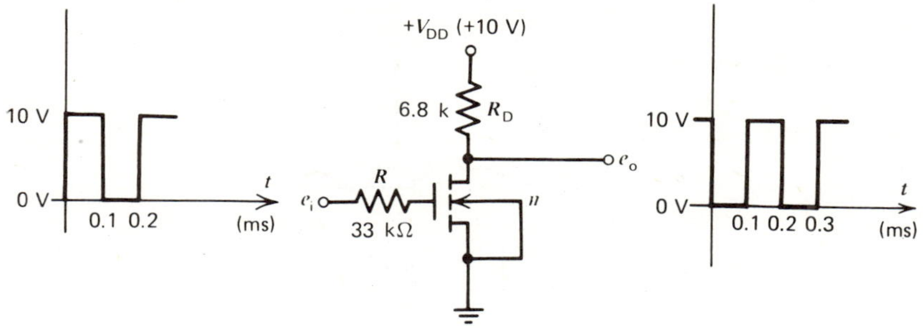

(a) Input (b) MOSFET switching circuit (c) Output

Figure 5-10 A MOSFET switch.

Solution

$$I_D(\text{ON}) = \frac{10 \text{ V}}{6.8 \text{ k}\Omega}$$
$$= 1.47 \text{ mA}$$

Thus, actual I_D, V_{DS} (10 V), and BV_{GSS} (10 V) are less than the maximum rated values of this MOSFET.

The MOSFET is ON when the input is at +10 V and therefore, the output voltage under the ON condition is $\cong$0 V.

The MOSFET is OFF when the input is at 0 V. The output voltage under this, the OFF condition, is $\cong$ 10 V (V_{DD}).

The output waveform is drawn in Fig. 5-10c.

5-10 The MOSFET Inverter

The MOSFET switch considered in Ex. 5-4 is essentially an inverter. But in practice, a MOSFET inverter is formed by connecting two MOSFETs, one an *n*-channel, the other a *p*-channel, as shown in Fig. 5-11. Usually, the whole unit (inverter) is fabricated on one chip. In fact, there is a whole family of logic circuits that use *complementary symmetry* (*p*-channel and *n*-channel MOSFETs). These are commonly called the COS/MOS or CMOS circuits. The inverter shown in Fig. 5-11 is the simplest form of COS/MOS circuitry.

Consider the inverter shown in Fig. 5-11. When the input E_i is 0 V (logic 0), then V_{GS} of the *p*-channel MOSFET is at $-V_{DD}$ and it is ON. Under this condition, the *p*-channel MOSFET behaves like a closed switch and thus the output is equal to V_{DD}. Note that when E_i is at 0 V, the *n*-channel MOSFET is OFF (open switch).

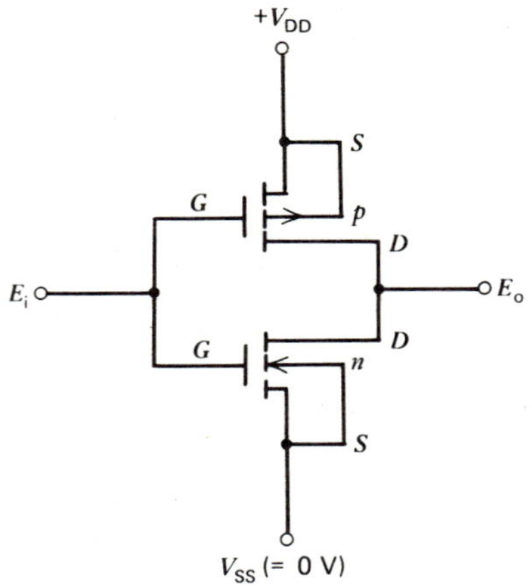

Figure 5-11 A COS/MOS inverter.

When the input is at V_{DD} (logic 1), the situation is reversed, i.e., the *p*-channel MOSFET is OFF and the *n*-channel MOSFET is ON, resulting in approximately 0 V at the output. In either logic state, one MOSFET is OFF while the other is ON.

5-11 TTL IC Inverter

The 7404 is a typical TTL IC inverter (ICs are covered in detail in Ch. 11.) It is called a hex inverter because there are six identical inverters on one 7404 IC chip. Figure 5-12 shows the pin configuration diagram of the whole chip and the circuit diagram of one complete inverter.

The 7404 inverter is designed for use in *positive* logic systems. The typical logic voltage levels are: HIGH = +4.5 V; LOW = + 0.3 V.

When a HIGH logic voltage level is applied at the input (pin 1):

1. D_1 turns OFF.
2. There is no base-emitter current in Q_1.
3. The base-collector diode of Q_1 conducts.
4. The collector current of Q_1 turns Q_2 ON.
5. The emitter current of Q_2 turns Q_3 ON.
6. The ON Q_3 gives rise to the LOW output voltage (pin 2).
7. Q_4 and D_2 are ON.

When a LOW logic voltage level is applied to the input terminal (pin 1):

1. D_1 turns OFF.
2. A base-to-emitter current exists in Q_1 and the voltage at the base of Q_1 is approximately 0 V.
3. 0 V at the base of Q_1 means that there is no collector current in Q_1.
4. Q_2 is OFF.

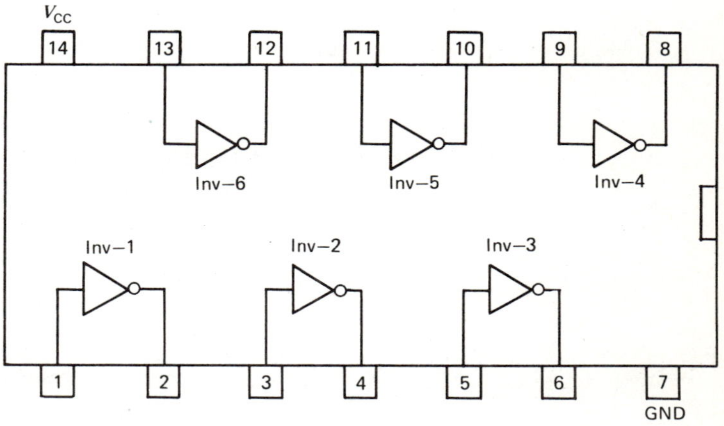

(a) Pin-configuration diagram

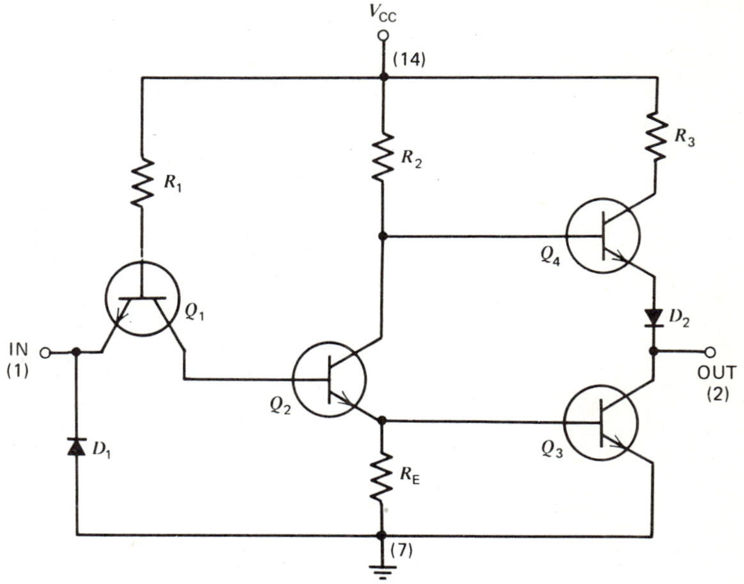

(b) Circuit diagram

Figure 5-12 The 7404, the TTL IC hex inverter.

5. Absence of emitter current in Q_2 turns Q_3 OFF.
6. Q_4 and D_2 are ON (through load).
7. Q_4 ON, D_2 ON, and Q_3 OFF result in HIGH logic voltage level at the output (pin 2).

The above discussion on the operation of 7404 shows how transistors and diodes are used as ON/OFF switches in ICs. Further discussion on transistor and diode switches used in digital ICs is left for Ch. 11.

5-12 Summary

1. Two types of transistors are used in switching circuits: one, the junction bipolar transistor; and the other, the field effect transistor (FET).
2. A bipolar transistor is usually called simply the transistor. There are two types of transistors, the *npn* and the *pnp*.
3. There are two types of transistor switches, the saturating type and the nonsaturating type.
4. The nonsaturating transistor switch operates only in the active region. Its t_{ON} and t_{OFF} times are very short, which is useful in high-speed switching circuits. However, any change in circuit operating conditions is immediately reflected in the output.
5. The saturating transistor switch is either ON (saturated) or OFF (cutoff) at any one time. Its switching speed is slower as compared to the nonsaturating transistor switch, but still quite satisfactory for most switching applications. It is the most common type of switch used, especially in integrated circuits (TTL).
6. When saturated, a transistor acts like a closed switch, and when cutoff, like an open switch.
7. A commutating or speed-up capacitor is used to improve the switching time of a transistor switch.
8. One application of a transistor switch is in the inverter circuit. An inverter is an electronic circuit used to invert or complement the logic state of the input.
9. The two logic voltages in an inverter are V_{CE}(ON) and V_{CC}.
10. An FET switch is much superior to a transistor switch in many ways. The major problem with an FET switch has been its slow switching time. This now has been practically overcome by the so-called *D*-type MOSFET.
11. From among the FET family, the MOSFET (enhancement) is used almost exclusively in pulse and switching circuits.
12. A MOSFET inverter consists of two enhancement MOSFETs, one the *p*-channel and the other the *n*-channel. Circuits that utilize this type of complementary symmetry are collectively called COS/MOS circuits.

13. Important equations:

$$I_C + I_B = I_E \tag{5-1}$$

$$\frac{I_C}{I_B} = \beta \tag{5-2}$$

$$\frac{I_C}{I_E} = \alpha \tag{5-3}$$

$$I_C(\text{ON}) = \frac{V_{CC} - V_{CE}(\text{ON})}{R_L} \tag{5-4}$$

$$I_B(\text{ON}) = \frac{V_{RB}}{R_B} \tag{5-5}$$

$$t_{ON} = t_d + t_1 \tag{5-6}$$

$$t_{OFF} = t_s + t_2 \tag{5-7}$$

$$C_s \leqq \frac{t_p}{5R_B} \tag{5-8}$$

$$C_s \geqq \frac{Q_T}{2E_{p\text{-}p}} \tag{5-9}$$

$$\left[\frac{V_{CC} - V_{BE}(\text{ON})}{R_s}\right] - \left[\frac{V_{BB} + V_{CE}(\text{ON})}{R_B}\right] = I_B(\text{ON}) \tag{5-10}$$

$$\left[\frac{V_{CE}(\text{ON}) + V_{BE}(\text{OFF})}{R_s}\right] - \left[\frac{V_{BB} - V_{BE}(\text{OFF})}{R_B}\right] = 0 \tag{5-11}$$

5-13 Glossary

Active region. The transistor's region of operation that lies between the saturation region and the cutoff region. The current relationships as given by β and α hold true only in this region.

Alpha (α). The ratio of the collector current to the emitter current in a transistor when it is operating in the active region.

Base. One type of semiconductor material sandwiched between the two opposite types of semiconductor materials in the transistor's physical construction. The base material is *n*-type in a *pnp* transistor and *p*-type in a *npn*-type transistor. Usually, the input signal to be amplified is fed to the base of a transistor.

Beta (β). The ratio of the collector current to the emitter current in a transistor when it is operating in the active region. It is also known as the forward current ratio or current gain of a transistor.

Bias. A term used to describe the **dc** base current level in a transistor amplifier circuit.

Collector. One of the two outside semiconductor materials (the other is the emitter) in the transistor's physical construction make-up. It is usually connected to the supply voltage.

Conduction channel. The ohmic connection between the drain and the source of an FET.

COS/MOS. A term used to describe MOSFET circuits that utilize *p*-channel and *n*-channel MOSFETs to accomplish a certain function.

Cutoff region. The region of transistor operation where no current flows in the transistor (except the leakage current).

Delay time. A part of the t_{ON} time. It is the time taken by BEJ capacitance to discharge to 0 V from its original bias voltage.

Depletion MOSFET. A MOSFET that conducts even when there is no potential charge on the gate.

Drain. A lead connected to one side of the conduction channel of an FET.

Emitter. One of the two outside semiconductor materials (the other is the collector) in the transistor's physical (construction) make-up. It is usually connected to ground.

Enhancement MOSFET. A MOSFET that does not conduct when there is no potential charge on the gate.

Field effect transistor (FET). A special type of a transistor. The current in its conduction channel is controlled by the potential charge on the gate, as opposed to the base current in a bipolar transistor. An FET is a unipolar device because the current in it consists of only one type of charge carriers (majority).

Gate. The control element in an FET. Potential charge on it controls the current in the conduction channel.

Inverter. An electronic circuit used to invert or complement the logic state of the input.

Junction field effect transistor (JFET). One type of an FET whose gate forms a *p-n* junction with the conduction channel. During normal operating conditions, this *p-n* junction is reverse-biased.

Metal oxide semiconductor field effect transistor (MOSFET). One type of an FET whose gate does not form an electrical junction with its conduction channel.

***n*-channel.** An FET whose conduction channel is made up of *n*-type semiconductor material.

***npn* transistor.** A transistor whose collector and emitter are made of *n*-type semiconductor material and whose base is made of *p*-type semiconductor material.

***p*-channel.** Conduction channel in an FET, consisting of *p*-type semiconductor material.

***pnp* transistor.** A transistor whose collector and emitter are made from the *p*-type semiconductor material and whose base is made from the *n*-type semiconductor material.

Saturation region. A maximum collector current flows in the transistor when it is operated in the saturation region. A further increase in the base current does not result in any increase in the collector current.

Source. One side of the conduction channel of an FET.

Speedup capacitor. A capacitor used to reduce the t_{ON} and t_{OFF} times of a transistor switch. It is also called a commutating capacitor.

Storage time. The time that the minority carriers take to be swept away from the base region as the input pulse switches the transistor OFF from ON.

Substrate. Base semiconductor material used to manufacture transistors.

Threshold voltage (V_{th}). The input voltage level required to turn a semiconductor device (transistor) ON.

Transistor. A current-controlled semiconductor device. It is also called a bipolar device because the current in it consists of two types of charge carrier, the minority carriers and the majority carriers.

Turn-off time (t_{OFF}). The time that a transistor switch takes to turn OFF from its previous ON state, after the switching pulse has been applied to turn it OFF.

Turn-on time (t_{ON}). The time that a transistor switch takes to turn ON from its previous OFF state, after the switching pulse has been applied to turn it ON.

TTL. Transistor–transistor logic is a family of IC logic circuits.

5-14 Recapitulation

Now return to the Objectives and Self-Evaluation questions given at the beginning of this chapter and see how well you can answer them. If you cannot answer certain questions satisfactorily, place a check mark (√) next to them and review appropriate parts of the chapter to find the answers. No attempt should be made to solve the Problems (Sec. 5-15) before answering the Self-Evaluation Questions. Ability to solve the Problems should serve as a test and feedback regarding your understanding of the material presented in this chapter. Answers to the Problems are given on p. 520.

5-15 Problems

5-1 Consider the simple transistor switch shown in Fig. 5-13. Assume that the transistor is ideal. Find:
a. The minimum input voltage level that will turn the transistor ON.
b. The output voltage level when the input is at 3 V.

5-2 Repeat Prob. 5-1 but this time consider Q to be a Si transistor, and its I_c(max) to be 70 mA.

5-3 Repeat Prob. 5-1 but this time consider Q to be a Ge transistor, and its I_C(max) to be 25 mA.

5-4 The input to the circuit of Fig. 5-13 is a 10-$V_{p\text{-}p}$, 100-kHz square wave. Find and sketch the output waveform. Assume Q to be ideal.

5-5 The input to the circuit of Fig. 5-13 is a 20-$V_{p\text{-}p}$, 25-kHz sine wave. Assume Q to be a Si transistor whose rated $I_C(\text{max}) = 50$ mA. Find and sketch the output voltage waveform.

5-6 The input to the circuit of Fig. 5-13 is a 20-$V_{p\text{-}p}$, 100-kHz square wave. Assume Q to be a Ge transistor whose rated $I_C(\text{max}) =$ 50 mA. Find and sketch the output waveform.

5-7 Repeat Prob. 5-6 if $t_{ON} = 2.0\ \mu s$, $t_{OFF} = 2.5\ \mu s$, $t_s = 1.5\ \mu s$, and $t_d = 1.5\ \mu s$.

5-8 Consider the circuit conditions described in Prob. 5-6. Assume $Q_T = 500$ pC for the transistor. Calculate and find the appropriate range of values for the speed-up capacitor.

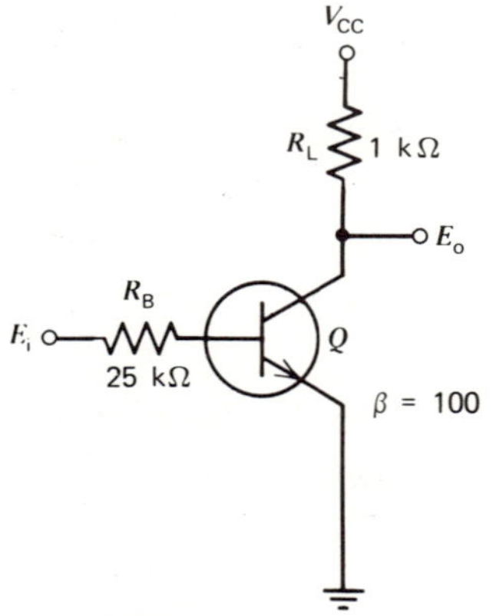

Figure 5-13 A transistor switch circuit for Prob. 5-1.

5-9 Design a simple transistor switch which is to be OFF when the input is -2 V and is to be ON when the input is $+10$ V. Overdrive the Si transistor (to be used) by a factor of 2 for saturation. Assume $V_{CC} = 15$ V, $I_C(\text{ON}) = 20$ mA, $\beta = 125$, and $Q_T = 500$ pC.

5-10 Repeat Prob. 5-9 except that this time, use a Ge transistor.

5-11 Design a transistor inverter for the two logic states 0 V and -5 V. Assume the following for the transistor to be used: $h_{fe} = 80$, $V_{BE}(\text{OFF}) = 1.0$ V, $V_{BB} = +5$ V, $I_C(\text{ON}) = 10$ mA, $V_{CE}(\text{ON}) = 0$ V, and $V_{BE}(\text{ON}) = 0$ V.

5-12 An inverter circuit is shown in Fig. 5-14. Check by calculations whether it will function properly or not.

5-13 Refer to the inverter circuit of Fig. 5-14. Describe various changes that you can make so that the circuit will function properly.

5-14 Design a transistor inverter circuit for the two logic states of 0.3 V and 4.5 V. Assume the following for the Si transistor to be used: $\beta = 150$; $I_C(\text{ON}) = 25$ mA. Also assume that $V_{BE}(\text{OFF}) = 0.5$ V, $V_{BB} = -3$ V, and $V_{CC} = 4.5$ V. Use only the standard (EIA) component values. Overdrive the transistor by 200 percent for the ON state.

5-15 Consider the inverter designed in Prob. 5-14. Find the minimum value of the input voltage that will still saturate the transistor.

5-16 Design a COS/MOS inverter for the two logic levels 0 V and $+5$ V.

5-17 Design a COS/MOS inverter for the two logic levels 0 V and -5 V

5-18 Design a COS/MOS inverter for the two logic levels $+5$ V and -5 V.

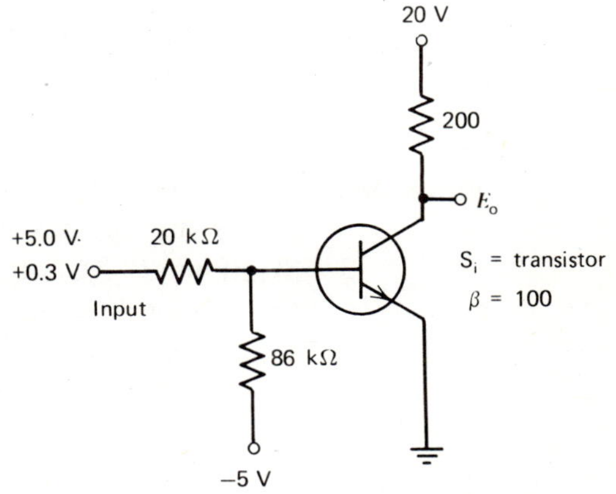

Figure 5-14 Inverter circuit for Prob. 5-12.

5-16 Experiment

Title **A. The Transistor Switch**

B. The MOSFET Switch

C. The Transistor Inverter

Theory A bipolar junction transistor (BJT), commonly called simply a transistor, may be made to act like a simple ON/OFF switch. In such an application, the transistor is operated in the *saturated* region to simulate the ON (closed) switch condition and in the *cutoff* region to simulate the OFF (open) switch condition. The input applied to the base of the transistor either turns it ON (saturates) or OFF (cuts off). When the transistor switch is ON, $V_{CE} = V_{CE}(\text{ON})$; and when the transistor is OFF, $V_{CE} = V_{CC}$. One common use of a transistor switch is in the inverter circuit. It basically is a simple transistor switch with a biasing voltage (V_{BB}) and a resistor (R_B) added to it. Addition of V_{BB} and R_B assures the OFF condition of the inverter.

The metal oxide semiconductor field effect transistor (MOSFET) may also be used as a switch in a manner similar to the BJT switch. In fact, the MOSFET switch is becoming more popular, especially since its switching speed is becoming favorably comparable to a BJT switch.

Equipment Dual beam oscilloscope: PHILIPS model 3230, or equal.
Power supply: ANATEK model 25-2.00, or equal.
Multimeter: RCA model WV 98C, or equal.
Multifunction generator: EXACT model 126, or equal.
Transistor: 2N4401.
MOSFET: 2N4351 (*n*-channel enhancement MOSFET).
Resistors: 330 Ω, 1 kΩ, 4.7 kΩ, 6.8 kΩ, 8.2 kΩ, 22 kΩ, and 47 kΩ.
Capacitor: 0.001 μF.

Procedure

1. Connect the transistor switch circuit shown in Fig. E5-1.
2. Apply 0 V (ground) at the input terminal E_i. Measure and record in Table E5-1 voltage appearing at the output terminal E_o.

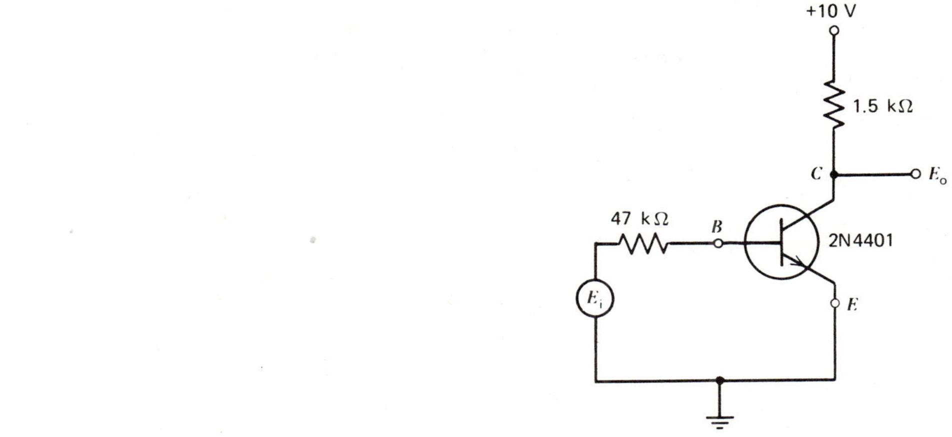

Figure E5-1 Bipolar transistor switch.

3. Apply 5 V **dc** at the input terminal E_i. Measure and record the voltage appearing at the output terminal E_o. Record your results in Table E5-1.
4. Apply a 10-$V_{p\text{-}p}$, 1-kHz sine wave at E_i.
5. Using CRO, observe and record in Table E5-2 the input as well as the output voltage waveforms.
6. From the waveforms observed in Procedure step 5, find and record in Table E5-3 V_{CE}(ON), V_{CE}(OFF), the minimum input voltage required to turn the transistor ON, and the minimum bias voltage required to turn the transistor OFF.
7. Change the input to a 10-$V_{p\text{-}p}$, 1-kHz square wave. Observe (CRO) and draw in Table E5-2 the input as well as the output waveforms.
8. Increase the input frequency slowly and observe its effects on the output wave. The output waveform should get progressively distorted as the frequency is increased.
9. Set the input frequency to 100 kHz. The output waveform should now appear considerably distorted. Draw this distorted waveform in Table E5-2.
10. Connect a 0.001-μF capacitor across the 47-kΩ resistor. The distortion in the output waveform should now disappear. Draw this output waveform in Table E5-2.
11. Connect the MOSFET switch circuit as shown in Fig. E5-2.
12. Apply 5 V at E_i. Measure and record in Table E5-4 the **dc** voltage appearing at E_o.
13. Decrease the input voltage slowly while monitoring the output voltage level. A point will be reached when the output becomes HIGH ($\simeq$10 V). Measure and record in Table E5-4 the input and output voltage levels at this point.

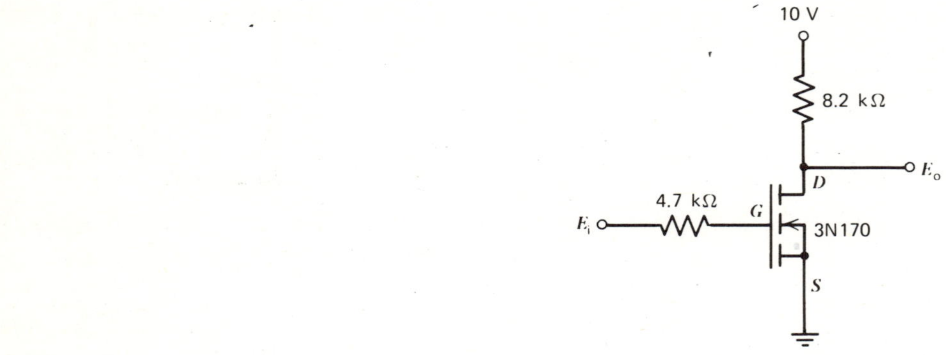

Figure E5-2 MOSFET switch.

14. Apply a 10-$V_{p\text{-}p}$, 1-kHz square wave at E_i. Observe and draw the output waveform in Table E5-5 (also the input waveform).
15. Increase the frequency slowly and note the distortion in the output waveform.
16. Connect the inverter circuit shown in Fig. E5-3.

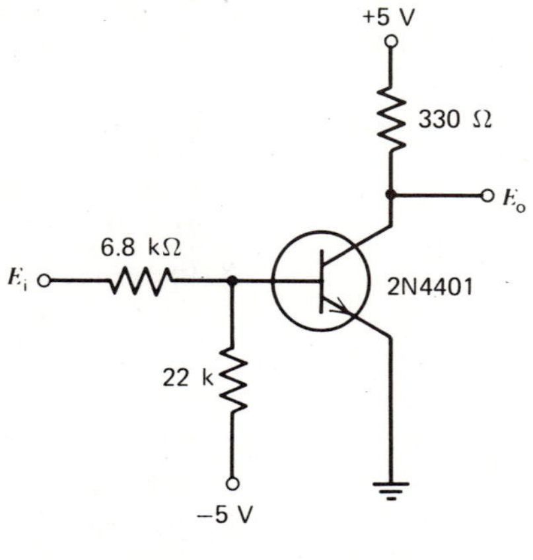

Figure E5-3 The inverter circuit for experiment 5.

17. Apply 5 V **dc** at E_i. Measure and record the voltage level appearing at E_o in Table E5-6.
18. Apply 0.4 volts at E_i. Measure and record in Table E5-6 the voltage level appearing at E_o.
19. Disconnect and return all the test equipment to its proper place.

Calculations

1. From the manufacturer's data sheets, it is noted that I_c(max) = 600 mA, V_{cc}(max) = 40 V, V_{CE}(ON) = 0.4 V, V_{BE}(ON) = 0.75 V, beta = 80 and Q_t = 40 μC for 2N4401.
2. Consider the simple transistor switch shown in Fig. E5-1. Find (calculate) and record in Table E5-1 the two output voltage levels for the two input voltage levels of 0 V and 5 V.
3. Calculate the percentage error between the experimentally observed and the theoretically expected output values in Table E5-1. Enter your results there.
4. From the manufacturer's data sheets, it is noted that I_D(max) = 30 mA, V_{DS}(max) = 25 V, V_{GS}(max) = 15 V, and P = 800 mW for the MOSFET 2N4351.
5. In light of the MOSFET ratings recorded in step 4 above, find whether the MOSFET switch will function properly and safely as a saturating-type switch.
6. The circuit shown in Fig. E5-3 is supposed to be an inverter. Check by calculations whether the ON/OFF conditions for the transistor are satisfied. The two logic levels are 5 V and 0.4 V.

Tabulation of Results

Table E5-1

E_i	E_o
0 V	Observed
	Theoretical
	Percentage error
5 V	Observed
	Theoretical
	Percentage error

Table E5-2

Procedure Step	Description	Waveform
5	Input Output	
7	Input Output	
9	Input Output	
10	Input Output	

Table E5-3

V_{CE}(ON)	Observed Theoretical Percentage error
V_{CE}(OFF)	Observed Theoretical Percentage error
$E_{i_{min}}$(ON)	
$E_{i_{min}}$(OFF)	

Table E5-4

E_i	E_o
+5 V	

Table E5-5

Description	Waveform
Input	______________
Output	______________

Table E5-6

E_i	E_o
+5 V	
+0.4 V	

Questions

1. The information recorded in Table E5-1 is intended to show that a transistor may be used as a switch. Explain whether that has been accomplished.
2. Examine the input/output waveforms recorded in Table E5-2. Explain briefly what happens in each case.
3. Explain how a speed-up capacitor accomplishes its task.
4. Explain the results observed in Procedure steps 8 and 15.
5. Examine the results recorded in Table E5-4. Assess whether or not the MOSFET functions as a switch.
6. Examine the results recorded in Table E5-6. Assess whether or not the inverter functioned properly (see Ex. 5-3).

Discussion

1. Summarize your conclusions from the observed and the theoretical values.
2. Account for discrepancies between the observed and the theoretically expected values.
3. Discuss and compare a transistor switch with a MOSFET switch.

Chapter 6 The Bistable Multivibrator

6-1 Objectives

The objectives of this chapter are to:

1. Explain the meaning of the term *multivibrator* (MV).
2. Describe the characteristics of various types of MVs.
3. Explain the CLEAR-SET (C-S) or RESET-SET (R-S) action of a flip-flop (F/F).
4. Explain the TOGGLE (T) or the COMPLEMENT action of an F/F.
5. Explain and describe the characteristics of a J-K F/F.
6. Discuss and explain the operation of a transistor F/F circuit.
7. Explain the operation of a MOSFET F/F circuit.
8. Discuss the use of speed-up capacitors for switching-time improvement.
9. Explain and describe the operation of R-S level triggering and edge-triggering circuits.
10. Explain and describe the operation of complement or T circuits.
11. Explain the operation of an emitter-coupled F/F circuit.
12. Discuss the triggering (R-T-S) techniques for an emitter-coupled F/F.

6-2 Self-Evaluation Questions

Watch for the answers to the following questions as you read the chapter. They should help you to understand the material presented. When you have finished the chapter, return to this section and be sure that you can answer all the questions. Then test your understanding by working out solutions to the Problems.

It is strongly recommended that you answer these questions satisfactorily before attempting to solve the problems (Sec. 6-21).

1. Explain the characteristics of an MV.
3. Give the characteristics of the three distinct types of MVs.
3. Explain the characteristics of a simple R-T-S F/F.
4. Describe the characteristics of a J-K F/F.
5. An F/F sometimes is also called a binary or a divide-by-two counter. Explain why.
6. Explain the operation of a transistor F/F circuit.
7. Design an F/F circuit for the two given logic levels V_1 and V_2.
8. Describe the function of speed-up (or commutating) capacitors used in F/F circuits.

9. Describe a procedure used to calculate the value of speed-up capacitors.
10. Explain the function of triggering networks in F/F circuits.
11. Describe the difference between level-triggering and edge-triggering.
12. Describe the design and operation of a base-triggered R-S F/F circuit using diodes.
13. Describe the operation of a collector-triggered R-S F/F circuit using diodes.
14. Describe the operation of a complement (T) triggering circuit.
15. Design a simple emitter-coupled F/F circuit for two given logic states at levels V_1 and V_2.
16. Describe the operation of a triggering circuit that may be used to COMPLEMENT, SET, or RESET an emitter-coupled F/F.
17. Show how F/Fs may be used as frequency dividers.
18. Explain the design and operation of a MOSFET F/F.

6-3 The Multivibrator

The term *multivibrator* (MV) is applied to a special class of electronic circuits. All MVs have two output terminals. The voltage level at one output terminal is always different (opposite) from the one at the other terminal. Assume that the MV shown in Fig. 6-1 has two output voltage levels of 0 V and 5 V. At any one time, one output is at 0 V while the other is at 5 V.

In logic circuits, the two output voltage levels of an MV correspond to the two logic states TRUE (1) and FALSE (0). One voltage level is made equivalent to the TRUE situation and the other to the FALSE situation. Thus at any one time, one output of an MV is TRUE while the other is false. These two output voltage levels are sometimes called the HIGH (V_H) and LOW (V_L).

The voltage levels that are assigned to TRUE and FALSE situations in a system are arbitrary. However, in one complete system, the two voltage levels (TRUE and FALSE) must be uniform throughout. The voltage levels commonly assigned to TRUE and

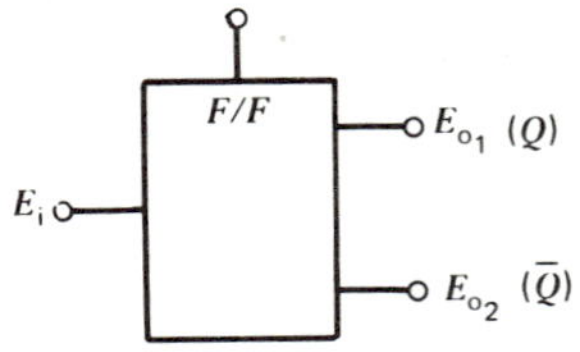

Figure 6-1 A multivibrator.

FALSE situations in modern-logic electronic circuits are 5 V and 0 V, 12 V and 0 V, +5 V and −5 V, respectively.

6-3.1 Positive Logic

It has already been indicated that two voltage levels are assigned to the two logic states (1 and 0) in a system. The logic system that uses a higher (algebraically) voltage level as 1 (TRUE) and the lower voltage level as 0 (FALSE) is known as the *positive logic* system. Assume that in a positive logic system, the two voltage levels are +5 V and −5 V. Then, the +5 V represents 1 (TRUE) and the −5 V 0 (FALSE).

6-3.2 Negative Logic

In a negative logic system, a lower (algebraically) voltage represents the 1 (TRUE) state and the higher voltage represents the 0 (FALSE) state. If +5 V and −5 V are the two logic levels in a negative logic system, then:

$$-5\text{ V} = 1 \quad \text{(TRUE)}$$

and

$$+5\text{ V} = 0 \quad \text{(FALSE)}$$

6-4 Types of Multivibrators

There are three main types of MVs: the *bistable*, the *monostable*, and the *astable*. These differ from one another in the manner in which they switch from one state to another. To understand the meaning of the word *state*, recall that at any one time, one output is TRUE while the other is FALSE. Then, the two states are:

$$E_{o1} = \text{TRUE and } E_{o2} = \text{FALSE} \cdots \text{state 1}$$

$$E_{o1} = \text{FALSE and } E_{o2} = \text{TRUE} \cdots \text{state 2}$$

6-4.1 Bistable MV (F/F)

A *bistable* MV is also commonly called a *flip-flop* (F/F). Other names given to it are the *binary* or the *divide-by-two* element. It is stable both in state 1 and in state 2. When an F/F is first energized, it may assume either state 1 or state 2. But once it has done so, it will stay in that state indefinitely, until the F/F is triggered. The state of an F/F may be changed (flipped) by an external trigger pulse. An external trigger is used to either flip the state or establish the state of an F/F. This action is further explained in Secs. 6-5 and 6-6.

6-4.2 Monostable MV

A *monostable* MV is stable only in one state and is unstable or quasi-stable in the other state. When a monostable MV is energized, it assumes its stable state. It will stay in its stable state indefinitely unless an external trigger pulse is applied to it. An external trigger pulse simply flips the monostable MV into its unstable state. After the trigger pulse has ended and a certain period of time has elapsed, it returns to its stable state on its own. The trigger pulse simply switches the state of the monostable MV temporarily. The time duration for which it stays in the unstable state is determined by the internal circuitry of the monostable MV. A monostable MV is also sometimes called a *one-shot* circuit. This circuit is discussed in detail in Ch. 7.

6-4.3 Astable MV

An *astable* MV is unstable in *both* of the states. It continously flips back and forth between the two states. Another name given to the astable MV is the *free-running* MV. The time duration that it stays in either state is determined by its internal circuitry. This circuit is also discussed in detail in Ch. 7.

6-5 The R-S Flip-Flop

A typical R-S F/F is shown in Fig. 6-2. R stands for RESET and S for SET. S and R simply correspond to the two stable states of an F/F. One state is called the SET state and the other the RESET or CLEAR (C) state. The states are arbitrarily assigned as SET and RESET. Either state could be named SET or RESET.

Assume that in the RESET state, E_{o_1} = TRUE and E_{o_2} = FALSE. Then in the SET state, E_{o_1} = FALSE and E_{o_2} = TRUE. When the power is turned ON for the first time, it is not known whether the F/F is in the SET or RESET state. Application of a trigger pulse at the S terminal turns the F/F to the SET state if it was in the RESET state before. If an F/F is already in the SET

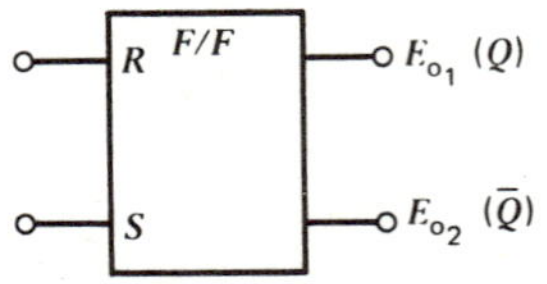

Figure 6-2 R-S flip-flop.

state, then the trigger pulse at the S terminal will not change anything. This means that the F/F becomes SET after a trigger pulse is applied at the S terminal, regardless of its previous state. Similarly, a trigger pulse applied at the R terminal turns the F/F to the RESET state regardless of its previous state. But if a trigger pulse is applied *simultaneously* to both the R and S terminals of an R-S F/F, an *indeterminate* situation results. The F/F may end up either in the SET or the RESET state. Table 6-1 summarizes the characteristics of an R-S F/F.

Table 6-1
States of an R-S F/F. True = 1, False = 0.

R	S	E_{o1}	E_{o2}
1	0	1	0
0	1	0	1
0	0	unchanged	
1	1	indeterminate	

6-6 The T Flip-Flop

The T (TOGGLE) type F/F has one input terminal marked T, as shown in Fig. 6-3. A trigger pulse applied at T simply flips the state of the F/F. Every time a trigger pulse is applied at T, the state of

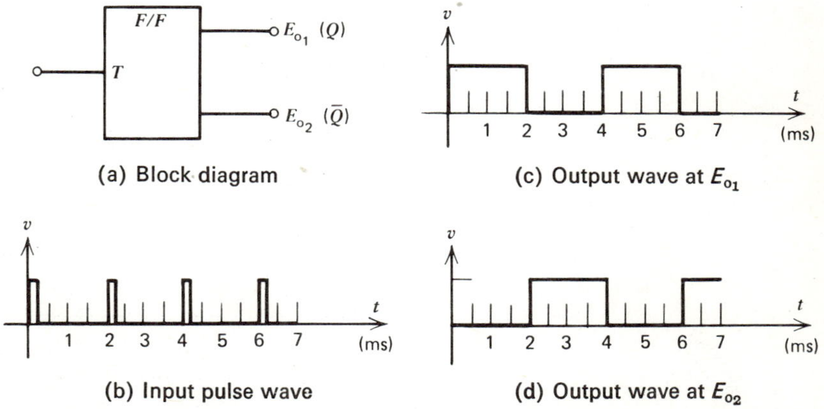

Figure 6-3 T-type F/F and input/output waveforms.

the F/F is flipped. If a series of trigger pulses are applied, as shown in Fig. 6-3b, then the outputs from the two output terminals are as shown in Figs. 6-3c and 6-3d. Note that one output is always opposite to the other.

6-7 The R-T-S Flip-Flop

The block diagram of an R-T-S F/F is shown in Fig. 6-4. An R-T-S F/F combines the capabilities of an R-S F/F and a T F/F in one unit. It can be used either as an R-S F/F or a T F/F. Table 6-2 summarizes the characteristics of an R-T-S F/F. Note that the R-S operation overrides the T operation. The T operation is accomplished by making $R = S = 0$ (FALSE) and $T = 1$ (TRUE).

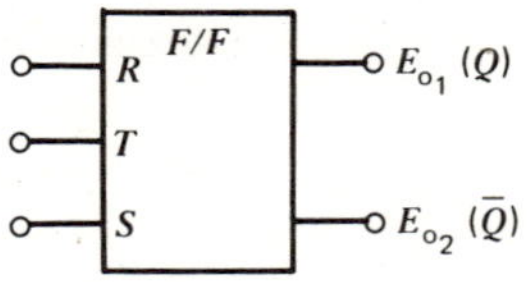

Figure 6-4 R-T-S flip-flop.

Table 6-2
Truth Table of an R-T-S F/F

R	S	T	E_{o1}	E_{o2}
0	0	0	unchanged	
0	0	1	TOGGLE	
0	1	1	0	1
1	0	0	1	0
1	0	1	1	0
1	1	0	indeterminate	
1	1	1		

6-8 The J-K Flip-Flop

Figure 6-5 shows a J-K F/F. It has only two input terminals, *J* and *K*. The terminals behave similarly to R-S terminals. However, when a trigger pulse is applied *simultaneously* to *both* the *J* and the *K* terminals, TOGGLE (*T*) action takes place. It is in this way that the *T* action is accomplished in a J-K F/F. Table 6-3 summarizes the characteristics of a J-K F/F.

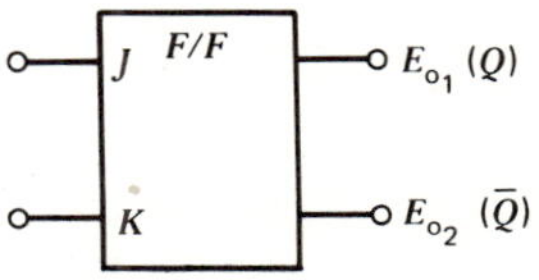

Figure 6-5 J-K flip-flop.

Table 6-3
Characteristics of a J-K F/F. True = 1, False = 0.

J	K	E_{01}	E_{02}
1	0	1	0
0	1	0	1
0	0	unchanged	
1	1	TOGGLE (*T*)	

6-9 The Flip-Flop Circuit

The basic circuit of an F/F is shown in Fig. 6-6a. It consists of two inverter circuits connected back-to-back. One inverter circuit is designed to be the mirror image of the other. The two outputs are taken at the two collectors. The circuit is designed such that at any one time, *one* transistor is *ON* while the *other* is *OFF*. Output voltage of the OFF transistor is approximately equal to the supply voltage V_{CC}, and that of the ON transistor is equal to V_{CE}(ON). Thus, the two output voltage levels are V_{CC} and V_{CE}(ON).

Some F/Fs are designed to be operated in the active region. In that case, the two output levels lie between V_{CC} and V_{CE}(ON).

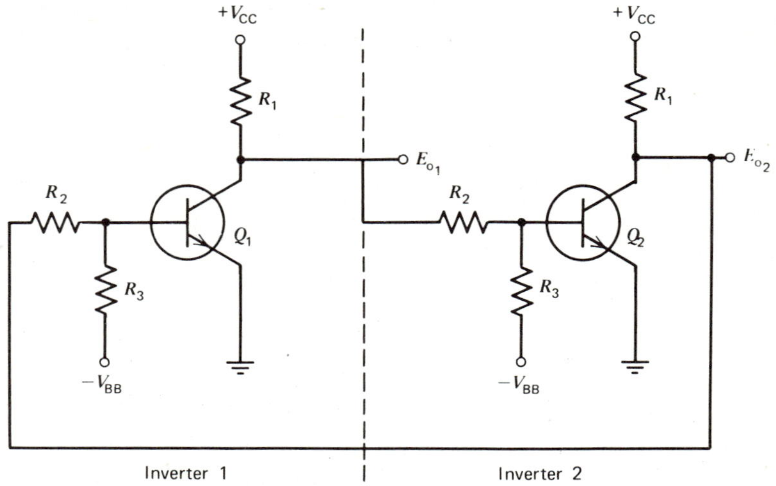

(a) Two inverters connected back-to-back to form a flip-flop circuit

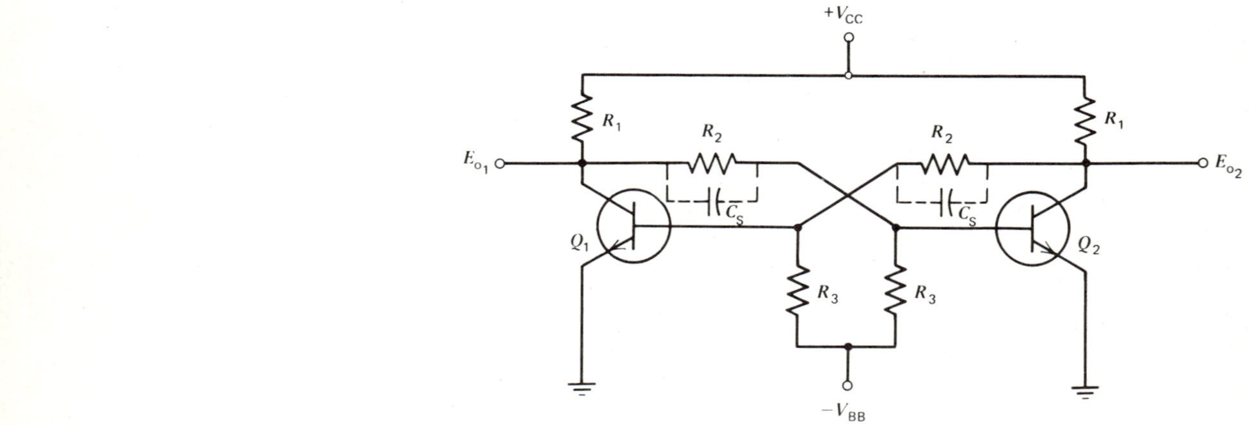

(b) Common form of a flip-flop circuit

Figure 6-6 Flip-flop circuit.

These are called the *nonsaturated*-type F/Fs. They are mainly used in high-speed switching circuits where the *turn-ON* time and the *turn-OFF* time become significant.

The *saturated*-type F/Fs are used in most switching applications. They are easy to design and their output voltage levels do not change with a small change in operating conditions. Only the saturated-type F/F (multivibrators) are discussed here.

The circuit of Fig. 6-6a is rearranged and shown in Fig. 6-6b in its most commonly encountered form. A basic design criterion is that at any one time, one transistor should be ON and the other should be OFF.

The two output voltage levels are V_{CC} and V_{CE}(ON). Thus, V_{CC} is chosen in accordance with the given HIGH logic level (V_H). R_1, the load resistor, is chosen such that the collector current at saturation, I_C(ON), is less than the maximum allowable collector-current, I_C(max), for the transistor. Thus, in the case of an ON transistor (Sec. 5-5):

$$R_1 \geqq \frac{V_{CC} - V_{CE}(\text{ON})}{I_C(\text{max})} \text{ ohms } (\Omega) \tag{6-1}$$

where:

V_{CC} is the supply voltage in volts
V_{CE}(ON) is the voltage at the collector with respect to the emitter when the transistor is saturated
I_C(max) is the maximum allowable collector-current

Usually, the next-higher standard value of resistance is selected to keep the collector current below I_C(max). After selecting a standard value for R_1, the actual I_C(ON) can be calculated:

$$I_C(\text{ON}) = \frac{V_{CC} - V_{CE}(\text{ON})}{R_1} \text{ amperes (A)} \tag{6-2}$$

Note that I_C(ON) should be less than I_c(max) and that

$$I_B(\text{ON}) = \frac{I_C(\text{ON})}{\beta} \text{ amperes (A)} \tag{6-3}$$

Resistors R_2 and R_3 are chosen such that one transistor is ON while the other is OFF. The *base-emitter* junction (BEJ) of the OFF transistor must be *reverse-biased.* For this purpose, a reverse bias of 0.5 V to 1.5 V is generally considered to be satisfactory.

To turn a transistor *ON*, its *base current* must be equal to or greater than the minimum base current needed for *saturation.* Usually, the circuit is designed such that at saturation, the base current is 150 percent to 200 percent of the minimum required for saturation. This assures that the transistor will saturate even if

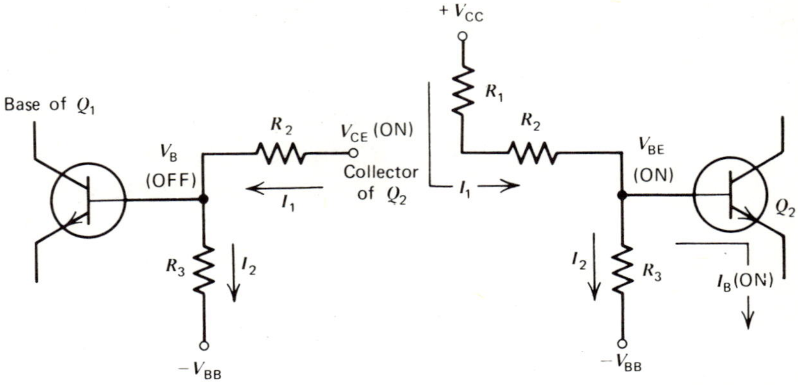

(a) Partial circuit showing Q_1 OFF (b) Partial circuit showing Q_2 ON

Figure 6-7 Partial circuits for Q_1 OFF and Q_2 ON.

some variation occurs in the circuit conditions. The two transistors Q_1 and Q_2 are chosen such that they are matched in every respect. This makes the circuit design simpler and easier. One inverter circuit is made identical to the other inverter circuit. ON/OFF conditions that apply to one inverter also apply to the other. This point will become clear in the following discussion.

Assume that Q_1 is OFF and Q_2 is ON (Fig. 6-6b). Figure 6-7a shows a partial circuit that controls the base current of Q_1. The base-emitter junction of Q_1 must be *reverse-biased* to turn Q_1 OFF [$I_B(Q_1) = 0$]. A reverse bias of 1.5 V to 0.5 V is deemed satisfactory for this purpose. Let us call this reverse-bias voltage V_{BE}(OFF). The partial circuit of Fig. 6-7a as applied to the OFF condition gives:

$$I_1 = I_2$$

$$\frac{V_C(\text{ON}) + V_{BE}(\text{OFF})}{R_2} = \frac{V_{BB} - V_B(\text{OFF})}{R_3} \tag{6-4}$$

where:

V_C(ON) is the collector voltage when the transistor is ON
V_{BB} is the reverse-bias supply voltage
V_B(OFF) is the base voltage when the transistor is OFF

Note that for the circuit configuration shown in Fig. 6-6:

$$V_C(\text{ON}) = V_{CE}(\text{ON})$$

and

$$V_B(\text{OFF}) = V_{BE}(\text{OFF})$$

A partial circuit that controls the base current to the ON transistor is shown in Fig. 6-7b. To turn Q_2 ON:

$$I_1 - I_2 \geqq I_B(\text{ON})$$

$$\frac{V_{CC} - V_B(\text{ON})}{R_1 + R_2} - \frac{V_{BB} + V_B(\text{ON})}{R_3} \geqq I_B(\text{ON}) \qquad (6\text{-}5)$$

Equations 6-4 and 6-5 are solved simultaneously for R_2 and R_3. Usually, R_1 is quite small compared with R_2 and can be neglected as far as Eq. 6-5 is concerned. This makes the process of solving for R_2 and R_3 much simpler (see Ex. 6-1 below).

6-10 Speed-up or Commutating Capacitors

The addition of two *speed-up capacitors* connected across the two base resistors improves the switching time of an F/F. These speed-up capacitors, shown in Fig. 6-6b as C_s, help to reduce the turn-ON time and the turn-OFF time. Their action is similar to that explained in Sec. 5-6. They are also sometimes referred to as *commutating capacitors*, because they direct the base storage charge into or out of the base when an F/F is triggered.

The value of these capacitors is usually chosen to be *small* so that their charge/discharge times are small (small τ). However, they must be large enough to be able to store the charge that accumulates at the transistor's junction. The maximum charge that may accumulate at a transistor's junction is usually given by the manufacturer of the transistor. It is normally in the order of pico coulombs (pC). A reasonable value for C_s may be calculated from Eq. 6-6.

$$C_s = \frac{Q_C}{V_{CC}} \text{ farads (F)} \qquad (6\text{-}6)$$

where:

C_s is the capacitance of the speed-up capacitor (F)
V_{CC} is the supply voltage (V)
Q_c is the charge-storage capacity of the transistor (C)

Example 6-1

Design an F/F circuit whose two logic levels are 0 V and +12 V. For the transistor to be used, assume:

$$I_C(\max) = 25 \text{ mA}$$
$$V_{BE}(\text{ON}) = 0$$
$$V_{BE}(\text{OFF}) = -1.5 \text{ V}$$
$$V_{BB} = -12 \text{ V}$$
$$Q_C = 400 \text{ pC}$$
$$\beta = 80$$
$$V_{CE}(\text{ON}) = 0$$

Use only standard (EIA) component values.
Calculate:

a. R_1.
b. $I_B(\text{ON})$.
c. Ratio of R_2 to R_3.
d. R_2.
e. R_3.
f. C_s.

Solution

a. The F/F circuit is shown in Fig. 6-8a. V_{CE} should be equal to +12 V when the transistor is OFF and it should be equal to 0 V when the transistor is ON.
Therefore, V_{CC} = **+12 V.**

$$R_1 \geqq \frac{V_{CC} - V_{CE}(\text{ON})}{I_C(\max)} \tag{6-1}$$

$$\geqq \frac{12 \text{ V} - 0 \text{ V}}{25 \text{ mA}}$$

$$\geqq 480 \ \Omega$$

b. Choose R_1 = **510 Ω**, the next-higher EIA standard value.

$$I_C(\text{ON}) = \frac{V_{CC} - V_{CE}(\text{ON})}{R_1} \tag{6-2}$$

$$= \frac{12 \text{ V} - 0 \text{ V}}{510 \ \Omega}$$

$$= 23.5 \text{ mA}$$

$$I_B(\text{ON}) = \frac{I_C(\text{ON})}{\beta} \tag{6-3}$$

$$\frac{23.5 \text{ mA}}{80}$$

$$= \mathbf{0.294 \text{ mA}}$$

c. Refer to the partial circuit of Fig. 6-8b. This is the OFF condition:

$$\therefore \frac{V_c(\text{ON}) + V_{BE}(\text{OFF})}{R_2} = \frac{V_{BB} - V_B(\text{OFF})}{R_3} \tag{6-4}$$

$$\frac{0 + 1.5}{R_2} = \frac{12 - 1.5}{R_3}$$

$$\frac{1.5}{R_2} = \frac{10.5}{R_3}$$

$$\mathbf{R_3 = 7.0R_2}$$

d. The partial circuit shown in Fig. 6-8c applies to the ON condition.

$$\therefore \left(\frac{V_{CC} - V_B(\text{ON})}{R_2}\right) - \left(\frac{V_{CC} + V_B(\text{ON})}{R_3}\right) \geqq I_B(\text{ON}) \tag{6-5}$$

$$\left(\frac{12 - 0}{510 + R_2}\right) - \left(\frac{12 + 0}{R_3}\right) \geqq 0.294 \text{ mA}$$

$$\left(\frac{12}{510 + R_2}\right) - \left(\frac{12}{R_3}\right) = 0.44 \text{ mA} \quad [150\% \text{ of } I_B(\text{ON})]$$

Substituting $7R_2$ for R_3 gives:

$$\left(\frac{12}{510 + R_2}\right) - \left(\frac{12}{7R_2}\right) = 0.44 \text{ mA} \tag{6-7}$$

Assuming that 510 Ω is small, compare R_2.

$$\left(\frac{12}{R_2}\right) - \left(\frac{12}{7R_2}\right) = 0.44 \text{ mA}$$

$$R_2 = \frac{12 - 1.7}{0.44 \times 10^{-3}}$$

$$= 23.4 \text{ k}\Omega$$

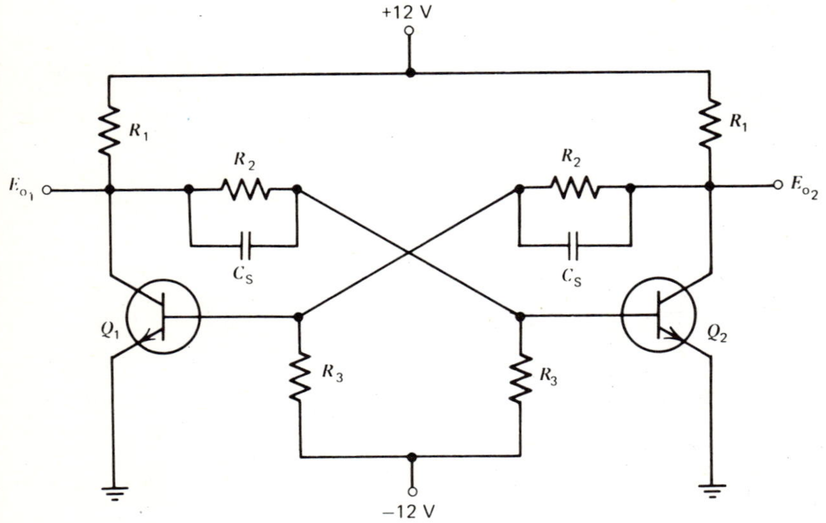

(a) The flip-flop circuit

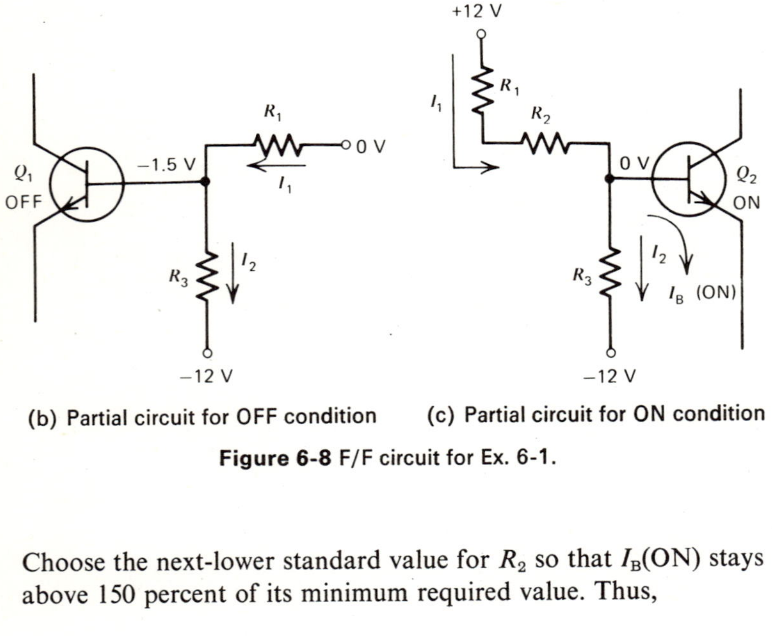

(b) Partial circuit for OFF condition (c) Partial circuit for ON condition

Figure 6-8 F/F circuit for Ex. 6-1.

Choose the next-lower standard value for R_2 so that I_B(ON) stays above 150 percent of its minimum required value. Thus,

$$R_2 = \mathbf{22\ k\Omega}$$

e. Therefore,

$$R_3 = 22\ \text{k} \times 7$$
$$= \mathbf{154\ k\Omega}$$

Choose next-higher standard value for R_3 for the same reason as R_2. Thus,

$$R_3 = \mathbf{160\ k\Omega}$$

f.

$$C_s \geqq \frac{Q_c}{V_{cc}} \tag{6-6}$$

$$\therefore C_s \geqq \frac{400 \times 10^{-12}}{12}$$

or

$$\geqq 33.3 \text{ pF}$$

Use:

$$C_s = \mathbf{40\ pF}$$

Let us see how much error is introduced by the assumption that $510 \ll R_2$. Starting from Eq. 6-7,

$$\left(\frac{12}{510 + R_2}\right) - \left(\frac{12}{7R_2}\right) = 0.44 \times 10^{-3}$$

$$\left(\frac{12R_2}{510 + R_2}\right) - \left(\frac{12}{7}\right) = 0.44 \times 10^{-3}R_2$$

$$(0.44 \times 10^{-3}R_2 + 1.7)(510 + R_2) = 12R_2$$

$$0.44R_2^2 - 10.1R_2 + 874 = 0$$

Solving this quadratic equation for R_2 yields:

$$R_2 = \frac{10.1 \pm \sqrt{(10.1)^2 - 1.54}}{2 \times 0.44 \times 10^{-3}}$$

$$= \frac{20.1}{0.88 \times 10^{-3}}$$

$$= \mathbf{22.8\ k\Omega}$$

$$\text{percentage error} = \frac{(23.4 - 22.8)\text{ k}\Omega}{23.4\text{ k}\Omega}$$

$$= \mathbf{2.5} \text{ percent}$$

Note that the error computed above is insignificant, especially when it is realized that:

1. Most standard resistors vary ±5 percent or ±10 percent of their rated values.
2. I_B(ON) is usually designed to be 150 percent or more of its minimum required value for saturation.
3. V_{BE}(OFF) is designed to be above 0.5 V, whereas any reverse bias across the bias emitter junction will turn the transistor OFF.
4. Closest standard EIA values of the components are used rather than the exact calculated values.

Example 6-2

Redesign the F/F circuit and repeat all parts of Ex. 6-1 assuming that the transistors used are standard Si transistors.

Solution

a.

$$R_1 \geqq \frac{V_{CC} - V_{CE}(\text{ON})}{I_C(\text{max})} \tag{6-1}$$

$$\therefore R_1 \geqq \frac{(12 - 0.3)\ \text{V}}{25\ \text{mA}}$$

or

$$\geqq 468\ \Omega$$

Choose:

$$R_1 = \mathbf{510\ \Omega.}$$

b. Now

$$I_C(\text{ON}) = \frac{V_{CC} - V_{CE}(\text{ON})}{R_1} \tag{6-2}$$

or

$$= \frac{(12 - 0.3)\ \text{V}}{510\ \Omega}$$

$$= 22.94\ \text{mA}$$

$$I_B(\text{ON}) = \frac{I_c(\text{ON})}{\beta} \tag{6-3}$$

$$\frac{22.94\ \text{mA}}{80}$$

$$= \mathbf{0.287\ mA}$$

c. OFF condition:

$$\frac{V_{CE}(\text{ON}) - V_{BE}(\text{ON})}{R_2} = \frac{V_{CC} - V_B(\text{OFF})}{R_3} \tag{6-4}$$

or

$$\frac{0.3 + 1.5}{R_2} = \frac{12 - 1.5}{R_3}$$

or

$$R_3 = \mathbf{5.83}\boldsymbol{R_2}$$

d. ON condition:

$$\left(\frac{V_{CC} - V_B(\text{ON})}{R_1 + R_2}\right) - \left(\frac{V_{BB} + V_B(\text{ON})}{R_3}\right) \geqq I_B(\text{ON}) \tag{6-5}$$

or

$$\left(\frac{12 - 0.7}{510 + R_2}\right) - \left(\frac{12 + 0.7}{R_3}\right) \geqq 0.287 \times 10^{-3} \times 1.5$$

[150 percent of I_B(ON)]

Assuming that 510 Ω ≪ R_2,

$$\left(\frac{11.3}{R_2}\right) - \left(\frac{12.7}{5.83R_2}\right) = 0.43 \times 10^{-3}$$

$$\left(\frac{11.3}{R_2}\right) - \left(\frac{2.18}{R_2}\right) = 0.43 \times 10^{-3}$$

$$R_2 = \frac{11.3 - 2.18}{0.43 \times 10^{-3}}$$

$$= 21.2\ \text{k}\Omega$$

Next-lower standard resistance value is **20 kΩ**.

e.

$$R_3 = 5.83 \times 20\ \text{k}\Omega$$

$$= 116.6\ \text{k}\Omega$$

Next-higher standard resistance value is **120 kΩ**.

f. C_s remains the same (40 pF) as in Ex. 6-1 because circuit conditions affecting its value have not changed.

Example 6-3

Check by calculations whether the F/F circuit previously designed in Ex. 6-2 will function properly. Do this by showing that:

a. V_{BE}(OFF) is below zero.

b. I_B(ON) exceeds the minimum value for saturation.

Solution

What is to be checked here is whether in fact one transistor is ON while the other is OFF. Consider the OFF condition (Fig. 6-9a):

a.
$$V_{BE}(\text{OFF}) = 0.3 - V_{20\text{ k}}$$
$$= 0.3 - \left[\frac{12 + 0.3}{(120 + 20)\text{ k}\Omega}\right] \times 20\text{ k}\Omega$$
$$= (0.3 - 1.76)\text{ V}$$
$$= -1.46\text{ V}$$

V_{BE}(OFF) is negative, which means that the transistor does go OFF.

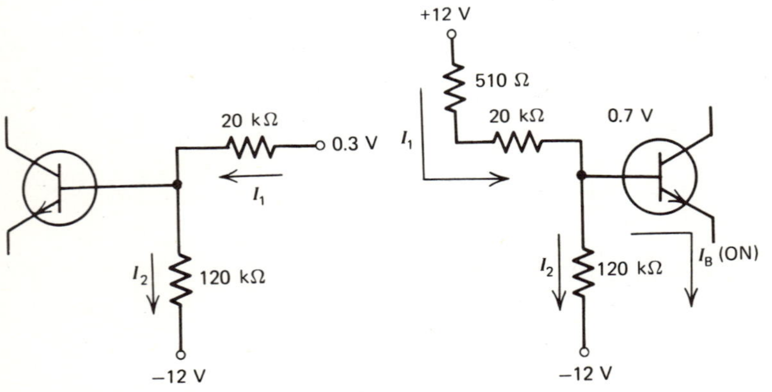

(a) Partial circuit for OFF condition (b) Partial circuit for ON condition

Figure 6-9 Partial circuits for Ex. 6-3.

b. Now consider the ON condition (Fig. 6-9b).

$$I_B = \left(\frac{12 - 0.7}{510 - 20\text{ k}}\right) - \left(\frac{12 + 0.7}{120\text{ k}}\right)$$
$$= 0.551\text{ mA} - 0.106\text{ mA}$$
$$= 0.445\text{ mA}$$

Minimum I_B required for saturation is:

$$I_B(\text{ON}) = \frac{(12.0 - 0.3)}{510} - \frac{1}{80}$$

$$= 0.277 \text{ mA}$$

I_B is actually greater than the minimum I_B required for saturation. Both ON and OFF conditions have been satisfied and therefore the circuit will function properly as a saturated type F/F.

Example 6-4

Consider the circuit shown in Fig. 6-10a. Assume $R_1 = 2\text{ k}\Omega$, $R_2 = 15\text{ k}\Omega$, $R_3 = 100\text{ k}\Omega$, $V_{CC} = -20$ V, $V_{BB} = +6$ V, $V_{CE}(\text{ON}) = -0.1$ V, $V_{BE}(\text{ON}) = -0.3$ V.

Calculate:

a. The two output voltage levels.

b. The minimum value of β required for the transistors.

c. $V_{BE}(\text{OFF})$.

Solution

a. The two output voltage levels are V_{CC} and $V_{CE}(\text{ON})$, namely **−20 V** and **−0.1 V**.

b.

$$I_C(\text{ON}) = \frac{V_{CC} - V_{CE}(\text{ON})}{R_1}$$

$$\therefore I_C(\text{ON}) = \frac{20 - 0.1}{2\text{ k}\Omega}$$

$$= 9.95 \text{ mA}$$

ON condition (Fig. 6-10b):

$$I_B(\text{ON}) = \frac{20 - 0.3}{2\text{ k}\Omega + 15\text{ k}\Omega} - \frac{6 + 0.3}{100\text{ k}\Omega}$$

$$= 1.159 \text{ mA} - 0.063 \text{ mA}$$

$$= 1.096 \text{ mA}$$

$$\therefore \beta(\text{min}) = \frac{9.95 \text{ mA}}{1.096 \text{ mA}}$$

$$= \mathbf{9.1}$$

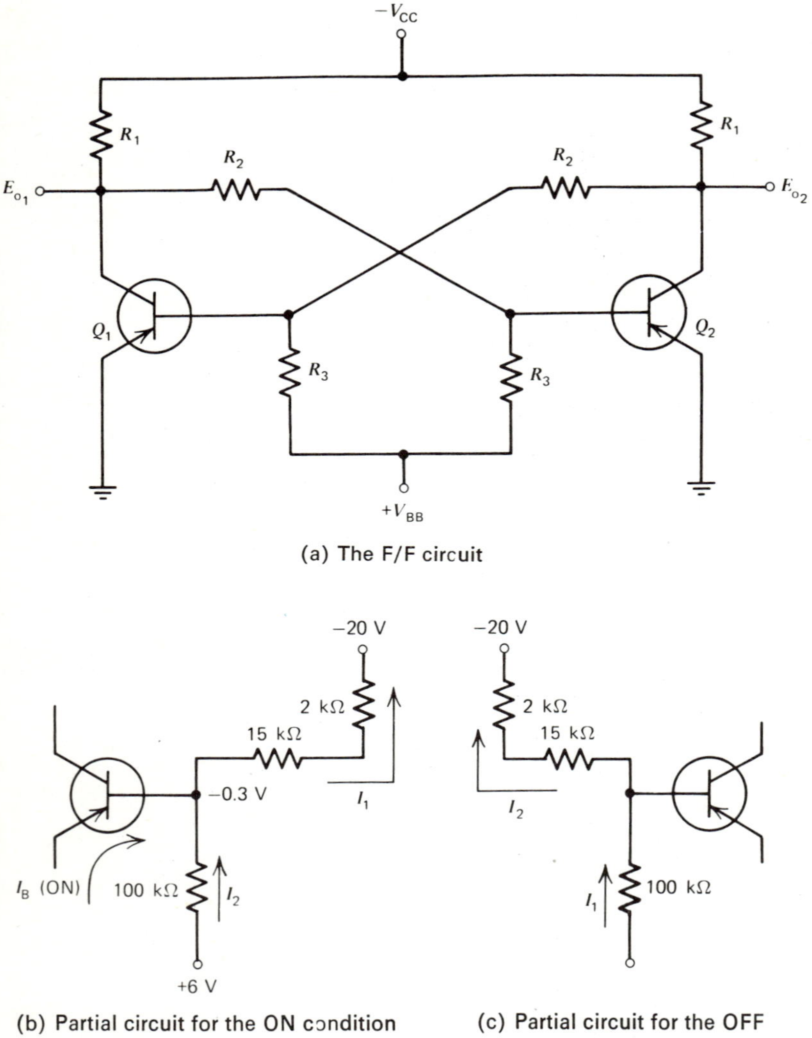

(a) The F/F circuit

(b) Partial circuit for the ON condition

(c) Partial circuit for the OFF condition

Figure 6-10 F/F circuit for Ex. 6-4.

c. OFF condition (Fig. 6-10c):

$$V_{BE}(\text{OFF}) = 0.1 + \frac{6 + 0.1}{(100 + 15)\text{ k}} = 0.15\text{ k}\Omega$$

$$= -0.1 + 0.796$$

$$= \mathbf{+0.696\ V}$$

6-11 Emitter-Coupled Flip-Flop

A typical circuit of an *emitter-coupled* flip-flop is shown in Fig. 6-11a. An *emitter resistor*, R_E, has been added to the basic flip-flop circuit of Fig. 6-6. This addition of R_E eliminates the bias supply V_{BB}. Other requirements for the circuit operation remain unchanged. At any one time, one transistor is ON and the other is OFF. The circuit is designed such that these ON/OFF conditions are fulfilled.

Consider the circuit of Fig. 6-11a. Regardless of whether Q_1 is

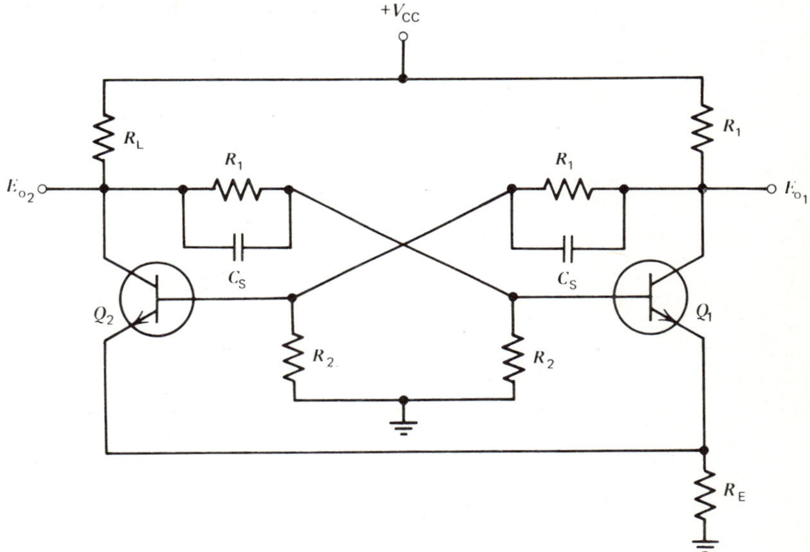

(a) F/F circuit

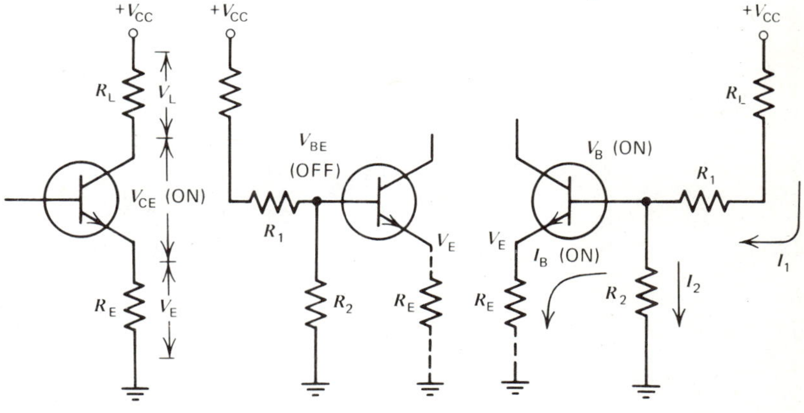

(b) ON transistor conditions

(c) Partial circuit for the OFF condition

(d) Partial circuit for the ON condition

Figure 6-11 Emitter-coupled F/F circuit.

ON or Q_2 is ON, current I_E in R_E always remains the same (assuming $Q_1 = Q_2$). Thus the emitter voltage (V_E) always remains the same. To satisfy the ON/OFF conditions it is necessary only to design the circuit such that the base-emitter junction of one transistor is reverse-biased while the base current in the other transistor is greater than or equal to its saturation current.

The two logic levels or output voltage levels are $+V_{CC}$ and $V_E + V_{CE}(\text{ON})$. Assume that voltages corresponding to the two logic levels are V_H(HIGH) and V_L (LOW). Then,

$$V_H = V_{CC}$$

$$V_L = V_E + V_{CE}(\text{ON}) \text{ volts (V)} \qquad (6\text{-}8)$$

A partial circuit of Fig. 6-11b shows conditions when the transistor is ON.

$$R_E = \frac{V_E}{I_E(\text{ON})} \text{ ohms } (\Omega) \qquad (6\text{-}9a)$$

$$= \frac{V_L - V_{CE}(\text{ON})}{I_C(\text{ON}) + I_B(\text{ON})}$$

$$= \frac{V_L - V_{CE}(\text{ON})}{I_C(\text{ON}) + [I_C(\text{ON})/\beta]} \text{ ohms } (\Omega) \qquad (6\text{-}9b)$$

and

$$R_L = \frac{V_{CC} - [V_E + V_{CE}(\text{ON})]}{I_C(\text{ON})} \text{ ohms } (\Omega) \qquad (6\text{-}10a)$$

$$= \frac{V_{CC} - V_L}{I_C(\text{ON})} \text{ ohms } (\Omega) \qquad (6\text{-}10b)$$

Values for R_1 and R_2 are calculated from the ON/OFF conditions. Figure 6-11c shows a partial circuit that applies to the *OFF condition.* A base-emitter reverse bias of 1.5 V is assumed to be enough to turn the transistor OFF [$V_{BE}(\text{OFF}) = -1.5$ V].

$$V_B(\text{OFF}) = V_E - V_{BE}(\text{OFF})$$

$$= V_E - 1.5$$

$$I_1 = I_2 \qquad (I_B = 0, \text{ OFF condition})$$

$$\frac{V_B\text{OFF}}{R_2} \quad \frac{V_L - V_B(\text{OFF})}{R_1}$$

or

$$\frac{V_E - 1.5}{R_2} = \frac{V_L - (V_E - 1.5)}{R_1}$$

A partial circuit for the *ON condition* is shown in Fig. 6-11d.

$$I_2 - I_1 \geqq I_B(\text{ON})$$

$$\frac{V_{CC} - V_B(\text{ON})}{R_L + R_1} - \frac{V_B(\text{ON})}{R_2} \geqq I_B(\text{ON}) \qquad (6\text{-}12a)$$

Since $V_B(\text{ON}) = V_E + V_{BE}(\text{ON})$,

$$\therefore \frac{V_{CC} - [V_E + V_{BE}(\text{ON})]}{R_L + R_1} - \frac{V_E + V_{BE}(\text{ON})}{R_2} \geqq I_B(\text{ON}) \qquad (6\text{-}12b)$$

Equations 6-11 and 6-12 are solved simultaneously to obtain values R_1 and R_2. Again, R_L is usually very small compared with R_1 and may be neglected (Eq. 6-12). This simplifies calculations.

Example 6-5

Design an emitter-coupled F/F circuit such that $V_L = -2$ V and $V_H = -12$ V. For the two transistors to be used, $h_{fe} = 100$, $I_C(\text{max}) = 15$ mA, $V_{BE}(\text{ON}) = 0.5$ V, and $V_{CE}(\text{ON}) = -0.2$ V. $V_{BE}(\text{OFF})$ is to be designed as 1.0 V.

Solution

The F/F circuit is shown in Fig. 6-12a. Values for V_{CC}, R_L, R_E, R_1, and R_2 need to be calculated.

$$V_{CC} = V_H$$
$$= \mathbf{-12\ V}$$

Consider the partial circuit shown in Fig. 6-12b.

$$R_L = \frac{V_{CC} - V_L}{I_C(\text{ON})} \qquad (6\text{-}10)$$

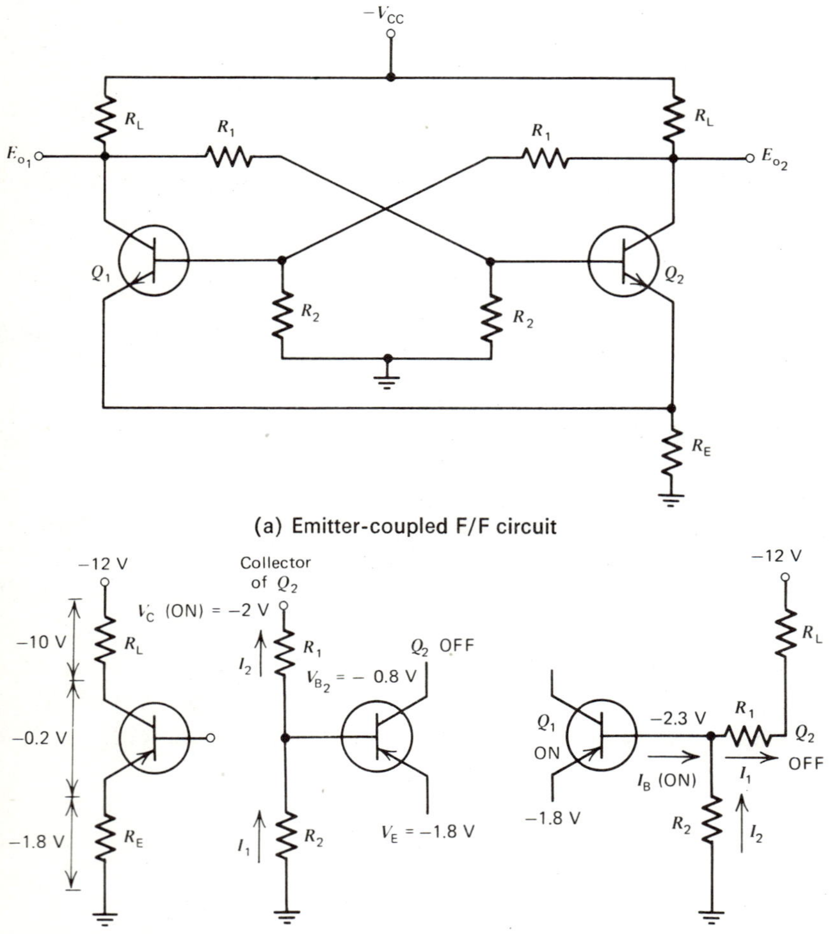

(b) Partial circuit of the ON transistor (c) Partial circuit for Q_2 OFF (d) Partial circuit for Q_1 ON

Figure 6-12 Circuits for Ex. 6-5.

or

$$= \frac{(12 - 2)\text{ V}}{15\text{ mA}}$$

$$= \mathbf{667\ \Omega}$$

$$R_E = \frac{V_L - V_{CE}(\text{ON})}{I_C(\text{ON}) - [I_C(\text{ON})/\beta]} \tag{6-9}$$

$$= \frac{(2 - 0.2)\text{ V}}{15 - (15/100)\text{ mA}}$$

$$= \mathbf{119\ \Omega}$$

Assume that Q_1 is ON and Q_2 is OFF. The partial circuit for the OFF condition is shown in Fig. 6-12c.

$$\frac{V_B(\text{OFF})}{R_L} = \frac{V_L - V_B(\text{OFF})}{R_1} \qquad (6\text{-}11)$$

or

$$\frac{0.8}{R_2} = \frac{2.0 - 0.8}{R_1}$$

$$R_1 = \mathbf{1.5}\boldsymbol{R_2} \qquad (6\text{-}13)$$

The partial circuit that applies to the ON condition is shown in Fig. 6-12d. Note that:

$$V_{BE}(\text{ON}) = -(1.8 \pm 0.5)$$
$$= -2.3\text{V}$$

$$\frac{V_{CC} - V_{BE}(\text{ON})}{R_L + R_1} - \frac{V_{BE}(\text{ON})}{R_2} \geqq \frac{I_c(\text{ON})}{\beta} \qquad (6\text{-}12)$$

or

$$\frac{12 - 2.3}{667 + R_1} - \frac{2.3}{R_2} = \frac{15 \times 10^{-3}}{100}$$

Assuming that $667 \ll R_1$, then,

$$\frac{9.7}{R_1} - \frac{2.3}{R_2} = 15 \times 10^{-5}$$

Substituting Eq. (6-13),

$$\frac{9.7}{1.5R_2} - \frac{2.3}{R_2} = 15 \times 10^{-5}$$

$$\frac{6.467}{R_2} - \frac{2.3}{R_2} = 15 \times 10^{-5}$$

$$R_2 = \frac{6.467 - 2.3}{15 \times 10^{-5}}$$
$$= \mathbf{27.8\ k\Omega}$$

$$R_1 = 1.5R_2 \qquad (6\text{-}13)$$
$$= 1.5 \times 27.8$$
$$= \mathbf{41.7\ k\Omega}$$

6-12 R-S Level-Triggering (dc)

Triggering F/F simply means flipping its state, that is, changing it either from SET to RESET or from RESET to SET. The triggering input to an F/F is also used to establish a known state of the F/F.

F/Fs may be triggered by applying certain **dc** voltage levels or pulses at the *R* and *S* terminals. Triggering circuits that operate according to the **dc** voltage levels applied at the *R* and *S* terminals are known as the *level-triggered* or ***dc**-triggered* circuits. Triggering circuits that operate as a result of pulses applied at the *R* and *S* terminals are known as the *edge-triggered* or ***ac**-triggered* circuits. The *edge-triggered* circuits are activated either by the *leading* edge or *trailing* edge of the input pulse. These are discussed further in Sec. 6-13.

The *level-triggered* or ***dc**-triggered* circuits get activated as soon as the trigger input voltages at the *R* and *S* terminals reach a certain *level*. The F/F shown in Fig. 6-13a is a level-triggered device. *R* and *S* are two momentary-contact-type switches. These correspond to the *R* and *S* terminals of Fig. 6-2.

Essentially, each switch (*S* and *R*), when activated (depressed) momentarily, makes the base voltage of the corresponding transistor negative, thus turning it OFF if it was not already OFF. Activating the switch has no effect on the corresponding transistor if it was already OFF. The F/F circuit shown in Fig. 6-13a is in the RESET state when Q_1 is OFF and Q_2 is ON; and it is in the SET state when Q_1 is ON and Q_2 is OFF.

The F/F circuit shown in Fig. 6-13b indicates another way of accomplishing the R-S action. In this case, when a switch (*R* or *S*) is momentarily activated, the collector voltage of the corresponding transistor becomes equal to 0 V. This reverse-biases the base of the other transistor and turns it OFF if it was not already OFF. Note that SET and RESET states for the two F/Fs shown in Figs. 6-13a and 6-13b are the same.

Figure 6-13c shows yet another level-triggered F/F. Momentarily activating switch *R* (RESET) saturates Q_3, lowering its collector voltage to $V_{CE}(\text{ON})$ ($\cong 0$ V). The collector voltage at Q_1 lowers along with the collector voltage of Q_3. The *lowered* collector of Q_1 *reverse-biases* the base-emitter junction of Q_2, turning it OFF if it was not already OFF. Momentarily activating switch *R* is equivalent to applying a pulse (of sufficient amptitude) at *R* so as to saturate Q_3. In practical circuits, that is exactly what is done. Pulsed transistor switches are used instead of mechanical switches. *Note* that the SET (*S*) operation works similar to RESET (*R*) operation for each case explained above.

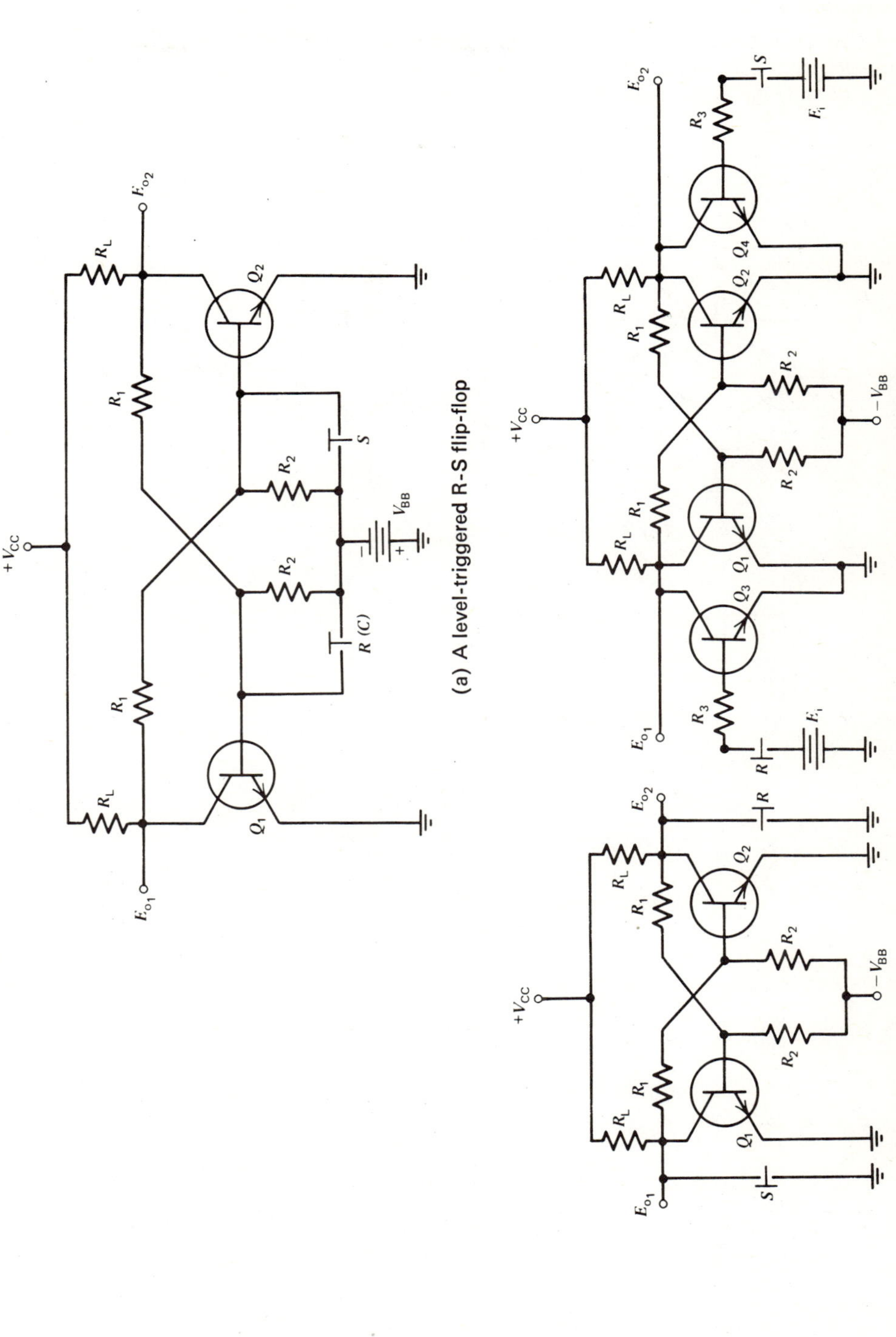

(a) A level-triggered R-S flip-flop

(b) A level-triggered R-S flip-flop

(c) A level-triggered R-S flip-flop

Figure 6-13 dc level-triggered R-S flip-flops.

Example 6-6

An R-S-type F/F is shown in Fig. 6-14a. Trigger input pulses applied at the R and S terminals are shown in Figs. 6-14b and 6-14c, respectively. Sketch the voltage waveforms that appear at:

a. E_{o_1}.

b. E_{o_2}.

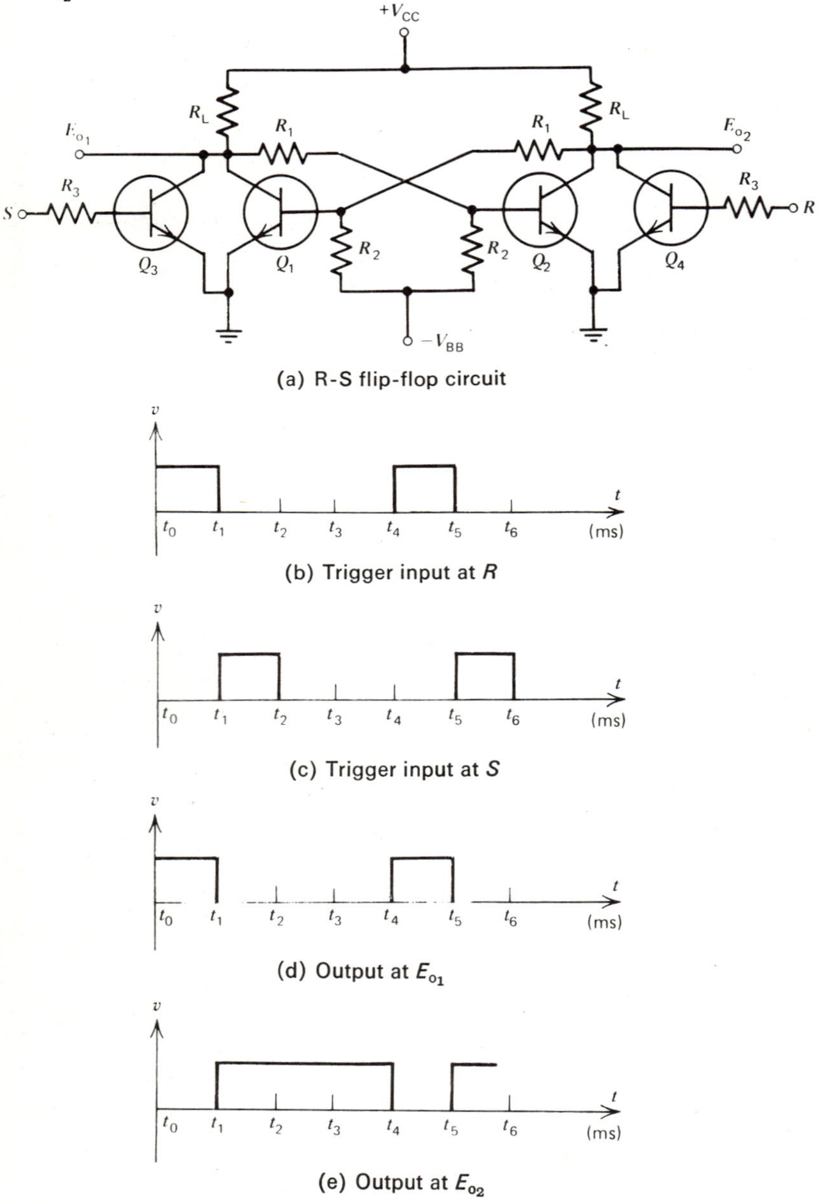

(a) R-S flip-flop circuit

(b) Trigger input at R

(c) Trigger input at S

(d) Output at E_{o_1}

(e) Output at E_{o_2}

Figure 6-14 Circuit and waveforms for Ex. 6-6.

Solution

At $t = t_{0+}$, the trigger pulse at R turns Q_4 ON, lowering the collector voltage of Q_2 and turning Q_1 OFF. At the same time, 0 V at S keeps Q_3 OFF, isolating it from the rest of the circuit. Thus, it is seen that a pulse at R and no pulse at S result in turning Q_1 OFF and Q_2 ON. E_{o_1} and E_{o_2} become equal to V_{CC} and V_{CE}(ON), respectively. This is the RESET state of the given F/F circuit.

At $t = t_{1+}$, the input trigger pulse at S turns Q_3 (and Q_1) ON and Q_2 OFF. No pulse at R isolates Q_4. Thus, after $t = t_1$, Q_2 is OFF and Q_1 is ON. E_{o_1} and E_{o_2} are V_{CE}(ON) and V_{CC}, respectively. This is the SET state of the given F/F.

From $t = t_{2+}$ to $t = t_{3+}$, there is no trigger pulse (voltage) at either S or R. Thus, both Q_3 and Q_4 stay OFF. The F/F circuit remains in the SET state.

At $t = t_{4+}$, there is a pulse at R and no pulse at S. The circuit goes into the RESET state as explained above ($t = t_{0+}$).

At $t = t_{5+}$, there is a pulse at S and no pulse at R. The circuit goes into the SET state as explained above ($t = t_{1+}$).

At $t = t_{6+}$, there is no pulse at either S or R. The circuit remains in the SET state.

The resulting output waveforms that appear at E_{o_1} and E_{o_2} are shown in Figs. 6-14c and 6-14d.

6-13 R-S Edge-Triggering (ac)

When the input is a rectangular pulse wave (*step function*), **ac**-triggering circuits are the ones that operate. A pulse is made up of two step-functions, a transition *up* and a transition *down*. When the pulse is *differentiated* by a short-time-constant circuit (Sec. 3-7), both positive and negative spikes are produced. The **ac**-triggering circuits may be designed to flip the state of the F/F, during either the *negative* spike or the *positive* spike. Because the F/F is triggered either by the leading edge or the trailing edge of the input pulse, it is called the *edge-triggered* F/F.

Figure 6-15 shows a *transistor edge-triggered* F/F. A pulse applied to the C (CLEAR) input terminal gets differentiated by the short-time-constant ($R_{B3}C_1$) circuit. The transistor Q_3 is turned ON by the positive spike. This lowers the collector voltage of Q_3 to $\simeq 0$ V, which turns Q_2 OFF. As soon as the positive spike disappears, Q_3 goes OFF. Q_1 gets turned ON because Q_2 was turned OFF. Any additional pulses at the C terminal will simply try to turn OFF Q_2, which is already OFF. Thus, this F/F is set into the CLEAR (C) state by applying a pulse input at the C terminal;

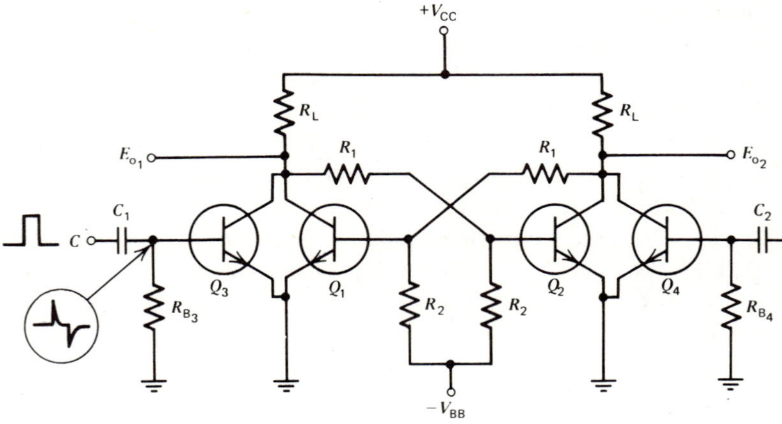

Figure 6-15 Edge-triggered F/F.

therefore, the F/F is set to the CLEAR state by the *positive*-going edge of the input pulse. The triggering capacitor (C_1) is usually made twice as large as the coupling capacitor ($C_1 \cong 2Q_T/V_{CC}$).

The *positive*-going edge of the input pulse applied at the S terminal will SET the F/F. The action is similar to the one explained above for CLEAR operation.

Figure 6-16 shows a *base* (*edge*)-*triggered* F/F circuit. The two triggering circuits consist of $C_1R_1D_1$ and $C_2R_2D_2$. Both triggering circuits are similar to each other. One input is designated as S (SET) and the other as R (RESET).

Let us examine the operation of the circuit as a pulse is applied at R. Assume that Q_1 is ON and Q_2 is OFF. The potential at the base of Q_1 is V_{BE}(ON), and therefore D_1 is ON. When applied at R, a pulse gets differentiated by C_1R_1, as shown in Fig. 6-16. The positive spike of the differentiated pulse turns D_1 OFF and does not reach Q_1. Thus, the positive spike has no effect on the state of the F/F. D_1 turns ON during the *negative* spike and the base of Q_1 becomes negative. This turns Q_1 OFF. As a result of the circuit design of the F/F, Q_2 turns ON as Q_1 is turned OFF. Thus a pulse applied at the R terminal turns Q_1 OFF and Q_2 ON. This is the RESET (R) state of the F/F.

Additional pulses applied at R cannot alter the state of the F/F. D_1 is OFF (OPEN) during the positive spike of the differentiated input pulse. The negative spike turns D_1 ON and is transferred to the base of Q_1. But Q_1 is already OFF. Thus, the negative potential at the base of Q_1 *changes nothing*.

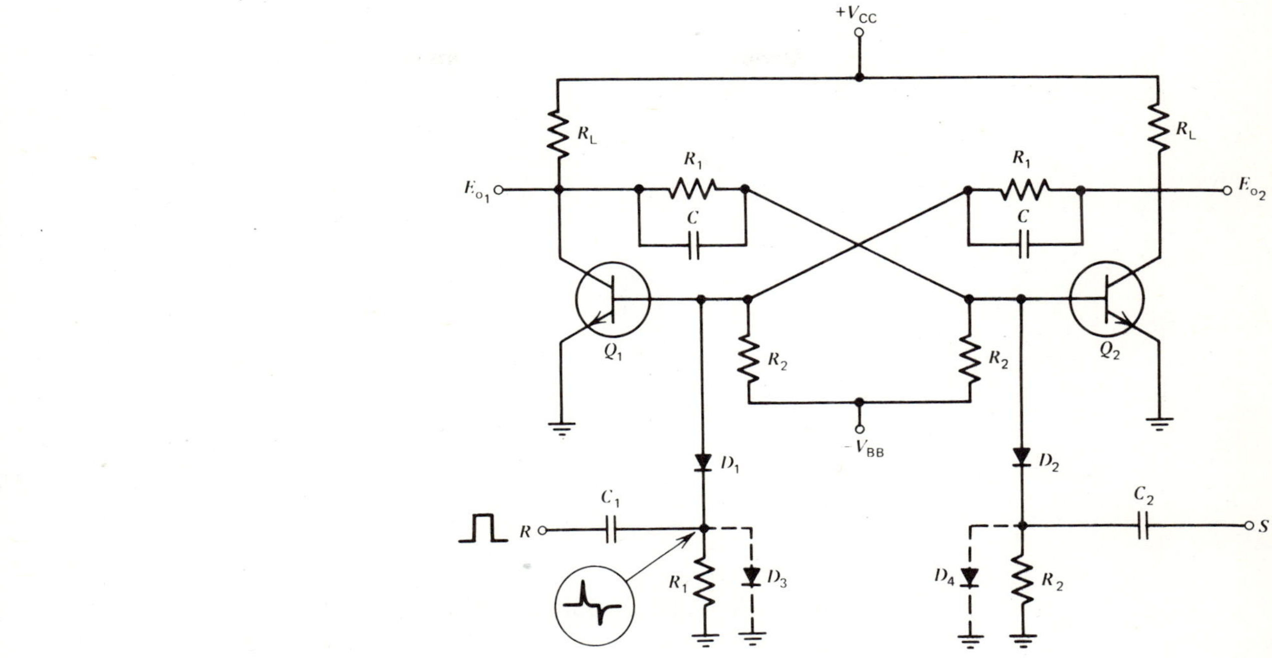

Figure 6-16 Base-triggered R-S flip-flop.

From the above discussion, it should be noted that the F/F of Fig. 6-16 is in the SET state when Q_1 is ON and Q_2 is OFF. When applied at the S terminal, an input trigger pulse will SET the F/F if it was not already SET. The process is similar to the RESET process explained above.

In the above-explained circuitry, the negative peak of the input trigger pulse was used to turn the transistor OFF. The positive peak of the differentiated input pulse may be used to turn the transistor ON if it was not already ON. This is easily accomplished simply by reversing the directions of diodes D_1 and D_2.

Note that C_1R_1 and C_2R_2 differentiate the input pulse waves that result in negative and positive spikes. In actual fact, all that is needed is only either the positive spike to turn the OFF transistor ON (*npn*) or the negative spikes to turn the ON transistor OFF (*npn*). The unnecessary spikes (positive or negative) may be *grounded* by connecting (pulse-steering) diodes instead of resistors R_1 and R_2 as shown by *dotted lines* in Fig. 6-16 (D_3 and D_4).

Figure 6-17 shows yet another method of R-S triggering, known as the collector-triggered R-S F/F. Assume that Q_1 is ON and Q_2

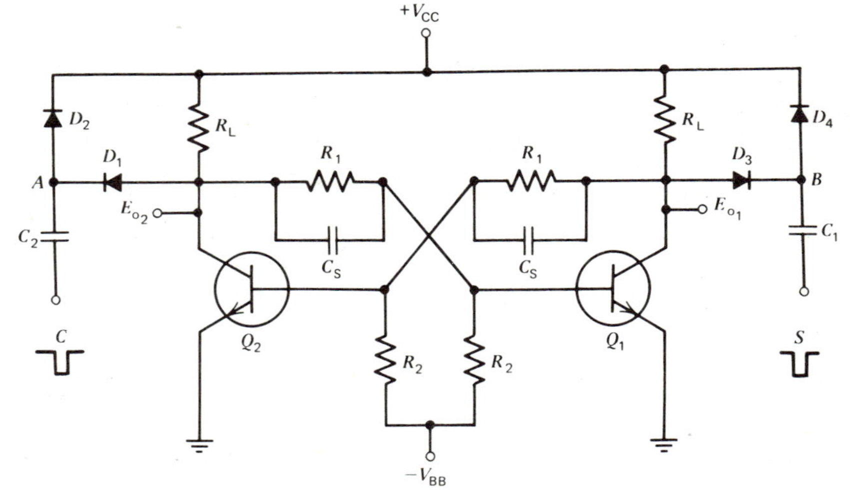

Figure 6-17 Collector-triggered R-S flip-flop.

is OFF. Voltage at point A is approximately $+V_{CC}$. A negative trigger pulse when applied at the C terminal lowers the voltage at point A by an amount equal to the amplitude of the negative trigger pulse. This forward-biases D_1 and the collector voltage of Q_2 is lowered. If it gets lowered to a point such that $V_{BE}(Q_1)$ becomes ≤ 0 V, Q_1 turns OFF, which in turn, turns Q_2 ON. If Q_1 was already OFF, then a negative trigger pulse at C will not change anything as far as the F/F is concerned. Thus, Q_1 OFF and Q_2 ON the CLEAR (C) state. In the SET (S) state, Q_1 is ON and Q_2 is OFF. A negative trigger pulse at the S terminal turns Q_2 OFF in the same way as it turned Q_1 OFF when applied to the C terminal.

6-14 T-Triggering

Toggle (T) *type triggering* is also known as *symmetrical triggering*. A trigger pulse applied to the T terminal simply flips the state of the F/F. This happens at every trigger pulse. Figure 6-18b shows the input trigger that is being applied to the T terminal. The trigger pulse that appears across D_1 is shown in Fig. 6-18c. Note that here the square wave has been converted to a series of negative spikes (C_1D_1 combination). Each spike (negative) occurs at the negative-going transition edge of the input wave.

Assume that Q_1 is ON and Q_2 is OFF before the trigger is applied. The base of Q_1 is at some positive voltage [V_{BE}(ON)] and the base of Q_2 is at some negative voltage [V_{BE}(OFF)], as shown in

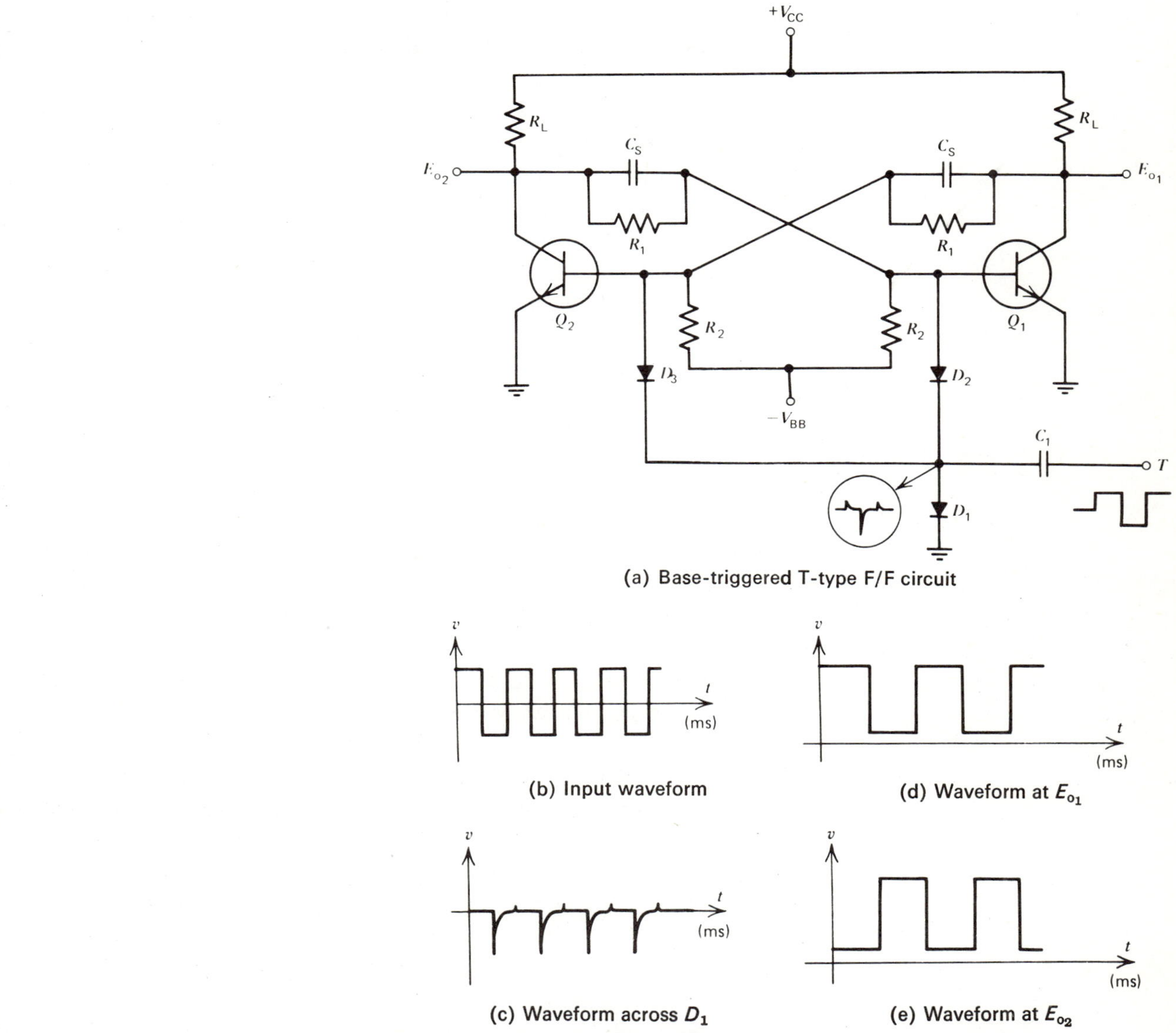

(a) Base-triggered T-type F/F circuit

(b) Input waveform

(d) Waveform at E_{o_1}

(c) Waveform across D_1

(e) Waveform at E_{o_2}

Figure 6-18 Base-triggered T-type F/F.

Fig. 6-18a. This means that D_1 is ON and D_2 is OFF (static conditions). As the negative spike appears across D_1, it gets transferred to the base of Q_1 because D_1 is ON. It is important that the amplitude of the negative spike should be less than V_{BE}(OFF), otherwise it will get transferred to the base of Q_2 as well as to the base of Q_1. Then it will be trying to turn both of the transistors OFF. This will result in an indeterminate situation as far as the resulting state of the F/F is concerned.

When operating properly, the negative spike turns the ON-transistor OFF by making its base momentarily negative (reverse bias). Note that if the first negative spike turns Q_1 OFF, the second spike will go to the base of Q_2, turning it OFF. The third spike will go to the base of Q_1, and so on. Thus, the trigger circuit really directs or *steers* the negative spike to the base of the ON transistor, which turns it OFF. D_2 and D_3 are called *steering* diodes.

Figures 6-18d and 6-18e show the voltage levels that will appear at E_{o_1} and E_{o_2} as a result of triggering pulses being applied at the T terminal. Note that the voltage wave at E_{o1} is 180 degrees out of phase with the one at E_{o2}. The frequency of the *output* waveform is *one-half* the frequency of the input waveform. Hence the name "divide-by-two or binary circuit." When connected in cascade, *two* F/Fs divide the input frequency by *four*. In this case the frequency of the output is *one-quarter* of that of the input. Figure 6-19 shows *three* cascaded (series-connected) F/Fs. The input frequency is eight times (2^3) the output frequency. Because of this frequency-dividing capability, on the order of 2^n, F/Fs are extensively used in digital binary counting circuits.

Figure 6-20 shows another method of accomplishing the TOGGLE action. The pulse-steering circuit consists of $C_1D_1R_5$ and $C_2D_2R_4$. This circuit *steers* the negative trigger pulse to the base of the *ON transistor*, which turns it OFF. Assume that Q_1 is ON and Q_2 is OFF. The input at T is at 0 V. Capacitor C_1 will have a very small voltage across it. Diode D_1 is ON. E_{o_2} is approximately equal to V_{CC}. C_2 is charged to approximately V_{CC} volts. D_2 is OFF.

A negative trigger pulse applied at T is transferred to the base of Q_1 through C_1 and D_1, which turns Q_1 OFF. It is important, however, that the amplitude of the negative input pulse should be less than V_{CC}; otherwise, the negative pulse, after getting transferred through C_2, may also turn D_2 ON. If this happens, the input negative pulse gets transferred to both of the bases, which results in an

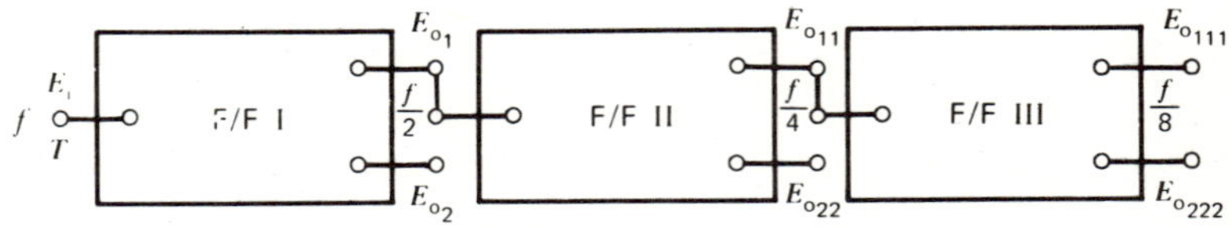

Figure 6-19 Divide by 8 counter.

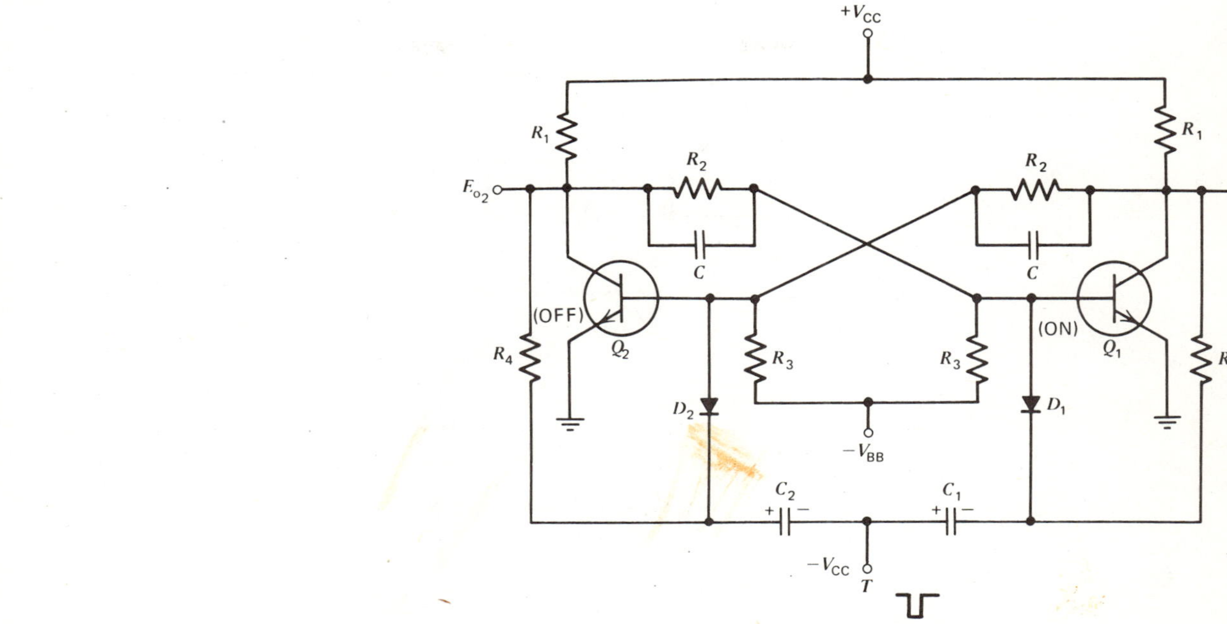

Figure 6-20 Collector-steered T-type F/F. Circuit showing how a negative-trigger input pulse turns the ON transistor OFF.

indeterminate output. Either one of the transistors may end up OFF. This, of course, is an undesirable situation. Each trigger pulse is supposed to flip the states.

Capacitors C_1 and C_2 are two *triggering* capacitors. Resistors R_4 and R_5 are chosen to be at least as high as 10 R_L so that they do not load the F/F circuit.

6-15 R-T-S–Triggering an Emitter-Coupled Flip-Flop

An R-T-S triggering circuit for an emitter-coupled F/F is shown in Fig. 6-21. To understand the R-S operation, consider that Q_1 is ON and Q_2 is OFF. Then D_1 and D_2 both are ON, and D_3 and D_4 are OFF. A negative trigger pulse applied to the R terminal is transferred to the base of Q_1 through D_1. It is important that the amplitude of the negative pulse should be *greater* than V_{BE}(ON) of Q_1. This reverse-biases the base-emitter junction of Q_1. Thus, Q_1 turns OFF. Any additional negative pulse at R will have no effect on Q_1, because it is already OFF. Thus, in RESET state, Q_1 is OFF and Q_2 is ON.

A negative trigger input pulse applied at the *S* terminal turns

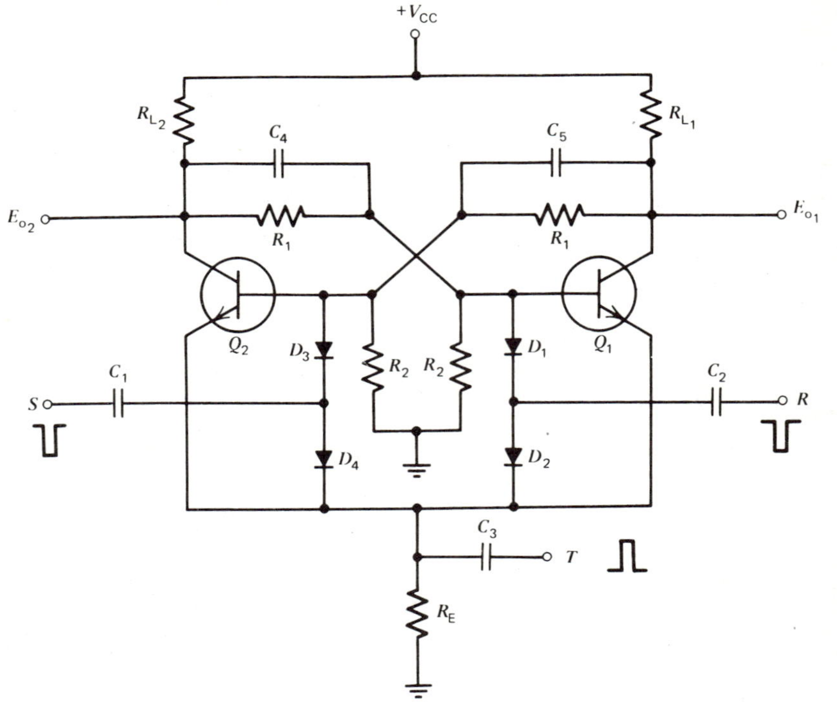

Figure 6-21 An R-T-S F/F circuit.

Q_2 OFF if it was not already OFF. The action is similar to the one explained above for the RESET operation. In the SET state, Q_1 is ON and Q_2 is OFF.

A positive trigger pulse applied at the T terminal TOGGLES the F/F. Assume that Q_1 is OFF and Q_2 is ON before the trigger pulse is applied. As the trigger pulse is applied, the differentiated wave (C_3R_3) that appears across R_E is superimposed on the static **dc** voltage across R_E. This additional positive spike also turns Q_2 OFF. Thus, momentarily, both of the transistors get cut OFF. As the positive spike disappears, both of the transitors start coming out of the cutoff region. Which transistor comes out of the cutoff first is controlled by the voltage across the commutating capacitors C_4 and C_5. Consider the situation before the trigger pulse was applied. Voltage across C_5 is $V_{CC} - V_E$, and across C_4 is approximately equal to V_E. The voltage across C_5 is larger than the one across C_4. Also note that the polarity of both voltages is such that it reverse-biases both transistors. Because the reverse-bias voltage across C_5 is significantly greater than the one across C_4, the positive trigger pulse puts Q_2 deep into the cutoff region as compared with Q_1. Hence, when both of the transistors start coming out of the cutoff region, Q_1 turns ON fairly quickly, thereby holding Q_2 OFF. Note that the state of the F/F has been flipped. C_4 and C_5 (capacitances) need to be considerably larger for the emitter-coupled F/F than the ones used for an ordinary F/F (Fig. 6-20).

6-16 The MOSFET Flip-Flop

A simple F/F circuit using two n-channel enhancement MOSFETs is shown in Fig. 6-22a. A practical yet equivalent MOSFET F/F is shown in Fig. 6-22b. In the practical MOSFET F/F, the drain current controlling resistor R_D is constructed by connecting the gate of a MOSFET to its drain terminal. When connected in this configuration, a MOSFET behaves like an ohmic resistor. The

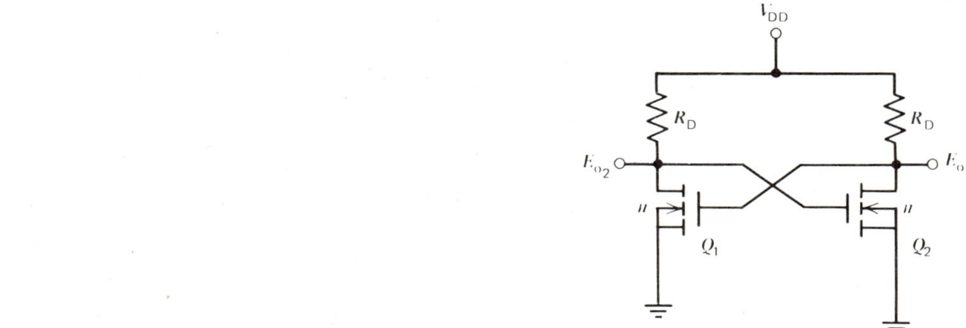

(a) A MOSFET F/F

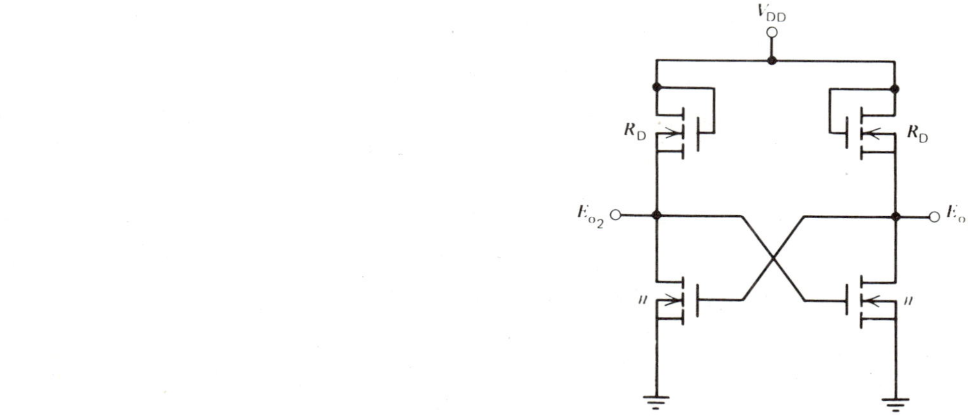

(b) A MOSFET F/F

Figure 6-22 Basic MOSFET F/F circuits.

operation of a MOSFET F/F is similar to the operation of a transistor F/F, described in Sec. 6-10. At any one time, one MOSFET is ON while the other is OFF. The two stable states are:

$$Q_1 \text{ OFF } (E_{o_1} = 0 \text{ V}), \quad Q_2 \text{ ON } (E_{o_2} = V_{DD}) \cdots \text{state 1}$$

$$Q_1 \text{ ON } (E_{o_1} = V_{DD}), \; Q_2 \text{ OFF } (E_{o_2} = 0 \text{ V}) \cdots \text{state 2}$$

The following conditions must be satisfied for safe operation.

1. When the MOSFET is ON, the drain current, I_D(ON), should be less than the maximum allowable value of the MOSFET's I_D(max).

$$I_D(\text{ON}) = \frac{V_{DD} - V_{DS}(\text{ON})}{R_D} \tag{6-14}$$

where:

V_{DD} is the supply voltage at drain (V)
R_D is the drain current limiting resistor (Ω)
V_{DS}(ON) is the drain-to-source voltage when MOSFET is ON (V)

2. V_{DD} should be less than the maximum allowable value of the drain-to-source voltage, V_{DS}(max).

3. V_{GS} (in this case, V_{DD}) should also be less than the breakdown voltage of the gate (BV_{GSS}).

Another version of a MOSFET F/F is shown in Fig. 6-23a. The negative supply voltage (V_{GG}) assures that the appropriate MOSFET does go OFF when required. Conditions that must be satisfied for proper and safe operation are:

1. I_D(ON) must be less than the rated I_D(max).
2. V_{DD} must be less than the rated V_{DS}(max).
3. V_{GS}(OFF), the voltage bias at the OFF gate, must be less than its breakdown voltage (BV_{GSS}).
4. The circuit design must assure that when one FET is ON, the other is OFF.

Let us investigate the ON/OFF condition as applied to the MOSFET F/F circuit shown in Fig. 6-23a. Assume that $V_{DS}(\text{ON}) = 0$ V, and that Q_1 is ON and Q_2 is OFF. Then,

1. $I_D(\text{ON}) = \dfrac{V_{DD}}{R_D}$

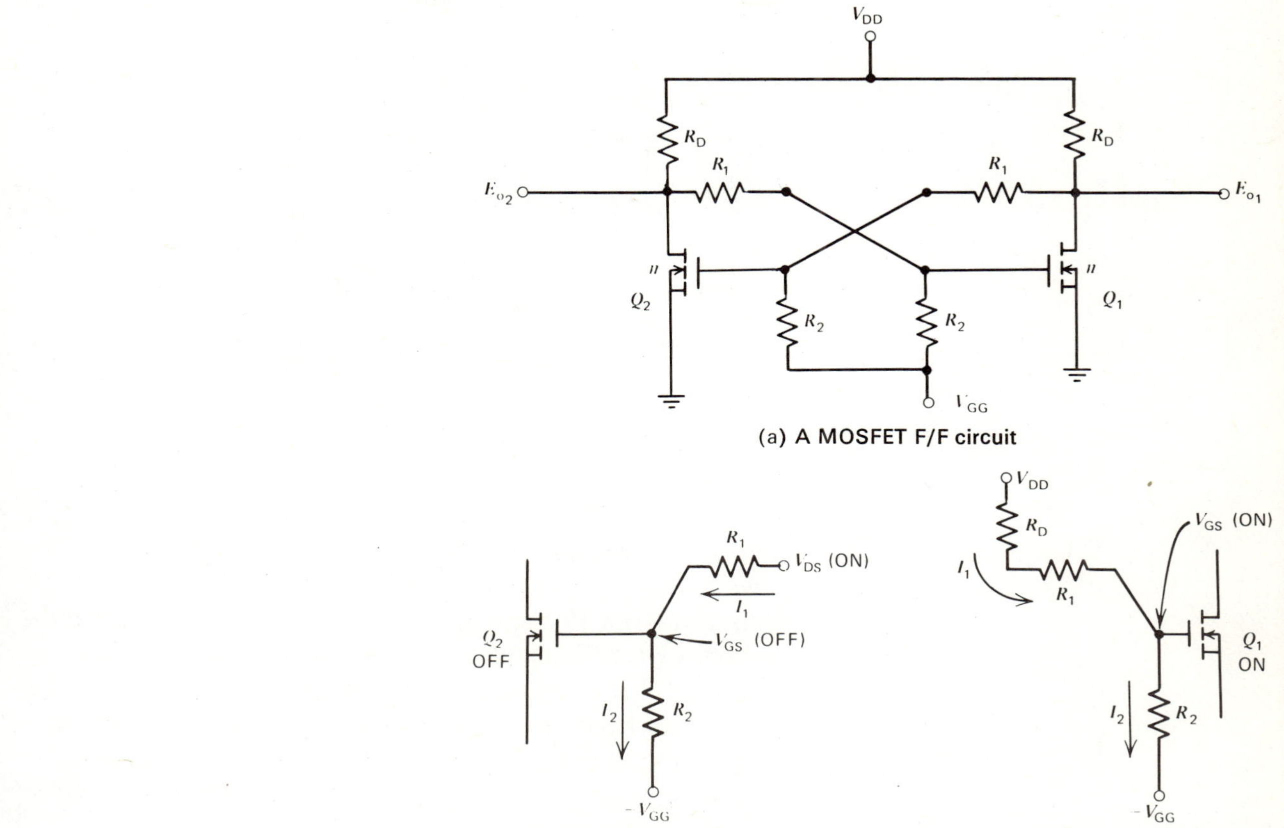

(b) Partial circuit for the OFF condition (c) Partial circuit for the ON condition

Figure 6-23 A MOSFET F/F and partial circuits for ON/OFF conditions.

2. The equivalent circuit for the OFF condition is shown in Fig. 6-23b. Note that the actual value of V_{GS}(OFF) must be lower than the minimum required to turn OFF the MOSFET.

$$I_1 = I_2$$

$$\frac{V_{DS}(\text{ON}) + V_{GS}(\text{OFF})}{R_1} = \frac{V_{GG} - V_{GS}(\text{OFF})}{R_2} \qquad (6\text{-}15)$$

3. The equivalent circuit for the ON condition is shown in Fig. 6-23c.

$$I_1 = I_2$$

$$\frac{V_{DD} - V_{GS}(\text{ON})}{R_D + R_1} = \frac{V_{GG} + V_{GS}(\text{ON})}{R_2} \qquad (6\text{-}16)$$

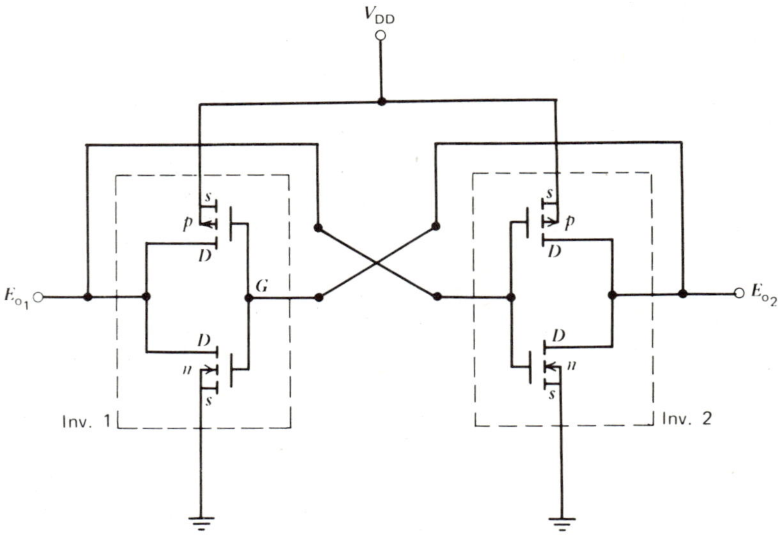

Figure 6-24 A COS/MOS F/F.

Equations (6-14), (6-15), and (6-16) are sufficient to design and analyze MOSFET F/F circuits of the type shown in Fig. 6-23a.

A COS/MOS F/F is constructed by connecting two COS/MOS inverters back-to-back, as shown in Fig. 6-24. At any one time, either $E_{o_1} = 0$ V and $E_{o_2} = V_{DD}$, or $E_{o_1} = V_{DD}$ and $E_{o_2} = 0$ V. Also, within each inverter at any one time, one MOSFET is ON while the other is OFF.

Assume that at a given time, $E_{o_1} = V_{DD}$ and $E_{o_2} = 0$ V. Under this state (condition):

1. The *p*-channel MOSFET in INV-1 is ON.
2. The *n*-channel MOSFET in INV-1 is OFF.
3. The *p*-channel MOSFET in INV-2 is OFF.
4. The *n*-channel MOSFET in INV-2 is ON.
5. $V_{GS1} = V_{DD}$ and $V_{GS2} = 0$ V.

The circuit conditions under the other stable state—that is, when $E_{o_1} = 0$ V and $E_{o_2} = V_{DD}$—are simply opposite to the ones listed above.

Example 6-7
A MOSFET (enhancement) F/F is shown in Fig. 6-25a. Assume $I_D(\text{max}) = 10$ mA, $V_{DS}(\text{max}) = 20$ V, $BV_{GSS} = \pm 30$ V, and $V_{DS}(\text{ON}) = 0.3$ V. Check by calculations whether the circuit functions safely and properly.

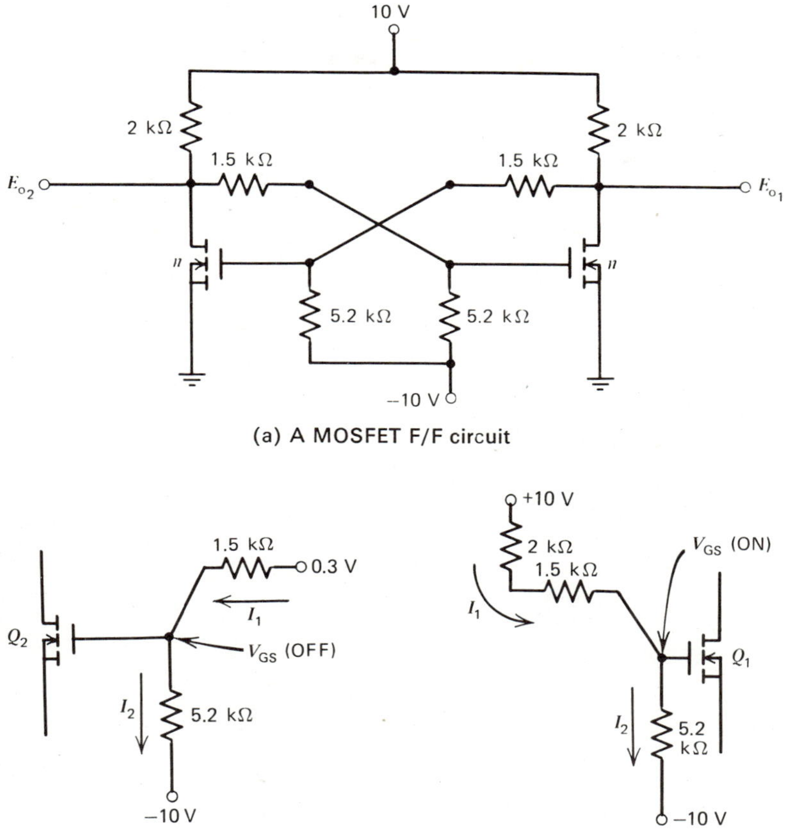

(b) Partial circuit for the OFF condition

(c) Partial circuit for the ON condition

Figure 6-25 A MOSFET F/F and partial circuits for the ON/OFF conditions for Ex. 6-7.

Solution

$$I_D(\text{ON}) = \frac{V_{DD} - V_{DS}(\text{ON})}{R_D} \tag{6-14}$$

$$= \frac{10 - 0.3}{2\text{ k}\Omega}$$

$$\cong 5 \text{ mA}$$

The two MOSFETs are safe because $I_D < 10$ mA, $V_{DD} < 20$ V, and $V_{DS} < 30$ V. To check for the ON/OFF conditions, assume that Q_1 is ON and Q_2 is OFF. A partial circuit for the OFF conditions is shown in Fig. 6-25b.

$$
\begin{aligned}
V_{GS}(\text{OFF}) &= V_{DS}(\text{ON}) - V_{1.5k} \\
&= 0.3 - \left[\frac{1.5\ \text{k}\Omega}{(1.5 + 5.2)\ \text{k}\Omega}\right](10 + 0.3) \\
&= 0.3 - 2.3 \\
&= -2.0\ \text{V}
\end{aligned}
$$

which means that Q_2 is OFF.

A partial circuit for the ON conditions is shown in Fig. 6-25c.

$$
\begin{aligned}
V_{GS}(\text{ON}) &= V_{DD} - (V_{2k+1.5k}) \\
&= 10 - (10 + 10)\left[\frac{(2 + 1.5)\ \text{k}\Omega}{(2 + 1.5 + 5.2)\ \text{k}\Omega}\right] \\
&= 10 - 8.08 \\
&= 1.92\ \text{V}
\end{aligned}
$$

which means that Q_1 is ON.

It is concluded from the above calculations that the F/F circuit is safe and functions properly, assuming that the MOSFET saturates when its $V_{GS} = 1.92$ V.

The triggering circuits discussed in Secs. 6-12, 6-13, 6-14, and 6-15 are applicable to MOSFET F/Fs as well (with minor modifications). F/Fs in COS/MOS integrated circuits are usually constructed from logic gates. For example, two NOR gates connected back-to-back form an R-S F/F. The reader is referred to Ch. 11 for a discussion of logic gates.

6-17 IC F/Fs

Integrated circuits (ICs) are covered in detail in Ch. 11. The F/F circuit discussed in Sec. 6-9 consists of two cross-coupled single-input RTL*NOR gates. Note that a single-input NOR gate behaves like an inverter (NOT gate). The 802 R-S F/F (IC) shown in Fig. 6-26 is similar to the F/F circuit discussed in Sec. 6-9. The R-S triggering arrangement is the level-triggering type discussed in Sec. 6-12.

The 7476 IC chip shown in Fig. 6-27 is a more versatile and a commonly used J-K F/F in TTL IC circuits. This dual J-K F/F

* See Sec. 11-9.

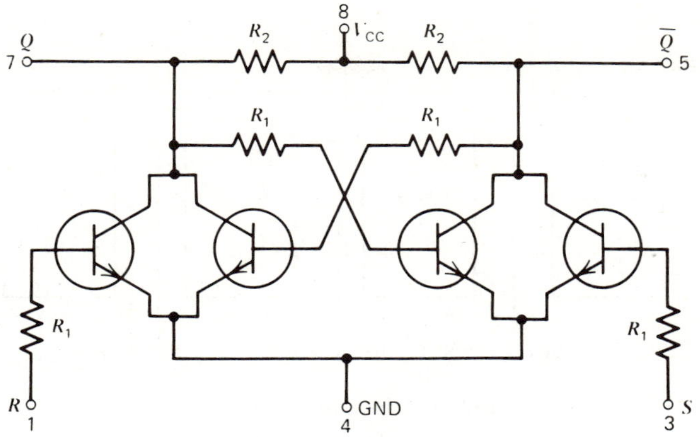

(a) Circuit diagram

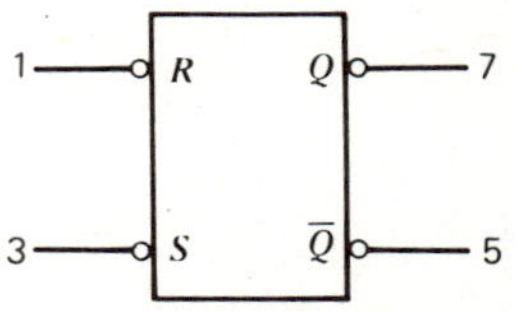

(b) Block diagram

R	*S*	*Q*
0	0	0
0	1	1
1	0	0
1	1	0

(c) State table

Figure 6-26 An IC R-S F/F (802).

operates on the *master-slave* principle. The positive-going (*leading*) edge of the input clock-pulse transfers the information (SET, RESET, or TOGGLE) from the input to the *master* section of the F/F. The negative-going (*trailing*) edge of the clock pulse transfers the information from the master to the *slave* part (output) of the F/F.

The two F/Fs in 7476 are identical. Each has *J-K* input terminals as well as *R-S* input terminals. The CLEAR and SET operations are accomplished via either *SET* and *RESET* terminals or *J-K*

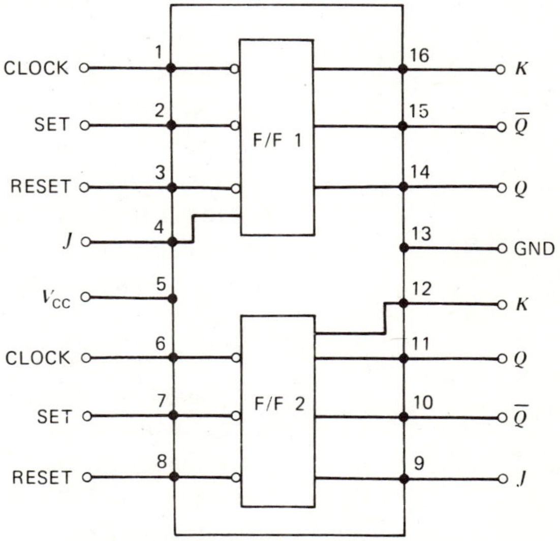

Figure 6-27 The 7476, a dual J-K F/F

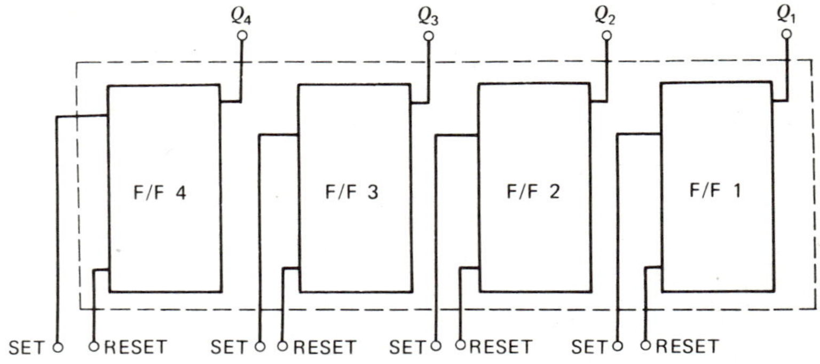

Figure 6-28 A 4-bit binary register.

terminals. The *R-S* operation takes precedence over the *J-K* operation. The F/F functions as a J-K F/F when $\overline{\text{RESET}}$ and $\overline{\text{SET}}$ terminals are connected to V_{cc}(HIGH).

The two most popular applications of F/Fs are in *registers* and *counters*. A register is simply a storage device for a number of bits in which each F/F stores one bit of information (1 or 0). Thus, a 10-bit register consists of 10 individual F/Fs, each able to store one bit of information at a time. Figure 6-28 shows a 4-bit register. Any number from 0000 to 1111 (binary)* may be stored in it simply by setting or resetting the appropriate F/Fs. Assume that F/F 1 is SET (1), F/F 2 is RESET (0), F/F 3 is RESET (0) and F/F 4 is SET (1). Therefore, the binary number stored in this register is 1001.

Another major application of F/Fs is in digital counters. A counter is used to count pulses or events. The TOGGLE capability of an F/F is used to perform this function. When a pulse wave is used to TOGGLE an F/F, the output frequency becomes one-half the input frequency, as already shown in Fig. 6-18. It is important to realize that an F/F TOGGLES (switches states) either at the positive-going or the negative-going edges of the input triggering pulse.

Figure 6-29 shows three F/Fs connected to form an 8-bit (111 in binary)† counter. In fact, it is one form of a register. The pulses to be counted are clocked into the F/F 1, which represents the least significant bit.

* See Appendix 5 for binary numbers.

† See Appendix 5 for binary numbers.

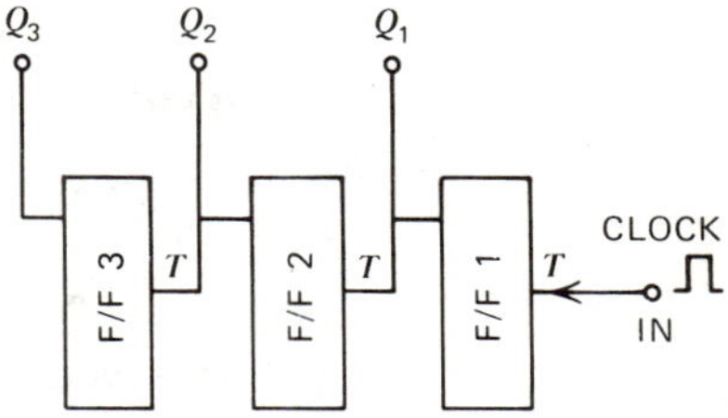

(a) Wiring diagram

Clock Pulse	Q_3	Q_2	Q_1
1	0	0	1
2	0	1	0
3	0	1	1
4	1	0	0
5	1	0	1
6	1	1	0
7	1	1	1
8	0	0	0
9	0	0	1

(b) State table for the three F/Fs as successive clock-pulses are applied.

Figure 6-29 A three-bit binary counter.

Assume that an F/F is TOGGLED by the *trailing* edge of the clock pulse and that it is a *positive* logic system. Each F/F TOGGLES when the input to it goes from 1 to 0.

The counter operates as follows:

1. Initially, all F/Fs are in the CLEAR state, that is, $Q_1 = 0$, $Q_2 = 0$, and $Q_3 = 0$. This initial condition is shown by the first line in Fig. 6-29b. The register reads 000.
2. The first clock pulse SETS F/F 1 while the other two F/Fs remain CLEAR. The register reads 001 as shown by the second line in Fig. 6-29b.
3. The second clock pulse CLEARS F/F 1 but SETS F/F 2, because Q_1 goes from 1 to 0. The register now reads 010, as shown by the third line in Fig. 6-29b.
4. A third pulse SETS F/F 1 while F/F 2 and F/F 3 remain unchanged. The register reads 011. Note that the counter is counting in binary—000, 001, 010, 011, and so on—as pulses are applied.
5. The seventh pulse SETS all three F/Fs and the register reads 111, as shown by the seventh line in Fig. 6-29b.

6. The eighth pulse CLEARS all three F/Fs.
7. The ninth pulse starts the counting process all over again.

Only one counter and a register are discussed above, in order to illustrate the basic principles involved. In practice, various types of registers and counters are constructed using F/Fs. ICs containing one to several F/Fs on one chip are now available. A particular IC containing a specific number of F/Fs is usually chosen to suit the requirements.

6-18 Summary

1. A multivibrator (MV) is one of the most important components of pulse, switching, and binary logic circuits.
2. An MV has two outputs; one at the 0 output voltage level (LOW) and the other at the 1 voltage level (HIGH).
3. The two output voltage levels of an MV represent the two logic situations, TRUE and FALSE, or "1" and "0", respectively.
4. When one output of an MV is at the 1 level, the other is at the 0 level. This is always true.
5. There are three types of MVs: the bistable (flip-flop), the monostable (one-shot), and the astable (free-running).
6. A flip-flop (F/F) is a bistable MV.
7. An F/F can be SET, RESET, or TOGGLED by an external trigger input pulse.
8. An F/F circuit consists of two inverter circuits connected back-to-back.
9. An F/F circuit is designed such that at any one time, one transistor is ON while the other is OFF.
10. An F/F circuit will function properly if one transistor is OFF when the other is ON.
11. A SET-RESET (S-R) operation may be accomplished using either transistor or diode triggering circuits.
12. An S-R operation may be accomplished either by **dc** input triggers or **ac** trigger pulses.
13. An F/F may be toggled using a pulse-steering network.
14. The speed-up (or commutating) capacitors used in a F/F improve the switching speed, as well as assisting in the switching process when a trigger input is applied.
15. IC F/Fs are used almost exclusively in electronic circuitry today. However, these are in fact miniaturized discrete-component circuits.
16. F/Fs are used in binary counters, registers, and memories.
17. Important equations:

$$R_L \geqq \frac{V_{CC} - V_{CE}(\text{ON})}{I_C(\text{max})} \qquad (6\text{-}1)$$

$$I_C(\text{ON}) = \frac{V_{CC} - V_{CE}(\text{ON})}{R_L} \tag{6-2}$$

$$I_B(\text{ON}) = \frac{I_c(\text{ON})}{\beta} \tag{6-3}$$

$$\frac{V_c(\text{ON}) + V_{BE}(\text{OFF})}{R_2} = \frac{V_{BB} - V_B(\text{OFF})}{R_3} \tag{6-4}$$

$$\frac{V_{CC} - V_B(\text{ON})}{R_L + R_2} - \frac{V_{BB} + V_B(\text{ON})}{R_3} \geqq I_B(\text{ON}) \tag{6-5}$$

$$C_s \geqq \frac{Q_t}{V_{CC}} \tag{6-6}$$

$$V_L = V_E + V_{CE}(\text{ON}) \tag{6-8}$$

$$R_E = \frac{V_E}{I_E(\text{ON})} \tag{6-9a}$$

$$R_E = \frac{V_L - V_{CE}(\text{ON})}{I_c(\text{ON}) + [I_C(\text{ON})/\beta]} \tag{6-9b}$$

$$R_L = \frac{V_{CC} - [V_E + V_{CE}(\text{ON})]}{I_C(\text{ON})} \tag{6-10a}$$

$$R_L = \frac{V_{CC} - V_L}{I_c(\text{ON})} \tag{6-10b}$$

$$\frac{V_B(\text{OFF})}{R_2} = \frac{V_L - V_B(\text{OFF})}{R_1} \tag{6-11}$$

$$\frac{V_{CC} - V_B(\text{ON})}{R_L + R_1} - \frac{V_B(\text{ON})}{R_2} \geqq I_B(\text{ON}) \tag{6-12}$$

6-19 Glossary

Astable (free-running) MV. Unstable in *both* of the MV states. The circuit continuously flips back and forth between the two states.

Bistable (flip-flop) MV. Stable in *both* of the MV states. Unless triggered by an external pulse, the circuit remains in its existing state.

Commutating capacitors. The capacitors (connected across the base resistors) that direct the current flow when an F/F is triggered. They help in the switching process of a flip-flop.

J-K flip-flop. A special type of F/F that has only two inputs, named *J* and *K*. A trigger pulse applied separately at *J* or *K* serves the same purpose as SET and RESET. The F/F gets TOGGLED when a trigger pulse is applied simultaneously at both the *J* and the *K* terminals.

Monostable (one-shot) MV. Stable in one state and unstable in the other state. Usually, a trigger pulse is used to switch it into the unstable state from the stable state. Then after a certain period of time, it switches back on its own to its stable state.

Multivibrator (MV). A circuit that has two outputs, each at a certain voltage level. There are only two voltage levels that either one of the outputs may have. When one output is at one voltage level, the other is always at the other voltage level.

Pulse steering circuit. The circuit that steers the input trigger pulse to the appropriate base (or collector) of a transistor so that the TOGGLE-type switching of an F/F takes place.

R-S flip-flop. The RESET and SET (*R-S*) type of F/F has two input terminals. A trigger input applied at one terminal (*R*) assures that the F/F is in state #1 and an input trigger applied at the other terminal (*S*) assures that the F/F is in state #2.

R-T-S flip-flop. Has three input terminals, labeled as RESET, TOGGLE, and SET (*R-T-S*). It has SET and RESET capabilities as well as the TOGGLE capability.

States of an MV. An MV has two states. In one state, the output voltage level at output terminal #1 is HIGH and that at output terminal #2 is LOW. In the second state, the output voltage level at output terminal #1 is LOW and that at output terminal #2 is HIGH.

T-type flip-flop. This has one input terminal labeled *T* (TOGGLE). An input trigger pulse applied at the *T* terminal switches the state of the F/F.

Triggering an MV. This is the process involved in switching the state of an F/F.

6-20 Recapitulation

Return to the Objectives and Self-Evaluation Questions given at the beginning of this chapter and see how well you can answer them. If you cannot answer certain questions satisfactorily, place a check mark (✓) next to them and review the appropriate parts of the chapter to find the answers. No attempt should be made to solve the Problems (Sec. 6-21) before answering the Self-Evaluation Questions. Ability to solve the Problems should serve as a test and feedback regarding your understanding of the material presented in this chapter. Answers to the Problems are given on p. 522.

6-21 Problems

6-1 In a simple F/F circuit, $V_{CC} = 24$ V, $V_{CE}(\text{ON}) = 0.2$ V, $I_C(\text{ON}) = 11.9$ mA, and $\beta = 150$. Find:
a. The value of the load resistor.
b. The minimum value of base current required for saturation ($I_B(\text{ON})$).

6-2 Consider the circuit shown in Fig. 6-6a. $V_{CC} = 12$ V, $V_{BB} = -6$ V, $V_{BE}(\text{OFF}) = -0.5$ V, $h_{fe} = 100$, and $I_C(\text{max}) = 20$ mA. Assume the transistors to be *ideal* as far as $V_{BE}(\text{ON})$ and $V_{CE}(\text{ON})$ are concerned. Find R_1, R_2, and R_3.

6-3 Repeat Prob. 6-2 for standard *silicon* transistors.

6-4 Repeat Prob. 6-3 and find *standard* resistance values for R_1, R_2, and R_3.

6-5 Consider the flip-flop circuit of Fig. 6-10a. $V_{CC} = -15$ V, $V_{BB} = +15$ V, $V_{BE}(\text{OFF}) = +1.0$ V, $\beta = 150$, and $I_C(\text{ON}) = 15$ mA. Assume the transistors to be *ideal* as far as $V_{CE}(\text{ON})$ and $V_{BE}(\text{ON})$ are concerned. Calculate R_1, R_2, and R_3.

6-6 Repeat Prob. 6-5 for standard *germanium* transistors.

6-7 Consider the F/F circuit of Fig. 6-8a. $R_1 = 470\ \Omega$, $R_2 = 50\ \text{k}\Omega$, and $R_3 = 330\ \text{k}\Omega$. Give specifications for the two germanium transistors Q_1 and Q_2 that will ensure that the F/F functions properly.

6-8 Refer to Prob. 6-7. For the two transistors to be used, assume $I_C(\text{max}) = 30$ mA, $V_{BE}(\text{ON}) = 0.5$ V, $\beta = 50$, and $V_{CE}(\text{ON}) = 0.2$ V. Analyze the circuit and check whether it will function properly as a saturated-type F/F.

6-9 Design an emitter-coupled F/F circuit similar to the one shown in Fig. 6-11a. $V_L = 4$ V and $V_H = 12$ V. For the two silicon transistors, $I_C(\text{ON}) = 20$ mA and $\beta = 50$. $V_{BE}(\text{OFF})$ is to be designed as -0.5 V.

6-10 The F/F circuit designed in Prob. 6-9 is being TOGGLED by the trigger pulse shown in Fig. 6-30. Sketch the voltage waveforms that appear at the two outputs.

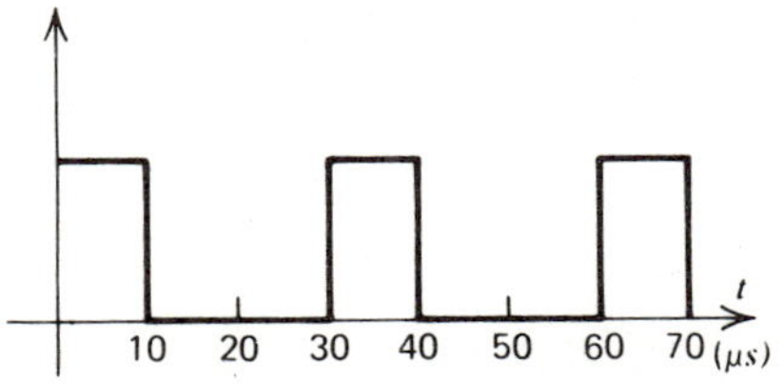

Figure 6-30 Input waveform for Prob. 6-10.

6-11 Waveforms shown in Fig. 6-31 are applied to the *R* and *S* terminals of an R-S F/F. Assume that in the SET state, E_{o_1} is HIGH and E_{o_2} is LOW. Sketch the two output waveforms.

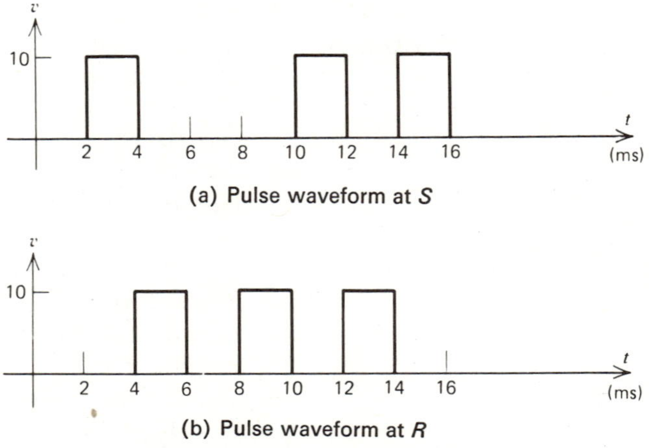

Figure 6-31 Waveforms for Prob. 6-11.

6-12 A 10-kHz square wave is applied to a binary counter circuit that uses two T-type F/Fs connected in series. Find the frequency of the output waveform.

6-13 Design a MOSFET F/F so that the two output voltage levels are +12 V and +0.3 V. For the two *n*-channel MOSFETs to be used, assume $I_D(\text{max}) = 20$ mA, $R_D(\text{ON}) = 50\Omega$, $V_{DS}(\text{max}) = +30$ V, $BV_{GSS} = \pm 30$ V, $V_{GG} = -12$ V, $V_{GS}(\text{OFF}) = -3$ V, and $V_{GS}(\text{ON}) = +2$ V.

6-14 Repeat Prob. 6-13. This time the two output voltage levels are to be +10 V and +0.5 V.

6-15 Design a COS/MOS F/F so that the two output voltage levels are +5 V and 0 V.

6-22 Experiment

Title **The R-T-S Operation of a Flip-Flop (F/F)**

Theory A multivibrator circuit has two outputs. At any one time, one output is HIGH and the other output is LOW. Let us call one output E_{o_1} and the other E_{o_2}. Then, at any one time, either E_{o_1} = HIGH and E_{o_2} = LOW (state 1) or E_{o_1} = LOW and E_{o_2} = HIGH (state 2). A bistable multivibrator, commonly known as a flip-flop (F/F), is stable in both of the states. As the circuit is energized, it assumes one of the two states. It stays in that state indefinitely unless triggered externally.

A transistor F/F consists of two inverters connected back-to-back. The two output levels are V_{CC} and V_{CE}(ON). At any one time, one transistor is ON while the other is OFF.

When designing or analyzing a discrete transistor F/F, one must assure that:

1. The two discrete transistors used in the F/F are a matched pair.
2. I_C(ON) is less than the rated I_C(max).
3. Conditions for one transistor being ON while the other is OFF, as stated in Eqs. (6-4) and (6-5), must be satisfied.
4. Proper values for the commutating capacitors (Eq. 6-6) and the coupling capacitors in the triggering circuitry are chosen.

IC F/Fs, similar to 7476, are used in most pulse circuits today. These are economical, readily available, and compatible with other IC circuits.

Equipment Dual-beam oscilloscope: PHILIPS model 3230, or equal.
±DC power supply: ANATEK model 25–2.00, or equal.
Multifunction generator: EXACT model 126, or equal.
Multimeter: RCA model WV 98C, or equal.
Four 2N4401 transistors.
Three switching diodes.
Two momentary contact switches.
Resistors: two each of 510 Ω, 22 kΩ, 27 kΩ, and 100 kΩ.
Capacitors: two each of 0.01 μF and 250 pF.
One 7476 IC F/F.

Procedure

1. Connect the F/F circuit shown in Fig. E6-1. Measure and record in Table E6-1 the two output voltage levels (E_{o_1} and E_{o_2}).

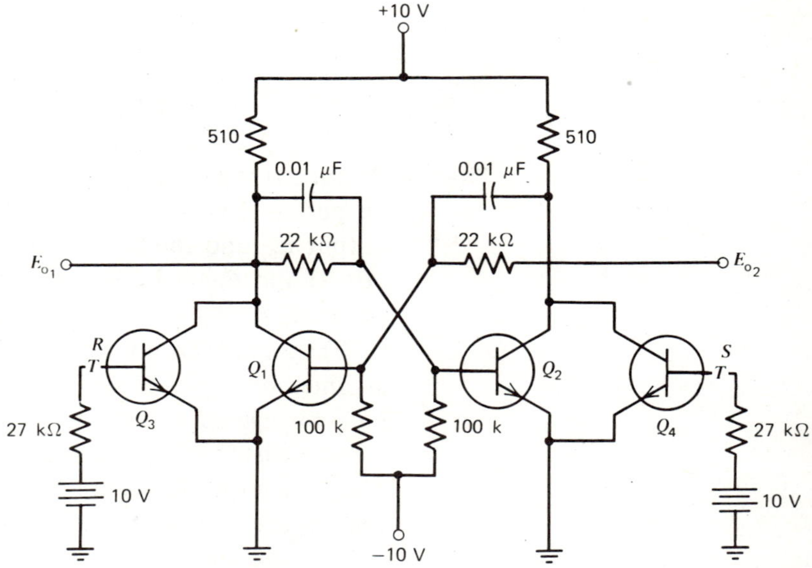

Figure E6-1 R-S F/F.

2. Measure and record in Table E6-1 the base voltages of both of the transistors [V_{BE}(ON) and V_{BE}(OFF)] of the F/F.
3. Depress the *R* switch momentarily. The F/F is now in RESET state. Measure and record in Table E6-2 the two output voltage levels.
4. Depress the *S* switch momentarily. Measure and record in Table E6-2 the two output voltage levels. The F/F is in the SET state.
5. Construct the TOGGLE steering circuit as shown in Fig. E6-2. Connect it to the F/F circuit of Fig. E6-1.

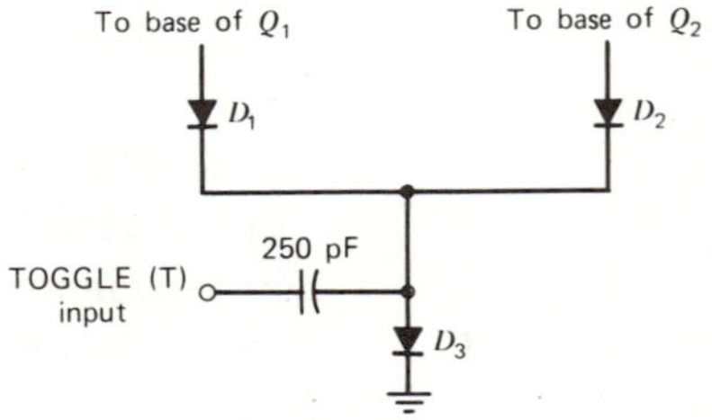

Figure E6-2 Pulse steering circuit for TOGGLE operation.

6. Apply a 10 kHz square wave of sufficient amplitude to the T terminal. Adjust the input amplitude until proper switching is obtained. Note that if proper switching is taking place, the output frequency is one-half the input frequency.
7. Observe (CRO) and draw in Table E6-3 the input as well as the output waveforms.
8. Increase the input frequency slowly. Find and record the maximum input frequency at which switching still takes place properly. This is the maximum switching frequency of the circuit (F/F). Record f_{max} = ————. Note that the input amplitude may have to be re-adjusted to find f_{max}.
9. Connect one-half of the 7476 F/F as shown in Fig. E6-3. For the remainder of this experiment, assume +5 V = 1 = HIGH and 0 V = 0 = LOW.

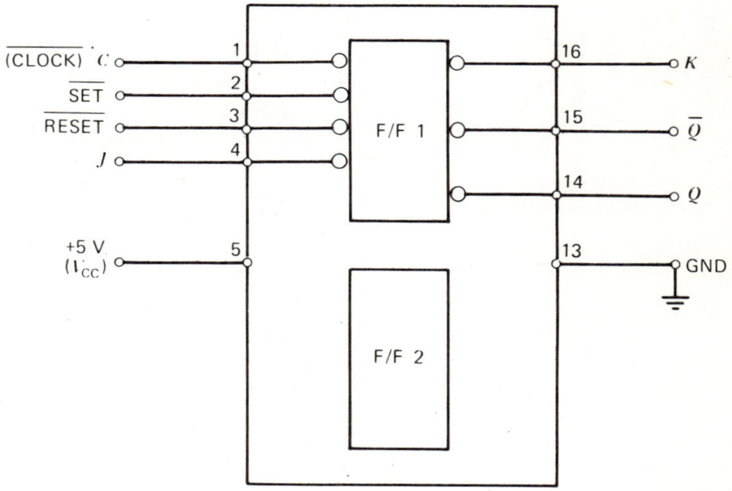

Figure E6-3 The 7476 F/F for Experiment 6.

10. Connect both $\overline{\text{SET}}$ and $\overline{\text{RESET}}$ to HIGH. Make J = 1 and K = 0. Apply a pulse (of +5 V) at C.
11. Measure and record in Table E6-4 the output levels at Q and $\overline{\text{Q}}$.
12. Repeat Procedure steps 10 and 11, but this time make J = 0 and K = 1.
13. With $\overline{\text{SET}}$ = $\overline{\text{RESET}}$ = J = K = 1, apply a 0 V to 5 V, 10-kHz square wave at C. Observe (CRO) and record in Table E6-5 the input as well as the two output (Q and $\overline{\text{Q}}$) waveforms.
14. Disconnect and return all test equipment to its proper place.

Calculations 1. From the transistor manual, find I_C(max), V_{CE}(ON), and V_{BE}(ON). Record V_{CE}(ON) and V_{BE}(ON) in Table E6-1.

2. Calculate V_{CE}(OFF) and V_{BE}(OFF) for the F/F circuit tested (see Ex. 6-4). Enter your results into Table E6-1.
3. Calculate the percentage error between the observed and the theoretical values recorded in Table E6-1. Record your results therein.

Tabulation of Results

Table E6-1
DC Collector and Base Voltage of an F/F

	Observed	Theoretical	Percentage Error
E_{o_1}			
E_{o_2}			
V_{BE}(ON)			
V_{BE}(OFF)			

Table E6-2
R-S States of an F/F

State	E_{o1}	E_{o2}
R		
S		

Table E6-3
Input/Output Waveforms of an F/F

	Waveforms
Input	
E_{o_1}	
E_{o_2}	

Table E6-4
J-K Operation

J	K	Q	$\bar{Q}$
1	0		
0	1		

Table E6-5
Input/Output Waveforms

Input Waveform	u	t	(ms)
Output Waveform at Q	u	t	(ms)
Output Waveform at Q			

Questions

1. Calculate I_C(ON) for the F/F circuit tested. Is I_C(ON) $<$ I_C(max)? Explain why this must be so for safe operation.
2. Examine Table E6-1. Explain whether or not the turn-ON and turn-OFF conditions have been satisfied.
3. Table E6-2 is intended to:
   a. Establish the two stable states (R-S) of the F/F.
   b. Test whether or not this circuit does function properly as an F/F. Explain whether or not this has been accomplished.
4. Examine the waveforms recorded in Table E6-3. Is this what you expected? Explain.
5. In Procedure step 8, you recorded the maximum switching frequency of the F/F. Explain its significance.

Discussion

1. Summarize your conclusions from the observed and the theoretical values in Table E6-1.
2. Account for discrepancies between the theoretical and the experimental values in Table E6-1.
3. Discuss why the output frequency is one-half the input frequency in Table E6-3.
4. Compare the discrete components F/F with its IC counterpart.

Chapter 7 Monostable and Astable Multivibrators

7-1 Objectives

The objectives of this chapter are to:

1. Describe the general characteristics of a monostable multivibrator (MV).
2. Explain the operation of a monostable MV circuit.
3. Show how to design a transistor monostable MV circuit.
4. Identify components that control the recovery time of a monostable MV.
5. Analyze a given monostable MV circuit for proper operation.
6. Explain the operation of an astable MV circuit.
7. Show how to design a simple astable MV circuit.
8. Analyze a given astable MV circuit for proper operation.
9. Discuss the synchronizing techniques of an astable MV.
10. Explain the operation of a MOSFET monostable MV.
11. Explain the operation of a MOSFET astable MV.
12. Discuss the IC monostable MV.

7-2 Self-Evaluation Questions

Watch for the answers to the following questions as you read the chapter. They should help you to understand the material presented. When you have finished the chapter, return to this section and be sure that you can answer all the questions. Then test your understanding by working out solutions to the Problems.

It is strongly recommended that you answer these questions satisfactorily before attempting to solve the Problems (Sec. 7-17).

1. Describe the basic characteristics of a monostable MV.
2. Describe, in detail, the operation of a monostable MV.
3. Design a sample monostable MV circuit.
4. Identify the components that control the output pulse duration.
5. Show a method of calculating the output pulse width.
6. What is meant by the term *recovery time*, and how is it calculated?
7. What is the significance of recovery time as far as the time interval between any two input trigger pulses is concerned?
8. Design a triggering circuit that may be used to trigger a monostable MV.
9. Describe the basic characteristics of an astable MV.
10. Describe, in detail, the operation of an astable MV.
11. Design a sample astable MV circuit.

12. Analyze a given circuit of an astable MV for proper operation.
13. Identify the components that control the duration of each half-cycle of the output waveform.
14. What is the significance of recovery time as far as the maximum possible frequency of the output of an astable MV is concerned?
15. Describe the process involved in synchronizing an astable MV.
16. Describe the operation of a monostable MOSFET MV.
17. Describe the operation of a MOSFET astable MV.
18. Describe the characteristics of monostable MVs available in TTL IC circuits.

7-3 A Monostable Multivibrator (MV)

A *monostable MV* is stable in one state and unstable in the other. Normally, the circuit is in its *stable* state. It is usually triggered into its unstable state by an external trigger pulse. After the input trigger pulse disappears, the circuit eventually returns to its stable state on its own. The time duration for which the circuit remains in the unstable state, as well as the voltage levels at the outputs, depend entirely upon the internal design of the monostable circuit and *not* on the input trigger pulse.

Consider the block diagram of a monostable multivibrator shown in Fig. 7-1a. Assume that in its stable state, $E_{o_1} = 0$ V and $E_{o_2} = +5$ V.

The output levels remain the same until a trigger input is applied at E_i. The input trigger (Fig. 7-1b) simply flips the state of the MV and switches it into its unstable state, that is, E_{o_1} becomes $+5$ V and E_{o_2} becomes 0 V (Figs. 7-1c and 7-1d). Time (t_p) for which E_{o_1} remains at $+5$ V and E_{o_2} at 0 V is determined by the internal circuit of the MV. This will be explained in Sec. 7-4. The circuit switches back to its stable state after t_p has elapsed.

A monostable MV may be a level-triggered or an edge-triggered circuit. A level-triggered MV switches into its unstable state as soon as the input trigger pulse reaches a certain voltage level. The input/output waveforms shown in Figs. 7-1b, 7-1c, and 7-1d display this condition. The outputs change states at the leading edge of the input trigger pulse.

An edge-triggered MV circuit may be designed to be switched by either the leading edge or the trailing edge of the input trigger pulse. The waveforms in Figs. 7-1b, 7-1c, and 7-1d show the *leading*-edge-triggered circuit behavior. The waveforms are the same as in the level-triggered circuit. If the circuit is designed to be

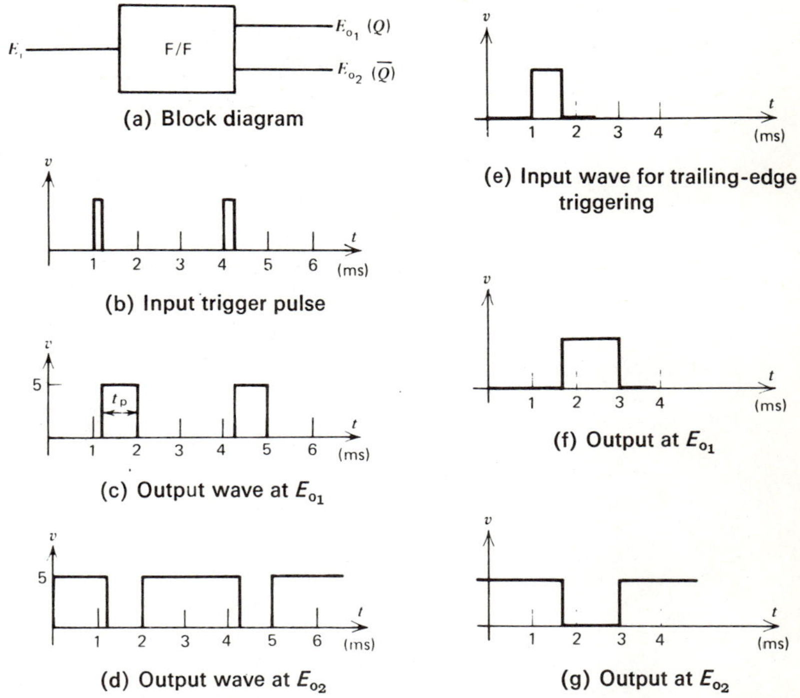

Figure 7-1 Monostable MV and input/output waveforms.

triggered by the *trailing* edge of the input trigger pulse, the output waveforms will be similar to the ones shown in Figs. 7-1e, 7-1f, and 7-1g.

Note from Fig. 7-1 that:

1. A level-triggered or a leading-edge-triggered monostable MV may be used as a *pulse stretcher* or a *pulse reshaper*.
2. A trailing-edge-triggered monostable MV may be used as a *pulse-delay* circuit.

7-4 Operation of a Monostable MV Circuit

Figure 7-2a shows a circuit of a monostable MV. Note the difference between it and the F/F circuit (Fig. 6-4a). The base of one of the transistors (Q_1) is connected to V_{cc} through a resistor (R) and is also coupled to the collector of the other transistor through a capacitor (C). At any one time, one transistor is ON and the other is OFF. An examination of the circuit shown in Fig. 7-2a indicates that in its stable state, Q_1 must be ON and Q_2 must be OFF. This observation is based upon the fact that there is a reverse-bias

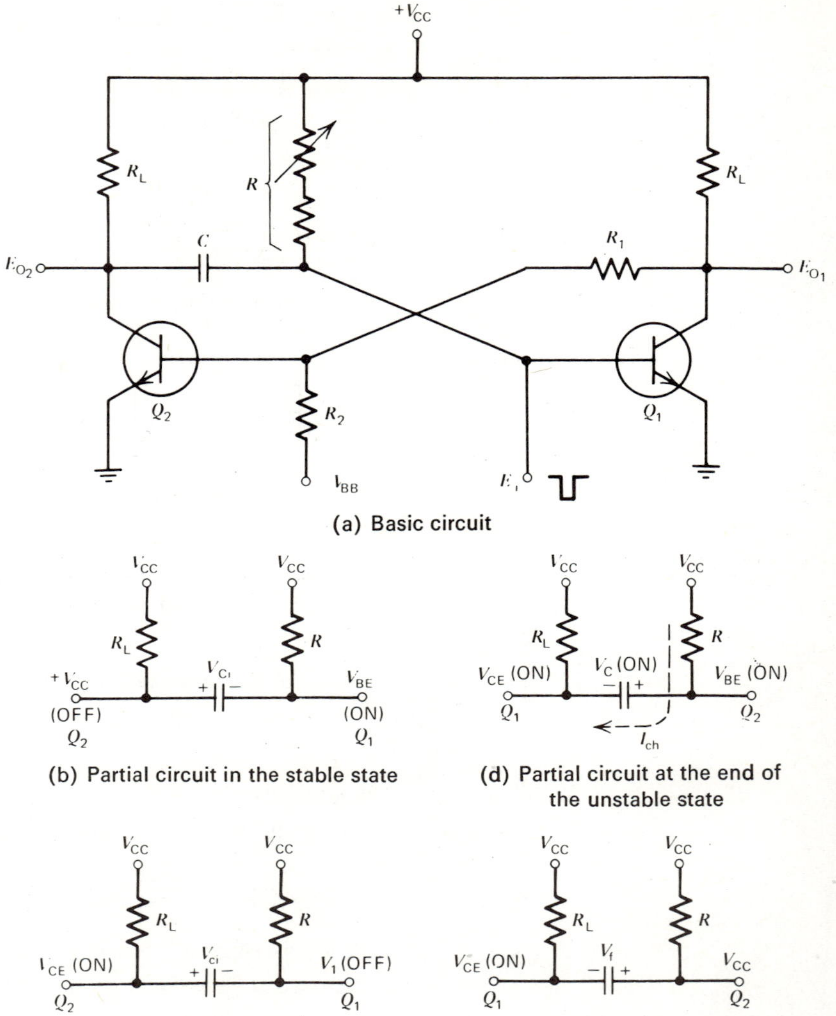

(a) Basic circuit

(b) Partial circuit in the stable state

(d) Partial circuit at the end of the unstable state

(c) Partial circuit at the start of the unstable state

(e) Partial circuit showing final conditions if the capacitor is allowed to become fully charged

Figure 7-2 The monostable MV circuit.

voltage ($-V_{BB}$) connected only to the base of Q_2. Therefore, Q_2 must be OFF. If Q_2 is OFF, then Q_1 must be ON.

To understand the operation, consider the static conditions as they exist around the capacitor C. One side of the capacitor is at V_{cc} and the other is at V_{BE}(ON). This is shown in Fig. 7-2b. Thus, the voltage across the capacitor V_{c_1} is:

$$V_{C_1} = V_{cc} - V_{BE}(\text{ON}) \text{ volts (V)} \tag{7-1}$$

A negative input trigger pulse applied to the base of Q_1 turns it OFF, which in turn, turns Q_2 ON. This immediately makes the voltage at the collector of Q_2 to be V_{CE}(ON). Thus, the input trigger switched the state of the MV, which in turn resulted in a voltage drop from V_{cc} to V_{CE}(ON) on one side of the capacitor C. This happens very quickly. The capacitor practically remains charged to V_{c_1}. The voltage drop [V_{cc} to V_{CE}(ON)] at the collector of Q_2 gets transferred to the base of Q_1 and reverse-biases its base-emitter junction. Thus, Q_2 turns OFF. The magnitude of the reverse bias is:

$$V_1 = V_{CE}(\text{ON}) - V_{c_i} \tag{7-2a}$$

or

$$= V_{CE}(\text{ON}) - [V_{cc} - V_{BE}(\text{ON})] \tag{7-2b}$$

$$= - [V_{cc} - V_{CE}(\text{ON}) - V_{BE}(\text{ON})] \tag{7-2c}$$

The condition stated in Eq. (7-2) is shown in Fig. 7-2c. It can be seen from the partial circuit of Fig. 7-2c that the capacitor will finally end up charged to $V_{cc} - V_{CE}$(ON) (Fig. 7-2e), if enough time (5τ) elapses without anything else occurring in the circuit. At the end of 5τ, the final voltage across capacitor C is:

$$V_f = V_{cc} - V_{CE}(\text{ON}) \text{ volts (V)} \tag{7-3}$$

But something does occur. As the base side of Q_1 reaches the turn-ON voltage, Q_1 of course turns ON. As a result of the circuit design, Q_1 ON turns Q_2 OFF. Thus, the circuit comes back to its stable state. Note that the capacitor was being charged at a rate determined by the time constant RC (Fig. 7-2d).

Voltages on and around the capacitor are shown in Fig. 7-2d just as Q_1 turns back ON. At this instant, voltage on the base of Q_1 is V_{BE}(ON) and the charge on the capacitor is:

$$V_c(\text{ON}) = V_{BE}(\text{ON}) - V_{CE}(\text{ON}) \text{ volts (V)} \tag{7-4}$$

The time duration (t_p) for which Q_1 remains OFF can now be calculated.

The capacitor has an initial charge of V_{ci}. It is charging toward

$[V_{cc} - V_{CE}(\mathrm{ON})]$. The switching takes place when the capacitor voltage reaches $V_{BE}(\mathrm{ON}) - V_{CE}(\mathrm{ON})$. The capacitor charge curve is shown in Fig. 7-3a. Writing it in equation form:

$$V_c(\mathrm{ON}) = V_f - (V_f + V_{ci})e^{-(t_p/RC)} \tag{3-8}$$

$$\begin{aligned} V_{BE}(\mathrm{ON}) - V_{CE}(\mathrm{ON}) = [V_{cc} - V_{CE}(\mathrm{ON})] \\ - \{[V_{cc} - V_{CE}(\mathrm{ON})] \\ + [(V_{CC} - V_{BE}(\mathrm{ON})]\}e^{-(t_p/RC)} \end{aligned}$$

or

$$e^{-(t_p/RC)} = \frac{[V_{cc} - V_{CE}(\mathrm{ON})] - [V_{BE}(\mathrm{ON}) - V_{CE}(\mathrm{ON})]}{[V_{cc} - V_{CE}(\mathrm{ON})] + [V_{cc} - V_{BE}(\mathrm{ON})]}$$

$$= \frac{V_{cc} - V_{BE}(\mathrm{ON})}{2V_{cc} - V_{CE}(\mathrm{ON}) - V_{BE}(\mathrm{ON})} \tag{7-5}$$

Alternately, let us look at the base of Q_1. The voltage drops from $V_{BE}(\mathrm{ON})$ to V_1 as the trigger pulse switches the monostable MV to its unstable state. It starts rising toward V_{cc} as the capacitor C charges. The transistor Q_1 turns back ON as its base voltage reaches $V_{BE}(\mathrm{ON})$. This is shown in Fig. 7-3b. Writing the above in equation form (using Eq. 3-8) yields,

$$V_{BE}(\mathrm{ON}) = V_{cc} - (V_{cc} + V_1)e^{-(t_p/RC)}$$

$$e^{-(t_p/RC)} = \frac{V_{cc} - V_{BE}(\mathrm{ON})}{V_{cc} + V_1}$$

But

$$V_1 = V_{cc} - V_{CE}(\mathrm{ON}) - V_{BE}(\mathrm{ON}) \tag{7-2c}$$

$$\therefore e^{-(t_p/RC)} = \frac{V_{cc} - V_{BE}(\mathrm{ON})}{V_{cc} + V_{cc} - V_{CE}(\mathrm{ON}) - V_{BE}(\mathrm{ON})}$$

$$= \frac{V_{cc} - V_{BE}(\mathrm{ON})}{2V_{cc} - V_{CE}(\mathrm{ON}) - V_{BE}(\mathrm{ON})}$$

Note that this is the same equation as Eq. (7-5); this must be so. Equation (7-5) indicates that the time constant RC controls the time duration t_p for the circuit to remain in the unstable state.

The monostable circuit was assumed to be in its stable state before the input trigger was applied. More important is to note that

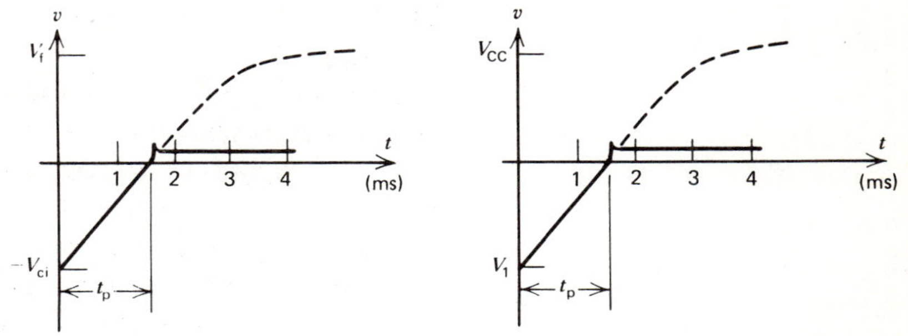

(a) Voltage across the capacitor (b) Voltage at the base of Q_1

Figure 7-3 The exponential charge curves.

the coupling capacitor was charged to V_{ci} (Eq. 7-1). As a result of this capacitive charge, the input trigger pulse sent the ON transistor deep into the OFF region. It is *this* charge (V_{ci}) that must be removed before the circuit returns to its stable state.

The above discussion is presented to emphasize that the coupling capacitor must have a charge (V_{ci}) on it *before* the circuit is triggered. It is only then that the equations presented above hold true. Thus, a certain minimum time must be allowed to elapse before the next input trigger pulse is applied. This minimum time is required for the coupling capacitor to charge to V_{ci} again or to *recover* back to V_{ci}. It is commonly known as the *recovery time* (t_{re}). During this time, the capacitor charges with the time constant $R_L C$.

The capacitor is considered to be fully charged when $t = 5\tau$. Thus, the recovery time is:

$$t_{re} = 5R_L C \tag{7-6}$$

The circuit will still function as a monostable device even if the capacitor did not quite get fully charged before the trigger pulse was applied. However, t_p of the output pulse will be shortened accordingly.

7-5 Design of a Monostable MV Circuit

A monostable circuit is usually designed such that, when triggered, it gives out a pulse of desired duration and amplitude. The peak-to-peak amplitude of the output pulse is:

$$V_{p\text{-}p} = V_{cc} - V_{CE}(\text{ON}) \text{ volts (V)} \tag{7-7}$$

Thus, essentially, the supply voltage (V_{cc}) sets the amplitude of the output pulse.

Equation (7-5) gives the output pulse width (t_p). It identifies the circuit components that control or influence t_p. After V_{cc} has been selected, R and C are the two remaining (t_p-controlling) components. R is selected such that enough saturation current flows to the base of Q_1 (Fig. 7-2a) when the circuit is in its stable state. A value for C is calculated using Eq. (7-5) for a given t_p.

For *ideal* transistors [$V_{CE}(ON) = V_{BE}(ON) = 0$], Eq. (7-5) simplifies to:

$$e^{-(t_p/RC)} = \frac{V_{cc}}{2V_{cc}}$$

$$= 0.5$$

$$\frac{t_p}{RC} = 0.693$$

$$t_p = 0.693RC \text{ seconds (s)} \tag{7-8}$$

For most practical purposes, the pulse width given by Eq. (7-8) is satisfactory, since $V_{CE}(ON)$ and $V_{BE}(ON)$ are usually small compared with V_{CC}.

Consider the circuit shown in Fig. 7-2a:

$$R_L = \frac{V_{cc} - V_{CE}(ON)}{I_C(max)} \text{ ohms } (\Omega) \tag{7-9}$$

If a standard value for R_L must be chosen, then choose the next higher EIA standard value to the one obtained by Eq. (7-9). This ensures that $I_C(ON)$ will be less than $I_C(max)$. Thus,

$$I_C(ON) = \frac{V_{cc} - V_{CE}(ON)}{R_L} \text{ amperes (A)} \tag{7-10}$$

The saturation base current $I_B(ON)$ is supplied through resistor R to the normally ON transistor (Q_1).

$$I_B(ON) = \frac{I_C(ON)}{\beta} \text{ amperes (A)} \tag{7-11}$$

$$R = \frac{V_{CE} - V_{BE}(ON)}{I_B(ON)} \text{ ohms } (\Omega) \tag{7-12}$$

If a standard value for R must be used, then a standard resistance value, next lower from the one obtained by Eq. (7-12), is chosen. This assures that the base current to the normally-ON transistor is greater than the minimum required for saturation.

Next, a value for C is calculated using Eq. (7-5). If the calculated value of C is not a standard capacitive value, then the next-higher standard value is chosen. This of course will result in a longer t_p than necessary. In practical circuits, the resistor R is made up of two resistors connected in series. One resistor is a fixed resistor and the other is a variable resistor. The fixed resistor keeps the base current well below the maximum allowed for the given transistor. The variable resistor is used to adjust t_p to the desired duration.

Values for R_1 and R_2 are calculated such that the ON/OFF conditions for the normally OFF transistor are satisfied. The process is similar to the one explained for an F/F in Ch. 6.

Consider Q_2 in Fig. 7-2a. A partial circuit for the OFF condition is shown in Fig. 7-4a. Assume that $V_{BE}(\text{OFF}) = 1.0$ V. Then,

$$I_1 = I_2$$

$$\frac{V_{CE}(\text{ON}) - V_{BE}(\text{OFF})}{R_1} = \frac{V_{BE}(\text{OFF}) - V_{BB}}{R_2} \tag{7-13}$$

Refer to the partial circuit shown in Fig. 7-3b for the ON condition.

$$I_1 - I_2 \geqq I_B(\text{ON})$$

$$\frac{V_{CC} - V_{BE}(\text{ON})}{R_L + R_1} - \frac{V_{BE}(\text{ON}) - V_{BB}}{R_2} \geqq I_B(\text{ON}) \tag{7-14}$$

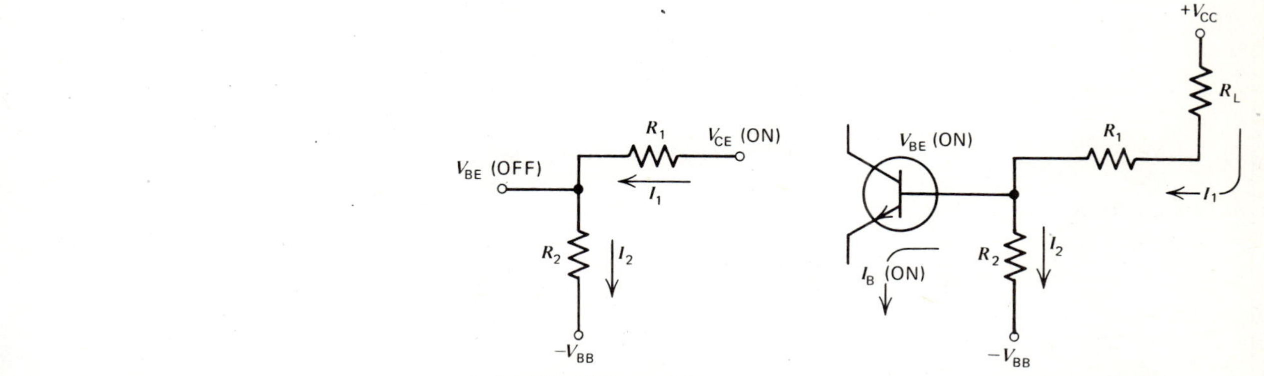

(a) The OFF condition (b) The ON condition

Figure 7-4 Partial circuits for ON/OFF conditions.

R_1 and R_2 are calculated by simultaneously solving Eqs. (7-13) and (7-14). R_L is usually quite small compared to R_1 and can safely be neglected in Eq. (7-14); this makes the calculation much simpler.

Example 7-1

Design a monostable MV circuit such that when it is triggered, its output pulse has $t_p = 100\ \mu s$ and $V_{p\text{-}p} = 10$ V (0 V to 10 V). Assume for the two npn transistors that $\beta = 125$, $V_{CE}(ON) = 0$, $I_C(max) = 20$ mA and $V_{BE}(ON) = 0$. Assume $V_{BB} = -10$ V.

Solution

The circuit is shown in Fig. 7-2a.

1. Determine V_{CC}:

$$V_{cc} = V_{p\text{-}p} + V_{CE}(ON) \quad (7\text{-}7)$$

$$= 10 - 0$$

$$= \mathbf{10\ V}$$

2. Calculate R_L:

$$R_L = \frac{V_{cc} - V_{CE}(ON)}{I_C(ON)} \quad (7\text{-}9)$$

$$= \frac{10 - 0}{20\text{ mA}}$$

$$= \mathbf{500\ ohms}$$

3. Find $I_B(ON)$:

$$I_B(ON) = \frac{I_C(ON)}{\beta} \quad (7\text{-}11)$$

$$= \frac{20\text{ mA}}{125}$$

$$= \mathbf{0.16\ mA}$$

4. Calculate R:

$$R = \frac{V_{cc} - V_{BE}(ON)}{I_B(ON)} \quad (7\text{-}12)$$

$$\frac{10}{0.16\text{ mA}}$$

$$= \mathbf{62.5\ k\Omega}$$

5. Find C:

$$V_{BE}(ON) = V_{CE}(ON) = 0$$

$$C = \frac{t_p}{0.693\ R} \tag{7-8}$$

$$= \frac{100\ \mu s}{0.693 \times 62.5\ k\Omega}$$

$$= 2.31 \times 10^{-9}\ F$$

or

$$= \mathbf{0.00231\ \mu F}$$

6. Evaluate the ratio of R_2 to R_1. Assume $V_{BE}(OFF) = -1.0$ V. OFF condition:

$$\frac{V_{CE}(ON) - V_{BE}(OFF)}{R_1} = \frac{V_{BE}(OFF) - V_{BB}}{R_2} \tag{7-13}$$

$$\frac{0 - (-1.0)}{R_1} = \frac{-1.0 - (-10)}{R_2}$$

$$\frac{1}{R_1} = \frac{9}{R_2}$$

$$\mathbf{R_2 = 9R_1} \tag{7-15}$$

7. Calculate R_2 and R_1. ON condition:

$$\frac{V_{cc} - V_{BE}(ON)}{R_L + R_1} = \frac{V_{BE}(ON) - V_{BB}}{R_2} \geqq I_B(ON) \tag{7-14}$$

$$\frac{10 - 0}{R_L + R_1} = \frac{0 - (-10)}{R_2} = 0.16\ mA$$

Assuming $R_L \ll R_1$,

$$\frac{10}{R_1} = \frac{10}{R_2} = 0.16\ mA \tag{7-16}$$

Solving Eqs. (7-15) and (7-16) simultaneously for R_1 and R_2 yields:

$$\frac{10}{R_1} - \frac{10}{R_2} = 0.16\ mA$$

$$R_1 = \frac{10 - (10/9)}{-0.6 \times 10^{-3}}$$

$$= \mathbf{56.25\ k\Omega}$$

$$\therefore R_2 = 9 \times 56.25\ k$$

$$= \mathbf{506.25\ k\Omega}$$

Example 7-2

Repeat Ex. 7-1 for standard Si transistors. All circuit components are to have standard EIA values. The two output levels are to be 12 V and 0.3 V.

Solution

1. Determine V_{cc}:

$$V_{cc} = V_{p\text{-}p} + V_{CE}(\text{ON}) \tag{7-7}$$
$$= 12.0 + 0.3$$
$$= \mathbf{12.3\ V}$$

2. Calculate R_L:

$$R_L = \frac{V_{cc} - V_{CE}(\text{ON})}{I_C(\text{ON})} \tag{7-9}$$

$$R_{L_1} = \frac{12.3 - 0.3}{20\text{ mA}}$$
$$= \mathbf{600\ \Omega}$$

The next higher and EIA standard resistor value is **680 Ω.**

3. Find $I_B(\text{ON})$:

$$\text{Actual } I_c(\text{ON}) = \frac{12.3 - 0.3}{680}$$
$$= 17.65\text{ mA}$$

$$I_B(\text{ON}) = \frac{I_C(\text{ON})}{\beta} \tag{7-11}$$
$$= \frac{17.65\text{ mA}}{125}$$
$$= 0.141\text{ mA}$$

4. Calculate R:

$$R = \frac{V_{cc} - V_{BE}(\text{ON})}{I_B(\text{ON})} \tag{7-12}$$
$$= \frac{12.3 - 0.7}{0.141\text{ mA}}$$
$$= \mathbf{82.27\ k\Omega}$$

The resistor R may be made up of two series-connected resistors, one a **50 kΩ** fixed resistor and the other a **50 kΩ** variable resistor.

5. Find C:

$$e^{-(t_p/RC)} = \frac{V_{cc} - V_{BE}(\text{ON})}{2V_{cc} - V_{CE}(\text{ON}) - V_{BE}(\text{ON})} \tag{7-5}$$

$$= \frac{12.3 - 0.7}{24.6 - 0.3 - 0.7}$$

$$= 0.491$$

$$\frac{t_p}{RC} = 0.71$$

$$C = \frac{t_p}{0.71\,R}$$

$$= \frac{100 \times 10^{-6}}{0.71 \times 82.27\text{ k}}$$

$$= 1.7 \times 10^{-9}\text{ F}$$

The next higher standard value of a capacitor is 2×10^{-9} F or **0.002 μF**.

If t_p is to be kept at 100 μs, a new value for R must be calculated.

$$R = \frac{100 \times 10^{-6}}{0.71 \times 2 \times 10^{-9}}$$

$$= \mathbf{70.42\ k\Omega}$$

This means that the variable resistor is to be set at **20.42 kΩ**.

$$R = 50\text{ k}\Omega\text{ (fixed)} + 20.42\text{ k}\Omega\text{ (variable)}$$

$$= 70.42\text{ k}\Omega$$

6. Evaluate the ratio of R_1 and R_2. OFF condition, [assume $V_{BE}(\text{OFF}) = -1.0$ V]:

$$\frac{V_{CE}(\text{ON}) - V_{BE}(\text{OFF})}{R_1} = \frac{V_{BE}(\text{OFF}) - V_{BB}}{R_2} \tag{7-13}$$

$$\frac{0.3 - (-1.0)}{R_1} = \frac{(-1.0) - (-10)}{R_2}$$

$$\frac{1.3}{R_1} = \frac{9}{R_2}$$

$$\mathbf{R_2 = 6.923R_1} \tag{7-17}$$

ON condition:

$$\frac{V_{cc} - V_{BE}(ON)}{R_L + R_1} - \frac{V_{BE}(ON) - V_{BB}}{R_2} \geqq I_B(ON) \qquad (7\text{-}14)$$

$$\frac{12.3 - 0.7}{680 + R_1} - \frac{0.7 - (-10)}{R_2} = 0.141 \text{ mA} \qquad (7\text{-}18)$$

Assuming $R_1 \gg 680\ \Omega$,

$$\frac{11.6}{R_1} - \frac{10.7}{R_2} = 0.141 \text{ mA}$$

7. Find R_1 and R_2. Solving Eqs. (7-17) and (7-18) yields:

$$\frac{11.6}{R_1} - \frac{10.7}{6.923R_1} = 0.141 \text{ mA}$$

$$R_1 = \frac{11.6 - (10.7/6.923)}{0.141 \text{ mA}}$$

$$= 71.3 \text{ k}\Omega$$

The next lower EIA standard resistor value is **68 kΩ.**

$$R_2 = 68 \text{ k}\Omega \times 6.923$$

$$= \mathbf{470\ k\Omega}$$

This is a standard EIA resistor value. If it was not so, then the next higher EIA standard resistance value should be chosen.

Example 7-3

Refer to the designed monostable MV circuit of Ex. 7-2. Calculate:

a. The recovery time.

b. The minimum time gap required between any two consecutive trigger pulses.

Solution

a.

$$t_{re} = 5R_LC \qquad (7\text{-}6)$$

$$= 5 \times 680 \times 2 \times 10^{-9}$$

$$= \mathbf{6.8\ \mu s}$$

b.

$$\Delta t_{min} = t_p + t_{re}$$

$$= (100 + 6.8)\ \mu s$$

$$= \mathbf{106.8\ \mu s}$$

Example 7-4

Refer to the designed monostable MV circuit of Ex. 7-2. Check whether this circuit will function satisfactorily.

Solution

$$I_C(\text{ON}) = \frac{12.3 - 0.3}{680}$$

$$= 17.65 \text{ mA}$$

Thus, $I_C(\text{ON})$ is less than $I_C(\text{max})$.

$$\min I_B(\text{ON}) = \frac{17.65}{125}$$

$$= 0.141 \text{ mA}$$

$$\text{Actual } I_B(\text{ON}) \text{ for } Q_1 = \frac{(12.3 - 0.7)\text{ V}}{70.42 \text{ k}\Omega}$$

$$= 0.165 \text{ mA}$$

Actual $I_B(\text{ON}) >$ min $I_B(\text{ON})$ required. Therefore, Q_1 does go ON in its stable state.

In the stable state, Q_2 is required to be OFF. This means that V_{BE} (OFF) must be a negative-bias voltage.

$$V_{BE_2}(\text{OFF}) = 0.3 - \frac{68 \text{ k}}{68 \text{ k} + 470 \text{ k}} \times (10 + 0.3)$$

$$= 0.3 - 1.65$$

$$= -1.35 \text{ V}$$

In the unstable state, Q_2 must go ON. The base current to Q_2 must be $\geqq$ min $I_B(\text{ON})$.

$$\text{Actual } I_{B_2}(\text{ON}) = \frac{12.3 - 0.7}{680 + 68 \text{ k}} - \frac{10 + 0.7}{470 \text{ k}} \qquad \text{(Fig. 7-3b)}$$

$$= 0.169 - 0.023 \text{ mA}$$

$$= 0.146 \text{ mA}$$

which is greater than the $I_{B_2}(\text{min})$ required for saturation. Thus, Q_2 does turn ON in the unstable state.

All of the ON/OFF conditions are satisfied. The circuit will function satisfactorily provided that the input trigger pulses are at least 106.8 μs apart (Ex. 7-3).

Example 7-5
Refer to the circuit designed in Ex. 7-2. Find:
a. The maximum output pulse-width obtainable.
b. The minimum output pulse-width obtainable.

Solution

a.
$$t_p(\text{min}) = 0.71 \times 50\ \text{k}\Omega \times 2 \times 10^{-9}$$
$$= \mathbf{71\ \mu s}$$

(50 kΩ variable resistor at 0 Ω)

b.
$$t_p(\text{max}) = 0.71 \times 82.27\ \text{k}\Omega \times 2 \times 10^{-9}$$
$$= \mathbf{116.8\ \mu s}$$

[82.27 kΩ is the maximum value of R that can be used and still keep Q_1 ON (Ex. 7-2)].

7-6 Triggering a Monostable MV Circuit

The monostable MV circuit by itself is of very little use unless triggered. It is mainly used to generate a pulse, and it does so only when triggered. Thus, there is always a triggering circuit associated with a monostable MV circuit.

The function of the triggering circuit is to switch the monostable MV to its unstable state whenever a trigger pulse is applied to it. Consider the triggered monostable MV circuit shown in Fig. 7-5. D_1, D_2, and C_2 make up the triggering circuit. A triggering pulse input applied at E_i gets differentiated by the C_2D_2 combination. Note that the positive spike of the differentiated input pulse gets grounded through D_2, leaving behind only the negative spike. This negative spike gets transferred to the base of Q_1 through D_1, making it momentarily negative. Negative voltage on the base of Q_1 cuts it OFF. As a result of the circuit design, when Q_1 goes OFF, Q_2 turns ON. The monostable MV is now in its unstable state. The process and manner in which this circuit comes back to its stable state was explained in Sec. 7-4.

Basically, the triggering circuit turns OFF the ON transistor. The SET-RESET triggering circuits discussed in Sec. 6-13 will accomplish the desired switching here. The basic idea is to switch the monostable MV to its unstable state. Figure 7-5 shows waveforms that appear at various points as the trigger input pulse is applied.

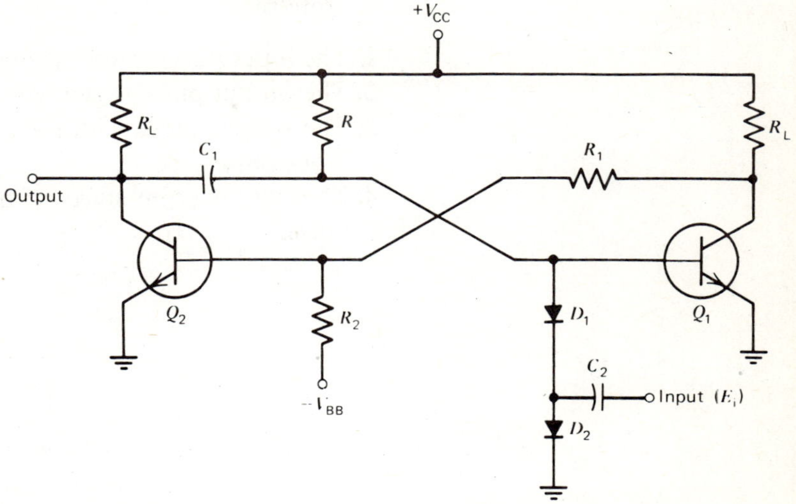

(a) Triggering circuit

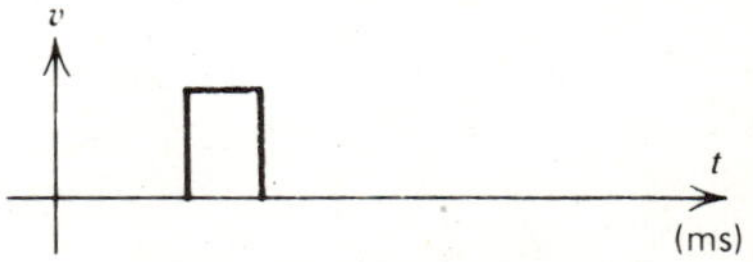

(b) Input trigger pulse

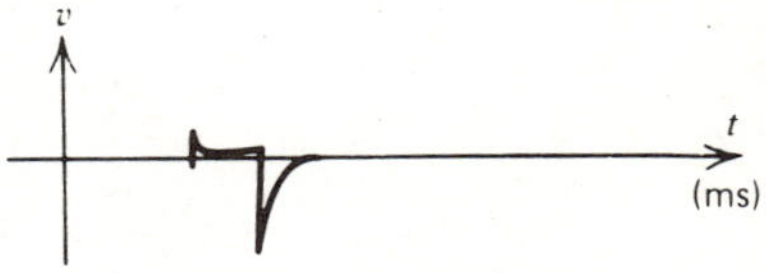

(c) Waveform across D_2

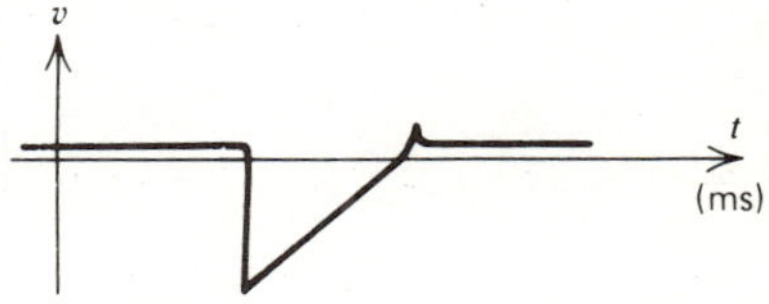

(d) Waveform at the base of Q_1

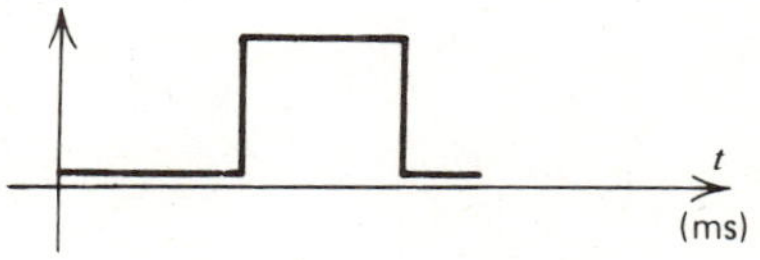

(e) Output pulse.

Figure 7-5 The triggering circuit and input/output waveforms.

Note that:

1. The input pulse simply switches the circuit into its unstable state.
2. The output pulse is generated by the monostable circuit.
3. The output pulse width and amplitude have no relation to the input pulse.
4. The output pulse starts at the negative-going edge of the input pulse.

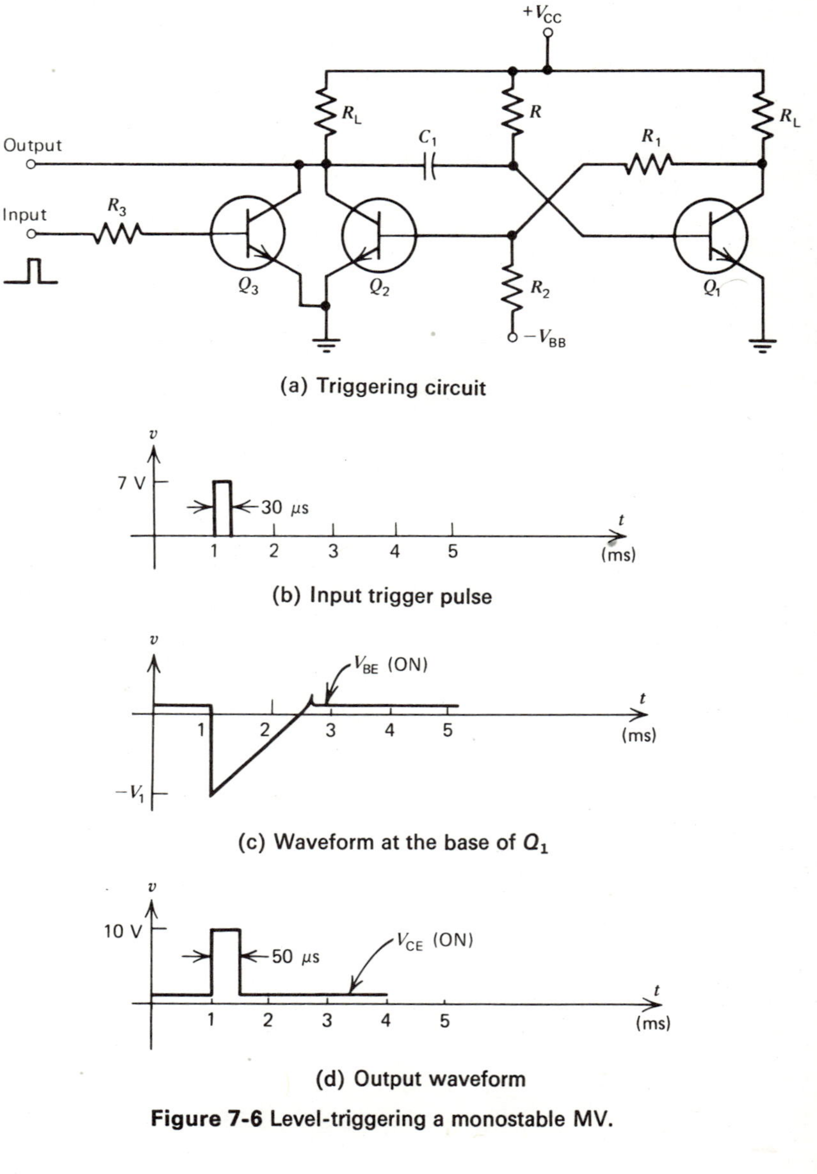

Figure 7-6 Level-triggering a monostable MV.

Waveforms shown in Figs. 7-5b and 7-5e indicate that a monostable MV may be used as a pulse delay circuit. If the circuit is designed such that the output pulse is similar to the input pulse, then the input pulse appears to have been delayed for time t_p by the monostable circuit.

A *level-triggering* circuit may be employed when a monostable MV is to be used as a pulse reshaper (or pulse stretcher). The circuit flips into its unstable state when the input trigger pulse reaches a certain voltage level. Figure 7-6 shows a typical pulse-reshaping circuit. The pulse to be reshaped is shown as being applied at the trigger input terminal E_i. Assume that originally it was a 10 $V_{p\text{-}p}$ (0 V to 10 V) 50 μs pulse. It now appears to be a 7 $V_{p\text{-}p}$, 30 μs distorted pulse. In the reshaping circuit of Fig. 7-6, the input pulse merely turns Q_3 ON, lowering the voltage at the collector of Q_2. This drop in voltage is transferred through C_1 to the base voltage of Q_1. Thus, Q_1 is turned OFF, which in turn, turns Q_2 ON.

The output pulse that appears at E_o is generated by the monostable circuit itself. Its amplitude and duration are set by the circuit designers as shown in Ex. 7-1. Essentially, a monostable circuit with $V_{cc} = 10$ V and $0.693RC \doteq 50$ μs will emit the desired pulse.

7-7 Astable MV

An *astable* or a *free-running* MV is unstable in both of the states. The circuit continuously switches back and forth between the two states on its own. Assume that the square block shown in Fig. 7-7a is a free-running MV. Waveforms that appear at the two outputs

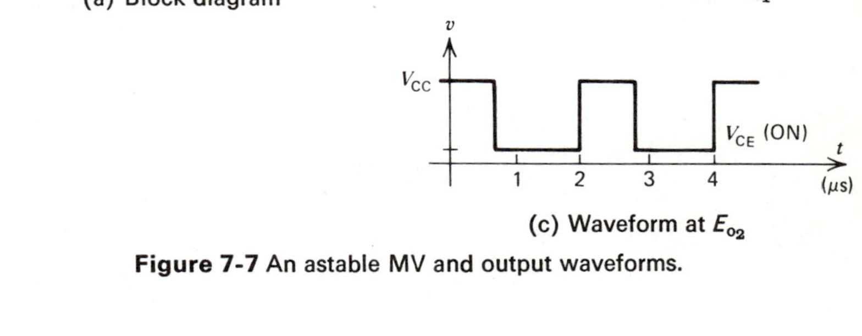

Figure 7-7 An astable MV and output waveforms.

E_{o_1} and E_{o_2} are shown in Figs. 7-7b and 7-7c. Note that when one output is at V_1, the other is at V_2. The output voltage levels interchange at the end of each unstable state. The time duration for which the circuit stays in either of the two unstable states (t_{p_1} or t_{p_2}) is controlled by the internal circuit design. The output waveform may be made either a square wave or a pulse wave by setting time durations t_{p_1} and t_{p_2}. Normally, the two output-voltage levels are V_{cc} and V_{CE}(ON). This is based upon the transistor being either OFF or ON.

7-8 Operation of an Astable MV

Figure 7-8a shows a circuit of an astable MV, which consists of two identical amplifiers coupled back-to-back. Both of the transistors automatically get switched ON and OFF turn-by-turn. The output voltage levels vary between V_{cc} and V_{CE}(ON).

To follow and understand the operation, assume that at some given time, Q_1 is ON and Q_2 is OFF. The voltage across C_2 is $V_i = V_{cc} - V_{BE}$(ON), as shown in Fig. 7-8f. It will become clearer later that at this time C_1 is charging toward $V_{cc} - V_{CE}$(ON) and the base of Q_2 is going toward $-V_{cc}$. As the base voltage (Q_2) reaches the turn-ON voltage, Q_2 turns ON. This makes the collector of Q_2 drop to V_{CE}(ON). The voltage drop at the collector of Q_2 is transferred through C_2 to the base of Q_1. C_2 still has a charge of V_i volts. As shown in Fig. 7-8g, the base of Q_1 becomes reverse-biased by $V = V_i - V_{CE}$(ON) immediately after it is turned OFF. It starts going toward $-V_{cc}$, with a time-constant of R_2C_2, immediately after Q_1 turns OFF. As it reaches the turn-ON voltage of Q_1 (Fig. 7-8h), Q_1 turns ON. This will switch Q_2 OFF in the same way as Q_1 was switched OFF. Now C_1 will start charging with a time-constant of C_1R_1, and the switching will take place when the base of Q_2 reaches its turn-ON potential. This back-and-forth switching between the two transistors goes on endlessly. Note that C_2 charges to V_{c_1} (Fig. 7-8e) through time-constant R_LC_2, while Q_2 is OFF. This is the recovery time for C_2. Thus,

$$t_{re_2}(C_2) = 5R_LC_2 \text{ seconds (s)} \tag{7-19}$$

Similarly,

$$t_{re_1}(C_1) = 5R_LC_1 \text{ seconds (s)} \tag{7-20}$$

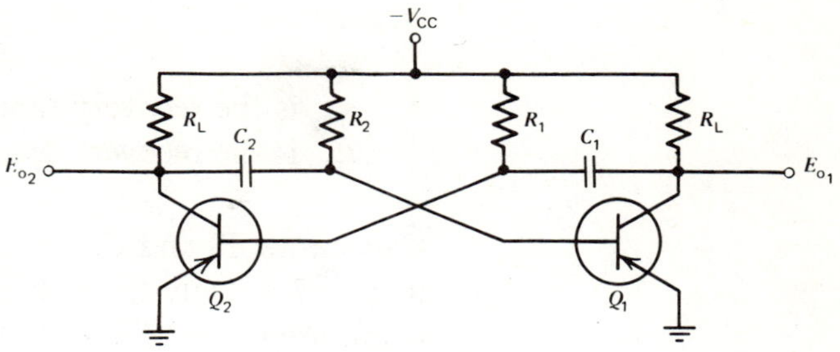

(a) Astable MV circuit

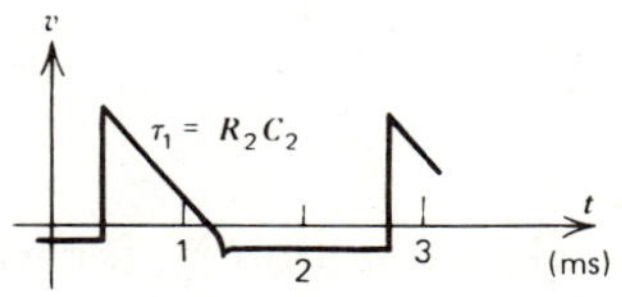

(b) Waveform at the base of Q_1

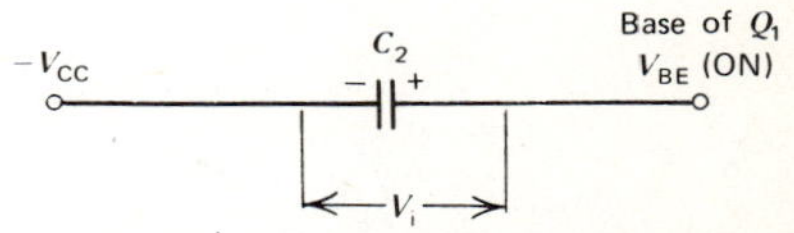

(f) Charge on C_2 when Q_1 is ON and Q_2 is OFF

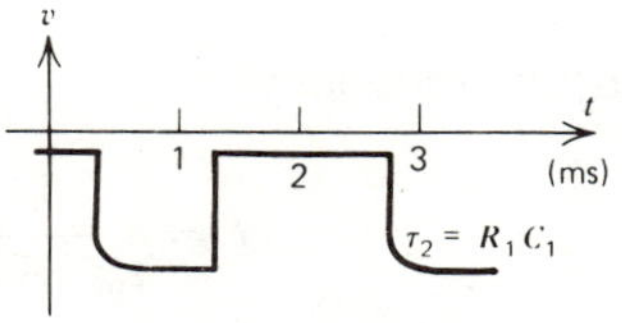

(c) Waveform at E_{o_1}

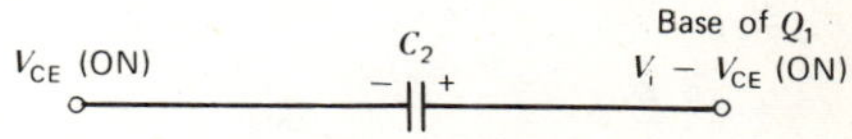

(g) Charge on C_2 right after Q_1 has turned OFF

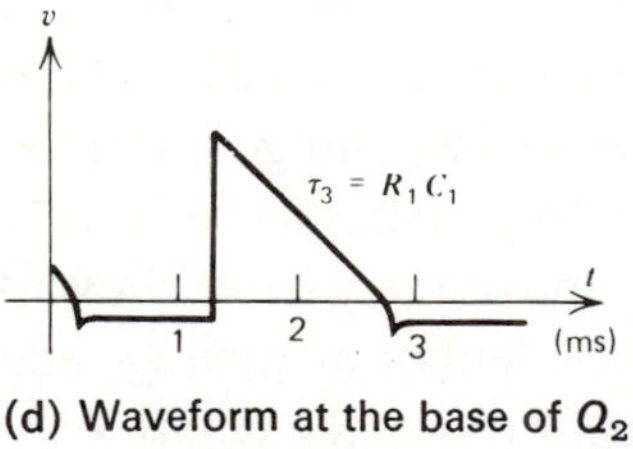

(d) Waveform at the base of Q_2

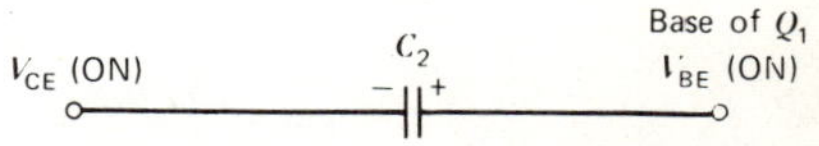

(h) Charge on C_2 just before Q_1 turns ON

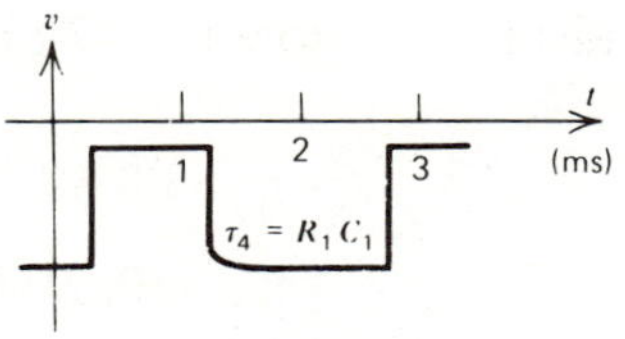

(e) Waveform at E_{o_2}

Figure 7-8 Astable MV circuit and waveforms.

where:

t_{re_2} is the recovery time for C_2
t_{re_1} is the recovery time for C_1

Capacitors C_1 and C_2 charge and discharge in a manner explained in Sec. 7-4. Only the polarity is different, since *pnp* transistors are being used here. Thus the time duration t_{p_1} for which Q_1 is OFF, is given by:

$$e^{-(t_{p_1}/R_2C_2)} = \frac{V_{CC} - V_{BE}(ON)}{2V_{CC} - V_{BE}(ON) - V_{CE}(ON)} \tag{7-5}$$

And the time duration t_{p_2} for which Q_2 is OFF, is given by:

$$e^{-(t_{p_2}/R_1C_1)} = \frac{V_{CCC} - V_{BE}(ON)}{2V_{CC} - V_{BE}(ON) - V_{CE}(ON)} \tag{7-5}$$

The output frequency is:

$$f = \frac{1}{t_{p_1} + t_{p_2}} \text{ hertz (Hz)} \tag{7-21}$$

Note that for proper operation, $t_{re}(C_1)$ should be less than t_{p_1}, and that $t_{re}(C_2)$ should be less than t_{p_2}. For proper operation, the capacitors should recover to the voltage charge as shown in Fig. 7-8b before switching takes place. Thus, $5R_LC_1$ should be less than t_{p_1} and $5R_LC_2$ should be less than t_{p_2}.

The above discussion shows that the time-constant R_1C_1 determines the length of time Q_2 stays OFF (t_{p_2}) and the time-constant R_2C_2 determines the length of time Q_1 stays OFF (t_{p_1}). The output will be a square wave if $R_1C_1 = R_2C_2$.

For ideal transistors (Eq. 7-8):

$$t_{p_1} = 0.693R_2C_2 \text{ seconds (s)}$$

$$t_{p_2} = 0.693R_1C_1 \text{ seconds (s)}$$

and

$$f = \frac{1}{0.693\ R_1C_1 + 0.693\ R_2C_2} \text{ hertz (Hz)}$$

7-9 Design of an Astable MV Circuit

An astable MV circuit is usually designed to produce a specified output voltage waveform. The output is specified in terms of its desired pulse repetition frequency (PRF) and desired amplitude.

Consider the astable circuit of Fig. 7-8a. Designing this circuit means that values for V_{cc}, R_L, R_1, R_2, C_1, and C_2 are to be calculated.

The output voltage is at V_{cc} when the transistor is OFF and it is at V_{CE}(ON) when the transistor is ON.

$$\therefore E_{o(p\text{-}p)} = V_{cc} - V_{CE}(\text{ON}) \tag{7-7}$$

Values for R_L, R_1 and R_2 are calculated using Eqs. (7-9), (7-10), (7-11), and (7-12). Note that values for R_1 and R_2 are chosen such that they provide enough current to saturate the corresponding transistors. Usually $R_1 = R_2$.

Values for C_1 and C_2 are calculated using Eq. (7-5). Note that they depend upon the time durations that each transistor is required to be OFF.

Example 7-6

Design an astable circuit that gives out a 12 $V_{p\text{-}p}$ (0 V to 12 V) output pulse waveform whose $t_{p_1} = 10\ \mu$s and PRF = 20 kHz. The two *npn* transistors that are to be used have $I_C(\max) = 20$ mA, $\beta = 50$, and $V_{BE}(\text{ON}) = V_{CE}(\text{ON}) = 0$ V.

Solution

The circuit is drawn in Fig. 7-9.

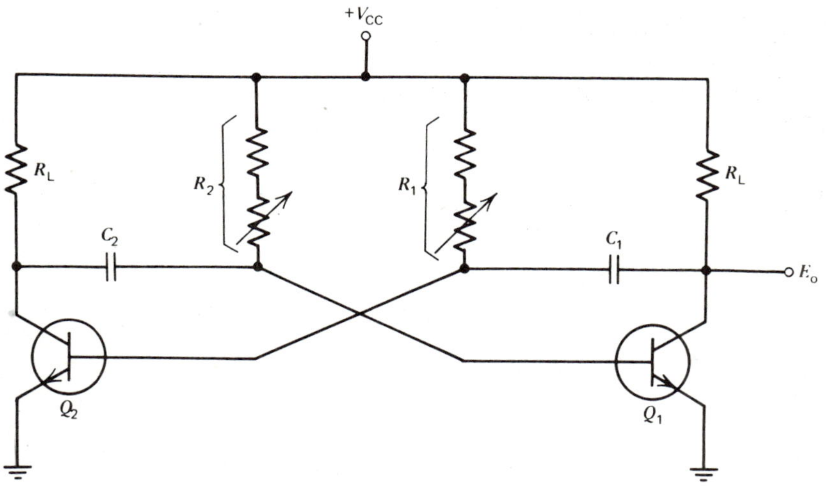

Figure 7-9 Circuit for Ex. 7-6.

1. Find V_{cc}:

$$V_{cc} = V_{\mathbf{p\text{-}p}} + V_{CE}(\mathrm{ON}) \qquad (7\text{-}7)$$
$$= \mathbf{12\ V}$$

2. Calculate R_L:

$$R_L = \frac{V_{cc} - V_{CE}(\mathrm{ON})}{I_C(\mathrm{ON})} \qquad (7\text{-}9)$$
$$= \frac{12\ \mathrm{V}}{20\ \mathrm{mA}}$$
$$= \mathbf{600\ \Omega}$$

3. Calculate I_B:

$$I_B(\mathrm{ON}) = \frac{I_C(\mathrm{ON})}{\beta} \qquad (7\text{-}11)$$
$$= \frac{20\ \mathrm{mA}}{50}$$
$$= 0.4\ \mathrm{mA}$$

4. Find R_1 and R_2:

$$R_1 = \frac{V_{cc} - V_{BE}(\mathrm{ON})}{I_B(\mathrm{ON})} \qquad (7\text{-}12)$$
$$= \frac{12\ \mathrm{V}}{0.4\ \mathrm{mA}}$$
$$= \mathbf{30\ k\Omega}$$

This is the maximum resistance value for R_1. The transistor will not saturate if R_1 is made greater than 30 kΩ.

Note from the circuit symmetry that Q_1 will also saturate if R_2 is made 30 kΩ or lower. Thus,

$$R_1 = R_2$$
$$= \mathbf{30\ k\Omega}$$

5. Calculate the rest period, t_r:

$$\mathrm{PRF} = 20\ \mathrm{kHz}$$
$$\therefore\ \mathrm{PRT} = \frac{1}{20\ \mathrm{kHz}} \quad \left(\text{or, } \mathrm{PRT} = \frac{1}{\mathrm{PRF}}\right)$$
$$= 50\ \mu\mathrm{s}$$

$$\text{Rest period} = 50\ \mu\text{s} - 10\ \mu\text{s}$$
$$= \mathbf{40\ \mu s}$$

6. Calculate C_2 and C_1. To generate the required waveform, Q_1 must stay OFF for 10 μs and ON for 40 μs. Thus,

$$C_2 = \frac{t_p}{0.693 \times R}$$
$$= \frac{10\ \mu\text{s}}{0.693 \times 30\ \text{k}\Omega}$$
$$= 0.481 \times 10^{-9}\ \text{F}$$
$$= \mathbf{481\ pF}$$

Similarly,

$$C_1 = \frac{40\ \mu\text{s}}{0.693 \times 30\ \text{k}\Omega}$$
$$= 1.924 \times 10^{-9}\ \text{F}$$
$$= \mathbf{0.00192\ \mu F}$$

Example 7-7

Repeat Ex. 7-6 for standard germanium transistors, that is, $V_{CE}(\text{ON}) = 0.1$ V and $V_{BE}(\text{ON}) = 0.3$ V. Use only standard values for the components.

Solution

1. Calculate R_L:

$$R_L = \frac{V_{cc} - V_{CE}(\text{ON})}{I_c(\text{ON})}$$
$$= \frac{12 - 0.1}{20\ \text{mA}}$$
$$= 595\ \Omega$$

Choose R = **680 Ω** (next higher EIA standard value).

$$\text{Actual } I_C(\text{ON}) = \frac{12 - 0.1}{680}$$
$$= 17.5\ \text{mA}$$

2. Calculate $I_B(\text{ON})$:

$$I_B(\text{ON}) = \frac{I_C(\text{ON})}{\beta} \qquad (7\text{-}11)$$

$$= \frac{17.5}{50}$$

$$= \mathbf{0.35\ mA}$$

3. Find R_1 and R_2:

$$R = \frac{V_{cc} - V_{BE}(\text{ON})}{I_B(\text{ON})} \qquad (7\text{-}12)$$

$$R_B(\text{max}) = \frac{12 - 0.3}{0.35\ \text{mA}}$$

$$= 33.43\ \text{k}\Omega$$

Choose

$$R_1 = R_2$$

$$= \mathbf{33\ k\Omega}\ \text{(next lower EIA standard value)}$$

4. Determine C_1:

$$e^{-(t_p/RC)} = \frac{V_{cc} - V_{BE}(\text{ON})}{2V_{cc} - V_{CE}(\text{ON}) - V_{BE}(\text{ON})} \qquad (7\text{-}5)$$

$$= \frac{12 - 0.3}{24 - 0.1 - 0.3}$$

$$= 0.496$$

$$\frac{t_p}{R_1C_1} = 0.70$$

$$C_1 = \frac{t_p}{0.7R_1} \qquad (7\text{-}22)$$

$$= \frac{(10 \times 10^{-6})\ \text{s}}{0.7 \times 33\ \text{k}\Omega}$$

$$= 0.433 \times 10^{-9}\ \text{F}$$

Choose

$$C_1 = \mathbf{470\ pF.}$$

5. Recalculate R_1 by using Eq. (7-22):

$$R_1 = \frac{10 \times 10^{-6}}{0.7 \times 0.47 \times 10^{-9}}$$

$$= \mathbf{30.4\ k\Omega}$$

A standard resistance value of **30 kΩ** is the closest.

6. Determine C_2 by using Eq. (7-22):

$$C_2 = \frac{40 \times 10^{-6}}{0.7 \times 33\ \text{k}\Omega}$$

$$= 1.732 \times 10^{-9}\ \text{F}$$

Choose C_2 = **0.002 μF** (next higher standard value).

7. Recalculate R_2:

$$R_2 = \frac{40 \times 10^{-6}}{0.7 \times 2 \times 10^{-9}}$$

$$= 28.57\ \text{k}\Omega$$

Choose R_2 = **27 kΩ** (closest EIA standard value).

Note that R_1 and R_2 are less than their maximum calculated value (33.43 kΩ).

If the pulse width and the rest period are required to be more exact or adjustable, then R_1 and R_2 consist of a fixed resistor and a variable (adjustable) resistor connected in series. This is shown in Fig. 7-9. Note that the value of the fixed resistor must be above the minimum value that is needed to limit the base current for safe operation.

7-10 Synchronizing an Astable MV

The PRF of the astable MV output is very much dependent upon the supply voltage (Eq. 7-5). Any variation or fluctuation in V_{cc} will be reflected in the PRF of the output waveform. The PRF may also change as a result of component aging or a change in operating temperature. Thus, as it stands, the astable circuit is not very stable.

Synchronization is the process of tying the PRF of the MV to an external source of stable PRF. This makes the astable MV as stable as the external source. Signals from the stable source are fed to the base of one of the transistors, as shown in Fig. 7-10a. These signals get superimposed on the base potential, as shown in Fig. 7-10d. If there was no synchronization signal, Q_1 would turn ON t_1 ms after it was turned OFF (Fig. 7-10b). The presence of a synchronization signal (Fig. 7-10d) forces Q_1 to turn ON in t_2 ms (which is less than t_1). Thus, the synchronization signal forces the transistor to turn ON a little *earlier* than it would have come ON on its own. Note that it is the leading edge of the third pulse that turns Q_1 ON in our example of Fig. 7-10.

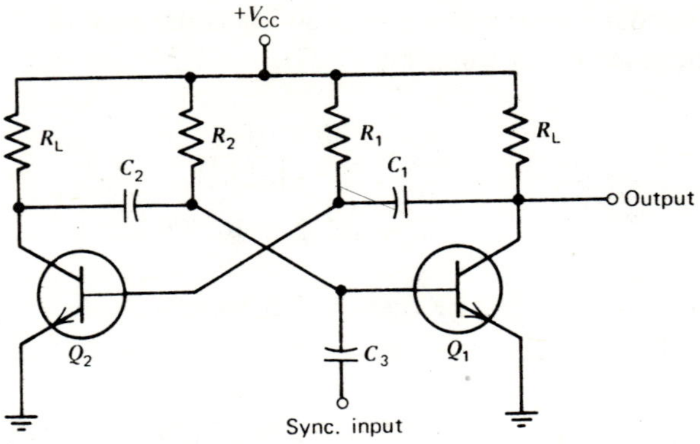

(a) Synchronizing circuit

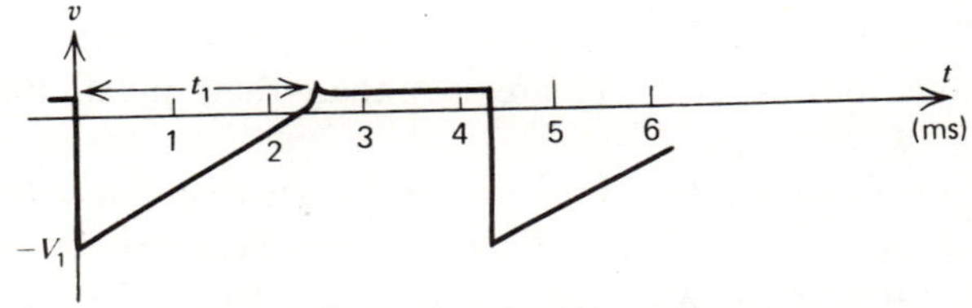

(b) Waveform at the base of Q_1 if unsynchronized

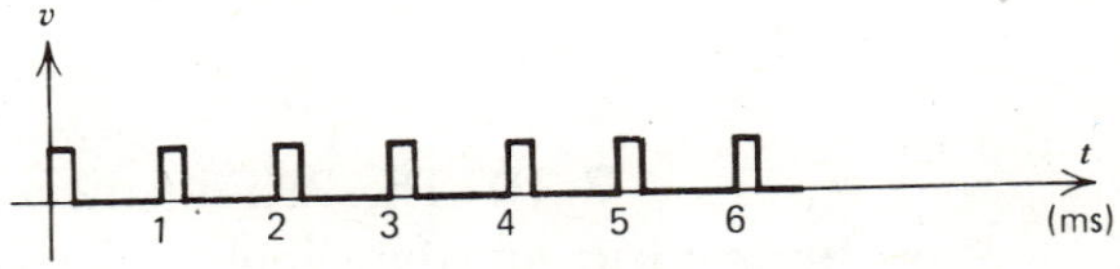

(c) Synchronizing input pulse

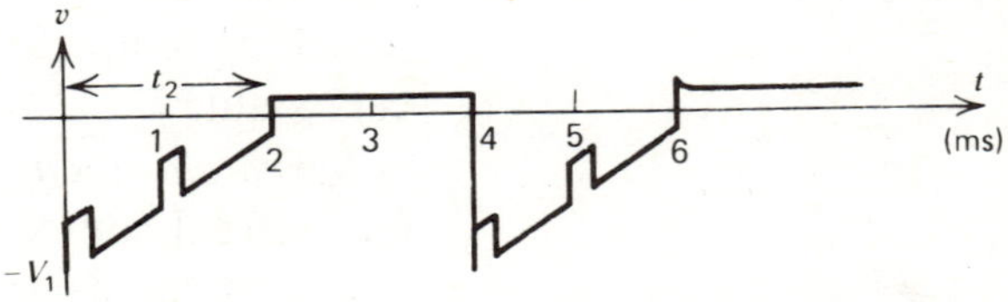

(d) Waveform at the base of Q_1 with synchronized pulse

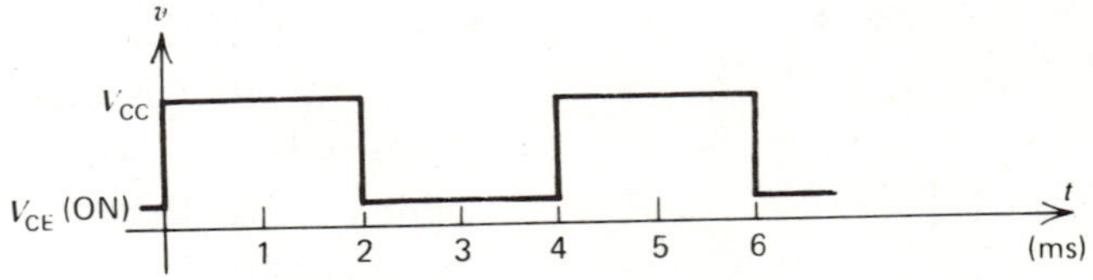

(e) Output waveform

Figure 7-10 Synchronizing circuit and waveforms for an astable MV.

The synchronizing clock input can be any waveshape. The only requirement is that its frequency must be a whole-number multiple of the frequency of the desired output waveform. In Fig. 7-10, the PRF of the clock input is 1 kHz and the frequency of the MV (output) is 250 Hz (whole-number multiple). This will make the astable MV circuit as stable as the synchronizing trigger input.

The amplitude level of the input trigger is also important. Note in Fig. 7-10d that the base voltage will rise above the transistor's turn-ON voltage only during the second pulse if its amplitude was big enough. The whole idea behind synchronization is to force the OFF-transistor ON after a certain set number of input pulses has gone by.

7-11 MOSFET Monostable MV

A monostable MV circuit using two enhancement-type *n*-channel MOSFETs is shown in Fig. 7-11a. Its operation is very similar to the bipolar-transistor monostable MV discussed in Sec. 7-4. In its stable state, Q_1 is OFF and Q_2 is ON. It is assumed that the threshold voltage (V_{TH}) of the MOSFETs is much greater than the drain-to-source voltage of the ON MOSFET. That is,

$$V_{TH} \gg V_{DS}(\text{ON})$$

When the circuit is in its stable state, the static conditions are:

1. $V_{G_1} = E_o = V_{DS}(\text{ON})$.
2. $V_{G_2} = V_{DD}$.
3. $V_{DS}(\text{OFF}) = V_{DD}$.
4. There is no charge on the capacitor C because both its ends are at V_{DD}.

A positive input pulse applied at E_i turns Q_1 ON. Immediately after Q_1 turns ON, the following occurs:

1. V_{DS_1} drops from V_{DD} to $V_{DS}(\text{ON})$.
2. Because C cannot charge instantaneously, this drop [from V_{DD} to $V_{DS}(\text{ON})$] is transferred to the gate of Q_2, making $V_{GS_2} = V_{DS}(\text{ON})$.
3. Q_2 turns OFF because now V_{GS_2} is less than V_{TH}.
4. The circuit assumes its unstable state, that is, $Q_1 = \text{ON}$ and $Q_2 = \text{OFF}$.

The partial circuit shown in Fig. 7-11b reflects circuit conditions

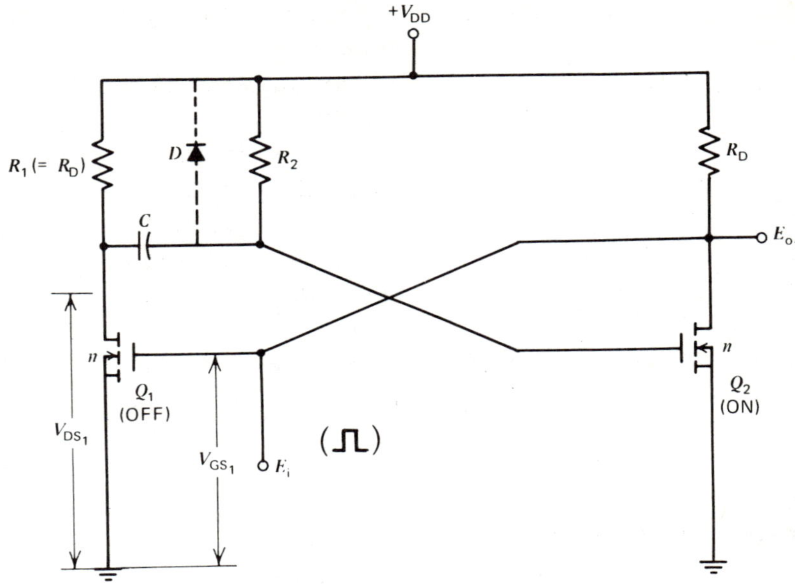

(a) Monostable F/F circuit

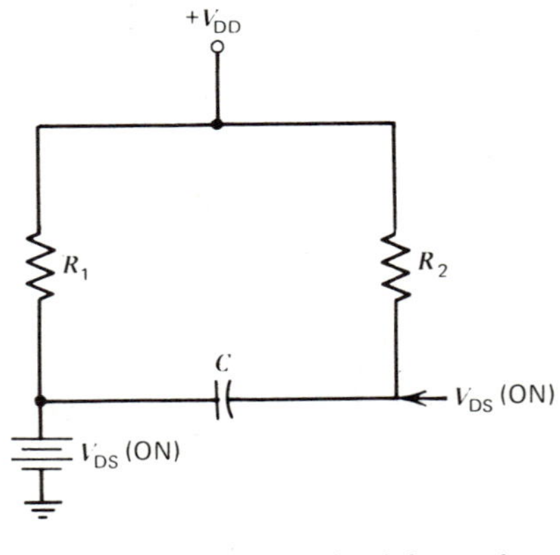

(b) Partial equivalent circuit instantly after Q_2 has been turned OFF

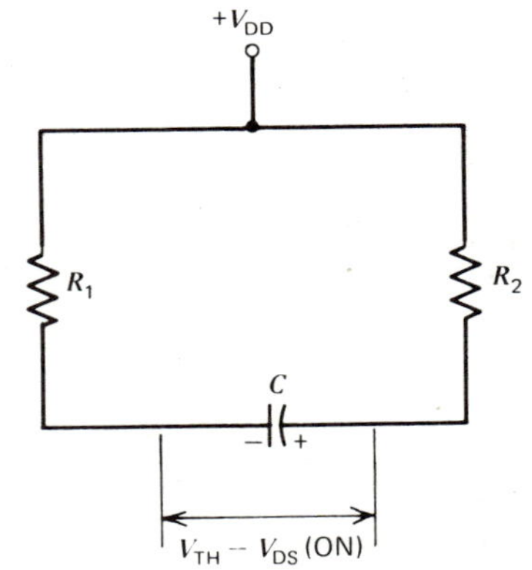

(c) Partial equivalent circuit instantly after Q_2 has returned to its stable state (ON)

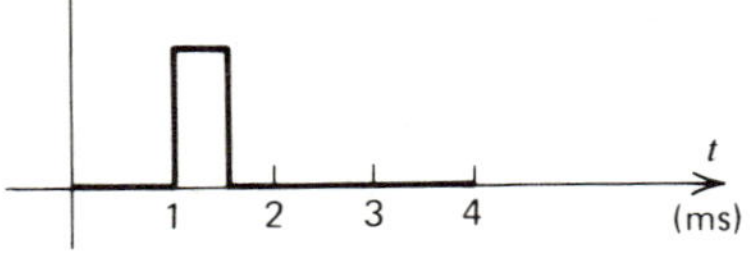

(d) Input pulse waveform

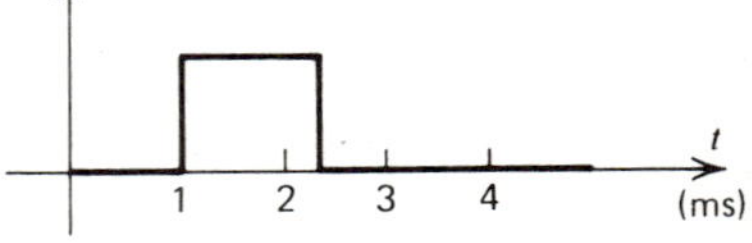

(e) Output pulse at E_{o_1}

Figure 7-11 Monostable F/F.

that exist immediately after the MV has switched to its unstable state. From then on, the following happens:

1. C starts charging toward $V_{DD} - V_{DS}(ON)$.
2. Consequently, V_{GS_2} rises exponentially from $V_{DS}(ON)$ toward V_{DD} with the time-constant R_2C.
3. As V_{GS_2} reaches V_{TH} (after time t_p), Q_2 turns ON.
4. As Q_2 turns ON, V_{GS_2} drops to $V_{DS_2}(ON)$ from V_{DD}.
5. The lowered $V_{DS_2}(ON)$ turns Q_1 OFF.
6. The circuit assumes its stable state.

The output pulse width t_p is calculated from the conditions stated in step 2 above, in other words, as V_{GS_2} goes from $V_{DS}(ON)$ to V_{TH}.

$$v_C = E_i - (E_i \pm V_{ci})e^{-(t/\tau)} \text{ volts (V)} \qquad (3\text{-}7)$$

$$V_{TH} = V_{DD} - [V_{DD} - V_{DS_2}(ON)]e^{-(t_p/R_2C_2)}$$

or

$$t_p = R_2C \ln \frac{V_{DD} - V_{DS_2}(ON)}{V_{DD} - V_{TH}} \text{ seconds (s)} \qquad (7\text{-}23)$$

Note that the charge on C must return to its static conditions ($V_C = 0$ V) before another pulse is applied. The capacitor discharge circuit is shown in Fig. 7-11c. The discharging time-constant is $(R_1 + R_2) \times C$. The time required to discharge the capacitor is:

$$\begin{aligned} t_{re} &= 5\tau \\ &= 5(R_1 + R_2)\cdot C \text{ seconds (s)} \end{aligned} \qquad (7\text{-}24)$$

The symbol t_{re} stands for the *recovery time*. The time space between any two consecutive input trigger-pulses must be greater than $t_{re} + t_p$. Note from Eqs. (7-23) and (7-24) that t_p is even smaller than t_{re}, or conversely, t_{re} is unusually large. In practical circuits, t_{re} is reduced by connecting a diode across R_2 (shown by dashed line in Fig. 7-11a). In such a case:

$$\begin{aligned} t_{re} &= 5\,(R_1 + R_f)\cdot C \\ &\cong 5\,R_1 \times C \quad (\text{since } R_1 \gg R_f) \end{aligned}$$

where:

R_f is the forward resistance of the diode, ohms (Ω)

The input pulse and the output pulse are shown in Figs. 7-11d and 7-11e. Note that the peak-to-peak amplitude of the output pulse is:

$$E_{o_{p\text{-}p}} = V_{DD} - V_{DS}(\text{ON}) \text{ volts (V)} \tag{7-25}$$

Example 7-8
Consider the monostable MV shown in Fig. 7-12. The two MOSFETs being used are the enhancement-type. For the MOSFETs, assume:

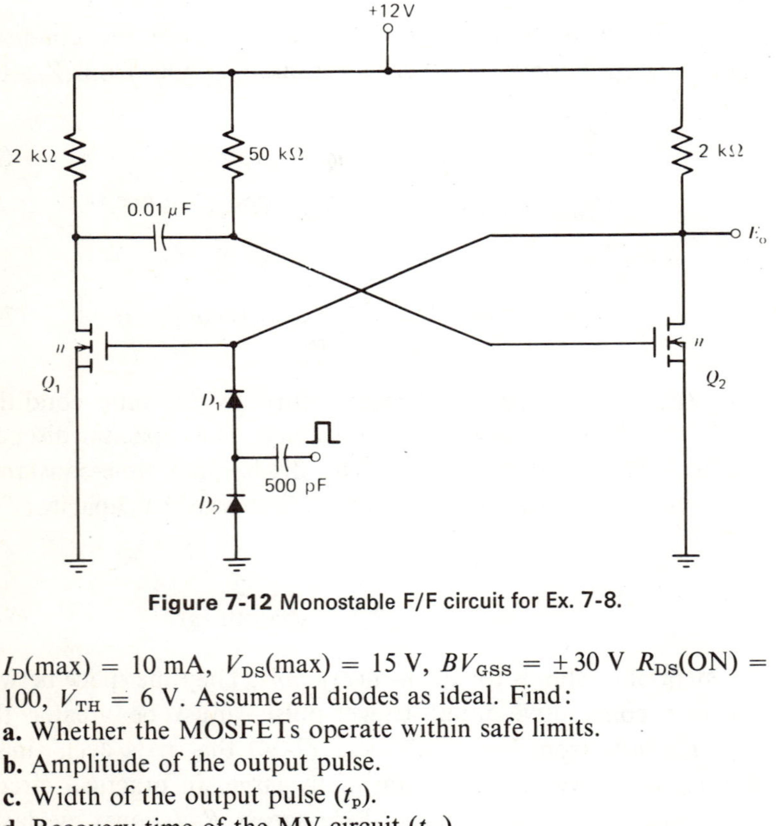

Figure 7-12 Monostable F/F circuit for Ex. 7-8.

$I_D(\text{max}) = 10$ mA, $V_{DS}(\text{max}) = 15$ V, $BV_{GSS} = \pm 30$ V $R_{DS}(\text{ON}) = 100$, $V_{TH} = 6$ V. Assume all diodes as ideal. Find:

a. Whether the MOSFETs operate within safe limits.
b. Amplitude of the output pulse.
c. Width of the output pulse (t_p).
d. Recovery time of the MV circuit (t_{re}).

Solution

a.

$$I_D(\text{ON}) = \frac{V_{DD}}{R_D + R_{DS}(\text{ON})}$$

$$= \frac{12}{2\text{ k}\Omega + 100}$$

$$= 5.714 \text{ mA}$$

I_D(ON), V_{DD}, V_{GS} are all below their respective maximum-rated values. Therefore, the MOSFETs operate within safe limits.

b.

$$V_{DS}(\text{ON}) = \frac{R_{DS}(\text{ON})}{R_D + R_{DS}(\text{ON})} \times V_{DD}$$

$$= \frac{100}{2\text{ k}\Omega + 100} \times 12$$

$$= 0.571 \text{ V}$$

$$E_{o\text{p-p}} = V_{DD} - V_{DS}(\text{ON}) \qquad (7\text{-}25)$$

$$= 12 - 0.571$$

$$= \mathbf{11.43\ V}$$

c.

$$t_p = R_2 C \ln \frac{V_{DD} - V_{DS}(\text{ON})}{V_{DD} - V_{TH}}$$

$$= 50\text{ k}\Omega \times 0.01\ \mu\text{F} \ln \frac{12 - 0.57}{12 - 6}$$

$$= 3.22 \times 10^{-4}$$

$$= \mathbf{0.322\ ms}$$

d.

$$t_{re} = 5(R_1 + R_f)\cdot C$$

$$= 5(2\text{ k}\Omega + 0) \times 0.01 \times 10^{-6}$$

$$= \mathbf{100\ \mu s}$$

7-12 MOSFET Astable MV

An astable MV using enhancement-type MOSFETs is shown in Fig. 7-13. Its operation is very similar to the bipolar-transistor astable MV discussed in Sec. 7-8. Note that for proper operation, the recovery time must be lower than the pulse width for each state. The two diodes must be connected across the gate resistors R_1 and R_2 for this purpose. Peak-to-peak amplitude of the output is:

$$E_{o(\text{p-p})} = V_{DD} - V_{DS}(\text{ON})$$

The time during which Q_1 is OFF is given by:

$$t_{p_1} = R_2 C_2 \ln \frac{V_{DD} - V_{DS}(\text{ON})}{V_{DD} - V_{TH}}$$

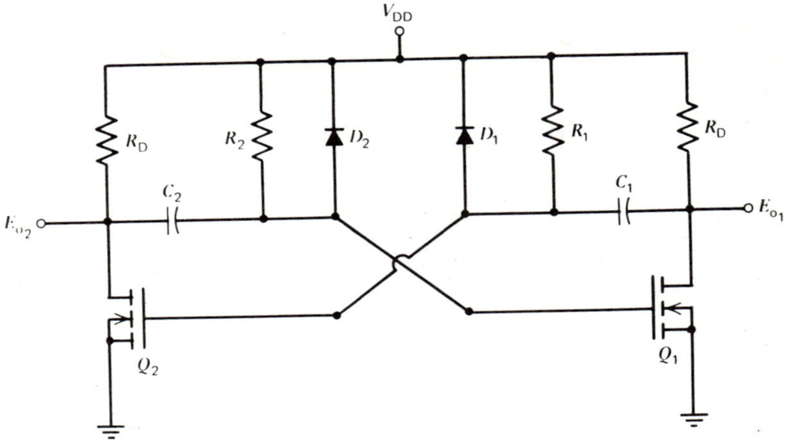

(a) Astable MV circuit

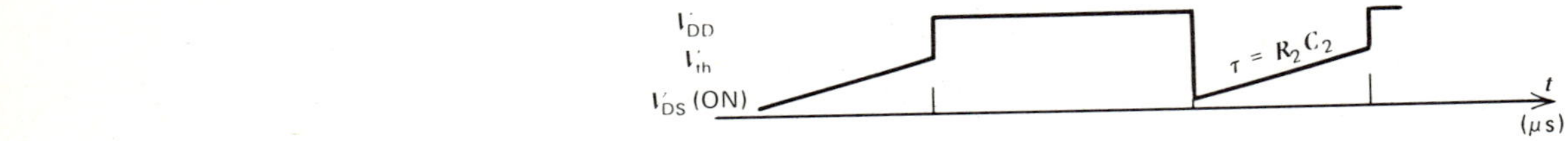

(b) Waveform at the gate of Q_1

(c) Waveform at the drain of Q_1 (E_{o_1})

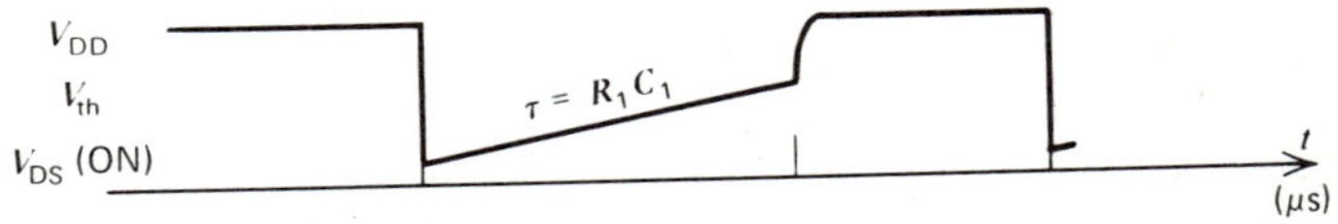

(d) Waveform at the gate of Q_2

(e) Waveform at the drain of Q_2 (E_{o_2})

Figure 7-13 MOSFET astable MV.

And the time during which Q_2 is OFF is given by:

$$t_{p_2} = R_1C_1 \ln \frac{V_{DD} - V_{DS}(ON)}{V_{DD} - V_{TH}}$$

The PRF of the output waveform is:

$$\text{PRF} = \frac{1}{t_{p_1} + t_{p_2}}$$

$$= \frac{1}{(R_1C_1 + R_2C_2)\ln \dfrac{V_{DD} - V_{DS}(\text{ON})}{V_{DD} - V_{TH}}} \text{ hertz (Hz)} \quad (7\text{-}26)$$

Waveforms that appear at both gates and both drains are shown in Fig. 7-13.

Example 7-9

Consider the free-running MV circuit shown in Fig. 7-13. Assume the following: $V_{DD} = 20$ V, $R_D = 1$ kΩ, $R_1 = 22$ kΩ, $R_2 = 33$ kΩ, $C_1 = C_2 = 0.01\ \mu$F, $I_D(\text{max}) = 25$ mA, $V_{DS}(\text{max}) = 20$ V, $BV_{GSS} = \pm 40$ V, $V_{TH} = 8$ V, $R_{DS}(\text{ON}) = 100\ \Omega$, and diodes as ideal. Find:

a. $V_{DS}(\text{ON})$.
b. $E_{o(p\text{-}p)}$.
c. PRF.

Solution

a.

$$V_{DS}(\text{ON}) = \frac{R_D(\text{ON})}{R_D + R_D(\text{ON})} \times V_{DD}$$

$$= \frac{100}{2\text{ k}\Omega + 100} \times 20$$

$$= \mathbf{0.95\ V}$$

b.

$$E_{op\text{-}p} = V_{DD} - V_{DS}(\text{ON}) \quad (7\text{-}25)$$

$$= 20 - 0.95$$

$$= \mathbf{19.05\ V}$$

c.

$$\text{PRF} = \frac{1}{(R_1C_1 + R_2C_2)\ln \dfrac{V_{DD} - V_{DS}(\text{ON})}{V_{DD} - V_{TH}}} \quad (7\text{-}26)$$

$$= \frac{1}{(22\text{ k}\Omega \times 0.01\ \mu\text{F} + 33\text{ k}\Omega \times 0.01\ \mu\text{F})\ln \dfrac{20 - 0.95}{20 - 8}}$$

$$= \frac{1}{(55 \times 10^{-5})(0.462)}$$

$$= \mathbf{3.93\ kHz}$$

7-13 IC Monostable MV

The 74121 is a typical monostable IC MV whose schematic and function tables are shown in Fig. 7-14. The OR gate with negated inputs constitutes a NAND gate. The block labeled *ST* represents a Schmitt trigger (see Sec. 9-4) AND gate. Both the input NAND

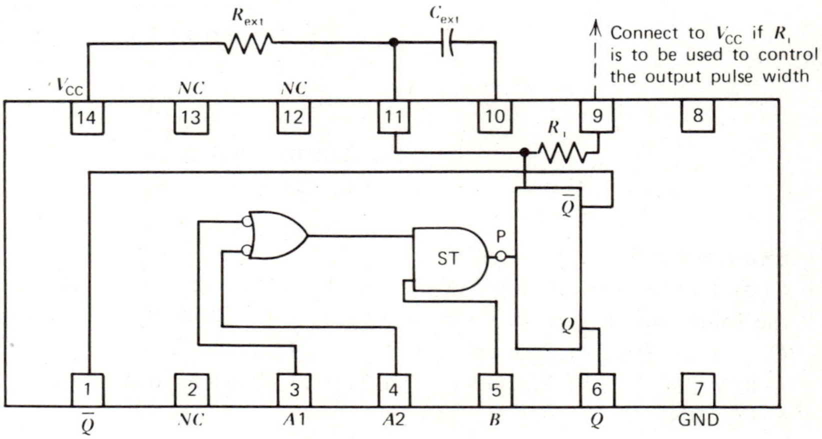

(a) Circuit diagram

Line	Inputs			Outputs	
	A1	A2	B	Q	$\bar{Q}$
1	L	X	H	L	H
2	X	L	H	L	H
3	X	X	L	L	H
4	H	H	X	L	H
5	H	↓	H	⊓	⊔
6	↓	H	H	⊓	⊔
7	↓	↓	H	⊓	⊔
8	L	X	↑	⊓	⊔
	X	L	↑	⊓	⊔

Note:
H = high level (steady state).
L = low level (steady state).
X = irrelevant (any input).
⊓ = one high-level pulse.
⊔ = one low-level pulse.
↓ = transition from high to low.
↑ = transition from low to high.

(b) Function table

Figure 7-14 The 74121 monostable MV and its function table.

gate and the Schmitt-trigger AND gate form a part of the monostable MV 74121. (See Ch. 11 for logic operations.)

A rising pulse edge at point P triggers the MV, switching it into its unstable state. The duration of time that it stays in the unstable state is controlled by the time constant RC. The output pulse width t_p is given by:

$$t_p = 0.7RC \text{ seconds (s)} \tag{7-27}$$

where:

R is either the externally-connected resistance or internal resistance when connected as shown in Fig. 7-14. In 74121 this internal resistance is a 2 kΩ.

and

C is the externally connected capacitance (F).

Note that the 74121 monostable MV will not function properly if the externally connected resistor R exceeds 40 kΩ. Also, a pulse width of 30 ns is obtained with no external capacitance and using the internal resistance R_i.

As mentioned above, the monostable MV triggers when a rising pulse edge appears at point P. This is accomplished in one of three ways:

1. One or both of the inputs to the NAND gate are LOW and the B input goes HIGH.
2. The B input is made HIGH and one of the A inputs goes LOW while the other A input stays HIGH.
3. The B input is made HIGH and both A inputs go LOW simultaneously.

The 74122 shown in Fig. 7-15 is a retriggerable monostable MV. It differs from 74121 in that:

1. It responds to a trigger pulse when it is in its stable state as well as when it is in its unstable state. If a trigger is applied when it is in its unstable state, it RESETS the timing (back to initial conditions). The monostable does not assume its stable state until one pulse duration has elapsed after the last trigger was applied. Therefore, a retriggerable MV may be kept in the unstable state indefinitely by an input pulse train whose PRT is less than the output pulse width (t_p).

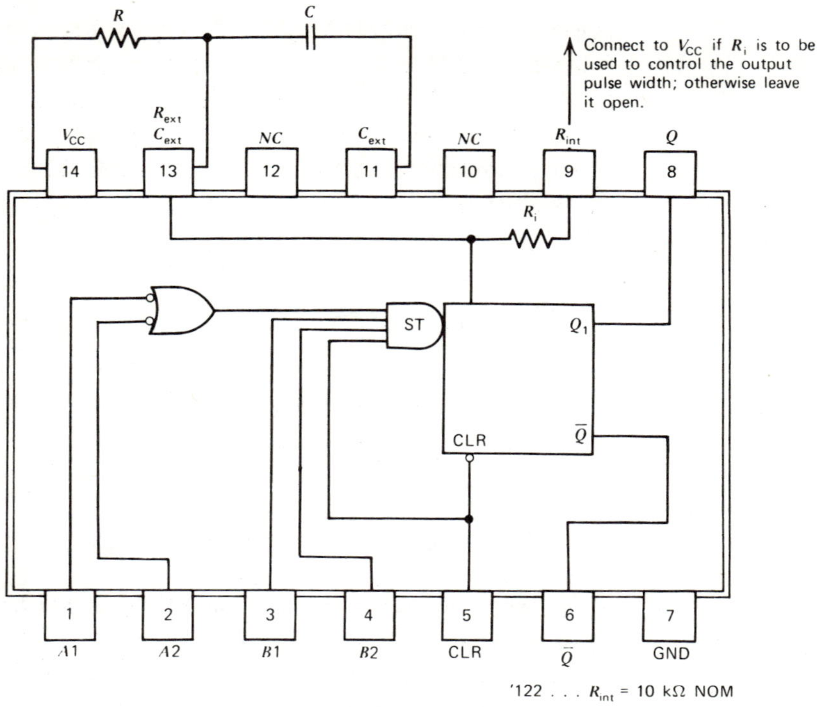

(a) Pin layout

Inputs					Outputs	
Clear	A1	A2	B1	B2	Q	$\bar{Q}$
L	X	X	X	X	L	H
X	H	H	X	X	L	H
X	X	X	L	X	L	H
X	X	X	X	L	L	H
X	L	X	H	H	L	H
H	L	X	↑	H	⊓	⊔
H	L	X	H	↑	⊓	⊔
H	X	L	H	H	L	H
H	X	L	↑	H	⊓	⊔
H	X	L	H	↑	⊓	⊔
H	H	↓	H	H	⊓	⊔
H	↓	↓	H	H	⊓	⊔
H	↑	H	H	H	⊓	⊔
↑	L	X	H	H	⊓	⊔
↑	X	L	H	H	⊓	⊔

Note:
H = high level (steady state).
L = low level (steady state).
X = irrelevent (any input).
⊓ = one high-level pulse.
⊔ = one low-level pulse.
↓ = transition from high to low.
↑ = transition from low to high.

(b) Function table

Figure 7-15 The 74122 monostable MV and its function table.

2. It has a RESET capability. The presence of a LOW logic level at the CLEAR input terminal (pin 5) RESETS the monostable immediately.

Example 7-10
Refer to the monostable MV shown in Fig. 7-14a. Assume $V_{cc} = 5$ V, $C = 0.01\ \mu$F, and $R = 20$ kΩ. Find the output pulse width (t_p).

Solution

$$t_p = 0.7RC \qquad (7\text{-}27)$$
$$= 0.7 \times 20\ \text{k}\Omega \times 0.01\ \mu\text{F}$$
$$= \mathbf{140\ \mu s}$$

Example 7-11
Consider the retriggerable monostable MV circuit shown in Fig. 7-15a. Assume $V_{cc} = 5$ V, $R = 22$ kΩ, and $C = 300$ pF. Find:
a. The output pulse width.
b. The minimum PRF of the input trigger pulse such that the monostable always stays in the unstable state.

Solution

a.
$$t_p = 0.7RC \qquad (7\text{-}27)$$
$$= 0.7 \times 22\ \text{k}\Omega \times 300\ \text{pF}$$
$$= \mathbf{4.62\ \mu s}$$

b.
$$\text{PRF}_{\text{min}} = \frac{1}{\text{PRT}_{\text{max}}}$$
$$= \frac{1}{4.62\ \mu\text{s}}$$
$$= \mathbf{216\ kHz}$$

7-14 Summary

1. A monostable MV (one-shot) is stable in one state and unstable in the other state.
2. An external trigger input pulse is used to switch the monostable MV into its unstable state.
3. The triggering circuit may be designed such that the monostable circuit is switched into its unstable state by either the leading edge or the trailing edge of the trigger pulse.

4. The time duration for which a monostable MV stays in its unstable state is controlled by the internal circuitry of the monostable (Eq. 7-5). This is usually the output pulse width (t_p).
5. The peak-to-peak amplitude of a pulse generated by a monostable MV is $V_{cc} - V_{CE}(\text{ON})$.
6. A monostable MV may be used as a pulse-reshaping circuit as well as a pulse-delay circuit.
7. The time interval between any two consecutive trigger pulses must be greater than $t_p + t_{re}$ where t_p = output pulse width, and t_{re} = recovery time.
8. An astable MV (free-running) is unstable in both of the states. The circuit flips back and forth between the two states on its own.
9. The output of a free-running MV is a rectangular waveform whose peak-to-peak amplitude is $V_{cc} - V_{CE}(\text{ON})$.
10. The time duration for which each transistor stays OFF in an astable MV is controlled by the corresponding time constant C_sR (Eq. 7-5) where C_s = coupling capacitor, and R = base-current-controlling resistor.
11. For proper operation, recovery time for one transistor (associated capacitor) must be greater than the time that the other transistor stays OFF.
12. A synchronizing circuit may be added to the basic astable circuit to improve stability.
13. The frequency of the synchronizing trigger input must be a whole-number multiple of the desired PRF of the astable output.
14. The operation of MOSFET MVs is similar to BJT MVs.
15. The most commonly-used TTL IC monostable MVs are the 74121 and the 74122.
16. *Important equations:*

$$e^{-(t_p/RC)} = \frac{V_{cc} - V_{BE}(\text{ON})}{2V_{cc} - V_{CE}(\text{ON}) - V_{BE}(\text{ON})} \tag{7-5}$$

$$t_{re} = 5R_LC \tag{7-6}$$

$$V_{p\text{-}p} = V_{cc} - V_{CE}(\text{ON}) \tag{7-7}$$

$$t_p = 0.693RC \tag{7-8}$$

$$R_L = \frac{V_{cc} - V_{CE}(\text{ON})}{I_c(\text{max})} \tag{7-9}$$

$$R = \frac{V_{cc} - V_{BE}(\text{ON})}{I_B(\text{ON})} \tag{7-12}$$

$$\frac{V_{CE}(ON) - V_{BE}(ON)}{R_1} = \frac{V_{BE}(OFF) - V_{DB}}{R_2} \tag{7-13}$$

$$\frac{V_{cc} - V_{BE}(ON)}{R_L + R_1} - \frac{V_{BE}(ON) - V_{BB}}{R_2} \geqq I_B(ON) \tag{7-14}$$

$$t_{re_2}(C_2) = 5R_LC_2 \tag{7-19}$$

$$t_{re_1}(C_1) = 5R_LC_1 \tag{7-20}$$

$$f = \frac{1}{t_1 + t_2} \tag{7-21}$$

$$t_p = RC \ln \frac{V_{DD} - V_{DS}(ON)}{V_{DD} - V_{TH}} \tag{7-23}$$

$$t_{re} = 5(R_1 + R_2)\cdot C \tag{7-24}$$

$$E_{o(p\text{-}p)} = V_{DD} - V_{DS}(ON) \tag{7-25}$$

$$\text{PRF} = \frac{1}{(R_1C_1 + R_2C_2) \ln \dfrac{V_{DD} - V_{DS}(ON)}{V_{DD} - V_{TH}}} \tag{7-26}$$

$$t_p = 0.7RC \tag{7-27}$$

7-15 Glossary

Recovery time (t_{re}). The time that the coupling capacitor in an MV circuit takes to recover to its steady-state situation right after the circuit has switched from one state to the other.

Synchronization. The process of tying the frequency of a free-running MV circuit to some external source of stable frequency.

7-16 Recapitulation

Now return to the Objectives and Self-Evaluation Questions given at the beginning of this chapter and see how well you can answer them. If you cannot answer certain questions satisfactorily, place a check mark (√) next to them and review appropriate parts of the chapter to find the answers. No attempt should be made to solve the Problems (Sec. 7-16) before the Self-Evaluation Questions are answered. Ability to solve the Problems should serve as a test and feedback regarding your understanding of the material presented in this chapter. Answers to the Problems are given on p. 524.

7-17 Problems

7-1 Design a monostable MV circuit such that it generates an output pulse of 10 $V_{p\text{-}p}$ amplitude and 20 μs duration. The transistors to be used have $I_C(ON) = 50$ mA and $\beta = 20$. Assume $V_{CE}(ON) = 0$, $V_{BE}(ON) = 0$ V, $V_{BB} = -10$ V, and $V_{BE}(OFF) = -1.0$ V.

7-2 Repeat Prob. 7-1 for $t_p = 50$ μs.

7-3 Repeat Prob. 7-1 for $V_{CE}(ON) = 0.2$ V and $V_{BE}(ON) = 0.5$ V.

7-4 Repeat Prob. 7-3 using only the EIA standard values for resistors and capacitors.

7-5 Design a one-shot MV circuit such that it generates an output pulse of 1.2 ms duration and -18 $V_{p\text{-}p}$ amplitude. The standard silicon transistors to be used have $I_C(ON) \leqq 20$ mA and $\beta = 75$. Assume that $V_{BB} = +10$ V and $V_{BE}(OFF) = 1$ V.

7-6 Repeat Prob. 7-5 using only the EIA standard values for resistors and capacitors.

7-7 Consider the monostable circuit shown in Fig. 7-2a. Assume:

$V_{cc} = -20$ V	$R_1 = 22$ kΩ	$V_{BB} = +20$
$R_L = 470$ Ω	$R_2 = 470$ kΩ	$\beta = 50$
$R = 22$ kΩ	$C = 325$ pF	Ideal transistors

Check by calculations whether this circuit will function satisfactorily as a saturated-type monostable MV.

7-8 Refer to Prob. 7-7. Find the minimum time interval that is required between any two successive trigger input pulses for proper operation.

7-9 Repeat Prob. 7-7 for $I_C(max) = 10$ mA, $V_{CE}(ON) = 0.3$ V, and $V_{BE}(ON) = 0.7$ V.

7-10 Design an astable MV circuit such that the output voltage waveform is as shown in Fig. 7-16. The two ideal transistors that are to be used have $\beta = 60$ and $I_C(ON) = 25$ mA. Assume $V_{BB} = -6$ V.

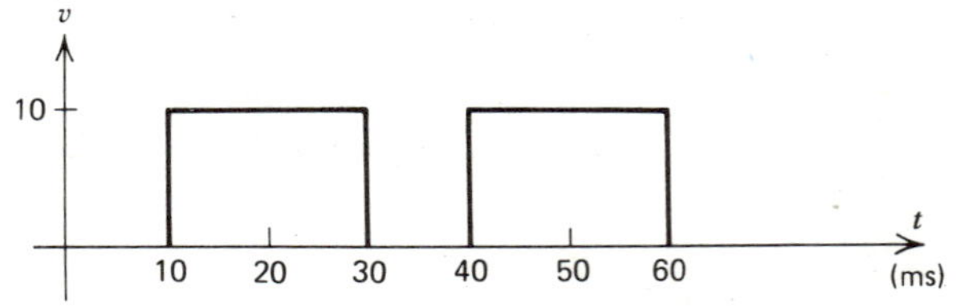

Figure 7-16 Waveform for Prob. 7-10.

7-11 Repeat Prob. 7-10 using only EIA standard values for resistors and capacitors.

7-12 Design an astable MV whose output is a 20 $V_{p\text{-}p}$, 50 kHz square wave. The two transistors to be used have $\beta = 150$, $I_C(max) = 15$ mA, $V_{CE}(ON) = 0.1$ V, and $V_{BE}(ON) = 0.3$ V. Use only the EIA standard values for all resistors and capacitors.

7-13 Refer to the astable MV circuit of Fig. 7-9.

$V_{cc} = 18$ V	$C_1 = 0.047\ \mu$F	$I_C(\max) = 20$ mA
$R_L = 1$ kΩ	$C_2 = 0.001\ \mu$F	$V_{CE}(\text{ON}) = 0.2$ V
$R_1 = 82$ kΩ	$\beta = 100$	$V_{BE}(\text{ON}) = 0.4$ V
$R_2 = 82$ kΩ		

Check by calculations whether this circuit functions properly as a saturated-type free-running MV.

7-14 Assume that the circuit of Prob. 7-13 is corrected so that it does operate properly ($C_2 = 0.1\ \mu$F). Draw the voltage waveforms that appear at both of the bases and collectors of the transistors.

7-15 The two diodes D_1 and D_2 shown in the astable MV circuit of Fig. 7-17 are sometimes used to protect the base-emitter junction of the

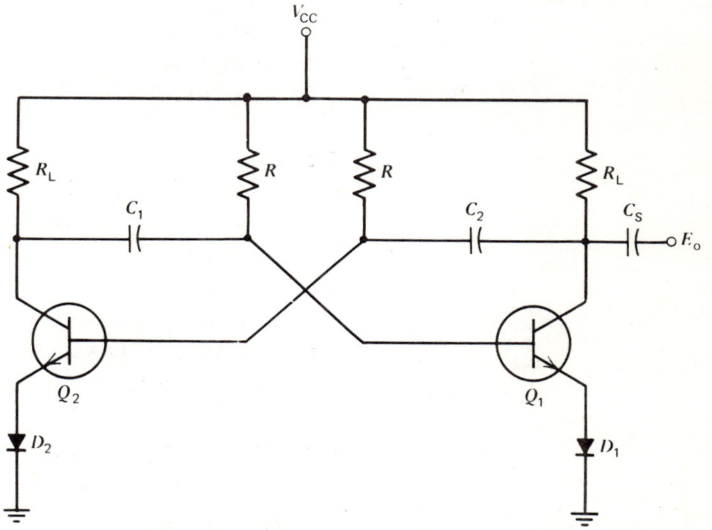

Figure 7-17 Astable MV circuit for Prob. 7-15.

transistor against excessive reverse-bias when the transistor turns OFF. Calculate the values for R_L, R_B, C_1, C_2, and V_{cc} if the desired output is a 12 $V_{p\text{-}p}$, 100 kHz square wave. The two silicon transistors have $\beta = 75$ and $I_C(\text{ON}) = 24$ mA. D_1 and D_2 are standard silicon diodes.

7-16 For the monostable circuit of Fig. 7-11, assume $R_D = 1.5$ kΩ, $R_2 = 470$ kΩ, $C = 300$ pF, and $V_{DD} = 10$ V. The two n-channel enhancement-type MOSFETs are rated as $I_D(\max) = 25$ mA, $R_{DS}(\text{ON}) = 75\ \Omega$, $V_{DS}(\max) = 30$ V, $V_{TH} = +5$ V and $BV_{SS} = \pm$ 30 V. Assume the diode to be ideal. Calculate and find:

a. Whether the MOSFETs operate within safe limits.
b. Amplitude of the output pulse.
c. Width (t_p) of the output pulse.
d. Recovery time (t_{re}) of the MV.

7-17 Repeat Prob. 7-16 for a new R_D value of 2.2 kΩ (instead of 1.5 kΩ).

7-18 Repeat Prob. 7-16 for a new C value of 0.001 μF (instead of 300 pF).

7-19 Consider the circuit described in Prob. 7-16. A 50 kΩ variable resistance is connected in series with the 470 kΩ resistor (R_2). Calculate the range of output pulse width (t_p) obtainable from this circuit.

7-20 For the astable MV shown in Fig. 7-13a, assume $R_D = 2.2$ kΩ, $C_1 = C_2 = 0.047$ μF, $R_1 = R_2 = 100$ kΩ, and $V_{DD} = +12$ V. The two n-channel enhancement-type MOSFETs are rated as $I_D(\text{max}) = 20$ mA, $R_{DS}(\text{ON}) = 100$ Ω, $V_{DS}(\text{max}) = 25$ V, $V_{TH} = 5$ V, and $BV_{SS} = \pm 25$ V. Assume D_1 and D_2 to be ideal. Calculate and find:

a. $V_{DS}(\text{ON})$.

b. $E_{o(p\text{-}p)}$.

c. The PRF of the output waveform.

7-21 Repeat Prob. 7-20 for a new R_D value of 1.5 kΩ (instead of 2.2 kΩ).

7-22 Repeat Prob. 7-20 for a new C_1 value of 0.025 μF (instead of 0.047 μF).

7-23 Consider the circuit described in Prob. 7-20. A 50 kΩ variable resistance is placed in series with R_1 (100 kΩ). Find the range of frequencies of the output waveform (PRF) obtainable from this circuit.

7-24 Consider the monostable circuit shown in Fig. 7-14a. Find a value of C such that when connected in conjunction with the internal resistance, R_i (2 kΩ) produces an output pulse of 56 μs.

7-25 Repeat Prob. 7-24, but this time use an externally connected resistance of 50 kΩ.

7-26 Refer to the retriggerable monostable circuit of Fig. 7-15a. Assume $V_{cc} = 5$ V, $C = 0.047$ μF, and $R = 10$ kΩ(R_1). Find the output pulse width t_p.

7-27 For the circuit described in Prob. 7-26, find the minimum PRF of the input trigger such that the monostable MV always stays in the unstable state.

7-18 Experiment

Title **Monostable and Astable MVs**

Theory A monostable MV is stable in one state and unstable in the other. Normally, the circuit is in its stable state. A trigger input pulse is applied to switch it into the unstable state. After the trigger pulse has disappeared, and a certain amount of time (t_p) has elapsed, the circuit comes back to its stable state on its own.

$$t_p = RC \ln \frac{2V_{CC} - V_{CE}(\text{ON}) - V_{BE}(\text{ON})}{V_{CC} - V_{BE}(\text{ON})}$$

$$\cong 0.693RC$$

and

$$E_{(\text{op-p})} = V_{CC} - V_{CE}(\text{ON})$$

A free-running MV is unstable in both the states. It continuously flips back and forth between the two states on its own.

$$t_{p_1} \cong 0.693R_1C_1 \quad \text{and} \quad t_{p_2} \cong 0.693R_2C_2$$

$$f = \frac{1}{t_{p_1} + t_{p_2}} = \frac{1}{0.693(R_1C_1 + R_2C_2)} = (\text{PRF})$$

$$E_{o(\text{p-p})} = V_{CC} - V_{CE}(\text{ON})$$

Equipment Dual-beam oscilloscope: PHILIPS model 3230 or equal.
Power supply: ANATEK model 25–2.00, or equal.
Multimeter: RCA model WV 98C, or equal.
Multifunction generator: EXACT model 126, or equal.
Two transistors: 2N4401 and two switching diodes.
Capacitors: 250 pF and two 0.01 μF.
Resistors: two each of 510 kΩ, 22 kΩ, 50 kΩ pot.; and one 120 kΩ.

Procedure

A. Monostable MV

1. Connect the monostable circuit shown in Fig. E7-1. Make:

$R_L = 510\ \Omega$	$R_1 = 22\ k\Omega + 50\ k\Omega$ pot.
$R_2 = 22\ k\Omega$	$R_3 = 120\ k\Omega$
$C_1 = 0.01\ \mu F$	$C_2 = 250\ pF$
$V_{CC} = 12\ V$	$V_{BB} = -12\ V$

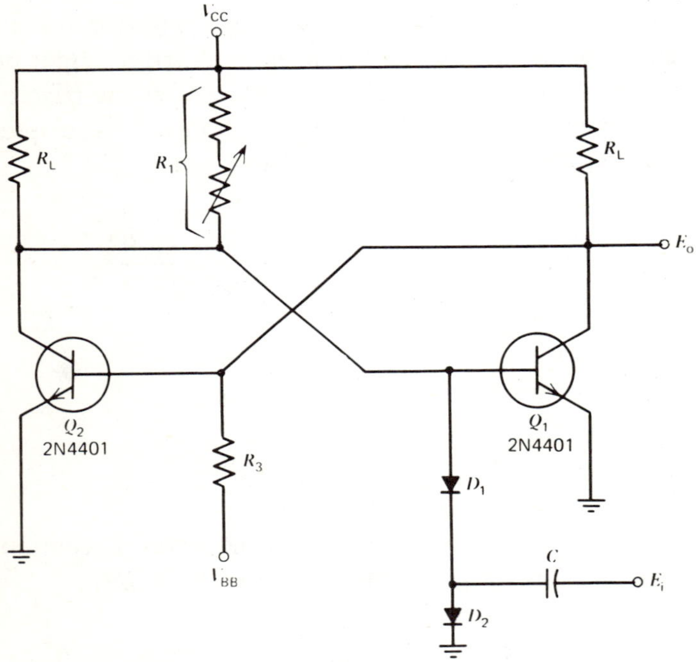

Figure E7-1 Monostable multivibrator.

2. Set the 50 kΩ potentiometer to the 0 Ω position. Measure and record in Table E7-1 the **dc** voltages appearing at both of the bases and collectors.
3. Set the potentiometer to a full 50 kΩ. Measure again the **dc** voltages appearing at both of the bases and collectors. They should remain the same as in Procedure step 2 above. If they are not the same, check with your instructor or your circuit to find why Q_1 does not go ON under this condition.
4. Set the potentiometer back to the 0 Ω position. Apply a 1 kHz square wave of appropriate amplitude (2 $V_{p\text{-}p}$) at E_i, the trigger input terminal. Observe (CRO) and record in Table E7-2 voltage waveforms appearing at the two bases, the two collectors (output), and the input. Note that $V_{BE}(Q_2)$ must be measured using a high-impedance probe to minimize the differentiating effect.

5. Increase the potentiometer's resistance slowly. Observe and record its effects on the output waveform, especially the pulse width (t_p). Measure and record in Table E7-1 the peak-to-peak amplitude and the maximum and minimum values of t_p obtainable from this monostable circuit (by varying the 50 kΩ potentiometer).

B. Free-running MV

6. Connect the free-running MV circuit shown in Fig. E7-2. Make $R_L = 510\ \Omega$, $R_1 = R_2 = 22\ k\Omega + 50\ k\Omega$ pot., $C_1 = C_2 = 0.01\ \mu F$, $V_{CC} = +12$ V.

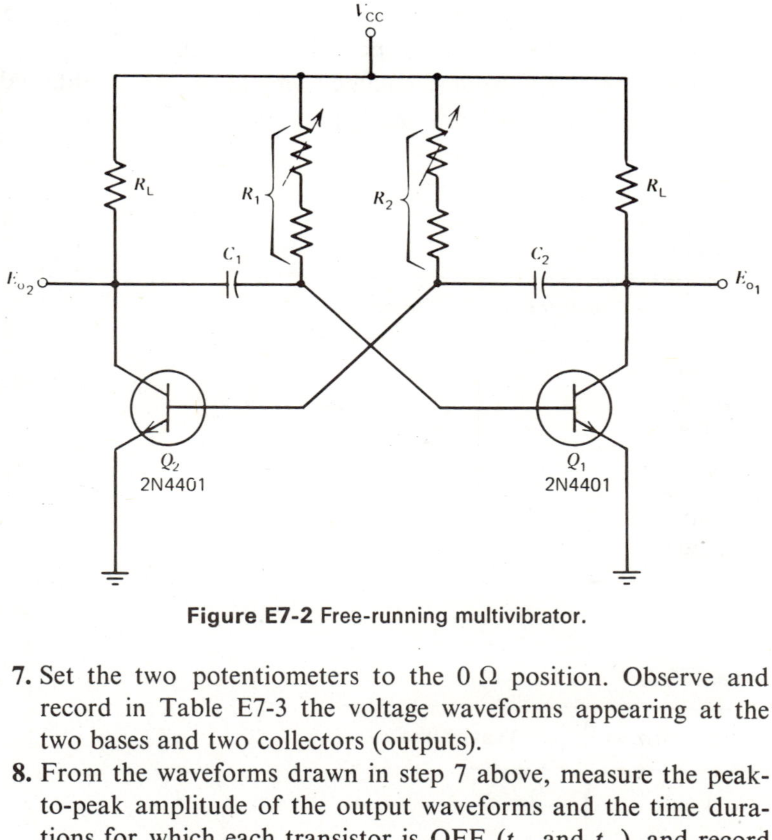

Figure E7-2 Free-running multivibrator.

7. Set the two potentiometers to the 0 Ω position. Observe and record in Table E7-3 the voltage waveforms appearing at the two bases and two collectors (outputs).
8. From the waveforms drawn in step 7 above, measure the peak-to-peak amplitude of the output waveforms and the time durations for which each transistor is OFF (t_{p_1} and t_{p_2}), and record them in Table E7-4.
9. Vary the resistance of the potentiometers, one at a time, and record the effect on the output waveforms. Measure and record in Table E7-4 the minimum and maximum output frequency obtainable from the circuit.
10. Disconnect and return all test equipment to its proper place.

Calculations

1. Find and record V_{CE}(ON) and V_{BE}(ON) from the data book for the 2N4401. Record your findings in Table E7-1 (See Exp. 5).
2. Check by calculations whether the monostable MV of Fig. E7-1 functions safely and properly (see Ex. 7-4).
3. Calculate the peak-to-peak amplitude value of the output voltage and the width of the output pulse. Enter your results in Table E7-1.
4. Consider the astable MV described in step 6 of the Procedure. Calculate:
   a. And check whether the circuit is safe and operates properly.
   b. And find t_{p_1}, t_{p_2}, t_{re_1} and t_{re_2} (Eqs. 7-5 and 7-6).
   c. The peak-to-peak amplitude of the output waveform and its frequency (PRF) when the pot. is set at 0 Ω.
   d. The range of output frequency (max. to min.) obtainable. Enter your results from **4b**, **4c** and **4d** in Table E7-4.

Tabulation of Results

Table E7-1

	Observed	Theoretical
V_{CE}(ON)		
V_{BE}(ON)		
V_{CE}(OFF)		
V_{BE}(OFF)		
$E_{o(p\text{-}p)}$		
t_p(min)		
t_p(max)		

Table E7-2

Waveform at	Waveform
Input	
$V_{BE}(Q_1)$	
$V_{CE}(Q_1)$	
$V_{BE}(Q_2)$	
$V_{CE}(Q_2)$	

Table E7-3

Waveform at	Waveform
$V_{BE}(Q_1)$	
$V_{CE}(Q_1)$	
$V_{BE}(Q_2)$	
$V_{CE}(Q_2)$	

Table E7-4

Item	Observed	Theoretical
t_{p1}		
t_{p2}		
f (max)		
f (min)		
$E_{o(p\text{-}p)}$		

Questions

1. Examine Table E7-1. How do the observed values compare with the theoretically expected ones?
2. What determines the width of the output pulse in a monostable MV?
3. Explain the operation of the triggering circuit used in Fig. E7-1.
4. What determines the minimum value of R_1 used in the monostable MV of Fig. E7-1?
5. Examine Table E7-4. Compare the experimental values with the theoretically observed ones.
6. How will the waveforms recorded in Table E7-3 change if C_1 and C_2 are both changed to 0.001 μF capacitors (Fig. E7-2)?
7. In Procedure step 3, you measured V_{BE} and V_{CE} for both transistors. Do your measured values still indicate that one transistor is ON while the other is OFF?

Discussion

1. Summarize your conclusions from the observed and the theoretical values.
2. Account for discrepancies between the observed and the theoretically expected values.
3. Compare the BJT monostable and astable MVs with their MOSFET counterparts.

Chapter 8 Sweep Generators

8-1 Objectives

The objectives of this chapter are to:

1. Show how a voltage *sweep* is generated by charging/discharging a capacitor.
2. Explain how to calculate the *slope error* of a sweep.
3. Explain the operation of a simple *astable sweep*-generator using a *four-layer* diode as an active element.
4. Describe the operation and the design procedure of a *transistor* sweep-generator.
5. Discuss the operation of a *Miller sweep* circuit.
6. Explain the operation of a *bootstrap sweep* circuit.
7. Describe the operation and the design procedure of a *constant-current* sweep circuit.
8. Explain the operation of a *current* sweep-generator.

8-2 Self-Evaluation Questions

Watch for the answers to the following questions as you read the chapter. They should help you to understand the material presented. When you have finished the chapter, return to this section and be sure that you can answer all the questions. Then test your understanding by working out solutions to the Problems.

It is strongly recommended that you answer these questions satisfactorily before attempting to solve the Problems (Sec. 8-13).

1. Draw an *ideal sawtooth* waveform.
2. Draw a *practical sawtooth* waveform. Label all parts including the sweep and the retrace times.
3. Explain how a voltage-sweep waveform is generated by charging a capacitor through a high resistance and discharging it through a low resistance.
4. Explain and identify important electrical characteristics of a *four-layer* diode.
5. Explain the operation of a simple astable-sweep-generator circuit that uses a *four-layer* diode as an active element.
6. Explain the operation of a *transistor* sweep-generator circuit.
7. Show how to design a simple transistor-sweep circuit and identify circuit conditions that will improve sweep linearity.
8. Show a method of calculating the *minimum retrace time* required for a given transistor-sweep-generator circuit.

9. Discuss the basis of *Miller* sweep generation.
10. Explain the operation of a simple *Miller* sweep circuit.
11. Explain the operation of a *bootstrap* sweep circuit.
12. Draw a simple *constant-current* sweep circuit.
13. Discuss the basis of *constant-current* sweep generation.
14. Discuss the basis of *current sweep* generation.
15. Describe the operation of a current sweep circuit.

8-3 Voltage Sweep Principles

The term *sweep* means a ramp function. *Voltage sweep* means that the voltage is increasing (or decreasing) with respect to time. Similarly, *current sweep* means that the current is increasing (or decreasing) with respect to time. The sawtooth waveform, discussed in Sec. 2-14, is an example of a sweep waveform.

Figure 8-1a shows an ideal sawtooth waveform. Each cycle is

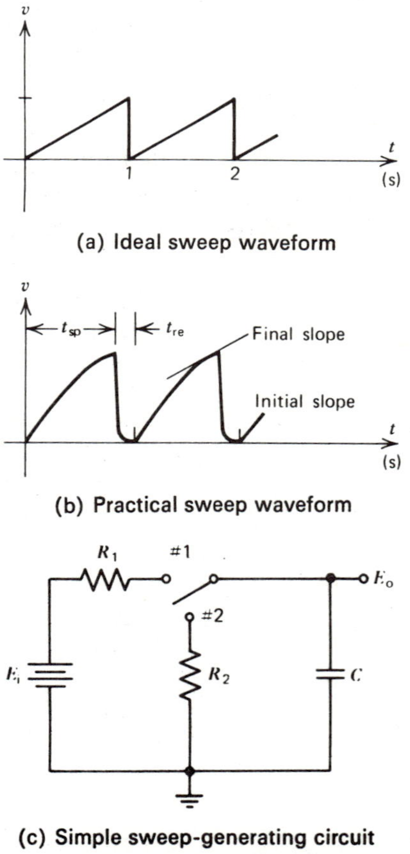

Figure 8-1 Voltage sweeps and a simple sweep-generating circuit.

made up of an ideal (rising) *ramp* function and an ideal (falling) *step* function. However, in practice it is impossible to generate ideal functions. A practical sawtooth wave is shown in Fig. 8-1b.

One way of generating a voltage sawtooth waveform is to charge a capacitor *slowly* through a *high* resistance and then discharge it *quickly* through a *low* resistance. The simple circuit arrangement shown in Fig. 8-1c will accomplish this. $R_1 \gg R_2$. C charges through a high resistance (R_1) with the switch in position #1. Usually the sweep-charging time (t_{sp}) is kept small in comparison to the charging time-constant CR_1. This allows the charge curve to be fairly linear. The capacitor C is then discharged through a low resistance (R_2) by switching the switch to position #2. C discharges fairly quickly (t_{re}) because the discharging time constant CR_2 is made very small.

Time shown as t_{sp} in Fig. 8-1b is known as the *sweep time*. It is also sometimes called the *trace time*. It is the time during which the function *increases* linearly. Time t_{re} is known as the *retrace time* or the *flyback time*. It is the time that the function takes to drop back down to the *initial* base voltage or the reference voltage.

When designing practical sweep-generating circuits, our goal is to generate a *linear* sweep as close to an ideal sweep (Fig. 8-1a) as possible. The practical (generated) sweep may not be so linear. Linearity of sweep is measured by its *slope error*.

$$\text{percentage slope error} = \frac{\text{Initial slope} - \text{Final slope}}{\text{Initial slope}} \times 100 \tag{8-1}$$

The initial slope and the final slope of the waveform are shown in Fig. 8-1b. The slope error measures the change in slope and the extent to which a *practical sweep* deviates from an *ideal* sweep. For a *perfectly* linear sweep, the initial and the final slopes are the same, which makes the slope error (Eq. 8-1) zero.

Sweep-generating circuits may be divided into two groups, the triggered type and the free-running type. Triggered-type circuits require an input trigger, whereas the free-running type are self-pulsing in nature. An astable sweep circuit using a four-layer diode (Sec. 8-4) is an example of a free-running sweep generator. The transistor sweep generator, the Miller circuits, and the constant-current sweep generator are examples of triggered sweep generators.

Example 8-1
If a voltage sweep has an initial slope of 1.2 and a final slope of 0.8, calculate the slope error.

Solution

$$\text{Slope error} = \frac{\text{Initial slope} - \text{final slope}}{\text{Initial slope}} \times 100 \qquad (8\text{-}1)$$

$$= \frac{1.2 - 0.8}{1.2} \times 100$$

$$= \mathbf{33.3\%}$$

8-4 An Astable Sweep Circuit (Relaxation Oscillator)

The simplest self-pulsing sweep generator uses a neon lamp and is shown in Fig. 8-2a. When the neon lamp is deionized (OFF), capacitor C connected across the neon lamp charges through a large circuit resistance R and gives rise to the sweep time, t_{sp}. As the capacitor charges to the ionization potential (V_2) of the neon lamp, the neon lamp fires (ionizes) and the capacitor discharges quickly to V_1 through the low resistance path provided by the ionized neon lamp. This is the retrace or the flyback time, t_{re}. The output sweep waveform is shown in Fig. 8-2b.

An astable sweep-generator circuit using a *four-layer* diode (FL diode) is shown in Fig. 8-3a. The FL diode turns ON only when the forward bias across it exceeds its *breakover voltage* (V_{BO}). From then on, it acts like an ordinary semiconductor diode (as long as it is conducting). Conversely, an ON FL diode turns OFF only if the forward current in it falls below its *holding current* (I_H). Holding current is the minimum forward current required to keep the FL

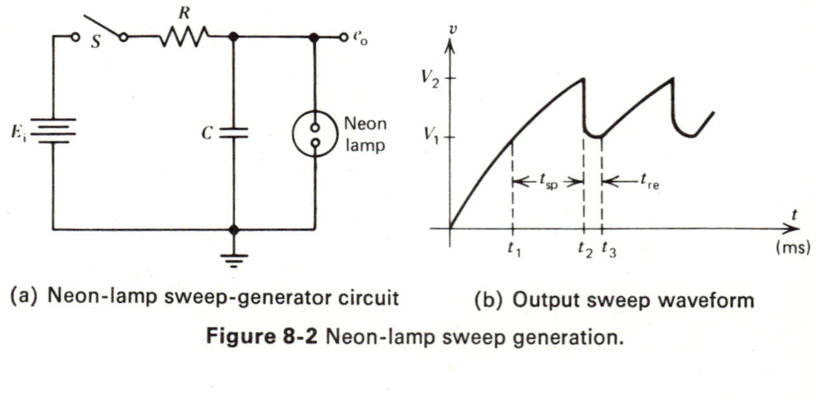

(a) Neon-lamp sweep-generator circuit (b) Output sweep waveform

Figure 8-2 Neon-lamp sweep generation.

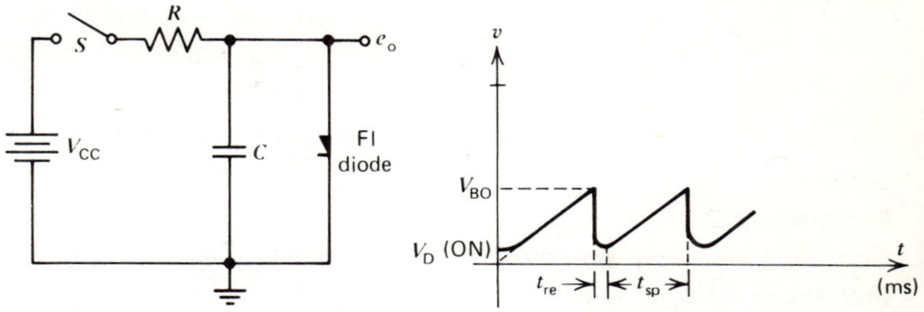

(a) Four-layer diode sweep-generator circuit

(b) Output sweep waveform

Figure 8-3 Four-layer diode sweep generator.

diode ON. Thus, the electrical behavior of a FL diode is very similar to a neon lamp.

As the circuit of Fig. 8-3a is energized, the following happens:

1. Capacitor C starts charging exponentially toward V_{cc} with the time constant $\tau_c(= RC)$.
2. When the capacitor voltage reaches V_{BO}, the FL diode turns ON. The capacitor charge time up to this instant is the sweep time, t_{sp}. It is given by:

$$V_{BO} = V_{cc} - [V_{cc} - V_D(ON)]e^{-(t_{sp}/\tau_c)} \quad \text{(using Eq. 3-8)}$$

or

$$t_{sp} = \tau_c \ln \frac{V_{cc} - V_D(ON)}{V_{cc} - V_{BO}} \text{ seconds (s)} \tag{8-2}$$

where V_D(ON) is the voltage across the FL diode when it is ON.

3. C discharges quickly from V_{BO} to V_D(ON) through the low resistance path provided by the ON FL diode. This capacitor discharge time is the retrace or the flyback time, t_{re}.
4. As the capacitor practically discharges fully, the circuit current falls below the holding current of the FL diode. This condition must be assured by the circuit design. For this purpose,

$$\frac{V_{cc} - V_D(ON)}{R} < I_H \text{ amperes (A)} \tag{8-3}$$

5. Whenever the circuit current falls below I_H of the FL diode, the FL diode turns OFF.

6. C again starts charging exponentially toward V_{cc} as in step 1 above. The whole process of C charging and discharging continues indefinitely until the circuit is de-energized. The resulting output sweep waveform is shown in Fig. 8-3b.

Example 8-2
Consider the sweep-generator circuit shown in Fig. 8-3a. Assume that $V_{cc} = 60$ V, $R = 33$ kΩ, $C = 0.01$ μF. For the *four-layer* diode assume that $V_{BO} = 15$ V, $V_D(\text{ON}) = 0$ V, $I_H = 5$ mA. Find t_{sp}.

Solution

$$t_{sp} = \tau_c \ln \frac{V_{cc} - V_D(\text{ON})}{V_{cc} - V_{BO}} \tag{8-2}$$

or

$$= (33 \times 10^3 \times 0.01 \times 10^{-6}) \ln \frac{60}{60 - 15}$$

$$= \mathbf{95\ \mu s}$$

8-5 Transistor Sweep Generator

Figure 8-4 shows a transistor-sweep-generator circuit. The input waveform (E_i) is used for turning the transistor ON and OFF. Capacitor C charges through R_L when the transistor is OFF and discharges through the transistor when the transistor is ON. The equivalent resistance of the ON transistor (R_T) is considerably

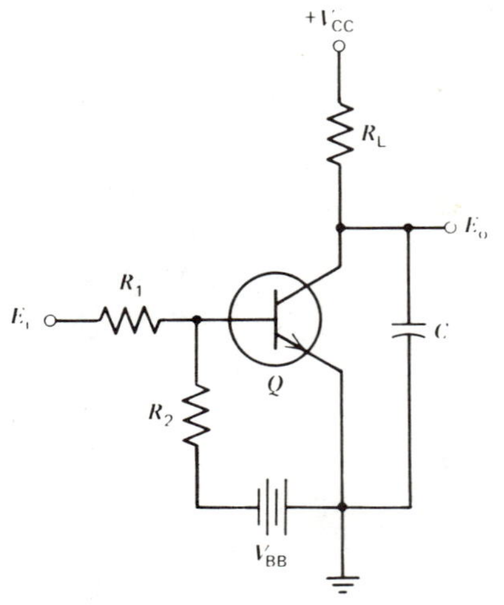

Figure 8-4 Transistor sweep generator.

lower than R_L. Thus, the charging time constant τ_c with which the capacitor charges, is considerably higher than the discharging time constant τ_d.

$$\tau_c = R_L C \text{ seconds (s)} \tag{8-4}$$

$$\tau_d = R_T C \text{ seconds (s)} \tag{8-5}$$

$$R_T = \frac{V_{CE}(\text{ON})}{I_C(\text{ON})} \text{ ohms } (\Omega) \tag{8-6}$$

where R_T is the equivalent CE resistance of an ON transistor.

The circuit shown in Fig. 8-4 is basically an inverter circuit with a grounded capacitor across the output. The transistor is ON when there is a positive pulse at the input and is OFF when there is no input pulse (rest period). Only during the *rest* period is the sweep generated. The retrace time occurs whenever the input pulse turns the transistor ON. Typical input and output waveforms are shown in Fig. 8-5. Note that V_{BB} and R_2 are unnecessary and may be removed from the circuit of Fig. 8-4 if the input pulse (Fig. 8-5a) goes negative to turn the transistor OFF during the rest period.

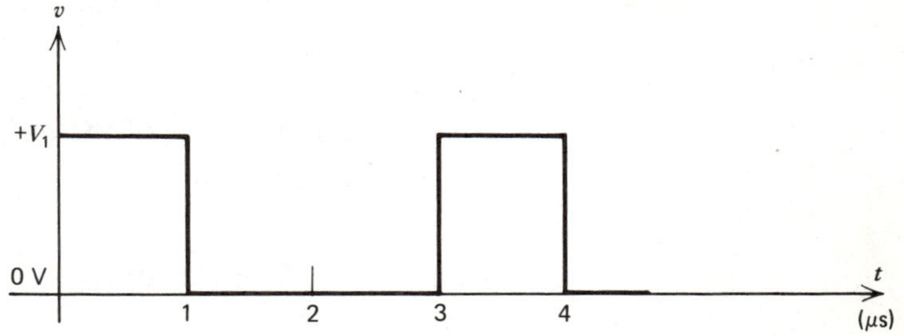

(a) Input pulse waveform

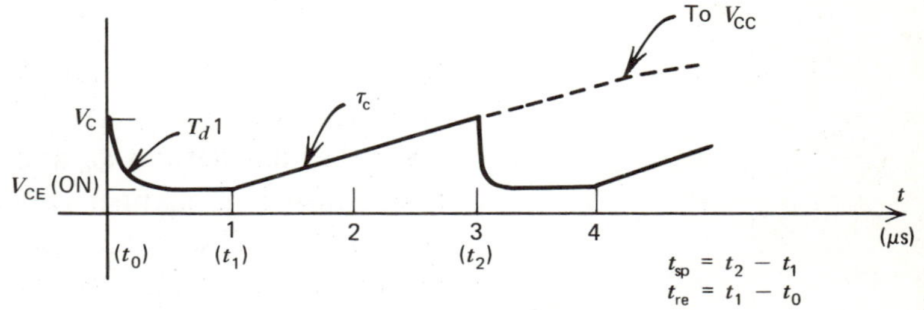

(b) Output sweep waveform

Figure 8-5 Input/output waveforms.

The sweep time (t_{sp}) is usually made smaller than τ_c so that a fairly linear sweep may be obtained. An expression for the sweep can be written using Eq. (3-8):

$$v_c = E_i - (E_i \pm V_{ci})e^{-(t/\tau)} \text{ volts (V)} \tag{3-8}$$

or

$$V_c = V_{cc} - [V_{cc} - V_{CE}(\text{ON})]e^{-(t_{sp}/\tau_c)} \text{ seconds (s)} \tag{8-7a}$$

or

$$t_{sp} = \tau_c \ln \frac{V_{cc} - V_{CE}(\text{ON})}{V_{cc} - V_C} \text{ seconds (s)} \tag{8-7b}$$

where:

t_{sp} is the sweep time (s)
τ_c is the charging time constant (s)
V_{cc} is the supply voltage (V)
V_c is voltage across the capacitor just before the transistor is turned ON (V)
V_{CE}(ON) is the collector-emitter junction voltage of the ON transistor (V)

The transistor ON time, t_{ON}, is made at least five times greater than the retrace time ($t_{ON} > 5\ \tau_d$) to assure that the capacitor is fully discharged before it starts charging again.

The slope error of the output sweep is calculated in the same way as previously indicated in Sec. 8-3:

$$\text{Slope error} = \frac{\text{Initial slope} - \text{Final slope}}{\text{Initial slope}} \times 100 \tag{8-1}$$

$$= \frac{V_c - V_{CE}(\text{ON})}{V_{cc} - V_{CE}(\text{ON})} \times 100 \tag{8-8}$$

(See Appendix 1 for detailed calculations.)

A transistor sweep circuit is designed to generate a sawtooth wave that has desired t_{sp} and amplitude. The basic inverter circuit is designed in a manner similar to the one explained in Sec. 5-7. A value for C is calculated using Eq. (8-7). Finally, the design insures that the transistor stays ON long enough for the capacitor to fully discharge. The design techniques are shown in Ex. 8-3 below.

Example 8-3

a. Design a transistor sweep generator such that the output sweep has an amplitude of 10 $V_{p\text{-}p}$ and a sweep time of 50 μs. Assume $V_{cc} = 25$ V, $V_{BB} = -10$ V, and $V_{BE}(\text{OFF}) = -1.0$ V. A standard silicon transistor is to be used whose $I_C(\text{ON}) = 20$ mA and $\beta = 100$. The input trigger pulse is shown in Fig. 8-6.

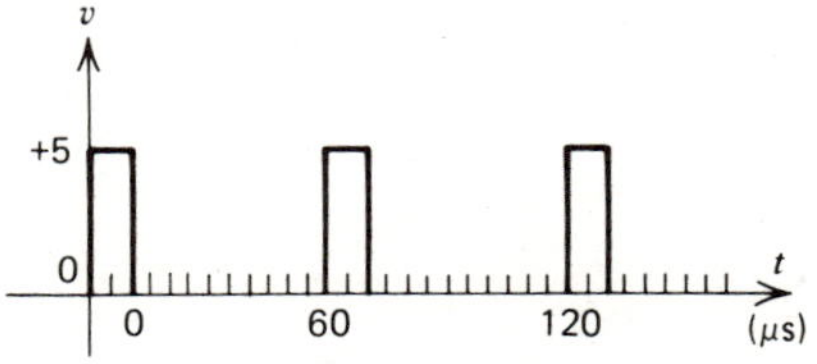

Figure 8-6 Input waveform for Ex. 8-3.

b. Find the minimum input pulse width required to provide adequate retrace time.

Solution

a. The circuit shown in Fig. 8-4 is that of a transistor sweep generator.

$$R_L = \frac{V_{cc} - V_{CE}(\text{ON})}{I_C(\text{ON})}$$

$$= \frac{25 - 0.3}{20 \text{ mA}}$$

$$= 1.235 \text{ k}\Omega$$

An output of 10 $V_{p\text{-}p}$ means that the waveform swings between 0.3 V and 10.3 V.

$$t_{sp} = \tau_c \ln \frac{V_{cc} - V_{CE}(\text{ON})}{V_{cc} - V_C} \qquad \text{(8-7b)}$$

$$50 \times 10^{-6} = R_L C \ln \frac{25 - 0.3}{25 - 10.3}$$

$$50 = 10^{-6} = 0.52\, R_L C$$

Solving for C yields:

$$C = \frac{50 \times 10^{-6}}{0.52 \times 1.235 \times 10^3}$$

$$= 0.0779\ \mu\text{F}$$

Using Eq. (5-11) for the OFF condition,

$$\frac{1}{R_1} = \frac{10 - 1.0}{R_2}$$

$$R_2 = 9R_1$$

Using Eq. (5-10) for the ON condition,

$$\frac{5 - 0.7}{R_1} - \frac{10 + 0.7}{R_2} \geqq \frac{20 \text{ mA}}{100}$$

$$\frac{4.3}{R_1} - \frac{10.7}{9R_1} = 0.2 \text{ mA}$$

$$R_1 = \frac{(4.3 - 1.189) \text{ V}}{0.2 \text{ mA}}$$

$$= \mathbf{15.56\ k\Omega}$$

$$R_2 = 9 \times 15.56 \text{ k}\Omega$$

$$= \mathbf{140\ k\Omega}$$

$$R_T = \frac{V_{CE}(\text{ON})}{I_C(\text{ON})} \tag{8-6}$$

$$= \frac{0.3 \text{ V}}{20 \text{ mA}}$$

$$= 15\ \Omega$$

$$C = 0.0779\ \mu\text{F}$$

$$t_p(\text{min}) \geqq 5\tau_d$$

$$= 5 \times 15\ \Omega \times 0.0779\ \mu\text{F}$$

$$= \mathbf{5.84\ \mu s}$$

8-6 Miller Sweep (Integrator) Circuit

The sweep-generating circuits discussed so far do not produce very linear sweeps. The slope error is quite high. The *Miller sweep circuit*, also known as the *Miller integrator*, produces a fairly linear sweep. The basic idea is that the voltage rise across a capacitor is always linear if it (the capacitor) is charged by a *constant current* (dc) source (Eq. 1-8).

Consider the circuit shown in Fig. 8-7a. V_f is a fictitious voltage generator. If V_f is made such that its terminal voltage is always

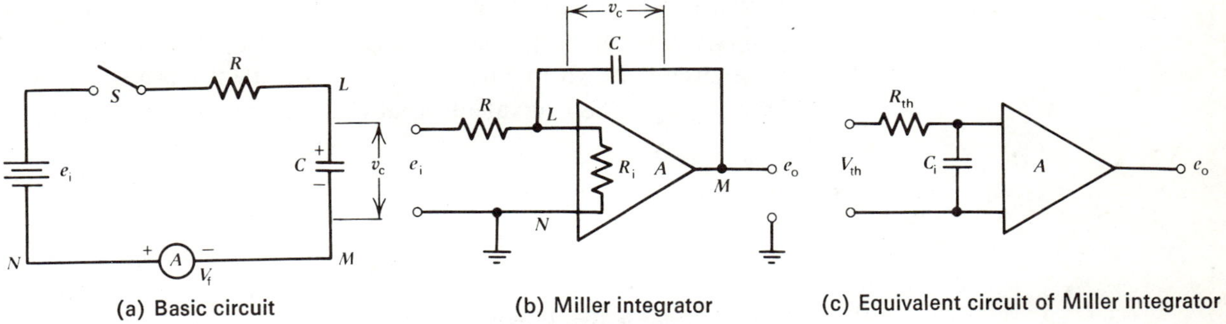

Figure 8-7 Miller sweep generator.

equal to the voltage across the capacitor C but *opposite* in polarity, then a constant current E_i/R will flow in the circuit. Constant **(dc)** current flowing through the capacitor results in a *linear voltage rise* across it. At any given time, the voltage across C is calculated in the following manner:

$$Q = CV_c$$

$$\frac{Q}{t} = C\left(\frac{V_c}{t}\right)$$

or

$$I_c = \frac{CV_c}{t}$$

or

$$\frac{E_i}{R} = C\left(\frac{V_c}{t}\right)\left(I_c = \frac{E_i}{R}\right)$$

or

$$V_c = \frac{E_i}{RC}\cdot t$$

$$= E_i\frac{t}{\tau} \text{ volts (V)} \tag{8-9}$$

The circuit configuration shown in Fig. 8-7b simulates conditions of Fig. 8-7a. The voltage and charge on capacitor C is used as a feedback element. The triangular block in Fig. 8-7b is an amplifier whose voltage gain is A. Note that the circuit is grounded at N

and the output is taken across MN, which is V_f. As v_c increases linearly, V_f or the output (E_o) decreases correspondingly.

The circuit shown in Fig. 8-7c is an equivalent circuit to that of Fig. 8-7b. C_i is the equivalent input capacitance, which is given by:

$$C_i = C(A + 1)$$

where:

C is the feedback capacitor (F)
A is the gain of the amplifier

V_{th} and R_{th} form the Thévenin's equivalent of E_i, R, and R_i.

$$V_{th} = E_i\left(\frac{R_i}{R + R_i}\right)$$

$$R_{th} = \frac{R_i R}{R_i + R}$$

C_i starts charging toward V_{th} with a time constant of $R_{th} \cdot C_i$ as the switch is closed or as E_i is applied. The capacitor charge equation is:

$$V_{C_i} = V_{TH} - V_{TH}e^{-(t/R_{TH} \cdot C_i)}, \text{ volts (V)} \qquad (8\text{-}10)$$

V_{C_i} is really the dynamic input (e_i) to the amplifier. Thus the output (e_o) is given by:

$$e_o = Ae_i$$

or

$$= AV_{C_i}$$

or

$$= AV_{TH} - AV_{TH}e^{-(t/R_{TH} \cdot C_i)}$$

For a high-gain amplifier (large A), the slope of the output is (see Appendix A-2):

$$\frac{e_o}{t} = \frac{E_i}{RC} \text{ volts/second (V/s)} \qquad (8\text{-}11a)$$

or

$$e_o = \frac{E_i}{RC} t, \text{ volts (V)} \tag{8-11b}$$

$$= E_i \frac{t}{\tau_c}, \text{ volts (V)} \tag{8-11c}$$

where $\tau_c = RC$, the charging time-constant (s).

Equation (8-11c) is similar to Eq. (8-9). This proves that the circuit shown in Fig. 8-7b approximately simulates the circuit conditions of Fig. 8-7a. The only condition is that the voltage gain (A) of the amplifier must be high.

It can be seen that the Miller sweep circuit converts a step function into a ramp function. This is exactly what happens when a step function is integrated. Thus the name *Miller integrator*.

Figure 8-8 shows a practical Miller sweep circuit. Essentially, Fig. 8-8b is a high-gain inverter amplifier. The output is coupled to the input through feedback capacitor C. The circuit is exactly the same as Fig. 8-7b.

When E_i is at $-V_2$, the transistor is OFF. Voltage across the capacitor is $V_{cc} - V_{BE}$(OFF). E_o is at V_{cc}. As the transistor is turned ON by $+V_1$ at e_i, e_o starts to drop. The voltage drop at the collector (e_o) is immediately transferred to the base of the transistor via C. This results in a lowered bias at the base, which tends to lower the base current. Accordingly, as a result of the feedback capacitor C being there, the collector voltage (e_o) tends to increase. This is the opposite of what $+V$ at e_i is trying to do to the collector voltage (e_o). This tug-of-war action, with the collector voltage trying

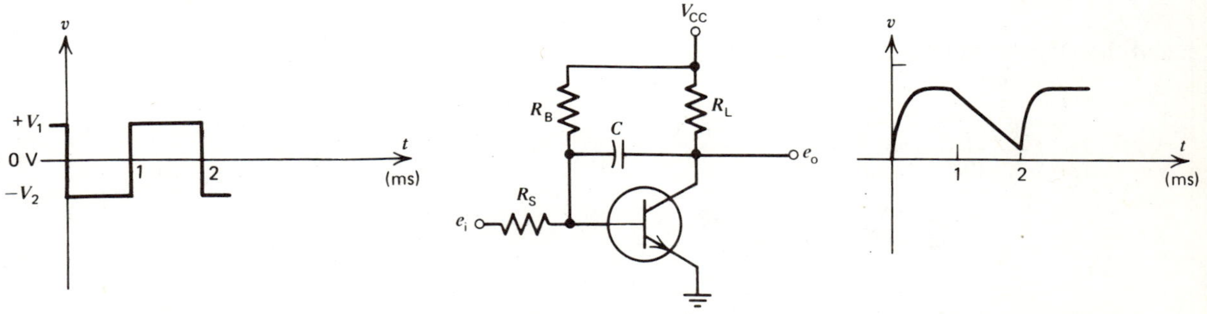

(a) Input waveform (b) Miller circuit (c) Output waveform

Figure 8-8 Miller sweep circuit.

to go down to V_{CE}(ON) on one side and the C transferring the drop to the base of the transistor on the other results in a linear drop in the collector voltage.

Example 8-4

Design a simple Miller sweep circuit to generate a 20-$V_{p\text{-}p}$ sweep of 100 μs duration. Assume $\beta = 100$ and I_C(ON) = 25 mA for a germanium transistor that is to be used in the amplifier. The input is a 10-$V_{p\text{-}p}$, 5-kHz square wave.

Solution

A basic Miller sweep circuit is shown in Fig. 8-8. The output voltage is to vary between 0.1 V [V_{CE}(ON)] and 20.1 V so that the required 20 $V_{p\text{-}p}$ output is obtained. Thus,

$$V_{cc} = 20.1 \text{ V}$$

$$R_L = \frac{(20.1 - 0.1)\text{ V}}{25 \text{ mA}}$$

$$= \mathbf{800\ \Omega}$$

Assume V_{BE}(OFF) = −0.5 V. The OFF condition yields (Eq. 5-11):

$$\frac{20.6}{R_B} = \frac{4.5}{R_s}$$

$$R_B = \frac{20.6}{4.5} R_s$$

$$= 4.58\ R_s$$

and the ON condition yields (Eq. 5-10):

$$\frac{20.1 - 0.3}{R_L} + \frac{5 - 0.3}{R_s} \geqq \frac{25 \text{ mA}}{100}$$

$$\frac{19.8}{4.58\ R_s} + \frac{4.7}{R_s} = 0.25 \text{ mA}$$

$$R_s = \frac{4.33 + 4.7}{0.25 \text{ mA}}$$

$$= \mathbf{36.12\ k\Omega}$$

and

$$R_B = 4.58 \times 36.12\ \text{k}\Omega$$
$$= \mathbf{165\ k\Omega}$$

$$C = \frac{e_i}{e_o} \times \frac{t_{sp}}{R} \qquad (8\text{-}11)$$

$$= \frac{5\ \text{V} \times 100\ \mu\text{s}}{20\ \text{V} \times 36.12\ \text{k}\Omega}$$

$$= \mathbf{692\ pF}$$

Another version of the Miller feedback idea is shown in Fig. 8-9. The IC linear operational amplifier (OP. AMP.) usually has a very high open-loop gain (several hundred thousand). Therefore, the high-gain condition mentioned in Eq. (8-11a) is adequately satisfied by an OP. AMP. Equation 8-11b now becomes:

$$e_o = \frac{e_i}{(R_1 + R_2)C} \cdot t_{sp},\ \text{volts (V)} \qquad (8\text{-}12)$$

where:

e_o is the output voltage at any time (V)
e_i is the amplitude of the input pulse (V)
t_{sp} is the sweep time (s) and R_1, R_2, and C are as shown in Fig. 8-9

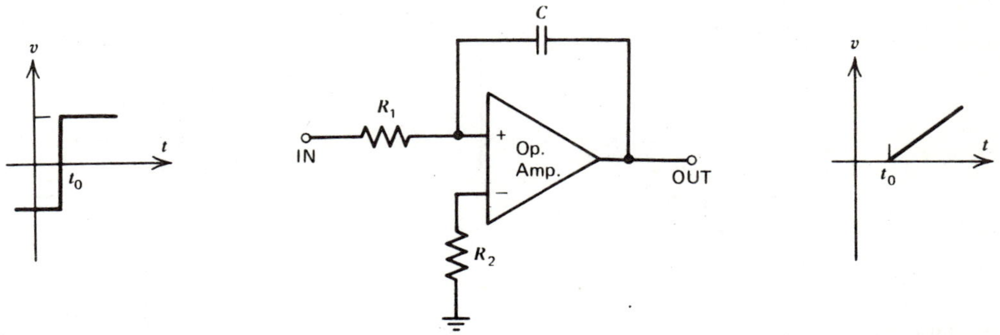

Figure 8-9 Miller sweep circuit using an OP. AMP.

8-7 Bootstrap Sweep Circuit

The bootstrap sweep circuit is just another version of the Miller integrator circuit. The basic idea is the same. In this case, however, the basic circuit (Fig. 8-7a) is grounded at M and the gain of the amplifier is made equal to unity. Figure 8-10a shows the basic configuration of the bootstrap circuit.

Capacitor C is charged by a constant current, E_i/R, when the switch (S) is open, and is quickly discharged by closing the switch. The linear voltage rise across C produces a linear output sweep when fed to an amplifier whose gain is unity.

Figure 8-10b shows an equivalent circuit to that of Fig. 8-10a. The only difference is that the fictitious voltage generator V_f has been replaced by an amplifier whose gain is unity. C is made to charge and discharge by opening and closing switch S. E_i becomes a part of the feedback circuit. The output, e_o, is given by:

$$e_o(t) = A\frac{E_i}{RC}\cdot t \text{ volts (V)} \tag{8-13}$$

If gain A of the amplifier is made equal to 1, then,

$$e_o(t) = \frac{E_i}{RC}\cdot t$$

$$= E_i\frac{t_{sp}}{\tau_c}$$

which is the same as Eq. (8-9), previously derived for a linear sweep.

The circuit shown in Fig. 8-10c is the same as that shown in Fig. 8-10b, except that the symbolic amplifier has been replaced by

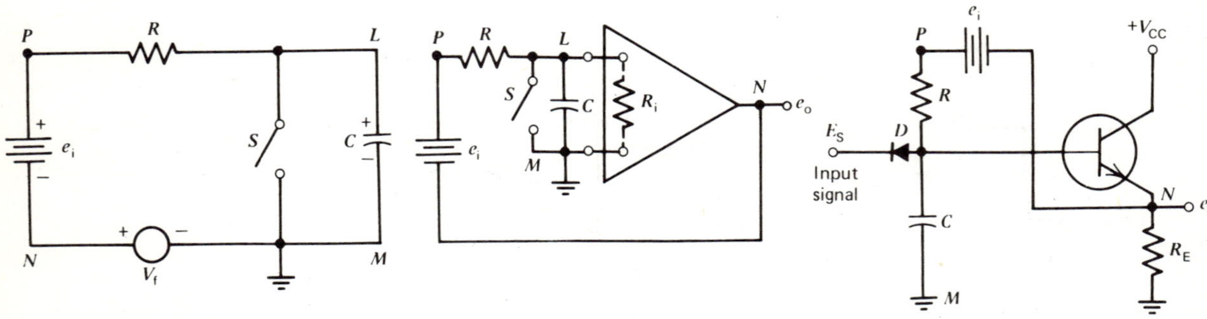

(a) Basic sweep circuit (b) Equivalent circuit (c) Bootstrap circuit

Figure 8-10 Bootstrap sweep generation.

an actual emitter-follower circuit and that the switch S has been replaced by a diode switch D. D is turned ON by applying a small negative voltage ($-V_1$) at E_S. An ON D transfers $-V_1$ to the base of the transistor turning it OFF. A slight current flowing through R_E, R, and D makes the voltage at e_o slightly negative.

A positive pulse of sufficient amplitude turns D OFF. C starts charging from $-V_1$ toward E_i. The charging time constant is:

$$\tau_c = C(R + R_E)$$
$$\approx C \cdot R \qquad (\because R_E \ll R)$$

The initial charging current is:

$$I_o = \frac{E_i + V_1}{R + R_E}$$
$$\approx \frac{E_i}{R} \qquad [\because E_i \gg V_1 \quad \text{and} \quad R \gg R_E]$$

The voltage rise across C serves as the input voltage as far as the emitter-follower is concerned. As C charges, the output voltage at e_o increases correspondingly. This increase in the output voltage gets transferred to point P. Thus, voltage at point P increases correspondingly (by the same amount) as the capacitor charges, keeping the voltage drop across R constant. A constant voltage drop across R results in a constant current, E_i/R, which is charging C. A linear voltage build-up across C results in a linear output voltage sweep.

$$I_c = \frac{E_i}{R}$$
$$v_c = I_c t_{sp}$$

or

$$= \frac{E_i}{RC} \cdot t_{sp}$$
$$e_o(t) = A v_c$$
$$e_o(t) = A \frac{E_i}{RC} \cdot t_{sp}$$
$$= E_i \times \frac{t_{sp}}{\tau_c} \text{ volts (V)} \quad (\tau_c = RC \text{ and } A = 1) \qquad \text{(8-14a)}$$

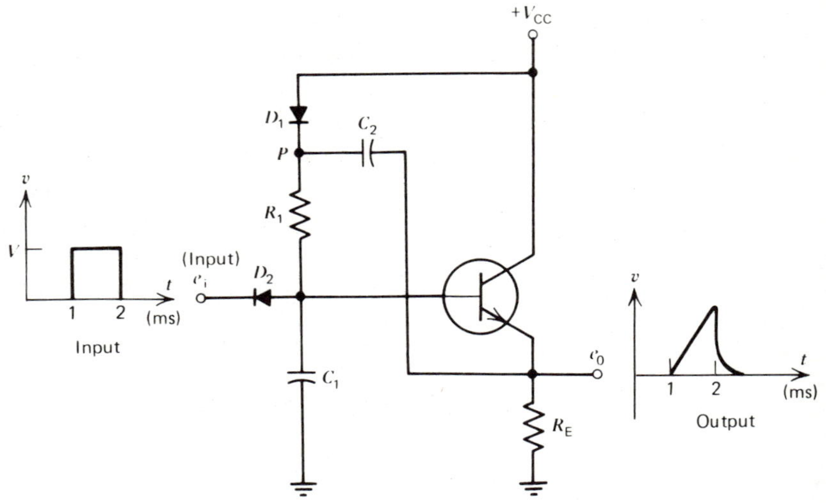

Figure 8-11 Bootstrap sweep circuit.

A practical problem with the circuit of Fig. 8-10c is that the voltage source E_i is floating (not grounded). This is easily overcome by connecting a large capacitor in its place. A more practical bootstrap circuit is shown in Fig. 8-11. C_2 is usually a very large capacitor. It does not discharge significantly at any time.

A slightly negative voltage at E_S grounds the base of the transistor, turning it OFF. Some current flows from V_{cc} through D_1, R, and D_2 to ground. The D_1R junction is approximately at $+V_{cc}$ volts. The other side of C_2 is at 0 V. Thus, C_2 charges to $+V_{cc}$, as shown in Fig. 8-11. It is in this way that E_i of Fig. 8-10 gets replaced by a $+V_{cc}$ charge on C_2 (Fig. 8-11).

A positive signal at E_S turns D_2 OFF and the capacitor starts charging with an initial current of V_{cc}/R. As C_1 charges, e_o increases. An increase in e_o immediately gets reflected in a corresponding increase in the D_1R junction voltage, because V_{c_2} remains constant. D_1 turns OFF and the voltage drop across R remains constant. As explained earlier, a constant voltage drop across R results in constant current charging C_1. Thus, a linear voltage build-up across C_1 produces a corresponding linear sweep at e_o. Equation (8-14a) now becomes:

$$e_o(t) = V_{cc} \times \frac{t_{sp}}{\tau_c} \text{ volts (V)} \qquad \text{(8-14b)}$$

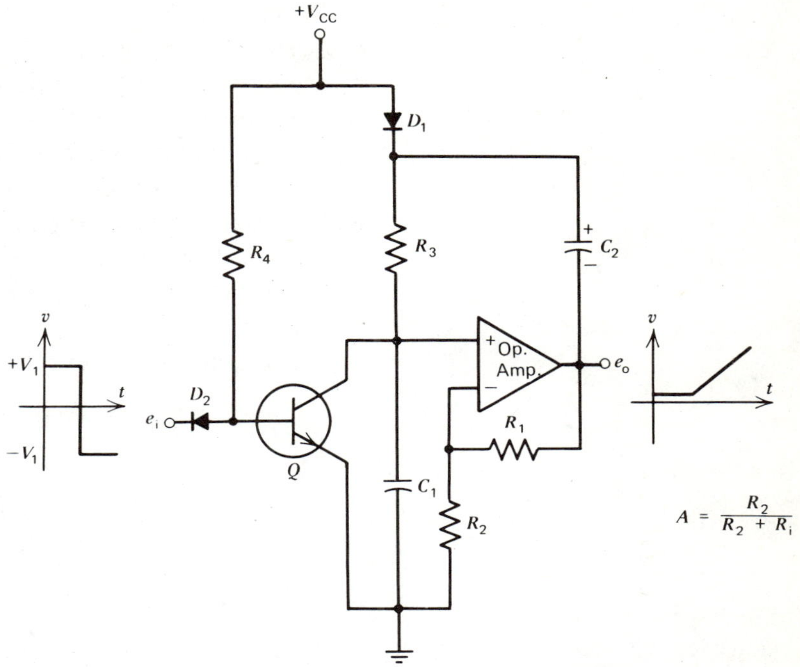

Figure 8-12 Bootstrap sweep circuit using an OP. AMP.

Another version of a practical bootstrap circuit is shown in Fig. 8-12. R_1 and R_2 are chosen such that the closed-loop gain of the OP. AMP. is unity.

$$A_v = \frac{R_2}{R_1 + R_2}$$

$$= 1, \qquad \text{if } R_2 \gg R_1 \text{ or } R_1 = 0$$

The circuit operates as follows (assume ideal diodes):

1. When a $+V_1$ is applied to the input:

  a. D_2 turns OFF.

  b. Q_1 turns ON.

  c. C_1 charges to V_{CE}(ON) through D_1 and R_3.

  d. The output is at V_{CE}(ON) because the closed-loop gain (A_v) of the OP. AMP. is unity.

  e. C_2, which was chosen to be large in its capacitive value, charges to $V_{cc} - V_{CE}$(ON).

2. When *a* $-V_1$ is applied to the input:

  a. D_2 turns ON.

b. Q_1 turns OFF.
c. C_1 starts charging through D_1 and R_3.
d. The output voltage (e_o) rises as the voltage across C_1 rises.
e. The charge on C_2 remains practically constant. Therefore, the rise in output is transferred to the junction of D_1R_3 through C_2.
f. Thus, the voltage drop across R_3 remains constant.
g. A constant voltage drop across R_3 results in a constant current through it that is also charging C_1.
h. A constant charging current results in a linear voltage rise across C_1.
i. The output is given by:

$$e_o = Av_{C_1}$$
$$= V_{C_1}$$
$$= \frac{V_{cc} - V_{CE}(\text{ON})}{R_3C_1} \cdot t \text{ volts (V)} \qquad (8\text{-}15)$$

Example 8-5

Consider the bootstrap sweep circuit shown in Fig. 8-10. Assume $V_{cc} = 15$ V, $R_E = 200\ \Omega$, $R_1 = 10$ kΩ, $C_1 = 680$ pF, $C_2 = 3300\ \mu$F. Consider D_1 and D_2 to be ideal diodes and Q an ideal transistor. Find and draw the output voltage waveform if the input is a 10-$V_{p\text{-}p}$, 500-kHz square wave.

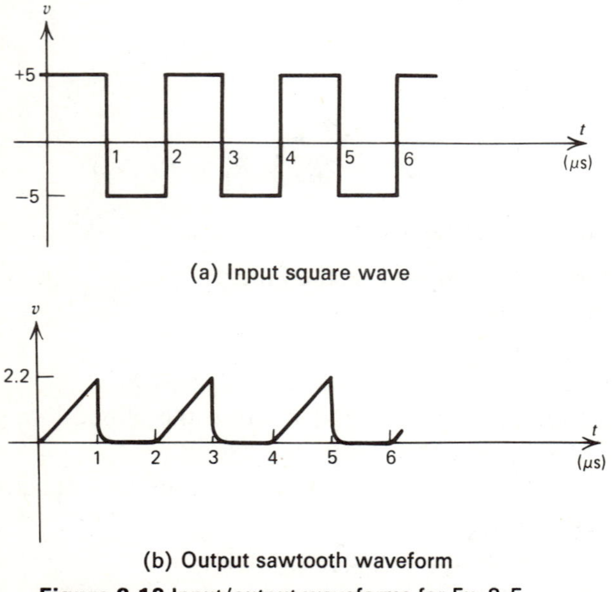

Figure 8-13 Input/output waveforms for Ex. 8-5.

Solution

$$t_{sp} = \frac{1}{2} \cdot \frac{1}{500\text{ kHz}}$$

$$= 1\ \mu\text{s}$$

$$e_o(t) = \frac{V_{cc}}{RC} \cdot t_{sp} \qquad \text{(8-14b)}$$

$$= \left(\frac{15\text{ V}}{10\text{ k}\Omega \times 680\text{ pF}}\right) 1\ \mu\text{s}$$

$$= \mathbf{2.21\ V}$$

The input and output waveforms are drawn in Fig. 8-13.

8-8 Constant-Current (Voltage) Sweep Generator

As mentioned earlier, a constant (**dc**) current flowing through a capacitor results in a linear voltage rise across it. Another way to obtain the constant current needed to charge a capacitor is to find a constant current source. The common base circuit configuration shown in Fig. 8-14 provides such a source. The constant emitter current I_E is given by:

$$I_E = \frac{V_{EE} - V_{BE}(\text{ON})}{R_E} \text{ amperes (A)} \qquad \text{(8-16)}$$

The collector current, I_c, charges the capacitor.

$$I_C = \alpha I_E$$

or

$$= \alpha \frac{V_{EE} - V_{BE}(\text{ON})}{R_E} \text{ amperes (A)} \qquad \text{(8-17)}$$

The capacitor C in Fig. 8-14 charges linearly with a constant current I_C when the switch S is open. When S is closed, C discharges quickly through it. Note that the discharge current, as well as I_C, flows through S when it is closed. The capacitor-charge equation when S is open is:

$$V_c = \frac{I_C}{C} \cdot t \text{ volts (V)} \qquad \text{(8-9)}$$

$$= \alpha \frac{V_{EE} - V_{BE}(\text{ON})}{CR_E} \cdot t \text{ volts (V)} \qquad \text{(8-18)}$$

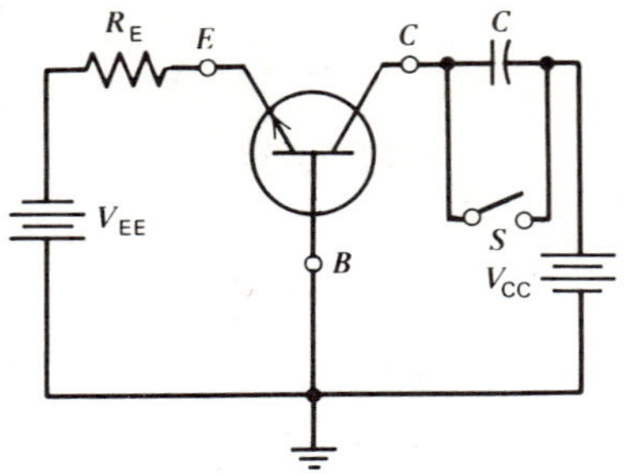

Figure 8-14 Constant-current voltage sweep generation.

Equation (8-18) gives voltage across the capacitor (V_c) at any time t. V_c is equal to 0 V when S is in the closed position. When S is open, V_c may charge up to a maximum of $V_{cc} - V_{CE}$(ON). Thus the maximum peak-to-peak swing of the sweep generated across C is $V_{cc} - V_{CE}$(ON). Usually, V_{CE}(ON) is very small. Then,

$$V_{c(p\text{-}p)} \approx V_{cc} \text{ volts (V)} \tag{8-19}$$

The maximum sweep time, t_{sp}(max) obtainable from a system is the time that the capacitor takes to charge from 0 V to approximately V_{cc}. Thus,

$$t_{sp}(\text{max}) = \frac{C}{I_c} \cdot V_{cc} \text{ seconds (s)} \tag{8-20}$$

In practical constant-current sweep circuits, the manual switch S is replaced by a transistor switch, as shown in Fig. 8-15. An input pulse is used to turn transistor Q_1 either ON or OFF. C charges linearly in accordance with Eq. (8-18) when Q_1 is turned OFF and discharges quickly (low resistance path) through Q_1 when Q_1 is turned ON. Note that when Q_1 is turned ON, the constant charging current (I_{ch}) and the capacitor discharge current (I_{dis}) both flow through it.

$$I_{C_1}(\text{ON}) = I_{ch} + I_{dis}$$

Usually, I_{dis} is made about five times the I_{ch}, in order to obtain a quick discharge (retrace) time. Thus,

$$\begin{aligned} I_{C_1}(\text{ON}) &= I_{ch} + 5I_{ch} \\ &= 6\, I_{ch} \text{ amperes (A)} \end{aligned} \tag{8-21}$$

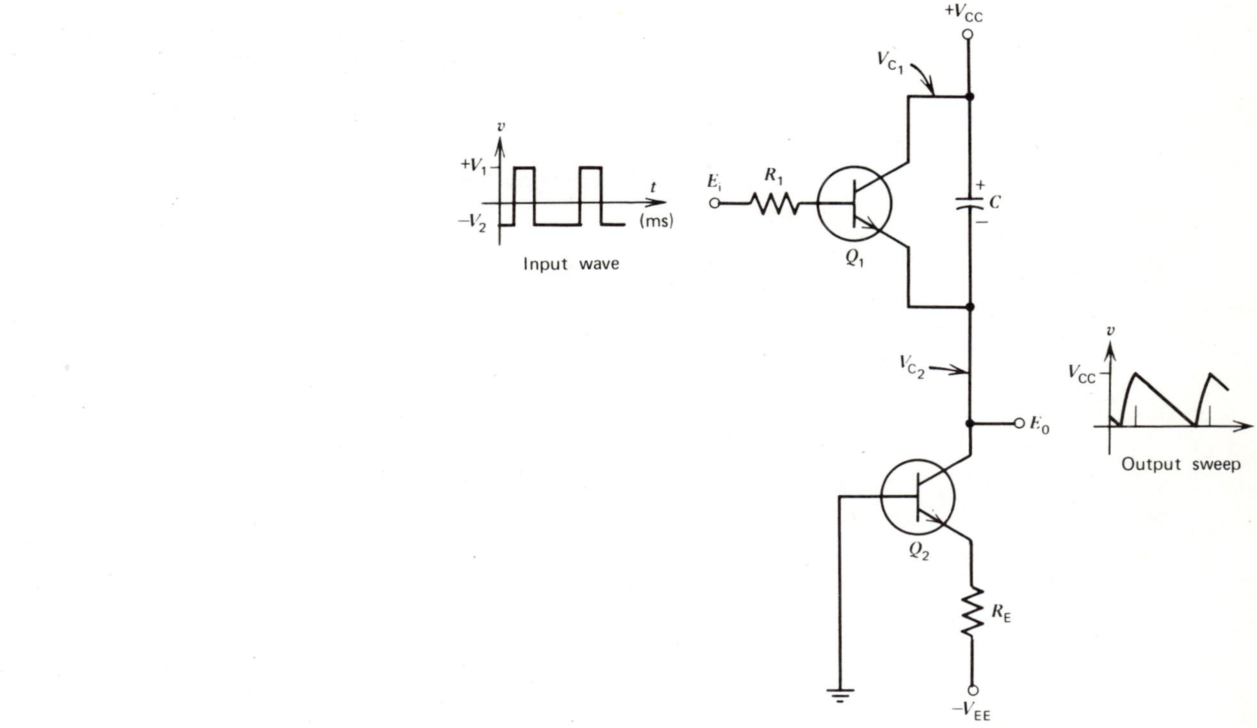

Figure 8-15 Constant-current voltage sweep generator.

Let us now calculate R_1. The function of R_1 is to provide the right amount of saturation base current to Q_1 when input is at $+V_1$.

$$R_1 = \frac{V_1 - V_{B_1}(\text{ON})}{I_{B_1}(\text{ON})}$$

where:

V_1 is the magnitude of the input voltage (V)
$V_{B_1}(\text{ON})$ is the base voltage of Q_1 when ON (V)
$I_{B_1}(\text{ON})$ is the base current of Q_1 when ON (A)

Assuming that the junction voltages of the ON transistors are approximately equal to 0 V, then,

$$V_{B_1}(\text{ON}) = V_{C_2}, \quad \text{the collector voltage of } Q_2$$

$$R_1 = \frac{V_1 - V_{C_2}}{I_{B_1}(\text{ON})} \text{ ohms } (\Omega) \tag{8-22a}$$

Equation (8-22a) is a general equation for calculating R_1. V_{C_2}, the collector voltage of Q_2, is approximately equal to 0 V when C is fully charged, and is equal to $V_{cc} - V_{CE}(\text{ON})$ when C has discharged. Q_1 must stay ON during and in-between these two conditions so long as the input is at $+V_1$.

When C is fully charged and the input of $+V_1$ has just been applied,

$$V_{C_2} \approx 0\text{ V}$$

$$R_1' = \frac{V_1}{I_{B_1}(\text{ON})} = \frac{V_1}{(6\,I_{ch})/\beta}\ \text{ohms }(\Omega) \qquad \text{(8-22b)}$$

When C has fully discharged, Q_1 should still be ON as long as $+V$ is there. Under this condition, though,

$$V_{B_1}(\text{ON}) = V_{cc}$$

and

$$I_{B_1}(\text{ON}) = (I_{ch})/\beta$$

or

$$R_1'' = \frac{V_1 - V_{cc}}{(I_{ch})/\beta}\ \text{ohms }(\Omega) \qquad \text{(8-22c)}$$

Equations (8-22b) and (8-22c) give two values for R_1. The lower of the two will assure that Q_1 saturates in each case. In fact, in practical circuits, Q_1 is driven 200 percent into saturation. This means that R_1 is usually chosen to be one-half the value calculated above.

Example 8-6

Design a constant-current sweep circuit similar to the one shown in Fig. 8-15. The input is a +15 V to −5 V, 10-kHz square wave. Assume $V_{EE} = -10$ V, $V_{CE}(\text{ON}) = V_{BE}(\text{ON}) = 0$ V, $I_C(\text{max}) = 25$ mA, $h_{fe} = 50$, and the constant charging current is 2 mA. The transistors are a matched pair. The output sweep is required to be 0 V − 10 V in amplitude and 20 μs long. Calculate appropriate values for V_{cc}, C, R_E, and R_1.

Solution

1. The amplitude condition:

$$V_{cc} = V_{c_{p\text{-}p}} \qquad (8\text{-}19)$$
$$= \mathbf{10\ V}$$

2. The sweep-time condition:

$$C = \frac{I}{V} \cdot t$$
$$= \frac{2 \times 10^{-3}}{10} \times 20 \times 10^{-6} \qquad (1\text{-}8)$$
$$= \mathbf{0.004\ \mu F}$$

3. The constant-charging-current condition:

$$I_{ch} = 2\text{ mA}$$
$$I_E = \frac{2 \times 10^{-3}}{0.98} \qquad \left(\alpha = \frac{50}{50 + 1} = 0.98\right)$$
$$R_E = \frac{V_{EE} - V_{BE}(\text{ON})}{I_E} \qquad (8\text{-}16)$$
$$= \frac{10\text{ V}}{2.041\text{ mA}} = \mathbf{4.9\ k\Omega}$$

4. Low retrace-time condition:

$$I_{c_1} = 6\, I_{ch} = 6 \times 2\text{ mA} = 12\text{ mA}$$
$$I_{B_1}(\text{ON}) = \frac{12\text{ mA}}{50}$$
$$= 0.24\text{ mA}$$
$$R_1' = \frac{V_1}{I_{B_1}(\text{ON})} \qquad (8\text{-}22b)$$
$$\therefore R_1' = \frac{15\text{ V}}{0.24\text{ mA}}$$
$$= 62.5\text{ k}\Omega$$

and

$$R_1'' = \frac{V_1 - V_{cc}}{I_{ch}/\beta} \qquad (8\text{-}22c)$$

$$R_1'' = \frac{15 - 10}{2 \text{ mA}/50}$$

$$= \frac{5 \times 50}{2 \text{ mA}}$$

$$= 125 \text{ k}\Omega$$

$\therefore R_1 =$ **62.5 kΩ** (lower of the two values)

8-9 Current Sweep Generation

The circuits previously considered are all *voltage* sweep generators. In some applications, a *current* sweep is required. One of the most common applications of current sweep is in the magnetic deflection coils of a cathode ray tube (CRT) in a TV receiver. In this and similar applications, a current is required to flow through the deflection coil to produce magnetic deflection of the electron beam.

Consider the simple circuit of Fig. 8-16a. After the switch is closed, current i_L in the circuit is given by:

$$i_L = \frac{E}{L} \cdot t \text{ amperes (A)} \qquad (8\text{-}23)$$

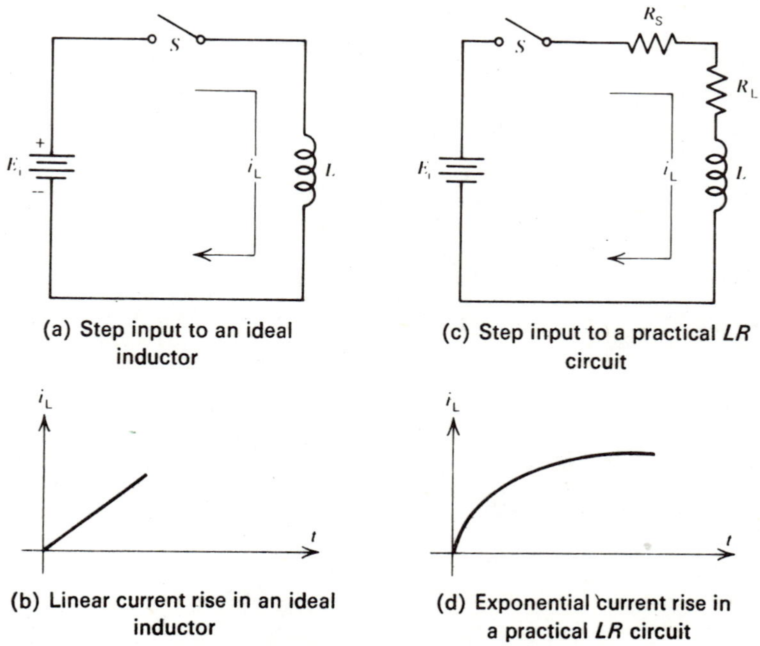

(a) Step input to an ideal inductor

(c) Step input to a practical *LR* circuit

(b) Linear current rise in an ideal inductor

(d) Exponential current rise in a practical *LR* circuit

Figure 8-16 Current-sweep generation.

where:

i_L is the current in the coil in amperes (A)
L is the inductance of the coil in henrys (H)
t is the time elapsed (seconds) since the switch was closed (s)

Equation (8-23) indicates that i_L increases linearly with time. Note that the internal resistance of both the source and the coil is assumed to be zero. In practical circuits, neither is the case. There is always some finite source- and coil-resistance in the circuit. Total resistance in the circuit determines the *maximum* current that may flow in the coil and of course, the *exponential* (rather than linear) manner in which it goes from 0 to its maximum value. Consider the more practical circuit of Fig. 8-16c. R_s is the equivalent resistance of the source and the switch (when closed), and R_L is the **dc** resistance of the coil. Total resistance in the circuit is:

$$R_t = R_s + R_L$$

The maximum value of the current, I_o (when $t > 5\tau$) is:

$$I_o = \frac{E}{R_t}$$

and

$$i_L = I_o - I_o e^{-(t/\tau)} \text{ amperes (A)} \qquad (8\text{-}24)$$

where:

i_L is the instantaneous current in the coil (A)
I_o is the maximum current in the coil at $t = s\tau$ (A)
τ is the time constant (L/R_t) in seconds (s)

Thus, we see that a step input applied to a *pure* inductor (no resistance) produces a linear current sweep (Eq. 8-23). In practical circuits, there always is some resistance present that tends to make the sweep *exponential.* The lower the resistance in the circuit, the more linear the current sweep.

Figure 8-17 shows a simple transistor current-sweep generator. A transistor is now being used as a switch. The transistor is turned ON and OFF by an input trigger (E_i). Inductance L represents the coil (deflection) in which a current sweep is needed.

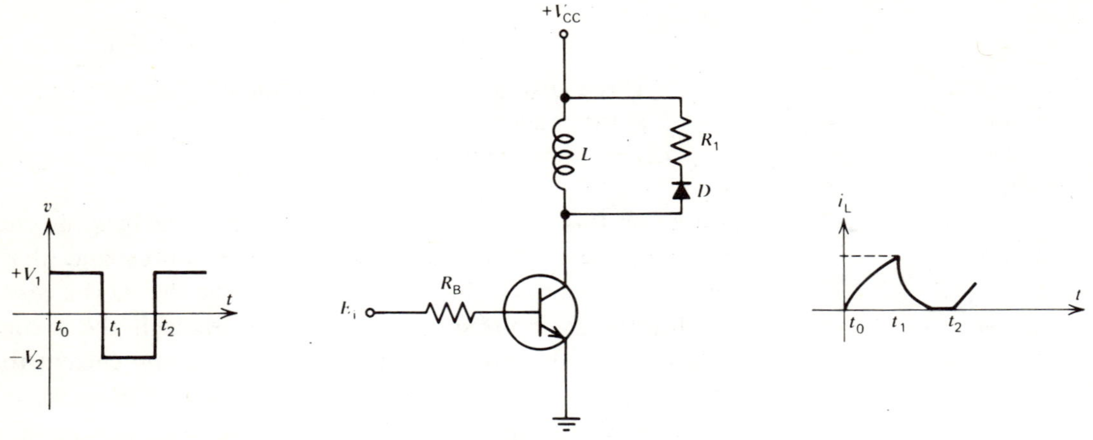

(a) Input voltage wave (b) Sweep generating circuit (c) Coil current

Figure 8-17 Transistor current-sweep generator.

At $t = t_{0+}$, $+V_1$ applied at E_i saturates the transistor. The inductor current i_L starts increasing from zero in accordance with Eq. (8-24). The rate of increase is fairly linear because the equivalent resistance of the ON transistor and the **dc** resistance of the coil are both very small. Essentially, it increases in accordance with Eq. (8-23). Diode D is reverse-biased (OFF) during this time. As E_i switches to $-V_2$ volts, the transistor turns OFF. The current i_L that was flowing at $t = t_1$ decays exponentially through D and R_1. The current decay is given by:

$$i_L = I_L e^{-(t/\tau)} \text{ amperes (A)} \tag{8-25}$$

where:

$$I_L \approx \frac{V_1}{L} \cdot t$$

and

$$\tau = \frac{L}{R_1 + R_D} \text{ seconds (s)} \tag{8-26}$$

R_D represents the forward resistance of the diode (Ω).

For all practical purposes, i_L will decay to zero if $t \geqq 5\,\tau$. Resistance $R_1 + R_D$ serves as the damping resistance as i_L decays. Damping is necessary to prevent undesirable oscillations (Sec. 3-10).

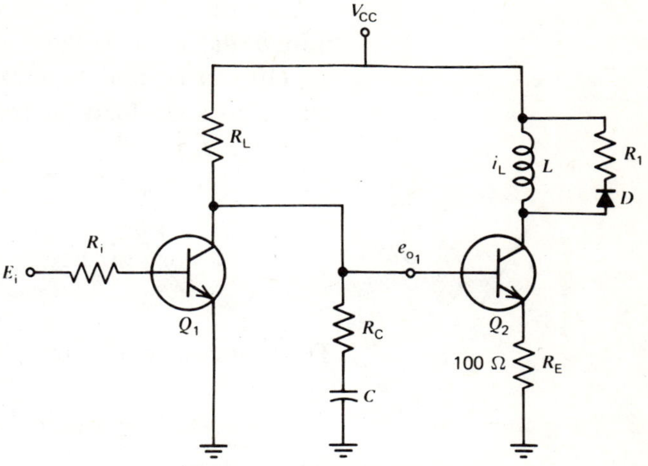

(a) Current sweep circuit

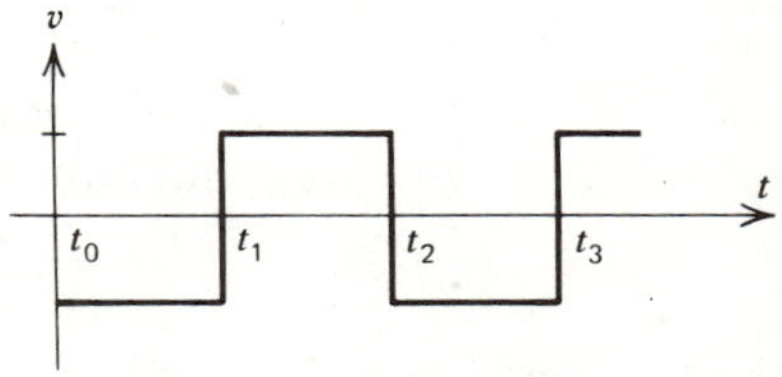

(b) Input waveform

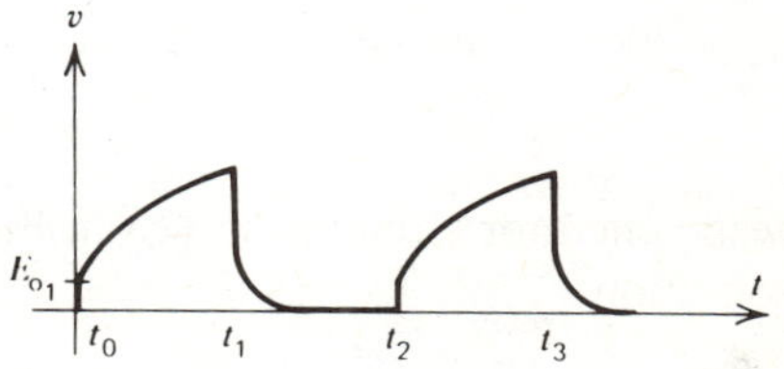

(c) Voltage waveform at e_{o_1}

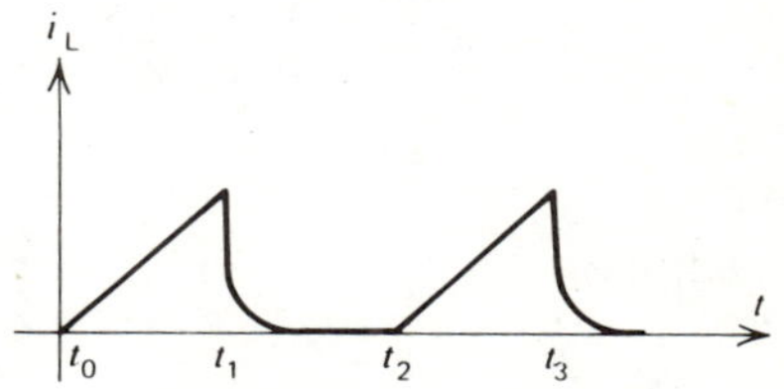

(d) Current sweep in coil

Figure 8-18 Current-sweep generator.

Linearity of the current sweep is vastly improved by applying a trapezoidal input voltage at e_{o_1}, as shown in Fig. 8-18a. Resistor R_E (100 Ω) is used to make the input impedance of Q_2 high so that it does not load the trapezoidal wave-producing circuit.

At $t = t_{0_+}$, $-V_1$ at E_i turns Q_1 OFF. Immediately, voltage at e_{o_1} jumps to V_{o_1} which is given by:

$$V_{o_1} = V_{cc}\left(\frac{R_c}{R_L + R_c}\right) \text{ volts (V)} \qquad (8\text{-}27)$$

Then, of course, C starts charging exponentially.

$$e_{o_1}(t) = V_{0_1} + V_{CC}\{1 - e^{-t/[C(R_L + R_c)]} \text{ volts (V)} \qquad (8\text{-}28)$$

In Eq. (8-28), t, the sweep time, is usually made small in comparison with $C(R_L + R_c)$, so that the voltage rise across C stays within the linear part of the exponential curve.

e_{o_1} acts as an input to Q_2. The small initial jump and then a linear voltage rise at e_{o_1} results in a more linear current sweep in L, as shown in Fig. 8-18d.

Example 8-7

Inductance of the horizontal deflection coil of a CRT is 5 mH. A current sweep of 0 mA to 20 mA is desired. The trace time is to be 20 μs and the flyback time to be 2 μs. Design an appropriate circuit. For the ideal transistor that is to be used as a switch, assume $I_C(\text{max}) = 20$ mA and $\beta = 300$. Assume $R_D = 10\ \Omega$.

Solution

The current-sweep-generator circuit is shown in Fig. 8-15.

1. Trace-time condition:

$$V_{cc} = \frac{i_L}{t}\cdot L \qquad (8\text{-}23)$$

$$\therefore V_{cc} = \frac{20\text{ mA} \times 5\text{ mH}}{20\ \mu\text{s}}$$

$$= \mathbf{5\ V}$$

2. Flyback-time condition:

$$2\ \mu\text{s} \geqq 5\tau$$

$$\geqq \frac{5L}{R_D + R_1} \qquad (8\text{-}26)$$

or

$$\geqq \frac{5 \times 5\,\text{mH}}{10\,\Omega + R_1}$$

$$\therefore R_1 \geqq \frac{25 \times 10^{-3}}{2 \times 10^{-6}} - 10$$

$$\geqq 12.5\,\text{k}\Omega - 10\,\Omega$$

$$\geqq \mathbf{12.5\,k\Omega}$$

The input voltage wave consists of a 20-μs positive pulse and a 2-μs rest period.

3. To assure transistor saturation:

$$R_B \leqq \frac{V_p}{20\,\text{mA}} \cdot (300)$$

where V_p is the peak input voltage.

8-10 Summary

1. A voltage sweep may be generated by charging a capacitor through a long time-constant circuit and discharging it through a short time-constant circuit.

2. Linearity of a sweep is measured by its slope error.

$$\text{Slope error} = \frac{\text{Initial slope} - \text{Final slope}}{\text{Initial slope}}$$

3. The slope error of a perfectly linear sweep (initial slope = final slope) is zero.

4. Usually, the retrace (flyback) time must be very small in comparison with the sweep time or the trace time.

5. In a simple self-pulsing, four-layer-diode sweep-generator circuit, a capacitor charges through a high resistance when the FL diode is OFF and then discharges through the low resistance of the ON diode after V_{BO} has been reached. The charging and discharging cycles continue indefinitely.

6. In a simple transistor sweep generator circuit, a capacitor is charged through the load resistor when the transistor is OFF and is discharged through the low resistance of the ON transistor. The transistor itself is turned ON and OFF by an input trigger pulse.

7. The Miller circuit uses a high-gain amplifier and a feedback capacitor to produce a fairly linear sweep. Essentially, a capacitor is charged by a constant current to produce a linear sweep.

8. The bootstrap circuit is another version of the Miller circuit. A feedback capacitor that acts like a power source and an amplifier of unity gain (emitter follower) are used to generate a linear sweep.
9. Still another way to charge a capacitor by a constant current is to use the common base configuration as a constant-current source. A transistor switch is connected across the capacitor. The capacitor is charged linearly by the constant current when the transistor switch is OFF and is discharged quickly through the low resistance of the transistor switch when it is ON.
10. The most common use of the current sweep is made in the deflection coils of a cathode ray tube. A current sweep is generated by applying a step function across a coil whose internal resistance is very low. At the end of the sweep, the coil current is made to decay through a resistor under critically damped circuit conditions.
11. Important equations:

$$\text{Percentage slope error} = \frac{\text{Initial slope} - \text{Final slope}}{\text{Final slope}} \times 100 \qquad (8\text{-}1)$$

$$t_{sp} = \tau_c \ln \frac{V_{cc} - V_D(\text{ON})}{V_{cc} - V_{RO}} \text{ seconds (s)} \qquad (8\text{-}2)$$

$$t_{sp} = \tau_c \ln \frac{V_{cc} - V_{CE}(\text{ON})}{V_{cc} - V_C} \text{ seconds (A)} \qquad (8\text{-}7b)$$

$$\text{Percentage slope error} = \frac{V_c - V_{CE}(\text{ON})}{V_{cc} - V_{CE}(\text{ON})} \qquad (8\text{-}8)$$

$$V_C = E_i \frac{t_{sp}}{\tau_c} \text{ volts (V)} \qquad (8\text{-}9)$$

$$e_o = E_i \frac{t_{sp}}{\tau_c} \text{ volts (V)} \qquad (8\text{-}11)$$

$$R_E = \frac{V_{EE} - V_{BE}(\text{ON})}{I_E} \text{ ohms } (\Omega) \qquad (8\text{-}16)$$

$$t_{sp}(\max) = \frac{C}{I_C} \cdot V_{cc} \text{ seconds (s)} \qquad (8\text{-}20)$$

$$V_{cc} = \frac{i_L}{t} \cdot L \text{ volts (V)} \qquad (8\text{-}23)$$

8-11 Glossary

Breakover voltage (V_{BO}). Forward voltage at which the four-layer diode breaks down (turns ON).

Deflection coil. Coil that is used to magnetically deflect the electron beam in a cathode ray tube (CRT).

Feedback capacitor. A capacitor that is connected between the output and the input of an amplifier.

Flyback time. The time duration taken by the signal to come back to its base or reference value.

Holding current (I_H). Minimum forward current required to keep the four-layer diode conducting (ON).

Linear sweep. A function whose rate of rise (slope) is *constant.*

Retrace time. The same as the *flyback time.*

Slope. Time rate of change of a function.

Slope error. Numerical measurement of linearity of a sweep.

Sweep time. The duration for which the function increases or decreases linearly.

Trace time. The same as the *sweep time.*

8-12 Recapitulation

Return to the Objectives and Self-Evaluation Questions given at the beginning of this chapter and see how well you can answer them. If you cannot answer certain questions satisfactorily, place a check mark (√) next to them and review the appropriate parts of the chapter to find the answers. No attempt should be made to solve the Problems (Sec. 8-13) before answering the Self-Evaluation Questions. Ability to solve the Problems should serve as a test and feedback regarding your understanding of the material presented in this chapter. Answers to Problems are given on p. 525.

8-13 Problems

8-1 Consider the circuit of Fig. 8-1c. $E_i = 100$ V, $R_1 = 50$ kΩ, $R_2 = 100$ Ω, and $C = 200$ pF. The switch is held in position #1 for 5 μs and in position #2 for 0.5 μs alternately for 3 cycles. Draw the voltage charge/discharge curve that appears across $C(E_o)$.

8-2 Repeat Prob. 8-1, but this time assume that the switch is held in position #1 for 15 μs and in position #2 for 1.0 μs during each cycle.

8-3 Calculate slope errors for:

a. Prob. 8-1.

b. Prob. 8-2.

8-4 Consider the relaxation oscillator shown in Fig. 8-3a. Assume that $V_{cc} = 100$ V, $R = 33$ kΩ, $C = 0.01$ μF, and for the FL diode, assume that $V_{BO} = 10$ V, $I_H = 5$ mA, and $V_D(ON) = 0.1$ V. Check by calculations whether the circuit functions satisfactorily as a relaxation oscillator.

8-5 Find sweep time, t_{sp}, for the circuit of Prob. 8-4.

8-6 Calculate the range of t_{sp} available if a 10-kΩ potentiometer is connected in series with the 33-kΩ resistor in the circuit of Prob. 8-4.

8-7 The 0.01-μF capacitor in the circuit of Prob. 8-4 is replaced by a 0.47-μF capacitor. Calculate the new t_{sp}.

8-8 Consider once again the circuit of Prob. 8-4. What value of resistance when connected in series with the 33-kΩ resistor will result in a t_{sp} of 50 μs?

8-9 Consider the transistor sweep circuit of Fig. 8-4. Assume $R_L =$ 5 kΩ, $R_1 =$ 50 kΩ, $R_2 =$ 100 kΩ, $C =$ 200 pF, $V_{cc} = +20$ V, and $V_{BB} = -5$ V. The input is a 10-$V_{p\text{-}p}$, 500-kHz square wave. A standard silicon transistor is used. Find:

a. I_C(max) rating of the transistor.
b. Minimum value of β required for proper operation.
c. Peak-to-peak amplitude of the output sweep.
d. The slope error.

8-10 Repeat Prob. 8-9 for a new R_L of 6.8 kΩ.

8-11 The input to the transistor sweep circuit of Fig. 8-4 is a pulse wave shown in Fig. 8-19. The output sweep is desired to be 10 μs long

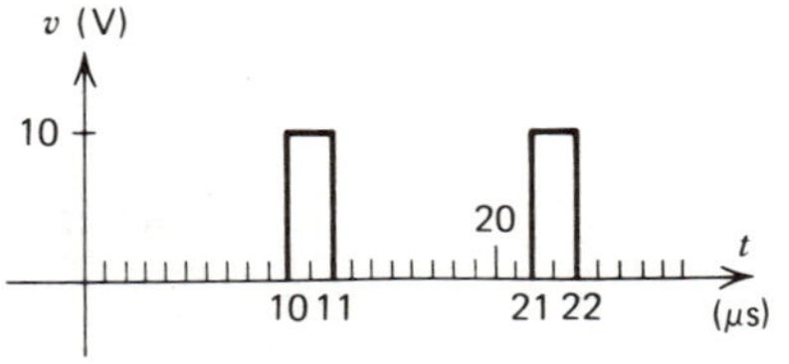

Figure 8-19 Input pulse waveform for Prob. 8-11.

and of 12 $V_{p\text{-}p}$ amplitude. Assume $V_{cc} = 24$ V, $V_{BB} = -2$ V, and $V_{BE}(OFF) = 0.5$ V. $I_C(ON) = 10$ mA and $\beta = 100$ for the germanium transistor being used. Find:

a. R_L.
b. R_1.
c. R_2.
d. C.

8-12 Consider the Miller integrator circuit of Fig. 8-8. Assume $R_L =$ 2 kΩ, $R_B =$ 100 kΩ, $R_s =$ 18 kΩ, $C =$ 300 pF, and $V_{cc} = +20$ V. Assume that the transistor is ideal. The input is a 10-$V_{p\text{-}p}$, 500-kHz square wave. Find:

a. I_C(max) for the transistor.

b. Minimum β required for the transistor.
c. Peak-to-peak amplitude of the output sweep.

8-13 Repeat Prob. 8-12, with a new R_L of 1.5 kΩ and a new R_s of 10 kΩ.

8-14 Consider the bootstrap sweep circuit of Fig. 8-11. D_1 and D_2 are two standard silicon diodes. $\beta = 150$ and $I_C(\max) = 20$ mA for the npn transistor being used. Other circuit components are $R_E = 100\ \Omega$, $R_1 = 10\ \text{k}\Omega$, $C_2 = 20\ \mu\text{F}$, $C_1 = 470$ pF, and $V_{cc} = +15$ V. The input is a 20-$V_{p\text{-}p}$, 500-kHz square wave. Find:
a. The constant current that charges C_1 during sweep generation.
b. Peak-to-peak amplitude of the sweep.

8-15 Refer to Prob. 8-14. The input is changed to a 20-$V_{p\text{-}p}$, 200-kHz square wave. Draw the output waveform.

8-16 Consider the constant-current sweep circuit shown in Fig. 8-15. $V_{cc} = 10$ V, $V_{EE} = -10$ V, $C = 0.002\ \mu\text{F}$, $R_E = 5\ \text{k}\Omega$, $R_1 = 68$ kΩ. The input is a 25-$V_{p\text{-}p}$, 100-kHz square wave. Find and sketch the output waveform. Assume $\beta = 100$ for the two otherwise ideal transistors.

8-17 Repeat Prob. 8-16 for a new value of $C = 0.001\ \mu\text{F}$.

8-18 Refer to Fig. 8-15. The input is a 20-$V_{p\text{-}p}$, 50-kHz square wave. For the two standard silicon transistors, $I_c(\max) = 10$ mA and $\beta = 125$. The output is a 8 $V_{p\text{-}p}$ sweep of 10 μs duration. Assume $V_{cc} = 8$ V, $V_{EE} = -4$ V and constant charging current = 1.5 mA. Find values for R_E, R_1, C.

8-19 Consider the current sweep circuit shown in Fig. 8-17. Assume $L = 2$ mH, $R_1 = 10\ \text{k}\Omega$, $R_B = 220\ \Omega$, $V_{cc} = 20$ V, for D $R_f = 25\ \Omega$, $R_r = 10\ \text{M}\Omega$, $\beta = 75$, and $I_C(\max) = 20$ mA for the silicon transistor. The input is a 5-$V_{p\text{-}p}$, 500 kHz square wave. Find and sketch the current waveform appearing in the coil.

8-20 Repeat Prob. 8-19 for an input of a 5-$V_{p\text{-}p}$, 100 kHz square wave.

8-14 Experiment

Part A

Title **Transistor Sweep Generator**

Theory The transistor sweep-generator circuit consists of a simple transistor switch with a capacitor as load. The load capacitor charges through a larger load resistor R_L (sweep time) when the transistor is OFF and discharges quickly (retrace time) through the lower resistance of the transistor when it (the transistor) is ON. Assuming V_{CE}(ON) to be negligible, the sweep voltage is given by:

$$v_C = V_{cc} - V_{cc}e^{-(t/R_L C)}$$

Equipment Dual-beam oscilloscope: PHILIPS model 3230, or equal.
Power supply: ANATEK model 25-2.00, or equal.
Multimeter: RCA model WV 98C, or equal.
Multifunction generator: EXACT model 126, or equal.
Transistor: 2N4401.
Resistors: 560 Ω, 680 Ω, and 22 kΩ.
Capacitors: 0.01 μF and 0.047 μF.

Procedure

1. Connect the transistor sweep circuit shown in Fig. E8-1.
2. Observe (CRO) and draw in Table E8-1 the input as well as the

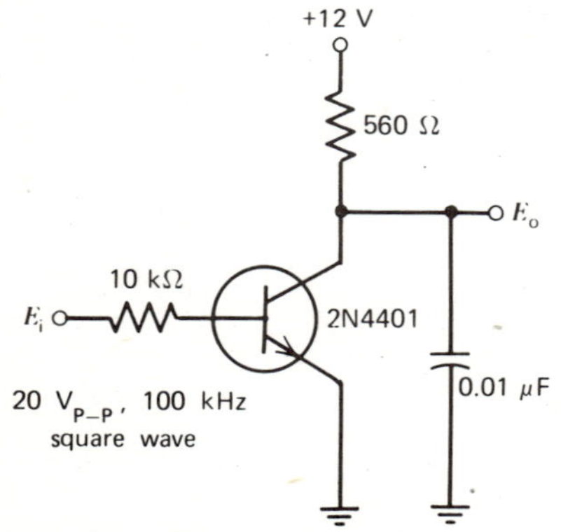

Figure E8-1 Circuit for Exp. 8-15A.

output voltage waveforms. Make sure you record the correct amplitude and the sweep time of the output sweep.

3. Replace the 0.01-μF capacitor with a 0.047-μF capacitor. Observe and draw in Table E8-1 the output voltage waveform.
4. Restore the load capacitance to 0.01 μF. Change the load resistor from 560 Ω to 680 Ω. Observe and draw in Table E8-1 the output voltage waveform.
5. Change the input frequency to 50 kHz but leave the rest of the circuit as shown in Fig. E8-1. Observe and record in Table E8-2 the output voltage waveform.
6. Change the input frequency to 500 kHz. Observe and record in Table E8-2 the output waveform.
7. Disconnect and return all test equipment to its proper place.

Calculations

1. From the output waveforms recorded in Table E8-1, find and record in Table E8-3 the experimentally observed peak-to-peak amplitude of the output sweep.
2. Calculate peak-to-peak amplitude (theoretical) of the output sweep for the circuit conditions given in Procedure steps 2, 3, and 4, respectively. Record the results in Table E8-3. Assume V_{CE}(ON) = 0 V.
3. Calculate the percentage error between the experimental and the theoretical values in Table E8-3. Record your results in the appropriate column in Table E8-3.

Tabulation of Results

Table E8-1

Procedure Step	Description	Waveform	Amplitude
2	Input		
	Output R_L = 560 Ω C = 0.01 μF		
3	Output R_L = 560 Ω C = 0.047 μF		
4	Output R_L = 680 Ω C = 0.01 μF		

Table E8-2

Procedure Step	Frequency	Output Waveform	Peak-to-Peak Amplitude
5	50 kHz		
6	500 kHz		

Table E8-3

Procedure Step	Observed $E_{o(p\text{-}p)}$	Theoretical $E_{o(p\text{-}p)}$	Percentage Error
2			
3			
4			

Questions

1. In Procedure step 3, the load capacitance was increased from 0.01 μF to 0.047 μF. How did this affect the output sweep? Explain.
2. In Procedure step 4, the load resistance was increased from 560 Ω to 680 Ω. How did this affect the output sweep? Explain.
3. Waveforms in Table E8-2 are intended to show the effect on the sweep linearity of a short charging-time and a longer charging time. Explain your results in this respect.

Discussion

1. Summarize your conclusions from the observed and theoretical values recorded in Table E8-3.
2. Account for discrepancies between the theoretical and the observed values.
3. Discuss and identify circuit conditions that will improve the linearity of the output sweep.

Part B

Title **Constant-Current Sweep Generator**

Theory The constant-current sweep generator circuit consists of a common base transistor (Q_2) with a capacitor in its collector circuit (Fig.

E8-2). The common-base configuration provides the constant current that charges the capacitor linearly.

A transistor switch (Q_1) is connected across the capacitor. The capacitor charges linearly when the transistor switch is OFF and discharges quickly (through the ON transistor) when the transistor switch is ON. The rate of capacitor charge is given by:

$$\frac{v_C}{t} = \frac{I_{ch}}{C}, \text{ volts/second (V/s)}$$

Equipment Dual-beam oscilloscope: PHILIPS model 3230, or equal.
Power supply: ANATEK model 25-2.00, or equal.
Multimeter: RCA model WV 98C, or equal.
Multifunction generator: EXACT model 126, or equal.
Resistors: 2.2 kΩ, 3.3 kΩ, and 22 kΩ.
Capacitors: 0.01 μF and 0.047 μF.
Transistors: two 2N4401.

Procedure

1. Connect the constant-current sweep generator circuit shown in Fig. E8-2.
2. Observe (CRO) and draw in Table E8-4 the input as well as the output voltage waveforms.

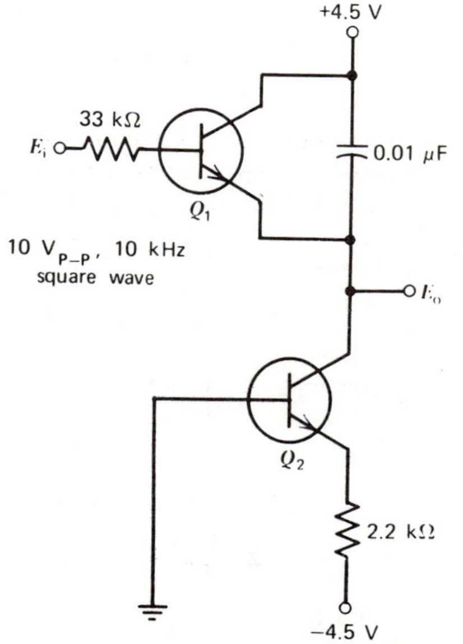

Figure E8-2 Circuit for Exp. 8-15B.

3. Change the 0.01-μF capacitor to a 0.047-μF capacitor. Observe and draw in Table E8-4 the output sweep waveform.
4. Insert back the 0.01-μF capacitor in place of the 0.047-μF capacitor. Change the emitter resistor from 2.2 kΩ to 3.3 kΩ. Again observe and draw in Table E8-4 the output sweep waveform.
5. Restore the circuit back to the one shown in Fig. E8-2. Change the input frequency from 10 kHz to 20 kHz. Observe and record in Table E8-5 the output sweep waveform.
6. Change the input frequency to 5 kHz. Observe and record in Table E8-5 the output sweep waveform.
7. Disconnect and return all test equipment to its proper place.

Calculations

1. From the waveforms recorded in Table E8-4, find and record in Table E8-6 the experimentally observed values of the amplitude and the sweep time of the output waveforms.
2. Calculate and record in Table E8-6 the theoretically expected values of the amplitude and sweep time of the output waveforms, for the circuit conditions of Procedure steps 2, 3, and 4.
3. Calculate the percentage error between the experimental and the calculated values recorded in Table E8-6.

Tabulation of Results Table E8-4

Procedure Step	Description	Waveform	$E_{o(p\text{-}p)}$	t_{sp}
2	Input			
	Output			
3	Output			
4	Output			

Table E8-5

Procedure Step	Frequency	Waveform	$E_{o(p\text{-}p)}$	t_{sp}
5	20 kHz			
6	5 kHz			

Table E8-6

Procedure Step	Observed	Theoretical	Percentage Error
2	$E_{o(p\text{-}p)}$		
	t_{sp}		
3	$E_{o(p\text{-}p)}$		
	t_{sp}		
4	$E_{o(p\text{-}p)}$		
	t_{sp}		

Questions

1. In Procedure step 3, the capacitance in the circuit was increased from 0.01 μF to 0.047 μF. How did this affect the output sweep? Explain.
2. In Procedure step 4, the emitter resistance was increased from 2.2 kΩ to 3.3 kΩ. How did it affect the output sweep? Explain.
3. Table E8-5 is intended to show the effects of a long charging-time and a short charging-time on the output sweep. Explain your observed results in this respect.

Discussion

1. Summarize your conclusions from the observed and the theoretical values recorded in Table E8-6.
2. Account for discrepancies between the observed and the theoretical values.
3. Discuss and identify circuit conditions that will improve the linearity of the output sweep.

Chapter 9 Other Pulse Circuits

9-1 Objectives

The objectives of this chapter are to:

1. Explain the operation of a Schmitt trigger circuit.
2. Show how to design a Schmitt trigger circuit.
3. Review the electrical characteristics of a unijunction transistor (UJT).
4. Explain the operation of a simple UJT relaxation oscillator.
5. Show how to design a UJT relaxation oscillator.
6. Show how a 555 timer may be used as a:
   a. Monostable MV.
   b. Astable MV.
   c. Sweep generator.

9-2 Self-Evaluation Questions

Watch for the answers to the following questions as you read the chapter. They should help you to understand the material presented. When you have finished the chapter, return to this section and be sure you can answer all the questions. Then test your understanding by working out solutions to the Problems.

It is strongly recommended that you answer these questions satisfactorily before attempting to solve the Problems (Sec. 9-14).

1. Draw the circuit diagram of a basic Schmitt trigger circuit.
2. Explain the terms *upper trigger potential* (UTP) and *lower trigger potential* (LTP) as used in Schmitt trigger circuits.
3. Show how to design a Schmitt trigger circuit for given LTP, UTP, supply voltage, input signal, and characteristics of the transistor.
4. Draw a typical Schmitt trigger circuit. Assume some reasonable values for circuit components. Find:
   a. Whether one transistor is OFF when the other's ON condition is satisfied.
   b. Values for LTP and UTP.
5. Identify and label the following points on the characteristic curve of a UJT:
   a. Peak point voltage (V_p).
   b. Peak point current (I_p).
   c. Valley point voltage (V_v).
   d. Valley point current (I_v).
6. Describe the operation of a UJT relaxation oscillator.

7. List and explain the conditions that are necessary for the UJT relaxation oscillator circuit to function properly.
8. For a UJT relaxation oscillator, find expressions for:
   a. Sweep time (t_{sp}).
   b. Retrace time (t_{re}).
   c. Frequency of the output waveform.
   d. Amplitude of the output voltage.
9. Explain how trigger pulses may be obtained from a UJT relaxation oscillator circuit.
10. Show that the frequency of the output waveform of a UJT relaxation oscillator is independent of the supply voltage.
11. What is a 555 timer? Describe its basic characteristics.
12. Show how a 555 timer is used as a:
    a. Monostable MV.
    b. Astable MV.
    c. Sweep generator.

9-3 Schmitt Trigger Operation

The Schmitt trigger is a bistable device. It contains two transistors connected in direct cascade. The circuit is designed such that at any one time, one transistor is ON while the other is OFF. An input signal is applied to the first (input) transistor and the output is taken at the collector of the second (output) transistor. The voltage level of the input signal controls the state (ON/OFF) of the input transistor.

A basic Schmitt trigger circuit is shown in Fig. 9-1a. The input signal (E_i) turns Q_1 ON or OFF. The circuit is designed such that when Q_1 is ON, Q_2 is OFF; and when Q_1 is OFF, Q_2 is ON. Typical input/output waveforms are shown in Figs. 9-1b and 9-1c. The two input voltage levels at which the circuit switches are known as the *upper trigger potential* (UTP) and the *lower trigger potential* (LTP). Q_1 switches from OFF to ON at UTP and it switches from ON to OFF at LTP.

Note that the emitter resistor R_E is common to both of the transistors. Thus, there is always an emitter current in R_E regardless of which transistor is ON. However, the emitter current of Q_1(ON) is *not* the same as the emitter current of Q_2(ON).

The Schmitt trigger shown in Fig. 9-1a operates as follows:

1. When $E_i = 0$ V (lower than, both UTP and LTP), V_{E_1} (as well as V_{E_2}) is at some positive potential. Thus, the BEJ of Q_1 is reverse-biased. Accordingly, under this condition, Q_1 is OFF and Q_2 is ON.

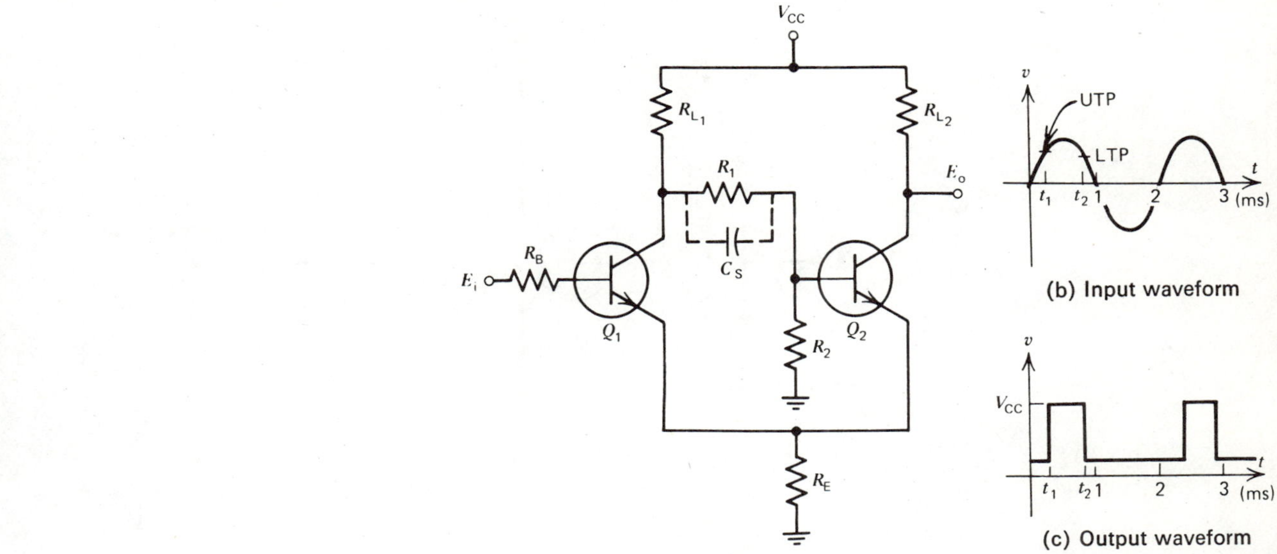

(a) Basic circuit

Figure 9-1 Schmitt trigger circuit and input/output waveforms.

2. As E_i increases, a point is reached when Q_1 turns ON and Q_2 turns OFF. The input voltage level at which Q_1 turns ON is called the *upper trigger potential* (UTP).
3. Q_1 stays ON and Q_2 OFF as long as $E_i \geq$ UTP.
4. As E_i decreases from above the UTP level, a point is reached when Q_1 turns OFF and Q_2 turns ON. The input voltage level at which this switching occurs is called the *lower trigger potential* (LTP).
5. The circuit is now in the same state as it was in step 1, above. Note that Q_1 requires higher potential (UTP) to turn ON from OFF (LTP) than to turn OFF from ON. This potential difference between UTP and LTP is an example of what is called *hysteresis*.

9-3.1 Schmitt Trigger Design A Schmitt trigger is designed for desired values of UTP and LTP. The step-by-step design procedure is:

1. Choose an appropriate value of the supply voltage V ... be the output voltage when $E_i \geqq$ UTP.
2. Choose two appropriate transistors (high ... their important characteristics: beta, ... V_{CC}(max), and I_C(max).

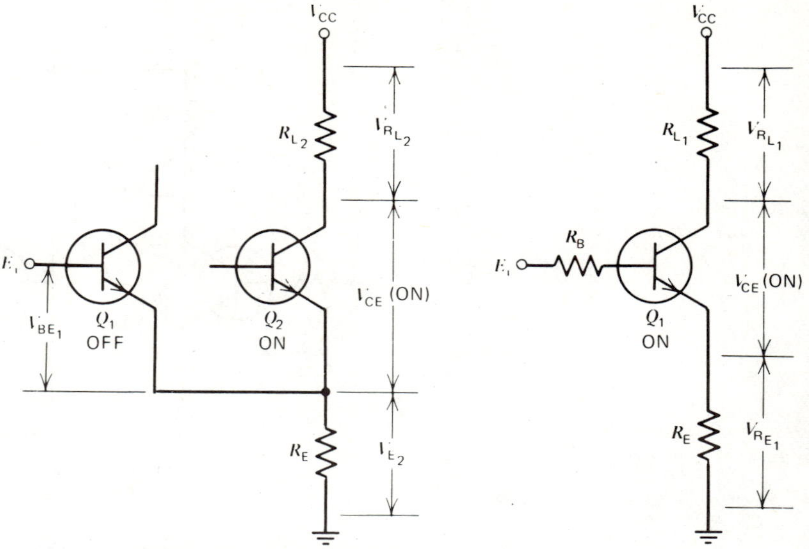

(a) Circuit condition when E_i < UTP

(b) Circuit condition when E_i > LTP

Figure 9-2 Partial circuits of a Schmitt trigger showing Q_1-OFF and Q_2-ON conditions.

3. Draw the equivalent circuit when E_i < UTP, as shown in Fig. 9-2a. Assume a safe value for $I_c(\text{ON}) < I_C(\text{max})$.

4. Calculate R_E:

a. $$I_{E_2}(\text{ON}) = \frac{I_{C_2}(\text{ON})}{\alpha} \text{ amperes (A)} \tag{9-1}$$

b. When Q_1 is on the verge of turning ON,

$$\text{UTP} = V_{BE_1}(\text{ON}) + V_{E_2}(\text{ON})$$

$$V_{E_2}(\text{ON}) = \text{UTP} - V_{BE_1}(\text{ON})$$

$$R_E = \frac{\text{UTP} - V_{BE_1}(\text{ON})}{I_{E_2}(\text{ON})} \text{ ohms } (\Omega) \tag{9-2}$$

5. Calculate R_{L_2}:

$$R_{L_2} = \frac{V_{RL2}}{I_{C_2}(\text{ON})}$$

$$= \frac{V_{cc} - [V_{CE_2}(\text{ON}) + V_{E_2}(\text{ON})]}{I_{C_2}(\text{ON})} \text{ ohms } (\Omega) \tag{9-3}$$

6. Draw the equivalent circuit when $E_i \geqq$ UTP, as shown in Fig. 9-2b. Under this condition, Q_1 is ON and Q_2 is OFF. When Q_1 is on the verge of turning OFF,

a. $$I_{E_1}(\text{ON}) = \frac{\text{LTP} - V_{BE_1}(\text{ON})}{R_E} \text{ amperes (A)} \qquad (9\text{-}4)$$

b. $$I_{C_1}(\text{ON}) = \alpha I_{E_1}(\text{ON}) \text{ amperes (A)} \qquad (9\text{-}5)$$

c. Calculate R_{L_1}:

$$R_{L_1} = \frac{V_{R_{L1}}}{I_{C_1}(\text{ON})}$$

$$= \frac{V_{cc} - [V_{CE_1}(\text{ON}) + V_{E_1}(\text{ON})]}{I_{C_1}(\text{ON})} \text{ ohms } (\Omega) \qquad (9\text{-}6)$$

7. Calculate R_1 and R_2:

a. Figure 9-3a shows a partial circuit when Q_1 is OFF and Q_2 is ON. Under this, the ON (Q_2) condition,

$$\frac{V_{cc} - V_{B_2}(\text{ON})}{R_{L_1} + R_1} - \frac{V_{B_2}(\text{ON})}{R^2} = \frac{I_{C_2}(\text{ON})}{\beta} \times 2 \quad \text{(overdriving 200 percent)}$$

but

$$V_{B_2}(\text{ON}) = V_{E_2}(\text{ON}) + V_{BE_2}(\text{ON})$$

$$= V_{E_2}(\text{ON}) + V_{BE_1}(\text{ON}) \quad [\text{usually } V_{BE_1}(\text{ON}) = V_{BE_2}(\text{ON})]$$

$$= \text{UTP}$$

$$\therefore \frac{V_{cc} - \text{UTP}}{R_{L_1} + R_1} - \frac{\text{UTP}}{R^2} = 2\, I_{B_2}(\text{ON}) \text{ amperes (A)} \qquad (9\text{-}7\text{a})$$

b. Figure 9-3b shows a partial circuit when Q_1 is ON and Q_2 is OFF. Under this, the OFF (Q_2) condition:

$$V_{c_1}(\text{ON}) = V_{E_1}(\text{ON}) + V_{CE_1}(\text{ON}) \text{ vol-}$$

and

$$V_{B_2}(\text{OFF}) = V_{E_1}(\text{ON}) + V_{BE_2}(\text{OFF}$$

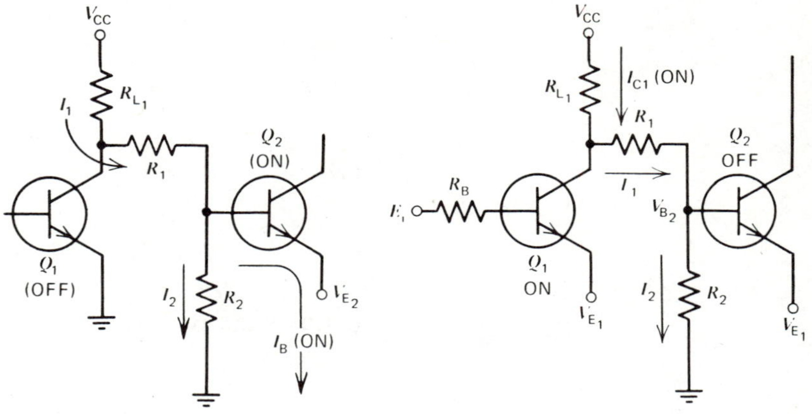

(a) Circuit conditions when Q_2 is ON (b) Circuit conditions when Q_2 is OFF

Figure 9-3 Partial circuits showing Q_2 ON/OFF conditions.

Now that $I_{B_2}(\text{OFF}) = 0$,

$$\frac{V_{R_1}}{R_1} = \frac{V_{R_2}}{R_2}$$

$$\frac{V_{C_1}(\text{ON}) - V_{B_2}(\text{OFF})}{R_1} = \frac{V_{B_2}(\text{OFF})}{R_2} \text{ amperes (A)} \qquad (9\text{-}10)$$

Solving Eqs. (9-7a) and (9-10) simultaneously yields appropriate values for R_1 and R_2. Note that as it stands, the solution will involve a quadratic equation, the solution of which is cumbersome at times. To avoid this and simplify the solving process, it may safely be assumed that R_{L_1} is considerably smaller than R_1. This assumption reduces Eq. (9-7a) to:

$$\frac{V_{cc} - \text{UTP}}{R_1} - \frac{\text{UTP}}{R_2} = 2\, I_{B_2}(\text{ON}) \text{ amperes (A)} \qquad (9\text{-}7b)$$

8. Resistor R_B is made equal to $\beta R_E/10$ so that it is insignificant as far as the total input impedance of Q_1 is concerned.

Quite often, a capacitor is connected across R_1 to improve the switching time of Q_2. It is, in fact, a speed-up capacitor, whose function was explained in Sec. 5-6.

Example 9-1

Design a Schmitt trigger circuit such that its UTP is +5 V and LTP is +4 V. Assume that $V_{cc} = +12$ V, and for the two transistors assume $h_{fe} = 100$, $V_{CE}(ON) = 0$ V, $I_C(ON) = 20$ mA, $V_{BE}(ON) = 0$ V.

Solution

The Schmitt trigger circuit is shown in Fig. 9-1a. Values for R_{L_1}, R_{L_2}, R_B, R_2, R_1, R_E are to be calculated.

1. Calculations for R_E:

$$I_{E_2}(ON) = \frac{I_{C_2}(ON)}{\alpha}$$

$$= \frac{20 \text{ mA}}{0.99} \qquad \left(\alpha = \frac{\beta}{1+\beta}\right)$$

$$= 20.2 \text{ mA}$$

$$R_E = \frac{\text{UTP} - V_{BE_1}(ON)}{I_{E_2}(ON)} \tag{9-2}$$

$$= \frac{(5.0 - 0) \text{ V}}{20.2 \text{ mA}}$$

$$= \mathbf{248\ \Omega}$$

2. Calculations for R_{L_2}:

$$V_{E_2}(ON) = \text{UTP} - V_{BE_1}(ON)$$

$$= 5.0 - 0$$

$$= 5 \text{ V}$$

$$R_{L_2} = \frac{V_{cc} - [V_{CE_2}(ON) + V_{E_2}(ON)]}{I_{C_2}(ON)} \tag{9-3}$$

$$= \frac{12 - (0 + 5)}{20 \times 10^{-3}}$$

$$= \mathbf{350\ \Omega}$$

3. Calculations for R_{L_1}:

$$I_{E_1}(ON) = \frac{\text{LTP} - V_{BE_1}(ON)}{R_E} \tag{9-4}$$

$$= \frac{4 - 0}{248}$$

$$= \mathbf{16.13\ mA}$$

$$I_{C_1}(\text{ON}) = 0.99 \times 16.13 \text{ mA}$$
$$= \mathbf{15.97\ mA}$$
$$V_{E_1}(\text{ON}) = I_{E_1}(\text{ON}) \times R_E$$
$$= 16.13 \text{ mA} \times 248\ \Omega$$
$$= \mathbf{4\ V}$$

$$R_{L_1} = \frac{V_{cc} - [V_{CE_1}(\text{ON}) + V_{E_1}(\text{ON})]}{I_{C_1}(\text{ON})} \tag{9-6}$$
$$= \frac{12 - (0 + 4)}{15.97 \times 10^{-3}}$$
$$= \mathbf{501\ \Omega}$$

4. Calculations for R_1 and R_2:
a. Q_2 OFF condition:

$$V_{C_1}(\text{ON}) = V_{E_1}(\text{ON}) + V_{CE_1}(\text{ON}) \tag{9-8}$$
$$= 4.0 + 0$$
$$= \mathbf{4\ V}$$

$$V_{B_2}(\text{OFF}) = V_{E_2}(\text{ON}) + V_{BE_2}(\text{OFF}) \tag{9-9}$$
$$= 4 + (-1.0)$$
$$= \mathbf{3.0\ V}$$

$$\frac{V_{C_1}(\text{ON}) - V_{B_2}(\text{OFF})}{R_1} = \frac{V_{B_2}(\text{OFF})}{R_2} \tag{9-10}$$
$$\frac{4.0 - 3.0}{R_1} = \frac{3.0}{R_2}$$
$$R_2 = 3R_1$$

b. Q_2 ON condition:

$$\frac{V_{cc} - \text{UTP}}{R_{L_1} + R_1} - \frac{\text{UTP}}{R_2} = 2\, I_{B_2}(\text{ON}) \tag{9-7a}$$
$$\frac{12 - 5}{501 + R_1} - \frac{5}{R_2} = 2\,\frac{20 \text{ mA}}{100}$$
$$\frac{7}{501 + R_1} - \frac{5}{3\,R_1} = 0.4 \times 10^{-3}$$

Assume that $501 \ll R_1$. Then,

$$R_1 = \frac{7 - (5/3)}{0.4 \times 10^{-3}}$$

$$= \mathbf{13.33\ k\Omega}$$

and

$$R_2 = 3 \times 13.33\ \text{k}\Omega$$

$$= \mathbf{40\ k\Omega}$$

5. Calculations for R_B:

$$R_B = \frac{\beta R_E}{10}$$

$$= \frac{100 \times 248}{10}$$

$$= \mathbf{2.48\ k\Omega}$$

Example 9-2

Repeat Ex. 9-1, but this time assume that Q_1 and Q_2 are standard Si transistors.

Solution

1. Calculations for R_E:

$$R_E = \frac{\text{UTP} - V_{BE_1}(\text{ON})}{I_{E_2}(\text{ON})} \tag{9-2}$$

$$= \frac{(5.0 - 0.7)\ \text{V}}{20.2\ \text{mA}}$$

$$= \mathbf{213\ \Omega}$$

2. Calculations for R_{L_2}:

$$V_{E_2}(\text{ON}) = \text{UTP} - V_{BE_1}(\text{ON})$$

$$= 5.0 - 0.7$$

$$= \mathbf{4.3\ V}$$

$$R_{L_2} = \frac{V_{cc} - [V_{CE_2}(\text{ON}) + V_{E_2}(\text{ON})]}{I_{C_2}(\text{ON})} \tag{9-3}$$

$$= \frac{12 - (0.3 + 4.3)}{20 \times 10^{-3}}$$

$$= \mathbf{370\ \Omega}$$

3. Calculations for R_{L_1}:

$$I_{E_1}(\text{ON}) = \frac{\text{LTP} - V_{BE_1}(\text{ON})}{R_E} \quad (9\text{-}4)$$

$$= \frac{4.0 - 0.7}{213}$$

$$= 15.49 \text{ mA}$$

$$V_{E_1}(\text{ON}) = I_{E_1}(\text{ON}) \times R_E$$

$$= 15.45 \times 10^{-3} \times 213$$

$$= 3.3 \text{ V}$$

$$I_{C_1}(\text{ON}) = 15.45 \times 0.99$$

$$= 15.34 \text{ mA}$$

$$R_{L_1} = \frac{V_{cc} - [V_{CE_1}(\text{ON}) + V_{E_1}(\text{ON})]}{I_{C_1}(\text{ON})} \quad (9\text{-}6)$$

$$= \frac{12 - (0.3 + 3.3)}{15.34 \times 10^{-3}}$$

$$= \mathbf{547.6\ \Omega}$$

4. Calculations for R_1 and R_2:

a. Q_2 OFF condition:

$$V_{C_1}(\text{ON}) = V_{E_1}(\text{ON}) + V_{CE_1}(\text{ON}) \quad (9\text{-}8)$$

$$= 3.3 + 0.3$$

$$= 3.6 \text{ V}$$

$$V_{B_2}(\text{OFF}) = V_{E_1}(\text{ON}) + V_{BE_2}(\text{OFF}) \quad (9\text{-}9)$$

$$= 3.3 + (-1.0)$$

$$= 2.3 \text{ V}$$

$$\frac{V_{C_1}(\text{ON}) - V_{B_2}(\text{OFF})}{R_1} = \frac{V_{B_2}(\text{OFF})}{R_2} \quad (9\text{-}10)$$

$$\frac{3.6 - 2.3}{R_1} = \frac{2.3}{R_2}$$

$$R_2 = 1.77 R_1$$

b. Q_2 ON condition:

$$\frac{V_{cc} - \text{UTP}}{R_1} - \frac{\text{UTP}}{R_2} = 2I_{B_2}(\text{ON}) \qquad (9\text{-}7b)$$

$$\frac{12 - 5}{R_1} - \frac{5}{1.77R_1} = 2\left(\frac{20 \text{ mA}}{100}\right)$$

$$\frac{7}{R_1} - \frac{2.825}{R_1} = 0.4 \times 10^{-3}$$

$$R_1 = \mathbf{10.44\ k\Omega}$$

and

$$R_2 = 10.44 \times 1.77 \text{ k}\Omega$$

$$= \mathbf{18.48\ k\Omega}$$

5. Calculations for R_B:

$$R_B = \frac{\beta R_E}{10}$$

$$= \frac{100 \times 213}{10}$$

$$= \mathbf{2.13\ k\Omega}$$

9-4 IC Schmitt Trigger and Applications

The 7413 is a typical IC Schmitt trigger. It in fact is a *dual* NAND Schmitt trigger, as shown in Fig. 9-4. It consists of two identical Schmitt trigger circuits on a single chip. Each circuit functions as a four-input NAND gate. However, as a result of the Schmitt action, the gate has different input threshold-switching levels for positive-going and negative-going signals. The hysteresis or the difference between two threshold levels (UTP − LTP) is typically 0.8 V.

One application of a Schmitt trigger is in the pulse generator. The connection diagram and the input/output waveforms for such an application are shown in Fig. 9-5. The output pulse width t_p is controlled by the **dc** level of the input waveform. Note that positive-going pulses may be obtained by simply connecting an inverter at the output.

Another application of a Schmitt trigger is in the pulse stretcher. The connection diagram and the input/output waveforms are shown in Fig. 9-6. A capacitor, C, is connected at the output of an inverter (with an open collector output) to obtain an exponential

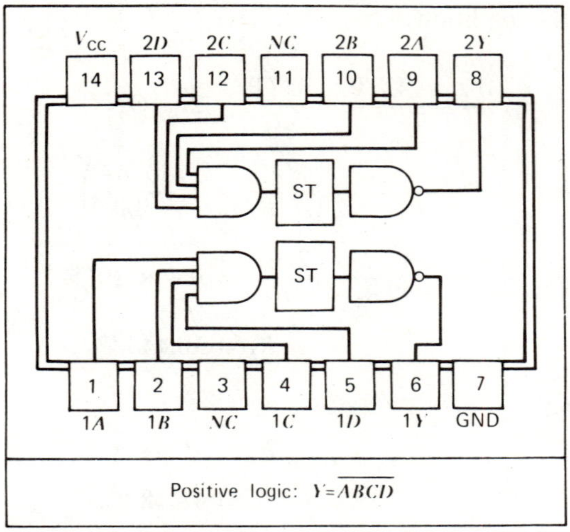

Figure 9-4 Logic diagram of the 7413, a dual NAND Schmitt trigger.

voltage rise across it when the input switches to LOW. The rate of the capacitor voltage rise is controlled by the magnitude of C ($\tau = RC$), which in turn controls the output pulse width.

Note that the waveforms shown in Figs. 9-6b, 9-6c and 9-6d show that an original input pulse is "stretched." It has a longer pulse duration at the output even though its repetition rate (PRR) is unchanged.

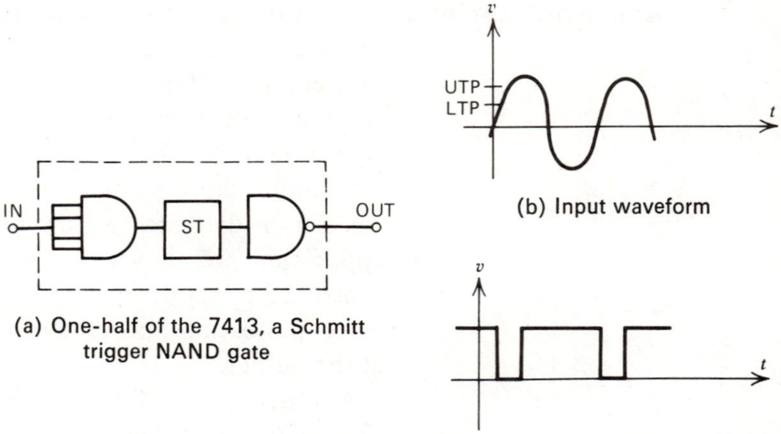

(a) One-half of the 7413, a Schmitt trigger NAND gate

(b) Input waveform

(c) Output waveform

Figure 9-5 Pulse generation, using an IC Schmitt trigger.

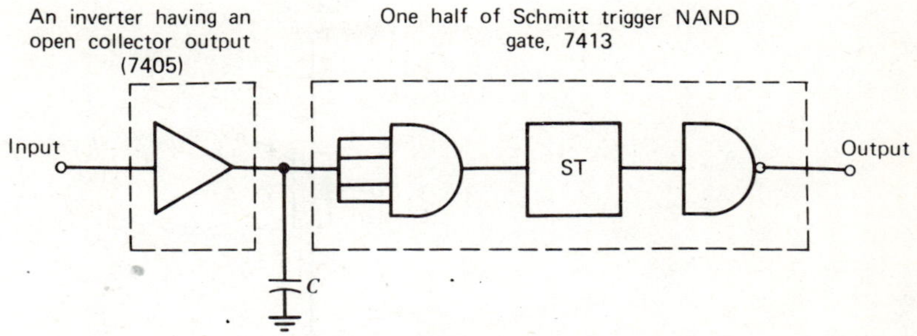

(a) Connection diagram

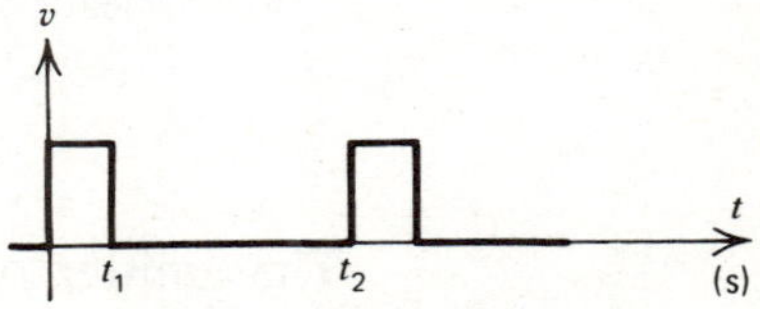

(b) Input pulse waveform

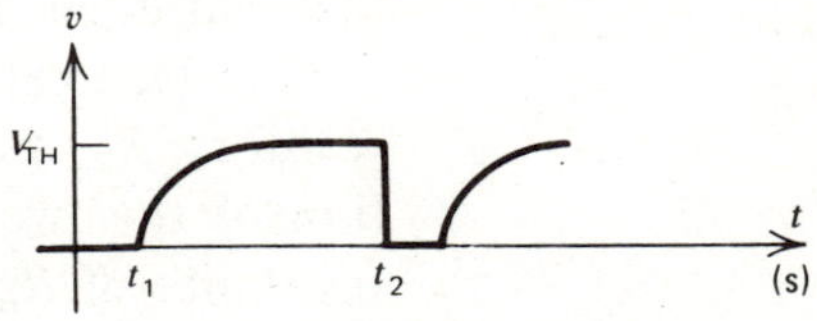

(c) Voltage across C

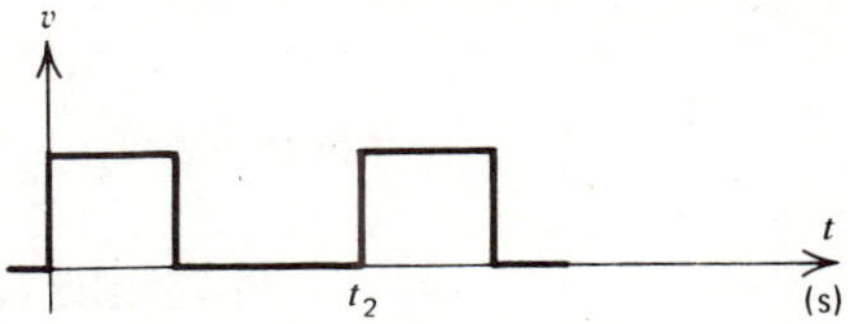

(d) Output pulse waveform

Figure 9-6 Pulse stretcher.

9-5 Characteristics of a UJT

The *unijunction transistor* (*UJT*) is primarily used in a self-pulsing relaxation oscillator circuit. A UJT relaxation oscillator produces very stable sharp-trigger pulses (spikes), which are used in timing circuits.

Figure 9-7 shows the physical makeup of (Fig. 9-7a), the standard symbol for (Fig. 9-7b), and the equivalent circuit of a UJT (Fig. 9-7c). Note that the three leads are called:

1. Base-1 (B_1).
2. Base-2 (B_2).
3. Emitter (E).

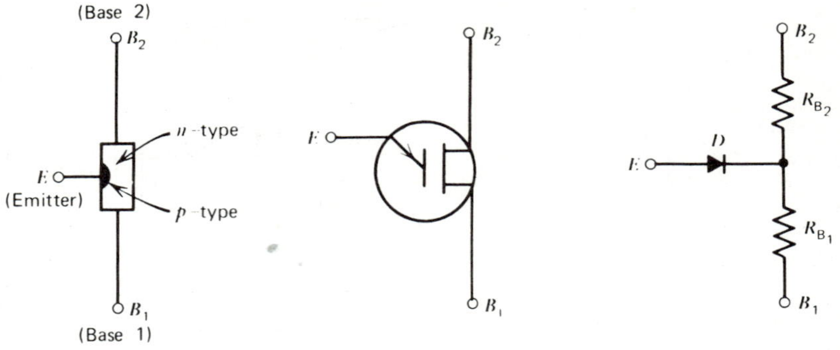

(a) Physical makeup (b) Standard symbol (c) Equivalent circuit

Figure 9-7 Unijunction transistor (UJT).

The emitter of a UJT may be made up of *p*-type material (Fig. 9-7a) or *n*-type material. If the emitter was made up of *n*-type material, then the emitter arrow in Fig. 9-7b will be reversed.

The electrical behavior of a UJT resembles two series-connected resistors, R_{B_1} and R_{B_2}, and a diode (emitter) connected to the junction of the two resistors, as shown in Fig. 9-7c. The equivalent resistance of $R_{B_1} + R_{B_2}$ is symbolized as R_{BB}.

$$R_{BB} = R_{B_1} + R_{B_2} \text{ ohms } (\Omega) \tag{9-11}$$

where:

R_{BB} is the equivalent resistance between base-1 and base-2 (Ω)
R_{B_1} is the equivalent resistance between base-1 and the cathode of the emitter diode (Ω)
R_{B_2} is the equivalent resistance between base-2 and the cathode of the emitter diode (Ω)

The ratio between R_{B_1} and R_{BB} is known as the *intrinsic stand-off ratio* and is symbolized by η(eta).

$$\eta = \frac{R_{B_1}}{R_{BB}} \tag{9-12}$$

The manufacturer usually lists values for R_{BB} and η at the zero emitter-current ($I_E = 0$) condition. Typical values for η range from 0.4 to 0.9.

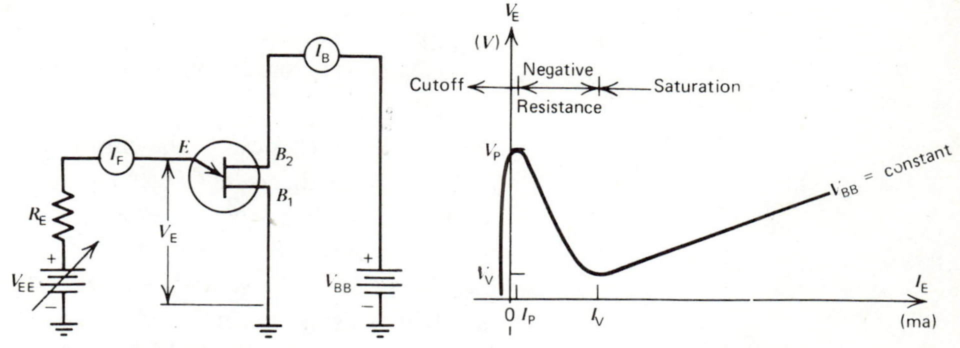

(a) Experimental circuit (b) Characteristic curve

Figure 9-8 Characteristics of a UJT.

The circuit of Fig. 9-8a may be used to determine the volt-ampere characteristic curve of a UJT. The volt-ampere characteristic curve of a UJT is the graph of the emitter current I_E versus emitter voltage V_E, as shown in Fig. 9-8b.

Initially, when $V_{EE} = 0$ and $I_E = 0$, the emitter diode is open. The emitter voltage V_E is:

$$V_E = V_{EE} \quad \text{(so long as } I_E = 0)$$

and

$$V_{R(B_1)} = V_{BB} \frac{R_{B_1}}{R_{BB}} \tag{9-13}$$

$$= \eta \, V_{BB} \text{ volt (V)} \tag{9-14}$$

V_E increases as V_{EE} is increased. When it reaches V_p (Fig. 9-8b), the emitter diode turns ON and the emitter current I_E begins to flow. Right at this point,

$$V_p = V_E \tag{9-15a}$$

$$= V_{R(B_1)} + V_D$$

$$= \eta V_{BB} + V_D \tag{9-15b}$$

$$\simeq \eta V_{BB} \text{ volts (V)} \tag{9-15c}$$

where:

V_{BB} is the supply voltage (V)

V_p is the *peak point voltage* (V)
V_D is the voltage drop across the emitter diode, which is usually very small compared to ηV_{BB} (V)

As soon as V_E reaches V_p, the emitter current I_E increases rapidly and V_E drops to the *valley point voltage* (V_v). Note that between V_p and V_v, V_E drops as I_E increases. This is opposite to what Ohm's law states (Sec. 2-4). That is why the region between V_p and V_v is called the *negative resistance* region. This very property of negative resistance is utilized to construct the UJT relaxation oscillator. Essentially, R_{B_1} is high when there is no I_E, and is low when there is an I_E. The UJT is considered to be ON when the emitter diode is conducting and OFF when the emitter diode is not conducting. R_{B_1} is low when UJT is ON and is high when UJT is OFF.

Beyond V_v, I_E rises proportionally to V_E and behaves like an ordinary semiconductor diode.

The standard symbols used in the industry to describe a UJT's characteristics are shown in Fig. 9-8b. They are:

V_p = the maximum emitter voltage with the emitter diode OFF, known as the peak point voltage (V).
V_v = the minimum emitter voltage with the emitter diode ON, known as the valley point voltage (V).
I_p = the emitter current when $V_E = V_p$, known as the peak point current (A).
I_v = the emitter current when $V_E = V_v$, known as the valley point current (A).

In addition to V_v, I_p, I_v, and η, the manufacturer usually gives a recommended value for V_{BB}. Note from Eq. (9-15c) that V_p is a function of V_{BB}.

The volt-ampere characteristic curve is redrawn in Fig. 9-9. The negative resistance characteristic of a UJT allows it to be used as a multivibrator for any of the three classes (astable, monostable, bistable). Location of the load line determines the class of multivibrator, as shown in Fig. 9-9. Conversely, the load line is chosen according to the desired class of the multivibrator circuit.

The UJT is basically used in an astable configuration known as the relaxation oscillator. The circuit is designed such that the load

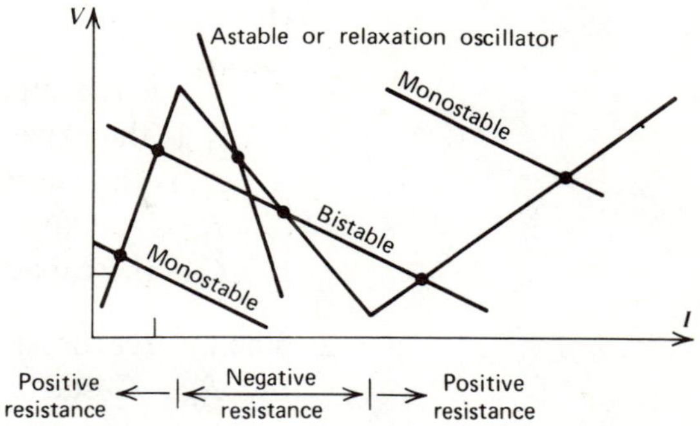

Figure 9-9 Load lines for UJT multivibrator circuits.

line crosses the characteristic curve only in the negative-resistance region. The negative-resistance region is an unstable region. Thus, the circuit becomes an astable multivibrator.

9-6 UJT Relaxation Oscillator

A UJT relaxation oscillator circuit is shown in Fig. 9-10a. It operates as follows:

1. As the circuit is energized, C starts charging toward V_{BB} through a large resistance (R). UJT is OFF at this time. The capacitor charge is given by:

$$v_C = V_{BB} - V_{BB}e^{-(t_{sp}/R_1C)} \text{ volts (V)}$$

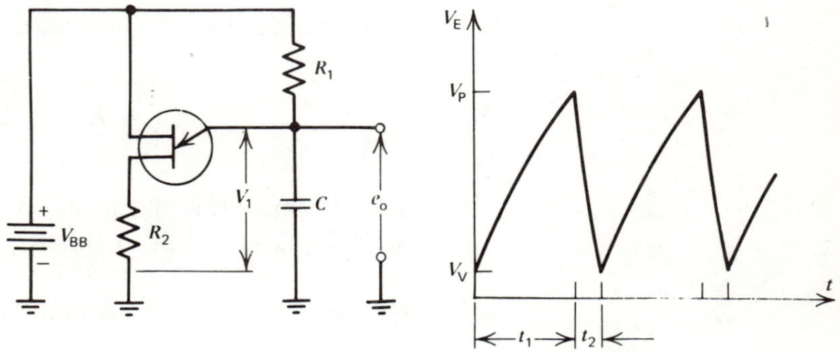

(a) Oscillator circuit (b) Output waveform

Figure 9-10 UJT relaxation oscillator.

where:

V_{BB} is the supply voltage (V)
v_C is the capacitor voltage (V)
t_{sp} is the charge time or the sweep time (s)
R_1 is the large circuit resistance (Ω)
C is the capacitance (F)

2. When v_C reaches V_p, the UJT turns ON. An ON UJT has a very low R_{B_1}. C starts discharging through R_{B_1}(ON) and R_2. The circuit resistance $[R_{B_1}(\text{ON}) + R_2]$, which forms a part of the discharging time constant, is kept to a minimum to achieve a quick retrace or discharge time, or sharp spikes across R_2.

3. C quickly discharges to V_v, at which time UJT turns OFF. The capacitor discharge equation is:

$$V_v = V_p e^{-\{t_{re}/[R_{B1}(\text{ON})+R_2]C\}} \text{ volts (V)} \qquad (9\text{-}16a)$$

or

$$t_{re} = [R_{B_1}(\text{ON}) + R_2]\cdot C \ln \frac{V_p}{V_v} \text{ seconds (s)} \qquad (9\text{-}16b)$$

4. As UJT turns OFF, C starts charging toward V_{BB} again (until V_p is reached). Note that this time, C has an initial charge of V_V. The capacitor charge equation is:

$$V_p = V_{BB} - (V_{BB} - V_v)e^{-(t_{sp}/R_1C)} \text{ volts (V)} \qquad (9\text{-}17a)$$

$$t_{sp} = R_1 C \ln \frac{V_{BB} - V_v}{V_{BB} - V_p} \text{ seconds (s)} \qquad (9\text{-}17b)$$

Assuming that $V_p = \eta V_{BB}$ and $V_V = 0$ V. Then,

$$t_{sp} = R_1 C \ln \left(\frac{1}{1-\eta}\right) \text{ seconds (s)} \qquad (9\text{-}17c)$$

5. This capacitor charge/discharge process continues indefinitely until the circuit is de-energized.

As a result of the charging/discharging process, the voltage waveform that appears at e_0 is shown in Fig. 9-10b.

Note from Eqs. (9-16b) and (9-17b) that t_{sp} and t_{re} are independent of the supply voltage, V_{BB}. Variations in V_{BB} do not affect the output

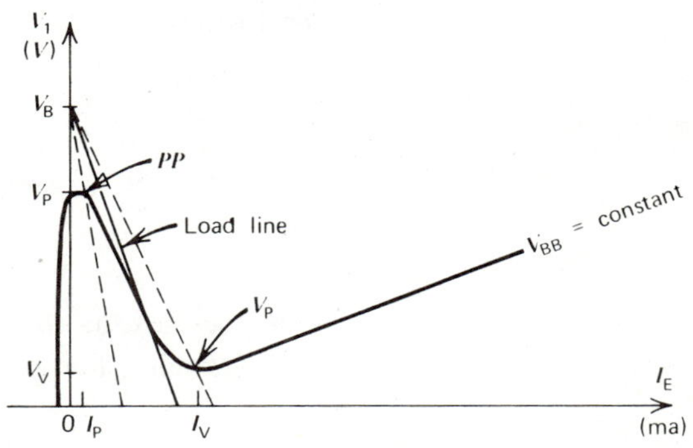

Figure 9-11 Load-line region of an astable circuit.

frequency significantly. Therefore, a UJT relaxation oscillator circuit is a stable circuit. The stable sharp spikes are frequently used as timing pulses.

It was pointed out earlier that the load line of an astable circuit should cross the UJT characteristic curve only in the negative resistance region. To satisfy this condition,

1. $V_{BB} > V_p$.
2. Load resistor $R(R_1$ in Fig. 9-10a) must be chosen such that $I_p > I_E > I_p$ when $V_E = 0$ (the current axis intercept).

Consider the load line shown in Fig. 9-11, which is the load line of an astable relaxation oscillator. The two dotted lines show the range of load-line location for a given V_{BB}. The two extremes are:

1. Crossing the characteristic curve at *peak point* (pp).
2. Crossing the characteristic curve at *valley point* (vp).

The load-line equation for the circuit of Fig. 9-11 is:

$$V_{BB} = I_E R + V_E \text{ volts (V)} \tag{9-18}$$

For the pp crossing,

$$V_{BB} = I_p R_{pp} + V_p$$

$$R_{pp} = \frac{V_{BB} - V_p}{I_p} \text{ ohms } (\Omega) \tag{9-19}$$

And for the vp crossing,

$$V_{BB} = I_v R_{vp} + V_v$$

$$R_{vp} = \frac{V_{BB} - V_v}{I_v} \text{ ohms } (\Omega) \qquad (9\text{-}20)$$

For oscillations to occur, R must be chosen such that $R_{pp} > R > R_{vp}$. Usually, the geometric mean of R_{pp} and R_{vp} is selected. Thus,

$$R = \sqrt{R_{pp} \times R_{vp}} \text{ ohms } (\Omega) \qquad (9\text{-}21)$$

Example 9-3

Design a UJT relaxation oscillator that produces a series of sharp positive spikes 50 μs apart.

Assume:

$V_{BB} = 12$ V	$I_p = 30\ \mu$A	$V_D = 0.4$ V
$V_v = 0.3$ V	$\eta = 0.70$	$R_{B_1}(\text{ON}) = 50\ \Omega$
$I_v = 30$ mA	$R_{BB} = 6$ kΩ	

Solution

The required circuit and the output are shown in Fig. 9-12. Values for R_1, R_2, and C need to be calculated.

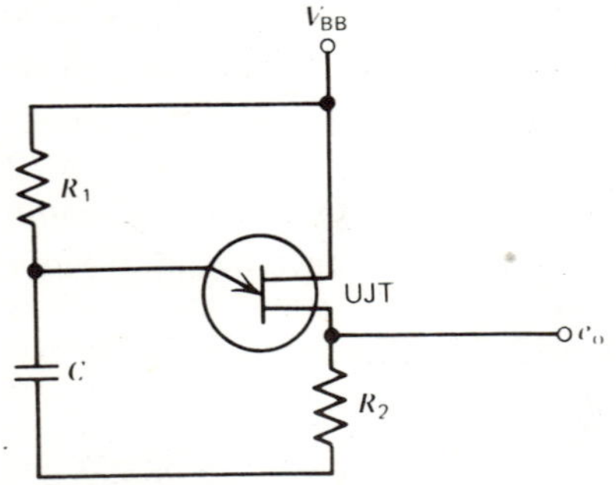

Figure 9-12 Circuit for Ex. 9-3.

$$V_p = \eta V_{BB} + V_D \qquad (9\text{-}15b)$$

$$= 0.7 \times 12 + 0.4$$

$$= 8.4 + 0.4$$

$$= \mathbf{8.8\ V}$$

$$R_{pp} = \frac{V_{BB} - V_p}{I_p} \qquad (9\text{-}19)$$

$$= \frac{12 - 8.8}{30\ \mu A}$$

$$= \mathbf{106.7\ k\Omega}$$

and

$$R_{vp} = \frac{V_{BB} - V_V}{I_V} \qquad (9\text{-}20)$$

$$= \frac{12 - 0.3}{30\ mA}$$

$$= \mathbf{390\ \Omega}$$

$$R_1 = \sqrt{R_{pp} \cdot R_{vp}} \qquad (9\text{-}21)$$

$$= \sqrt{106.7\ k\Omega \times 390\ \Omega}$$

$$= \mathbf{6.45\ k\Omega}$$

$$e^{-(t/RC)} = \frac{V_{BB} - V_p}{V_{BB} - V_V} \qquad (9\text{-}17)$$

$$e^{-[50us/(6.45 \times 10^3 \times C)]} = \frac{12 - 8.8}{12 - 0.3} \quad \text{(Assume that } t_{sp} \gg t_{re},\ \therefore\ t_{sp} \cong T\text{)}$$

$$= 0.2735$$

$$\frac{50\ \mu s}{6.45 \times 10^3 \times C} = 1.296$$

$$C = \frac{50\ \mu s}{6.45 \times 10^3 \times 1.296}$$

or

$$= 5.98 \times 10^{-9}\ F$$

or

$$= \mathbf{0.00598\ \mu F}$$

$$R_{B_1} = \eta R_{BB} \qquad (9\text{-}12)$$

$$= 0.7 \times 6\ k\Omega$$

$$= \mathbf{4.2\ k\Omega}$$

Choose R_2 less than 0.1 R_{B_1} and less than 0.1 R_1, so as not to load the circuit regardless of whether UJT is ON or OFF.

$$\therefore R_2 = \mathbf{47\ \Omega}$$

Example 9-4

A UJT trigger-pulse-generating circuit is shown in Fig. 9-10. $V_{BB} =$ 12 V, $R_1 = 20$ kΩ, $R_2 = 50$ Ω, and $C = 0.047$ μF. The UJT characteristics are the same as given in Ex. 9-3. Find the PRT of the output trigger pulses.

Solution

$$V_p = 8.8 \text{ V} \qquad \text{(from Ex. 9-3)}$$

$$t_{sp} = R_1 C \ln \frac{V_{BB} - V_V}{V_{BB} - V_p} \qquad \text{(9-17b)}$$

$$= 20 \times 10^3 \times 0.047 \times 10^{-6} \ln \left(\frac{12 - 0.3}{12 - 8.5}\right)$$

$$= 1.296 \text{ ms}$$

and

$$t_{re} = [R_{B_1}(\text{ON}) + R_2] \cdot C \ln \frac{V_p}{V_v} \qquad \text{(9-16b)}$$

$$= (50 + 50) \times 0.047 \times 10^{-6} \ln \frac{8.8}{0.3}$$

$$= 0.016 \text{ ms}$$

$$\text{PRT} = t_{sp} + t_{re}$$

$$= 1.306 \text{ ms} + 0.016 \text{ ms}$$

$$= \mathbf{1.312\ ms}$$

9-7 The 555 Timer

The 555 IC timer is a very useful and versatile device. Its schematic diagram is shown in Fig. 9-13a, and its pin-configuration diagram in the dual-in-line package is shown in Fig. 9-13b. Functions of the pin terminals, as labeled on Fig. 9-13b, are:

1. **GND**. This is the common ground point of the circuit. The ground terminal of the external circuit as well as of the power supply (V_{cc}) is connected to it.
2. **Trigger**. When a negative-trigger pulse of amplitude less than

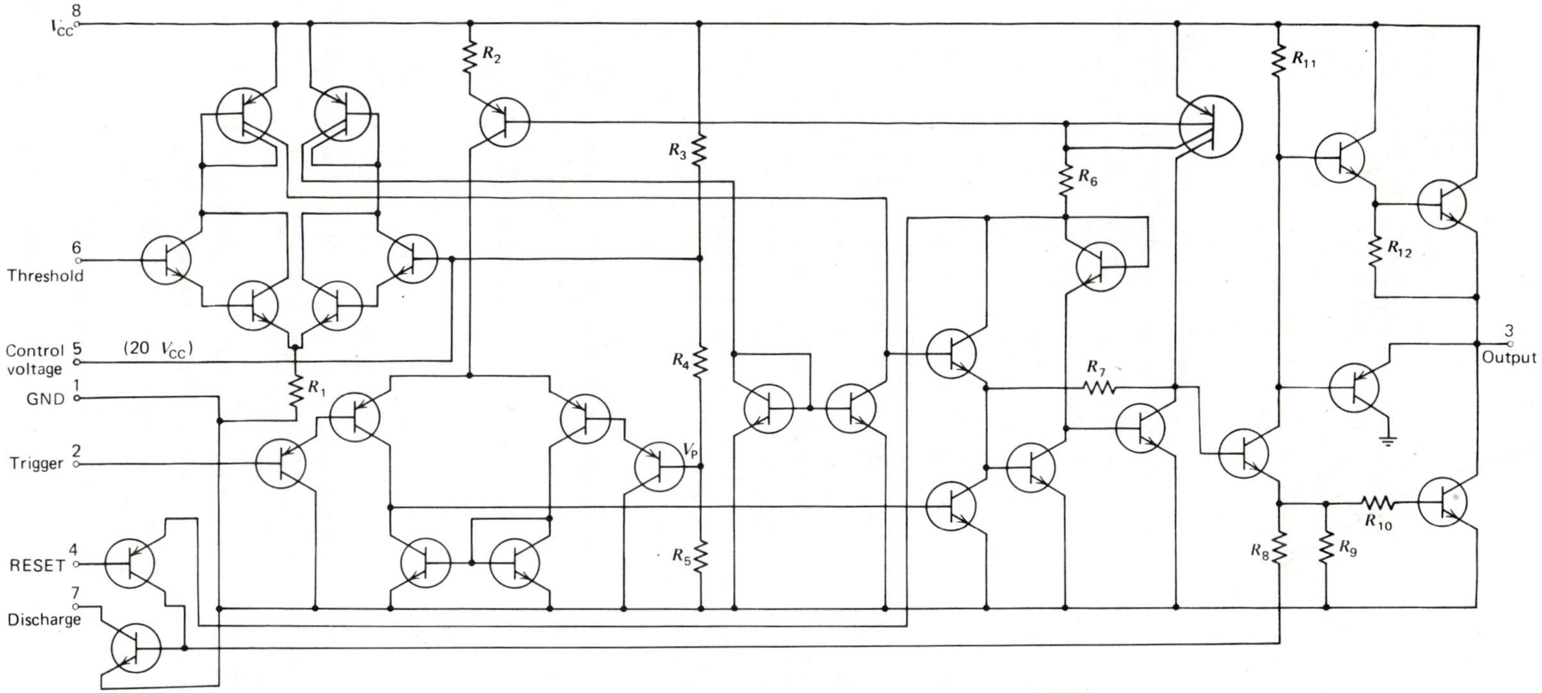

(a) Wiring schematic

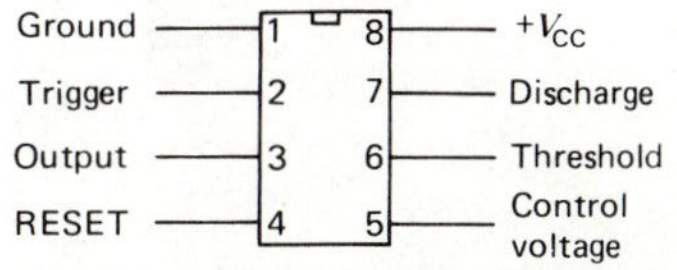

(b) Pin-configuration diagram of the dual-in-line package

Figure 9-13 Wiring-schematic and pin-connection diagrams of a 555 timer.

one-third of V_{cc} is applied to this terminal, the output switches from the LOW to the HIGH state.

3. **Output**. The output is taken from this terminal (which is connected to the load). At any one time, it is either LOW or HIGH.
4. **RESET.** Regardless of the previous state of the output, a trigger pulse applied to this terminal resets the output to LOW. It overrides the trigger-input terminal.
5. **Control voltage.** A positive voltage, equal to two-thirds of V_{cc}, must be present at this terminal. It forms a part of the comparator circuit. Usually, a capacitor is connected between this terminal and ground.
6. **Threshold voltage.** The threshold voltage and the control voltage are the two inputs to the comparator circuit. The circuit compares the voltage level present at this terminal with the reference voltage present at the control terminal. If the voltage at pin 6, the threshold voltage, is *above* the control voltage (two-thirds V_{cc}), the output is LOW; and if the voltage at pin 6 is *below* the control voltage, the output is HIGH.
7. **Discharge**. This terminal provides a low-resistance discharge path for the externally connected capacitor when the output is LOW. It acts like an open circuit when the output is HIGH.
8. $+V_{cc}$: The supply voltage is connected to this terminal.

9-8 The 555 Timer as a Monostable MV

Figure 9-14a shows the connection diagram of a 555 timer in the one-shot mode. The circuit operates as follows:

1. The circuit has been energized:
   a. The output is LOW.
   b. There is no charge on capacitor C.
   c. The discharge circuit (pin 7) is LOW and closed.
2. A negative-going trigger input pulse, shown in Fig. 9-14b, switches the output to HIGH and *opens* the discharge path.
3. Capacitor C charges exponentially toward V_{cc} with the time constant RC.
4. When the capacitor voltage, which is also connected to the threshold terminal, reaches two-thirds V_{cc} (control voltage), the comparator sets the output to LOW. The discharge path closes again.
5. The closed discharge path allows C to discharge quickly.

The voltage waveforms that appear at the input (trigger), the

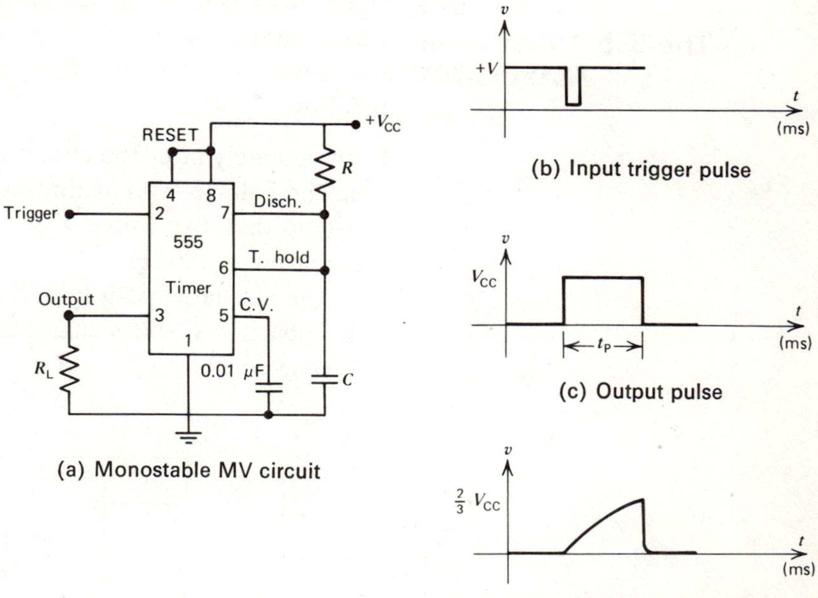

(a) Monostable MV circuit

(b) Input trigger pulse

(c) Output pulse

(d) Voltage across the capacitor *C*

Figure 9-14. A 555 timer connected in monostable mode.

output, and across the capacitor are shown in Figs. 9-14b, 9-14c and 9-14d. The output pulse duration t_p is calculated as follows:

$$v_C = E_i - E_i e^{-(t/\tau)} \text{ volts (V)} \tag{3-4}$$

$$\tfrac{2}{3}V_{CC} = V_{CC} - V_{CC}e^{-(t_p/RC)}$$

$$t_p = 1.1RC \text{ seconds (s)} \tag{9-22}$$

Example 9-5

Consider the monostable circuit shown in Fig. 9-14a. Assume that $V_{cc} = 10$ V, $R_L = 33$ kΩ, $R = 10$ kΩ, and $C = 300$ pF. Calculate the output pulse width t_p.

Solution

$$t_p = 1.1\,RC \tag{9-22}$$

$$= 1.1 \times 10^4 \times 300 \times 10^{-12}$$

$$= \mathbf{3.3\ \mu s}$$

9-9 The 555 Timer as an Astable MV

Figure 9-15a shows the connection diagram of a 555 timer in the *astable* mode. Note that the THRESHOLD terminal is directly connected to the input TRIGGER terminal. The circuit operates as follows:

1. Immediately after the circuit is energized,

a. The voltage level at threshold and the input terminal is LOW (less than two-thirds V_{cc}).

b. Output is HIGH.

c. The discharge path (pin 7) is open.

d. Capacitor C starts charging exponentially toward V_{cc} with the time constant $(R_1 + R_2)C$.

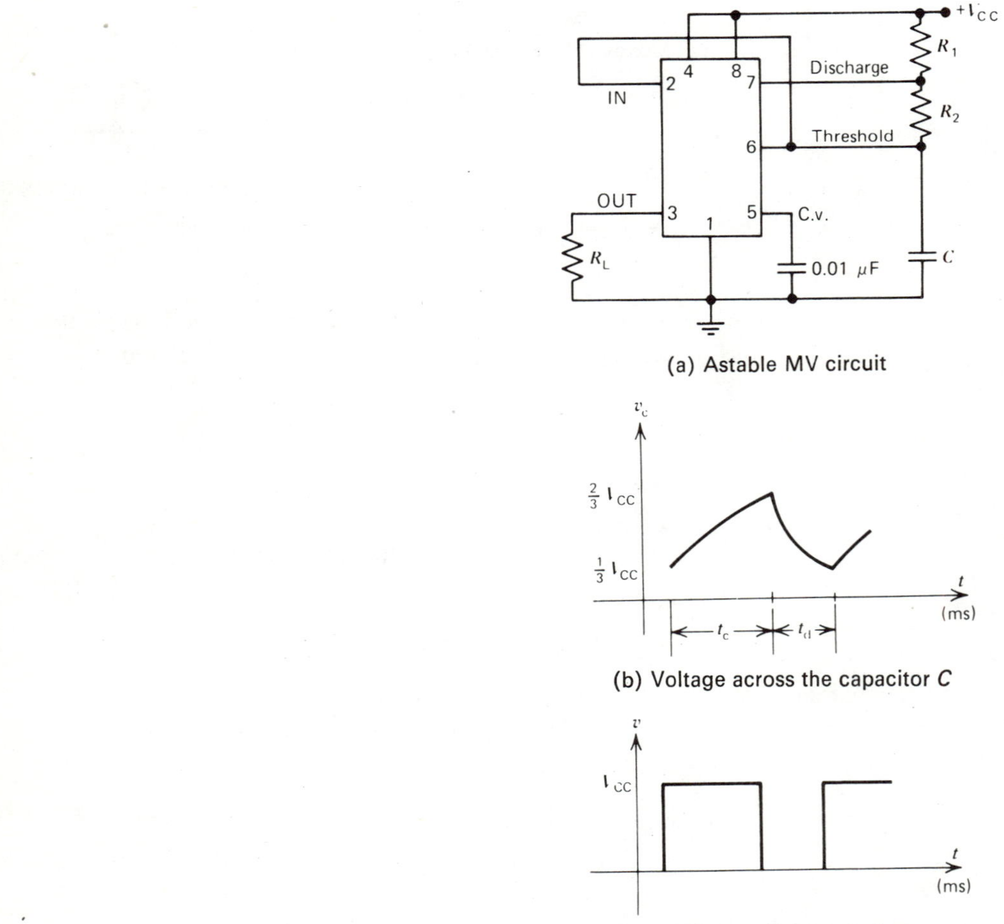

(a) Astable MV circuit

(b) Voltage across the capacitor C

(c) Output voltage waveform

Figure 9-15 A 555 timer connected in the astable mode.

2. When voltage across C, which is also connected to the threshold terminal, reaches two-thirds V_{cc} (control voltage), the comparator sets the output to LOW. The discharge path closes.
3. The closed discharge path allows C to discharge quickly through R_2. The discharge time constant is R_2C.
4. When C discharges down to one-third of V_{cc}, the comparator circuit switches the output to HIGH. The discharge path opens.
5. C again charges exponentially toward V_{cc}, with the charging time constant of $(R_1 + R_2)C$.
6. After this, steps 2, 3, 4, and 5 are repeated indefinitely or until the circuit is de-energized.

Voltage waveforms appearing at the output and across the capacitor are shown in Figs. 9-15b and 9-15c. The capacitor charge time, t_c, and the discharge time, t_d, are calculated as follows:

1. Capacitor charge time, t_c:

$$v_C = E_i - (E_i - V_{ci})e^{-(t/\tau)} \text{ volts (V)} \qquad (3\text{-}7)$$

$$\tfrac{2}{3}V_{cc} = V_{cc} - (V_{cc} - \tfrac{1}{3}V_{cc})e^{-[t_c/(R_1+R_2)C]}$$

$$t_c = 0.693(R_1 + R_2)C \text{ seconds (s)} \qquad (9\text{-}23)$$

2. Capacitor discharge time, t_d:

$$v_C = E_i e^{-(t/\tau)} \qquad \text{(exponential decay)}$$

$$\tfrac{1}{3}V_{cc} = \tfrac{2}{3}V_{cc}e^{-[t_d/(R_1C)]}$$

$$t_d = 0.693R_1C \text{ seconds (s)} \qquad (9\text{-}24)$$

The pulse repetition frequency (PRF) of the output is:

$$\text{PRF} = \frac{1}{t_c + t_d}$$

$$= \frac{1}{0.693(R_1 + 2R_2)C} \text{ hertz (Hz)} \qquad (9\text{-}25)$$

Note from Eq. (9-25) that the PRF of the output waveform is independent of the supply voltage V_{cc}, and is therefore quite stable.

Example 9-6
Consider the astable circuit shown in Fig. 9-15a. Assume that $V_{cc} = 8$ V, $R_L = 50$ kΩ, $R_1 = 10$ kΩ, $R_2 = 2$ kΩ, and $C = 470$ pF. Calculate the PRF of the output waveform.

Solution

$$\text{PRF} = \frac{1}{0.693(R_1 + 2R_2)C} \qquad (9\text{-}25)$$

$$= \frac{1}{0.693(10\ \text{k}\Omega + 2\ \text{k}\Omega) \times 470\ \text{pF}}$$

$$= \mathbf{256\ kHz}$$

9-10 The 555 Timer as a Ramp Generator

Figure 9-16a shows a 555 timer connected in the *ramp generator* mode. Resistors R_1, R_2, and R_E and the transistor Q constitute a constant current source. The capacitor C charges linearly (when the discharge path is open) by the constant current and discharges quickly through the closed discharge path. The circuit goes through following step-by-step operation:

1. Initially, the output is LOW, the discharge path is closed, and there is no voltage charge on the capacitor C.

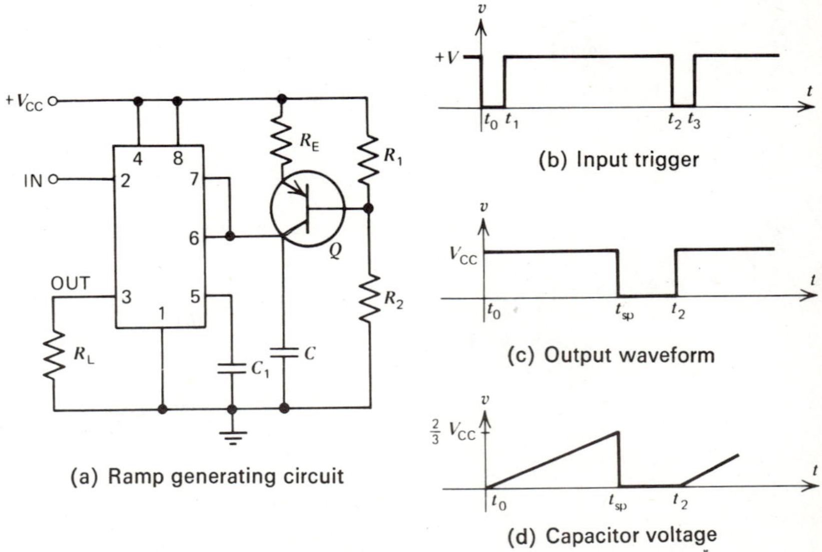

Figure 9-16 A 555 timer connected in the ramp-generating mode.

2. A negative (or 0 V) input trigger pulse, when applied to pin 2, switches the output to HIGH, which in turn opens the discharge path.

3. C charges linearly by a constant current provided by the constant-current-source circuit.

4. When C has charged to two-thirds V_{cc}, the comparator circuit resets the output to LOW, which in turn closes the discharge path. The sweep time, t_{sp}, is given by:

$$t_{sp} = \frac{\frac{2}{3}V_{CC}R_E(R_1 + R_2)C}{R_1V_{CC} - V_{BE}(R_1 + R_2)} \text{ seconds (s)} \qquad (9\text{-}26a)$$

The letter symbols in Eq. (9-26a) correspond to the circuit shown in Fig. 9-16a. See Appendix A-3 for the detailed derivation of Eq. (9-26a). The timing diagram of the input trigger pulse, the output pulse, and the linear sweep across C is shown in Figs. 9-16b, 9-16c, and 9-16d.

Assuming Q to be an ideal transistor, Eq. (9-26a) may be simplified to:

$$t_{sp} = \frac{\frac{2}{3}V_{CC}R_E(R_1 + R_2)\cdot C}{R_1V_{CC}} \text{ seconds (s)} \qquad (9\text{-}26b)$$

Note that Eq. (9-26b) may be further simplified by cancelling out V_{CC}.

Example 9-7

Consider the 555 timer ramp-generating circuit shown in Fig. 9-16a. Assume $R_1 = 30\ \text{k}\Omega$, $R_2 = 4.7\ \text{k}\Omega$, $C = 0.1\ \mu\text{F}$, $V_{cc} = 12$ V, and Q is an ideal transistor. Find the sweep time, t_{sp}, of the circuit when R_E equals:

a. 1 kΩ.
b. 2 kΩ.

Solution

a.

$$t_{sp} = \frac{\frac{2}{3}V_{CC}R_E(R_1 + R_2)\cdot C}{R_1V_{CC}} \qquad (9\text{-}26b)$$

$$= \frac{\frac{2}{3} \times 12 \times 1\ \text{k}\Omega\ (30\ \text{k}\Omega + 4.7\ \text{k}\Omega)\ 0.1\ \mu\text{F}}{30\ \text{k}\Omega \times 12}$$

$$= \mathbf{77.11\ \mu s}$$

b.

$$t_{sp} = 77.11\ \mu\text{s} \times 2 \quad \text{[in Eq. (9-26b), only } R_E \text{ has doubled]}$$

$$= \mathbf{145.22\ \mu s}$$

9-11 Summary

1. A Schmitt trigger is basically a flip-flop circuit; in other words, when one transistor is ON, the other is OFF.
2. The input transistor turns ON when the input potential is above the upper trigger potential (UTP), and it turns OFF when the input potential is below the lower trigger potential (LTP).
3. The 7413 is the most commonly used TTL IC Schmitt trigger.
4. A unijunction transistor differs from an ordinary bipolar transistor in that it exhibits a negative-resistance region at the lower end of its volt-ampere characteristic curve.
5. The negative-resistance characteristics of a UJT are utilized to construct bistable, monostable, or astable circuits, by placing the load line in an appropriate position.
6. Where a UJT is used, the most common circuit is the UJT relaxation oscillator (astable).
7. The output frequency of a UJT relaxation oscillator is practically independent of the supply voltage. Thus, any minor fluctuations in the supply voltage do not affect the output frequency.
8. A 555 timer may be used as a monostable MV, an astable MV, or a ramp generator.
9. Important equations:

$$R_E = \frac{\text{UTP} - V_{BE_1}(\text{ON})}{I_{E_2}(\text{ON})} \tag{9-2}$$

$$R_{L_2} = \frac{V_{cc} - [V_{CE_2}(\text{ON}) + V_{E_2}(\text{ON})]}{I_{C_2}(\text{ON})} \tag{9-3}$$

$$I_{E_1}(\text{ON}) = \frac{\text{LTP} - V_{BE_1}(\text{ON})}{R_E} \tag{9-4}$$

$$R_{L_1} = \frac{V_{cc} - [V_{CE_1}(\text{ON}) + V_{E_1}(\text{ON})]}{I_{C_1}(\text{ON})} \tag{9-6}$$

$$I_{B_2}(\text{ON}) = \left[\frac{V_{cc} - \text{UTP}}{R_{L_1} + R_1}\right] - \frac{\text{UTP}}{R_2} \tag{9-7}$$

$$\frac{V_{B_2}(\text{OFF})}{R_2} = \frac{V_{C_1}(\text{ON}) - V_{B_2}(\text{OFF})}{R_1} \tag{9-10}$$

$$V_p = \eta V_{BB} \tag{9-15c}$$

$$t_{re} = [R_{B_1}(\text{ON}) + R_2] \cdot C \ln \frac{V_p}{V_V} \tag{9-16b}$$

$$t_{sp} = R_1 \cdot C \ln \frac{V_{BB} - V_V}{V_{BB} - V_p} \tag{9-17b}$$

$$R = \sqrt{R_{pp} \times R_{vp}} \tag{9-21}$$

$$t_p = 1.1RC \tag{9-22}$$

$$PRF = \frac{1}{0.693(R_1 + 2R_2)C} \tag{9-25}$$

$$t_{sp} = \frac{\frac{2}{3}V_{cc}B_E(R_1 + R_2)C}{R_1V_{cc} - V_{BE}(R_1 + R_2)} \tag{9-26a}$$

9-12 Glossary

Intrinsic stand-off ratio. The ratio of R_{B_1} to R_{BB} when $I_E = 0$.

Lower trigger potential (LTP). The input voltage level below which the input transistor turns OFF.

Negative resistance region. The region between the valley point and the peak point on the characteristic curve. In this region, the current in a device increases as the voltage across it decreases.

Peak point. A point located on the characteristic curve of a UJT just before the negative resistance region starts.

Peak point current (I_p). The value of the forward current at the peak point.

Peak point voltage (V_p). The value of the forward voltage at the peak point.

Schmitt trigger. A triggered bistable circuit.

Unijunction transistor (UJT). A specially constructed transistor that exhibits negative-resistance characteristics.

Upper trigger potential (UTP). The input voltage level above which the input transistor turns ON.

Valley point. A point located on the characteristic curve of a device, at the end of the negative-resistance region.

Valley point current (I_V). The value of forward current at the valley point.

Valley point voltage (V_V). The value of the forward voltage at the valley point.

9-13 Recapitulation

Return to the Objectives and Self-Evaluation Questions given at the beginning of this chapter and see how well you can answer them. If you cannot answer certain questions satisfactorily, place a check mark (√) next to them and review the appropriate parts of the chapter to find the answers. No attempt should be made to solve the Problems (Sec.

9-14) before answering the Self-Evaluation Questions. Ability to solve the Problems should serve as a test and feedback regarding your understanding of the material presented in this chapter. Answers to the Problems are given on p. 527.

9-14 Problems

9-1 Refer to the Schmitt trigger circuit shown in Fig. 9-1a. Assume:

$R_{L_1} = 400\ \Omega$	$R_2 = 1.5\ \text{k}\Omega$	$I_C(\text{max}) = 50\ \text{mA}$
$R_{L_2} = 1\ \text{k}\Omega$	$R_B = 2.5\ \text{k}\Omega$	$V_{BE}(\text{ON}) = 0\ \text{V}$
$R_E = 200\ \Omega$	$V_{CC} = 12\ \text{V}$	$V_{CE}(\text{ON}) = 0\ \text{V}$
$R_1 = 1\ \text{k}\Omega$	$\beta = 80$	

a. Check by calculations whether the circuit functions properly.
b. Find UTP and LTP.

9-2 Repeat Prob. 9-1 if $V_{BE}(\text{ON}) = 0.7$ V and $V_{CE}(\text{ON}) = 0.3$ V for the two transistors.

9-3 Design a Schmitt trigger circuit using two transistors whose $\beta = 75$, $I_C(\text{max}) = 25$ mA, $V_{CE}(\text{ON}) = 0$ V, and $V_{BE}(\text{ON}) = 0$ V. $V_{BE}(\text{OFF})$ is to be designed as -0.5 V. The LTP is to be designed as 2.0 V and UTP as 3.0 V. Available V_{CC} is 20 V.

9-4 Repeat Prob. 9-3 if $V_{BE}(\text{ON}) = 0.7$ V and $V_{CE}(\text{ON}) = 0.3$ V for the two transistors.

9-5 Consider the UJT relaxation oscillator circuit of Fig. 9-10a. Assume

$R = 10\ \text{k}\Omega$	$V_{BB} = 20\ \text{V}$	$I_v = 5\ \text{mA}$
$C = 0.001\ \mu\text{F}$	$V_D = 0$	$I_p = 10\ \mu\text{A}$
$R_{BB} = 10\ \text{k}\Omega$	$V_v = 0.04\ \text{V}$	$\eta = 0.75$

Find:
a. V_p.
b. R_{B_1}.
c. R_{B_2}.
d. t_{sp}.

9-6 Repeat Prob. 9-5 if a new value of η is 0.60.

9-7 Design a simple UJT relaxation oscillator circuit. Characteristics of the UJT to be used are:

$V_v = 0.05\ \text{V}$	$I_p = 50\ \mu\text{A}$	$R_{BB} = 6.8\ \text{k}\Omega$
$I_v = 10\ \text{mA}$	$\eta = 0.70$	$V_D = 0.3\ \text{V}$

The output sweep is desired to be of 0.3 V to 10.3 V amplitude and of 5 μs duration.

9-8 Repeat Prob. 9-7 for the UJT used in Prob. 9-5. The output peak is to be 10 V.

9-9 Consider the UJT trigger-pulse-generator circuit shown in Fig. 9-10a. Assume $C = 0.01\ \mu$F, $R_1 = 33\ \text{k}\Omega$, $R_2 = 100\ \Omega$, $R_{B_1}(\text{ON}) = 120\ \Omega$, and $V_{BB} = 20$ V. Characteristics of the UJT are described in Prob. 9-7. Find the PRT of the output pulse.

9-10 Repeat Prob. 9-9, but this time use $C = 100$ pF.

9-11 Consider the monostable MV circuit of Fig. 9-14a. Assume that $R = 12$ kΩ and $C = 0.025$ μF. Calculate t_p of the output pulse.

9-12 Consider again the circuit conditions described in Prob. 9-11. Find the value of resistance that when connected in series with the existing 12 kΩ resistor, will produce an output pulse of 10 ms duration.

9-13 Refer to the astable circuit shown in Fig. 9-15a. Assume that $C = 300$ pF, $R_1 = 10$ kΩ, and $R_2 = 5$ kΩ. Calculate the PRF of the output waveform.

9-14 Repeat Prob. 9-13, but this time assume $R_1 = R_2 = 10$ kΩ.

9-15 Consider the sweep generator circuit shown in Fig. 9-16a. Assume that $R_1 = 15$ kΩ, $R_2 = 22$ kΩ, $R_E = 1.5$ kΩ, $C = 47$ μF, $V_{cc} = 4.5$ V, and $V_{BE}(ON) = 0.5$ V. Calculate the output sweep time, t_{sp}.

9-16 Refer to the circuit conditions described in Prob. 9-15. The 47 μF capacitor is replaced by an unknown capacitor, C. Calculate the capacitance of C if the output sweep-time is 25 ms.

9-15 Experiment

Part A

Title **The Schmitt Trigger**

Theory A Schmitt trigger circuit is basically a switched bistable device, that is, at any one time, one transistor is OFF while the other is ON. The input voltage level determines whether Q_1 is ON and Q_2 is OFF, or Q_1 is OFF and Q_2 is ON.

The upper trigger potential (UTP) refers to the input voltage level above which Q_1 stays ON and Q_2 stays OFF. The lower trigger potential (LTP) refers to the input voltage level below which Q_1 stays OFF and Q_2 stays ON.

A Schmitt trigger circuit is usually designed such that it switches from one stable state to the other at given values of UTP and LTP. A detailed design procedure for a typical Schmitt trigger is demonstrated and explained in Ex. 9-2.

Equipment Multifunction generator: EXACT model 126, or equal.
Multimeter: RCA model WV 98C, or equal.
DC power supply: ANATEK model 25-2.00, or equal.
Dual-beam oscilloscope: PHILIPS model 3230, or equal.
Resistors: 100 Ω, 220 Ω, 390 Ω, 560 Ω, 2.2 kΩ, two 1.8 kΩ.
Potentiometer: 100 Ω.
Transistors: two 2N4401.
One IC 7413, Schmitt trigger.

Procedure

1. Connect the Schmitt trigger circuit shown in Fig. E9-1.
2. With E_i at 0 V (ground), measure and record the **dc** voltages appearing at collectors, bases, and emitters of both transistors. Enter your readings in Table E9-1.
3. Connect a **dc** power supply at E_i. Increase the input voltage (E_i) slowly while constantly monitoring the output voltage at E_o. The output voltage should suddenly jump to about V_{cc} when the input voltage reaches a certain level. This is UTP. Record it in Table E9-2.
4. Now decrease the input **dc** voltage (from above UTP) slowly

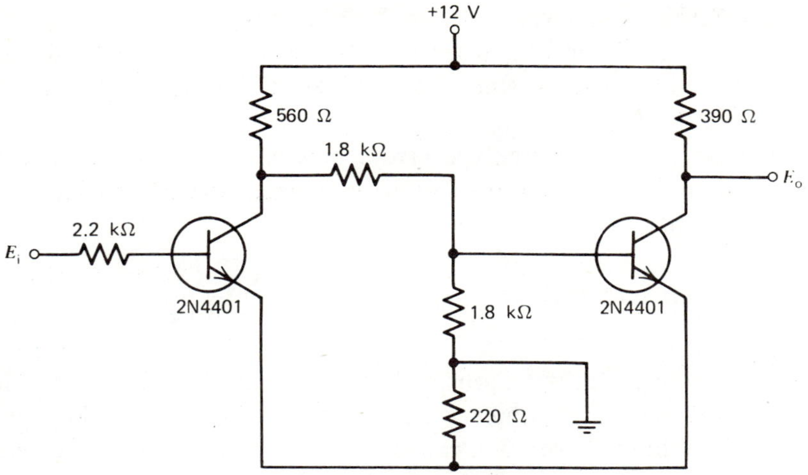

Figure E9-1 Schmitt trigger circuit.

while monitoring the output voltage level (E_o). The output voltage level should drop suddenly as the input reaches a certain level (LTP). Record your observed LTP in Table E9-2.

5. Replace the **dc** input power supply by a 10-$V_{p\text{-}p}$, 5-kHz sine wave. Observe and draw in Table E9-3 the input as well as the output waveforms.
6. Again find UTP and LTP from the waveforms drawn in Table E9-3. Record these observed values of UTP and LTP in Table E9-3.
7. Replace the 220 Ω resistor by a 100 Ω resistor. Repeat steps 5 and 6. Record your observed results in Table E9-3.
8. Reinsert the 220-Ω emitter resistor (in place of 100-Ω) in the circuit. Insert a 100-Ω potentiometer in between the emitter of Q_1 and the emitter resistor (220 Ω).
9. Vary the 100-Ω potentiometer back and forth, and observe its effects on the output waveform, especially insofar as it affects the UTP and LTP values. Measure and record (in Table E9-4) the UTP and LTP values for the potentiometer settings of 0 Ω, 50 Ω, and 100 Ω.
10. Connect one half of the 7413 Schmitt trigger, as shown in Fig. 9-5a. Use +6 V as V_{cc}.
11. Apply a 6-$V_{p\text{-}p}$, 2-kHz sine wave at the input (IN).
12. Observe and record in Table E9-5 UTP, LTP, and the input and output waveforms.
13. Shift the **dc** level of the input such that it now swings between +4 V and −2 V. Observe and record in Table E9-5 UTP, LTP, and the input and output waveforms.
14. Disconnect and return all test equipment to its proper place.

Calculations

1. Consider the circuit shown in Fig. E9-1. Calculate V_{C_1}, V_{C_2}, V_{B_1}, V_{B_2}, V_{E_1} and V_{E_2} when $E_i = 0$ V. Enter your calculated values in Table E9-1. Assume $V_{BE}(ON) = 0.7$ V and $V_{CE}(ON) = 0.3$ and $\beta = 80$.
2. Calculate the percentage errors between the measured and the calculated values in the following manner and enter your results in Table E9-1:

$$\text{Percentage error} = \frac{\text{Theoretical value} - \text{Observed value}}{\text{Theoretical value}}$$

3. For the Schmitt trigger circuit shown in Fig. E9-1, calculate UTP and LTP. Enter your results in Table E9-2. For value recorded in Table E9-2, calculate and record the percentage error as was done in step 2 above.

Tabulation of Results

Table E9-1

Voltage at:	Observed	Theoretical	Percentage Error
V_{C_1}			
V_{C_2}			
V_{B_1}			
V_{B_2}			
V_{E_1}			
V_{E_2}			

Table E9-2

	Observed	Theoretical	Percentage Error
UTP			
LTP			

Table E9-3

$R_E = 220\ \Omega$	$R_E = 100\ \Omega$
Input waveform	Input waveform
Output waveform	Output waveform
UTP = LTP =	UTP = LTP =

Table E9-4

Pot. at	Circuit UTP and LTP
0 Ω 50 Ω 100 Ω	

Table E9-5

Procedure Step 12	Procedure Step 13
Input waveform	Input waveform
Output waveform	Output waveform
UTP = LTP =	UTP = LTP =

Questions

1. From the calculated percentage errors in Tables E9-1 and E9-2, describe how your calculated results compare with the experimentally observed ones.
2. In step 7 of the Procedure, the emitter resistance was decreased from 220 Ω to 100 Ω. How did this affect the UTP and LTP of the circuit?
3. In step 8 of the Procedure, a resistance (100 Ω pot.) was inserted in

the basic Schmitt trigger circuit. Examine Table E9-4 and describe how this additional resistance in the emitter circuit of Q_1 affected the UTP and LTP of the circuit.

4. Examine the results recorded in Table E9-5. Explain how the UTP, LTP, and the output are affected when the input **dc** level is changed.

Discussion

1. Summarize your conclusions from the observed and calculated results of this experiment.
2. Account for any serious discrepancies between the observed and calculated values in Tables E9-1 and E9-2.
3. Describe briefly, in your own words, the operation of a Schmitt trigger circuit.
4. Compare the discrete-component Schmitt trigger with its IC counterpart.

Part B

Title

The UJT Relaxation Oscillator and Trigger Pulse Generator

Theory

The UJT is a device that exhibits negative resistance or an unstable region as a part of its characteristics. A circuit designed such that its load line crosses the UJT's characteristic curve only in the unstable region becomes an astable circuit and is commonly known as the UJT relaxation oscillator.

A simple UJT relaxation oscillator is shown in Fig. E9-2. When the circuit is energized, capacitor C starts charging toward V_{BB} with time constant RC. As the capacitor voltage V_C (also V_E) reaches the peak point voltage V_p, the UJT starts conducting, thereby lowering the resistance R_{B_1}. The capacitor discharges fairly rapidly through the lower R_{B_1} until it reaches the valley point voltage V_V, at which time the UJT turns OFF (ceases to conduct). After the UJT turns OFF, C starts charging again through resistance R toward V_{BB}. The charging of capacitor C through R and discharging through R_{B_1}(ON) continues indefinitely.

Sharp timing spikes of stable frequency may be obtained by inserting a low resistance between the ground and B_1. Example 9-3 shows how to design and numerically analyze a UJT relaxation oscillator circuit.

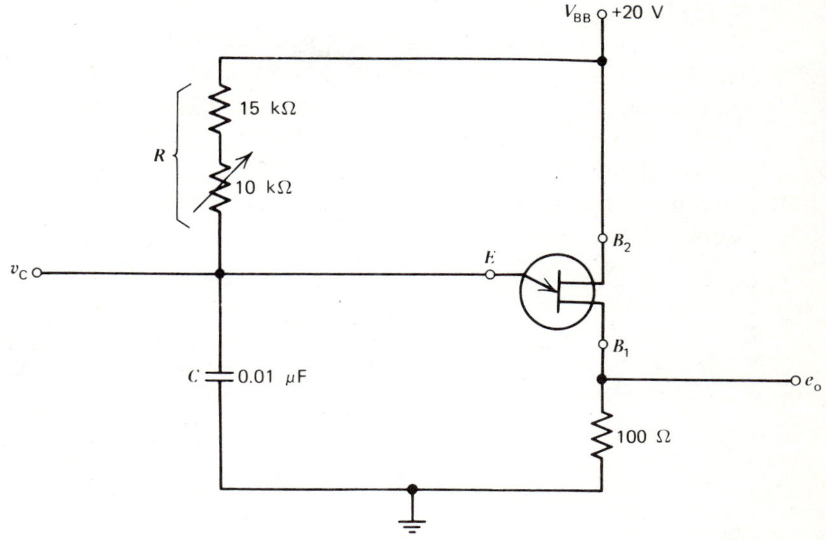

Figure E9-2 UJT relaxation oscillator.

Equipment Power supply: ANATEK Model 25-2.00, or equal.
Multimeter: RCA model WV 98C, or equal.
Dual-beam oscilloscope (CRO): Philip PM 3230, or equal.
UJT: 2N2646.
Resistors: 100 Ω, 15 kΩ.
Potentiometer: 10 kΩ.
Capacitors: 0.01 μF and 0.047 μF.

Procedure

1. Connect the UJT relaxation oscillator circuit shown in Fig. E9-2.
2. Adjust the 10-kΩ potentiometer to the 0-Ω position.
3. Observe (CRO) and record in Table E9-5 the voltage waveforms appearing at v_C and e_o. Record directly on the waveforms drawn, the amplitudes as well as t_{re} and t_{sp}.
4. Adjust the 10-kΩ potentiometer (pot.) to full 10-kΩ position.
5. Repeat Procedure step 3.
6. Keep the pot. setting at 10 kΩ. Replace the 0.01-μF capacitor with a 0.047-μF capacitor.
7. Repeat Procedure step 3.
8. Vary the 10-kΩ pot. back and forth, and note its effect on the PRF of the trigger pulses that appear across the 100-Ω resistor.

NOTES:
The PRF is ____________ when the pot. resistance is increased.
The PRF is ____________ when the pot. resistance is decreased.

9. Set the pot. at 10 kΩ. Vary the supply voltage V_{BB} from 16 V to 24 V in two-volt steps. Measure and record in Table E9-7 the PRF of the trigger pulses at each V_{BB} setting.
10. Disconnect and return all equipment and components to their proper places.

Calculations

1. From the manufacturer's data sheets, the following parameters for the UJT 2N2646 can be noted: $I_V = 4$ mA, $V_V = 3$ V, $I_p = 0.4\ \mu$A, $R_{BB} = 7$ kΩ, and $\eta = 0.65$.
2. Record the theoretical values of V_V and η in Table E9-7.
3. From the waveforms drawn in Table E9-5, find the observed (measured) values of V_V and η and enter them into Table E9-7. Note that the measured value of $\eta = V_p/V_{BB}$.
4. For the circuit conditions described in Procedure steps 1 and 2, calculate the sweep time of the output waveform and enter it in Table E9-8.
5. Find the measured value of the sweep time for the circuit conditions of Procedure steps 1 and 2 from the waveform drawn in Table E9-5, and enter it into Table E9-8.
6. Repeat Calculation steps 4 and 5 for the circuit conditions described in Procedure steps 4 and 6.
7. Calculate the percentage deviation for Table E9-6 and enter your results there.

$$\text{Percentage deviation} = \frac{(\text{PRF})_{18\text{V}} - (\text{PRF})_{V_{BB}}}{(\text{PRF})_{18\text{V}}}$$

Tabulation of Results Table E9-5

Circuit Conditions	$R = 4.7$ kΩ $C = 0.01\ \mu$F	$R = 14.7$ kΩ $C = 0.01\ \mu$F	$R = 14.7$ kΩ $C = 0.047\ \mu$F
Waveform at e_o			
Waveform at v_c			

Table E9-6

V_{BB} (volts)	PRF	Percentage Deviation
16 18 20 22 24		

Table E9-7

Circuit Parameter	Observed	Theoretical	Percentage Error
V_V η			

Table E9-8

Procedure Step	Observed t_{sp}	Theoretical t_{sp}	Percentage Error
2 4 6			

Questions

1. From the waveforms recorded in Table E9-5, describe what happens to the output waveform when:
   a. Resistance R is increased or decreased.
   b. Capacitance C is increased or decreased.
2. From the calculated percentage errors in Tables E9-7 and E9-8, describe how your calculated results compare with the experimentally observed ones.
3. The waveforms drawn in Table E9-5 are intended to confirm that sharp trigger spikes are obtained by inserting a small resistor in the emitter circuit (100 Ω) of the UJT relaxation oscillator, and that these spikes are generated during the UJT's ON time. Have these things been confirmed? Explain.
4. The tabulation of Table E9-6 is intended to confirm that the PRF of a UJT relaxation oscillator is practically independent of the supply voltage. Has this been confirmed? Explain.
5. Explain the results observed in Procedure step 8.

Discussion

1. Summarize your conclusions from the observed and calculated results of this experiment.
2. Account for any serious discrepancies between the observed and the calculated results.
3. Explain the load-line location requirements for a UJT relaxation oscillator.
4. Explain what determines the PRT of a UJT trigger pulse generator.
5. Explain briefly the operation of a UJT relaxation oscillator.

Part C

Title

Monostable and Astable MVs Using the 555 Timer

Theory The 555 timer may be used to construct either a monostable or an astable circuit. In the monostable mode, the output stays HIGH as long as voltage across an externally connected capacitor to the threshold terminal is below two-thirds of V_{cc}. When this capacitor voltage reaches two-thirds V_{cc}, the comparator (built inside the 555 timer) resets the output to LOW, which in turn closes the discharge path for the capacitor (stable state).

$$t_p = 1.1RC \text{ seconds (s)} \tag{7-27}$$

In the astable mode, the threshold terminal is connected to the input trigger terminal and a resistance is inserted between the threshold and the discharge terminal. The capacitor (0.001 μF in Fig. E7-4) now charges and discharges between one-third V_{cc} and two-thirds V_{cc}.

$$t_1 = 0.693(R_1 + R_2)\cdot C \text{ seconds (s)} \tag{7-28}$$

$$t_2 = 0.693R_2C \text{ seconds (s)} \tag{7-29}$$

Equipment Dual-beam oscilloscope: PHILIPS model 3230, or equal.
Power supply: ANATEK model 25-2.00, or equal.
Multimeter: RCA model WV 98C, or equal.
Multifunction generator: EXACT model 126, or equal.
Resistors: 33 kΩ, 47 kΩ, 100 kΩ, and 100 kΩ potentiometer.
Capacitor: 0.001 μF.
555 timer.

Procedure

1. Connect the monostable circuit shown in Fig. E9-3. Set the 100-kΩ pot. to 0 Ω.

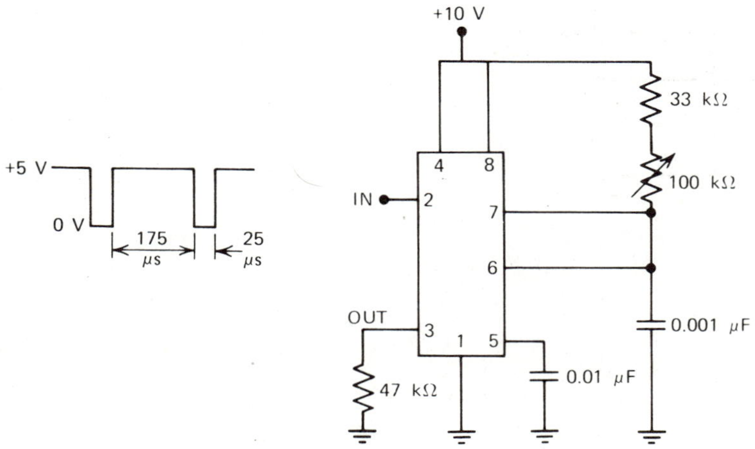

Figure E9-3 A 555 timer in the monostable mode.

2. Measure and record in Table E9-9, the **dc** voltage levels appearing at the output, the control voltage, and across the 0.001-μF capacitor when the input is HIGH (5 V).
3. Apply the pulse wave shown in Fig. E9-3 to the trigger input terminal.
4. Observe and record in Table E9-10 the input, the output, and the threshold voltage waveforms.
5. Set the 100-kΩ pot. to the full 100-kΩ position. Repeat step 4 above.
6. Vary the 100-kΩ pot. back and forth; observe its effects on the output waveform.
   - **a.** The pulse width increases when the pot. resistance is ____________?
   - **b.** The pulse width decreases when the pot. resistance is ____________?
7. Connect the astable circuit shown in Fig. E9-4. Set the 100-kΩ pot. to the 0-Ω position.
8. Observe and record in Table E9-11 the voltage waveforms that appear at the output, the discharge, and across the 0.001-μF capacitor.
9. Set the 100-kΩ pot. to the full 100-kΩ position. Repeat step 8 above.
10. Vary the 100-kΩ pot. back and forth; observe its effects on the output waveform.

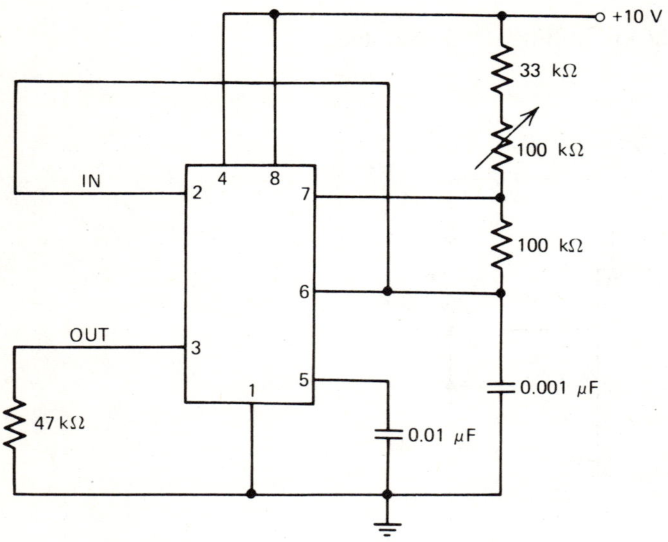

Figure E9-4 A 555 timer in the astable mode.

a. The output PRF increases when the pot. resistance is _________?

b. The output PRF decreases when the pot. resistance is _________?

11. Disconnect and return all test equipment to its proper place.

Calculations

1. Calculate the percentage error between the observed and expected values of Table E9-9. Record your results in Table E9-9.

2. From Table E9-10, note t_p values of the output waveform for both pot. at 0 Ω and at 100 kΩ. Enter these experimental values of t_p in Table E9-12.

3. Calculate the theoretical values of t_p for Table E9-12 and enter your results there.

4. From Table E9-11, note the experimental values of t_1 and t_2 and record these in Table E9-13.

5. Calculate the theoretical values for t_1 and t_2 for Table E9-13 and enter your results there.

6. Calculate the percentage error between the experimental and the theoretically expected values of Tables E9-12 and E9-13. Record your results in the respective Tables.

Tabulation of Results Table E9-9

Description	Observed Voltage	Theoretical Voltage	Percentage Error
Output			
Control			
0.001-μF Cap.			

Table E9-10

Description	Pot. at 0 Ω	Pot. at 100 kΩ
Input		
Output		
Discharge		

Table E9-11

Description	Pot. at 0 Ω	Pot. at 100 kΩ
Output		
Discharge		
Across		
0.001-μF Cap.		

Table E9-12

Description		Observed	Theoretical	Percentage Error
t_p	Pot. at 0 Ω			
	Pot. at 100 kΩ			

Table E9-13

Description		Observed	Theoretical	Percentage Error
Pot. at 0 Ω	t_1 t_2			
Pot. at 100 k Ω	t_1 t_2			

Questions

1. Examine the waveforms recorded in Tables E9-10 and E9-11. Is that what you expected? Explain.
2. Explain the results observed in Procedure steps 6 and 10.
3. Consider the output pulse drawn in Table E9-10. How would it change if the 0.001-μF capacitor were replaced by a 0.002-μF capacitor?
4. Consider the t_1 and t_2 values recorded in Table E9-13. How would these change if the 0.001-μF capacitor were replaced by a 0.01-μF capacitor?

Discussion

1. Summarize your conclusions from the experimental and the theoretical values.
2. Account for discrepancies between the observed and the theoretically expected values.
3. List some practical applications in which a 555 timer can gainfully be utilized.
4. Compare the 555 timer astable and monostable circuits with the corresponding discrete BJT circuits.

Chapter 10 Pulse-Delaying Circuits

10-1 Objectives

The objectives of this chapter are to:

1. Review the general characteristics of a transformer.
2. Identify the desired characteristics of an ideal pulse transformer.
3. Discuss the characteristics of a practical pulse transformer.
4. Explain the function of a delay line.
5. Identify various types of delay-line circuits used in pulse work.
6. Discuss electromagnetic delay lines.
7. Explain the terms *characteristic impedance* (Z_o), *delay time* (t_d), and *cutoff frequency* (f_c).
8. Show how a monostable multivibrator may be used as pulse delay circuit.

10-2 Self-Evaluation Questions

Watch for the answers to the following questions as you read the chapter. They should help you to understand the material presented. When you have finished the chapter, return to this section and be sure that you can answer all the questions. Then test your understanding by working out solutions to the Problems.

It is strongly recommended that you answer these questions satisfactorily before attempting to solve the Problems (Sec. 10-13).

1. Describe in detail the physical makeup and operation of a transformer.
2. Explain the following terms:
   a. Primary winding.
   b. Secondary winding.
   c. Transformation ratio.
   d. Iron core.
   e. Primary inductance.
   f. Secondary inductance.
3. A load of R ohms is connected to the secondary of a transformer whose transformation ratio is a and whose secondary terminal voltage is V volts. Find:
   a. Load current.
   b. Current in the secondary.
   c. Current in the primary.
   d. Primary voltage.
   e. Power rating of the transformer.
4. How does a pulse transformer differ from an ordinary transformer?

5. What is a delay line and where is it used?
6. Draw an equivalent circuit of a transmission line.
7. A lumped-parameter delay line has a total inductance L henrys and total capacitance C farads. Write expressions for its:
   a. Characteristics impedance (Z_o).
   b. Time delay (t_d).
   c. Cutoff frequency (f_c).
8. A two-conductor distributed-parameter delay line is to be used to delay a pulse for t seconds. Write expressions that may be used to find:
   a. Required length.
   b. Characteristic impedance.
   c. Cutoff frequency.
9. Repeat Question 8 for a two-conductor coaxial cable.
10. Show how a monostable multivibrator may be used as a pulse delay circuit.

10-3 Review of Basic Transformer Characteristics

A transformer consists of two magnetically-linked windings. One winding is called the *primary* and the other the *secondary*. The primary and the secondary are usually linked by a soft iron core because it magnetizes and demagnetizes easily.

A basic transformer is shown in Fig. 10-1. The input power source is connected to the primary and the load is connected to the secondary of the transformer. A *changing* input voltage/current in the primary produces a *changing* flux in the iron core (in accordance with Lenz's law).

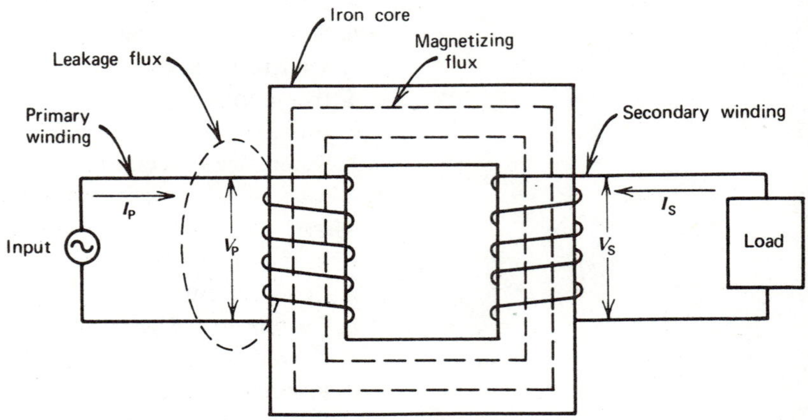

Figure 10-1 Transformer.

Also, a changing flux in the core induces voltage/current in the secondary winding. The magnitudes of the primary voltage, the primary current, the secondary voltage, and the secondary current are related and are influenced by the *transformation ratio* (a) of the two windings.

$$a = \frac{N_p}{N_s} \tag{10-1}$$

where:

a is the transformation ratio
N_p is the number of turns in the primary winding
N_s is the number of turns in the secondary winding

Also

$$a = \frac{V_p}{V_s} = \frac{I_s}{I_p} \tag{10-2}$$

or

$$P = V_p \cdot I_p = V_s \cdot I_s \text{ volt-amperes (VA)} \tag{10-3}$$

where:

a is the transformation ratio
V_p is the primary voltage (V)
I_p is the primary current (A)
V_s is the secondary voltage (V)
I_s is the secondary current (A)
P is the apparent power rating (VA)

Note that ideally, the primary volt-amperage is equal to the secondary volt-amperage, as given by Eq. (10-3). In practical circuits, losses do occur in the transformer. They are usually small and for most practical purposes, are neglected.

Equation (10-2) indicates that the output voltage (V_s) of a transformer is equal to the primary input voltage (V_p) divided by the transformation ratio (a). Thus, depending upon the transformation ratio, V_s may be greater than, equal to, or less than the input, V_p.

A transformer whose transformation ratio a is less than unity has $V_s < V_p$ and is known as a *step-down* transformer. A transformer whose transformation ratio is unity has $V_s = V_p$ and is known as an *isolation* transformer. A *step-up* transformer is the one whose transformation ratio is less than unity and has $V_s > V_p$.

Note that only a *changing* voltage/current produces changing flux in the magnetic core. A constant (**dc**) voltage/current will *not* produce a changing flux. Only a changing flux will induce voltage/current in the secondary, and a **dc** will not. In fact, the magnitude of the induced voltage is directly proportional to the rate of change of flux. Therefore, a **dc** voltage/current transformer using magnetically coupled circuits is impossible.

Figure 10-2a shows the symbolic representation of a transformer with a load, R_L, connected to its secondary. Let us examine how does this transformer-load combination look from the primary terminals insofar as the input impedance (as presented to the source) is concerned.

$$V_p = a \cdot V_s \text{ volts (V)} \tag{10-2}$$

and

$$I_p = (I_s)/a \text{ amperes (A)} \tag{10-2}$$

$$R_i = (V_p)/(I_p) = \frac{a \cdot V_s}{(I_s)/a} = a^2 \cdot (V_s/I_s)$$

But

$$(V_s/I_s) = R_L$$

$$\therefore R_i = a^2 \cdot R_L \text{ ohms } (\Omega) \tag{10-4}$$

where R_i is the reflected secondary load as seen by the primary (Ω).

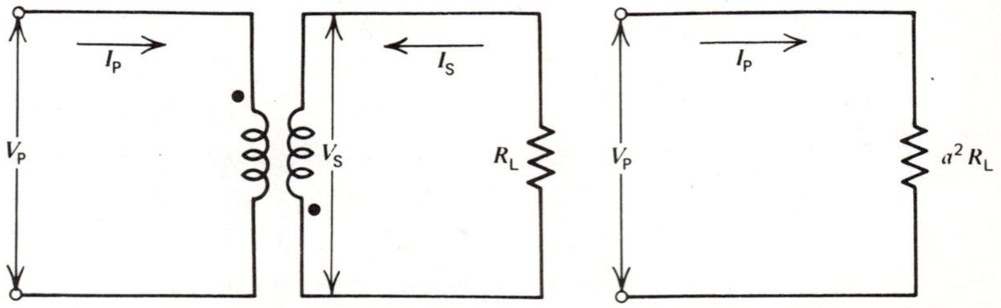

(a) Transformer and load (b) Equivalent circuit

Figure 10-2 Input impedance of a transformer.

Equation (10-4) shows that the load R_L appears a^2 times bigger (or smaller) when viewed from the primary side. This is shown in Fig. 10-2b. This makes the transformer useful as a *load-matching* (impedance-matching) device. Transformers are also used for coupling, isolation, and step-up or step-down purposes.

The polarity of the induced secondary voltage depends upon the direction in which the secondary is wound relative to the primary. It may be made to appear 180 degrees out of phase or in phase with the primary by the choice of the appropriate side of the secondary as reference point. The polarity of the secondary with respect to primary is indicated by dots on the transformer diagram (Fig. 10-2a). When the voltage at the dotted end of the primary is positive, the induced voltage at the dotted end of the secondary is also positive. Similarly, a negative voltage at the dotted end of the primary induces a negative voltage at the dotted end of the secondary coil.

Example 10-1

A 15-ohm load is connected to a 120-V, 60-Hz source through a transformer. Voltage across the load is 24 V. Find:

a. Transformation ratio (a).
b. Current in the secondary.
c. Current in the primary.
d. Power rating required for the transformer.
e. Whether it is a step-up or a step-down transformer.

Solution

a.

$$a = \frac{V_p}{V_s} \qquad (10\text{-}2)$$

$$= \frac{120\ \text{V}}{24\text{V}}$$

$$= \mathbf{5}$$

b.

$$I_s = \frac{V}{R} \qquad \text{(Ohm's law)}$$

$$= \frac{24\ \text{V}}{15\ \Omega}$$

$$= \mathbf{1.6\ A}$$

c.
$$I_p = \frac{I_s}{a} \qquad (10\text{-}2)$$
$$= \frac{1.6}{5}$$
$$= \mathbf{0.32\ A}$$

d.
$$P = V_s \cdot I_s \qquad (10\text{-}3)$$
$$= 24 \times 1.6$$
$$= \mathbf{42.4\ VA}$$

e. The secondary voltage (24 V) is lower than the primary voltage (120 V). Therefore, it is a *step-down* transformer.

10-4 An Ideal Pulse Transformer

A pulse transformer is a specially designed transformer used in pulse and switching circuits. It may be used as a coupling, isolating, step-up, step-down, or impedance-matching device. An ideal pulse transformer serves its purpose without distorting the pulse wave. Equations (10-1) apply to an ideal pulse transformer as well.

Basically, an ideal pulse transformer does not:

1. Have any internal losses.
2. Affect the rise- and fall-times of the pulse waveform.

Example 10-2

A pulse shown in Fig. 10-3a is applied to the ideal pulse transformer shown in Fig. 10-2a. Assume $a = 0.2$ and $R_L = 10\ k\Omega$. Find:

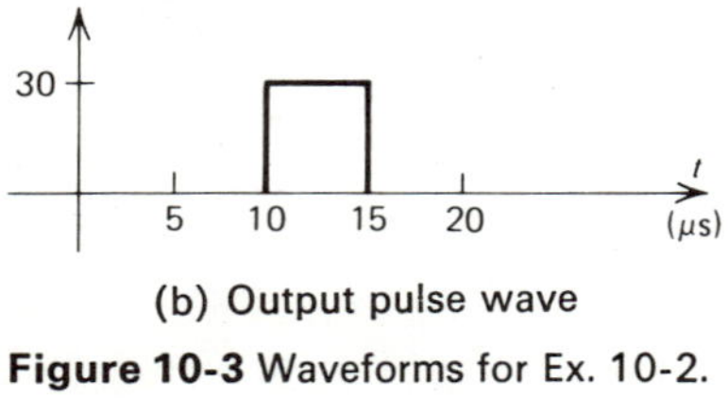

Figure 10-3 Waveforms for Ex. 10-2.

a. Amplitude of the pulse as seen by the load R_L.
b. Current in the load (R_L) during pulse.
c. Current drain from the source during pulse.
d. Input impedance as seen by the source.

Solution

a.
$$V_s = \frac{V_p}{a} \qquad (10\text{-}2)$$
$$= \frac{6}{0.2}$$
$$= \mathbf{30\ V}$$

b.
$$I_s = \frac{V_s}{R_L} \qquad \text{(Ohm's law)}$$
$$= \frac{30}{10\ \text{k}\Omega}$$
$$= \mathbf{3\ mA}$$

c.
$$I_p = \frac{I_s}{a} \qquad (10\text{-}2)$$
$$= \frac{3\ \text{mA}}{0.2}$$
$$= \mathbf{15\ mA}$$

d.
$$Z_i = a^2 R_L \qquad (10\text{-}4)$$
$$= (0.2)^2\ 10\ \text{k}\Omega$$
$$= \mathbf{400\ \Omega}$$

Also note that:

$$Z_i = \frac{V_p}{I_p}$$
$$= \frac{6}{15\ \text{mA}}$$
$$= 400\ \Omega$$

10-5 A Practical Pulse Transformer

A complete equivalent circuit of a practical pulse transformer is shown in Fig. 10-4. Various parameters shown in Fig. 10-4 represent the following:

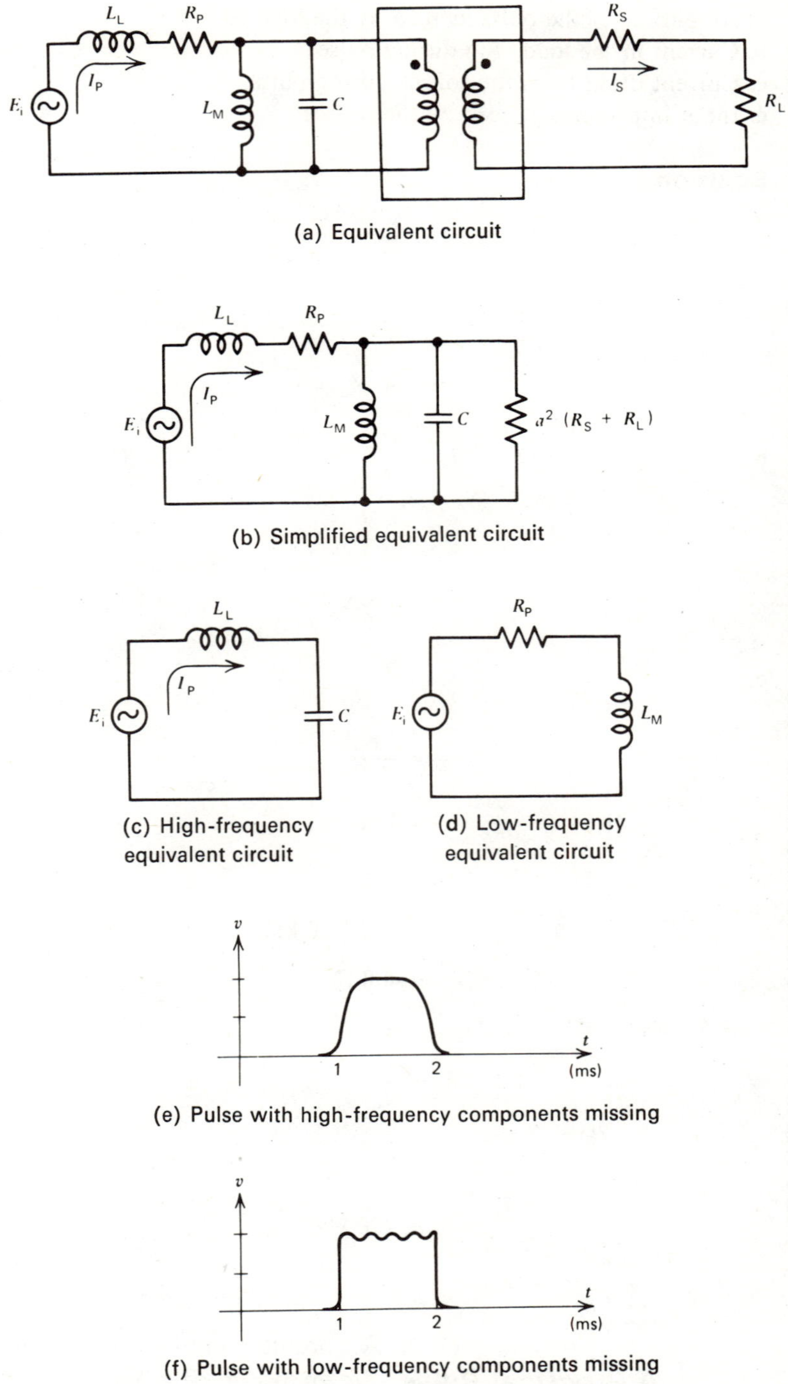

(a) Equivalent circuit

(b) Simplified equivalent circuit

(c) High-frequency equivalent circuit

(d) Low-frequency equivalent circuit

(e) Pulse with high-frequency components missing

(f) Pulse with low-frequency components missing

Figure 10-4 Transformer equivalent circuit and pulse distortion.

E_i is the input voltage (power) source (V)
L_L is the leakage inductance portion of the total primary inductance (H)
R_p is the resistance of the primary winding (Ω)
L_M is the magnetizing inductance portion of the total primary inductance (H)
C is the capacitance between primary and secondary windings (F)
R_s is the resistance of the secondary winding (Ω)
R_L is the load resistor (Ω)

Note that L_p, the total primary inductance, is:

$$L_p = L_L + L_M$$

An equivalent circuit to Fig. 10-4a is shown in Fig. 10-4b. Note that the secondary winding resistance and the load have been replaced by their equivalent resistances (Eq. 10-4), as seen by the source (E_i).

The response of a practical pulse transformer to a step input (pulse) is rather complicated and its analysis involves the use of calculus. However, certain characteristics of the circuit shown in Fig. 8-4b may easily be identified.

George E. Fourier demonstrated that each cyclical waveform is made up of three components, namely:

1. A **dc** component.
2. A *fundamental* sinusoidal component whose frequency is the same as the cyclical waveform in question.
3. Harmonics whose frequencies are whole-number multiples of the fundamental frequency.

It is important to note here that a pulse waveform is made up of a fundamental and its harmonics. It can be shown using the Fourier series-analysis method (not explained in this book) that the rising or falling edges of a pulse wave contain the higher harmonics and the flat or constant portion contains the lower harmonics. In other words, the high-frequency components of a pulse control its rise and fall times, whereas the low-frequency components control its flatness. Thus, the rise and fall times will become significant if the transformer filters or blocks the high-frequency components. On the other hand, if the transformer blocks the low-frequency components, the pulse will not be flat or constant at the top (middle).

Now consider the equivalent circuit shown in Fig. 10-4b. Note that the capacitive reactance (X_c) of C decreases as the frequency increases. This means that at *high* frequencies, X_c becomes very small and shunts the high-frequency signals. High-frequency components are thus prevented from reaching the load, R_L. Therefore, the pulse at R_L will have rounded edges (Fig. 10-4e).

Again, consider the magnetizing inductance L_M in Fig. 10-4b. Its inductive reactance, X_L, is low at *low* frequencies. Therefore, at low frequencies the current is shunted away from the load by L_M. Missing low-frequency components (harmonics) in a pulse wave distort the flat portion of the pulse (Fig. 10-4f).

We can conclude from the above discussion that a *lower* value of C will assure smaller rise and fall times and a *higher* value of L_M will assure a flatter pulse at the peak.

Another very important point that must be considered for the circuit shown in Fig. 10-4b is its *damping characteristics*. It was pointed out in Sec. 3-10 that a pulse wave input to an *RLC* circuit will result in oscillations if the circuit was underdamped. On the other hand, an overdamped circuit will result in long rise and fall times. What is needed is the critically damped circuit condition.

At high frequency, X_c becomes very small and X_{LM} becomes very high. The parallel combination of X_c, X_{LM}, and $a^2(R_s + R_L)$ becomes approximately equal to X_c. An approximate equivalent circuit for this condition (*high frequency*) is shown in Fig. 10-4c. Now its approximate damping characteristics may be identified (Fig. 10-5c).

$$Q = \frac{1}{R}\sqrt{\frac{L}{C}} \tag{3-33}$$

$$= \frac{1}{R_p} \cdot \sqrt{\frac{L_L}{C}} \tag{10-7}$$

and

$$f = f_0\sqrt{\frac{4Q^2 - 1}{4Q^2}} \tag{3-35}$$

$$\cong f_0 \text{ (for a high } Q\text{) hertz (Hz)} \tag{10-8}$$

$$\therefore f \cong \left(\frac{1}{2\pi}\right)\left(\frac{1}{\sqrt{L_L \cdot C}}\right) \text{ hertz (Hz)} \tag{10-9}$$

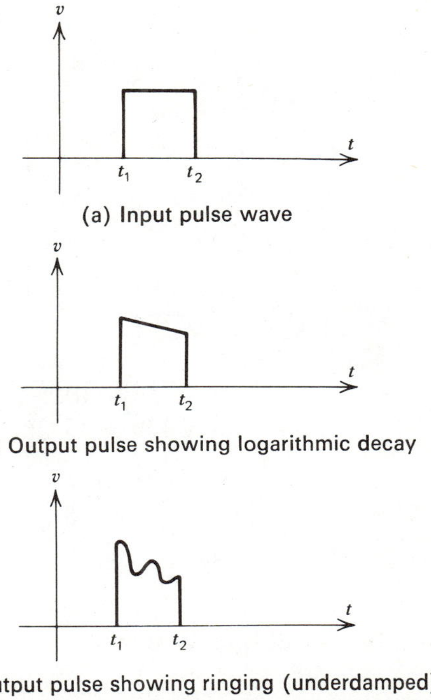

Figure 10-5 Typical input-output waveforms of a pulse transformer.

At *low frequencies*, the inductive reactance of the magnetizing inductance X_{L_M} is considerably lower than the capacitive reactance of the interwinding capacitance C and the reflected load. Therefore, the parallel combination of X_{L_M}, X_c, and $a^2(R_s + R_L)$ is essentially equal to X_{L_M}. The equivalent circuit under this condition is drawn in Fig. 10-4d.

Note that the low-frequency equivalent circuit (Fig. 10-4d) is a differentiator (Sec. 3-7). Its time constant (τ) is $(L_M + L_L)/R_p$. When a step input is applied to this circuit, the voltage across L_M(output) decays to 0 V in 5τ. Typical input-output waveforms are shown in Fig. 10-5.

Example 10-3

The pulse waveform shown in Fig. 10-6a is applied to a step-down transformer (Fig. 10-4b) where:

$a = 10$ $\quad$ $R_s = 100\ \Omega$
$L_L = 30\ \mu H$ $\quad$ $C = 150$ pF
$L_M = 150\ \mu H$ $\quad$ $R_L = 2.5\ k\Omega$
$R_p = 150\ \Omega$

Sketch the approximate voltage waveform that will appear across R_L.

Solution

$$V_s = \frac{V_p}{a} \qquad (10\text{-}2)$$

$$= \frac{20}{10}$$

$$= 2.0 \text{ V}$$

$$Q = \frac{1}{R_p}\sqrt{L_L/C} \qquad (10\text{-}7)$$

$$= \frac{1}{150}\sqrt{\frac{30 \times 10^{-6}}{150 \times 15^{-12}}}$$

$$= \frac{1}{150}\sqrt{20 \times 10^4}$$

$$= 2.98$$

Since $Q > 0.5$, the circuit is *underdamped.*

$$f = \left(\frac{1}{2\pi}\right)\left(\frac{1}{\sqrt{LC}}\right) \qquad (10\text{-}9)$$

$$= \left[\frac{1}{6.28}\right]\left[\frac{1}{(30 \times 10^{-6} \times 150 \times 10^{-12})^{1/2}}\right]$$

$$= \frac{1}{6.28 \times 6.71 \times 10^{-8}}$$

$$= 2.37 \text{ MHz}$$

$$T = \frac{1}{2.37 \text{ MHz}}$$

$$= 0.42\ \mu\text{s}$$

$$\text{Oscillations decay time} = \frac{10L}{R} \qquad (3\text{-}37)$$

$$= \frac{10 \times 30 \times 10^{-6}}{150}$$

$$= 2\ \mu\text{s}$$

The output waveform is shown in Fig. 10-6b.

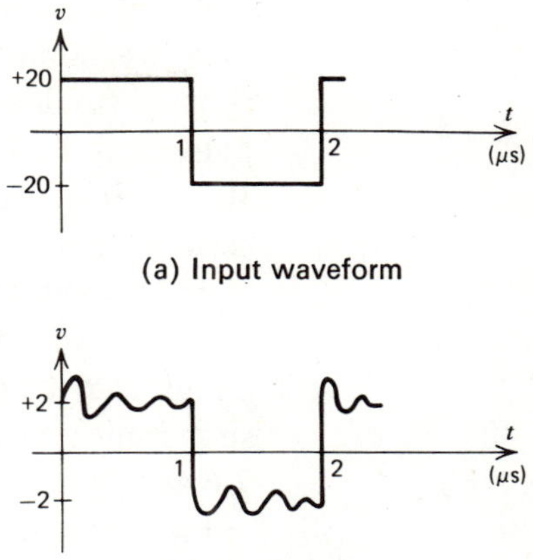

Figure 10-6 Input-output waveforms for Ex. 10-3.

10-6 Time Delay Lines

A *delay line* is used to delay a pulse for a predetermined period of time without distorting it. Assume that the block diagram of Fig. 10-7a represents a delay line. The pulse input to it is shown in Fig. 10-7b. The output is shown in Fig. 10-7c. Note that the delay line simply delays the input pulse by time t_d, as shown in Fig. 10-7c.

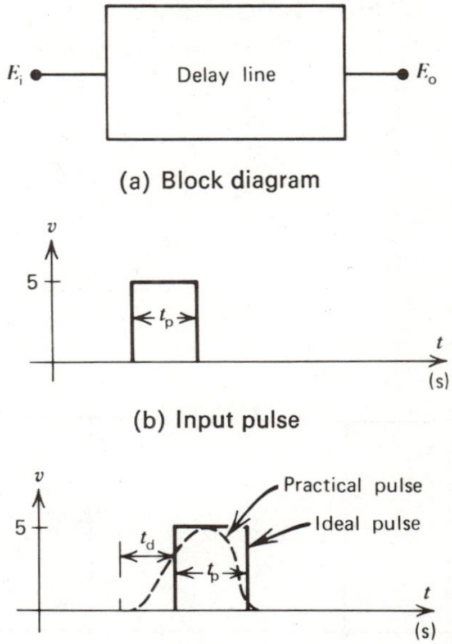

Figure 10-7 Delay line and input-output waveforms.

A pulse going through a delay line sometimes gets distorted. It may lose some amplitude. Rise and fall times may deteriorate and ringing (underdamped) may be introduced. A practical delayed pulse may look similar to the one shown as a dotted line in Fig. 10-7c. However, the distortion may be kept to a minimum under certain design conditions.

There are two types of pulse delay lines used in switching circuits, the *electronic* type and the *electromagnetic* type. Any monostable circuit may be used as an electronic pulse-delay line. Use of a one-shot multivibrator as a pulse-delay circuit was explained in Sec. 7-6 and will be discussed further in Sec. 10-9.

Electromagnetic delay lines use actual transmission lines or equivalent *LC* circuits. The fact that electromagnetic waves travel at the speed of light (3×10^8 m/s in space) is utilized in this case. The actual conductor length, when used as a delay line, is called the *distributed-parameter* delay line. When a transmission line is simulated by an equivalent *LC* circuit, it is called a *lumped-parameter* delay line.

10-7 Lumped-Parameter Delay Lines

Fig. 10-8a shows an equivalent circuit of a certain length of a lossless transmission line. C represents the interconductor capacitance and L the total equivalent inductance. Figure 10-8b shows the transmission line divided into several equal lengths. A step

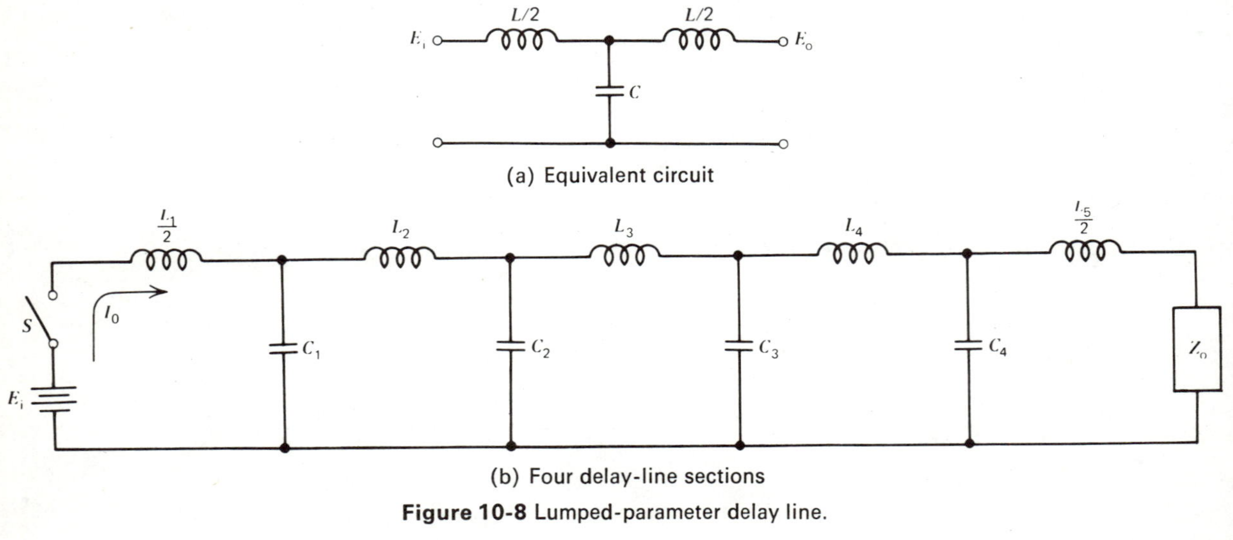

Figure 10-8 Lumped-parameter delay line.

input applied at E_i takes a certain finite time, t_d, to reach the load Z_o; t_d is called the *delay time*.

Let us examine the lumped-parameter delay line closely and identify its important features as applied to pulse work. Consider the circuit of Fig. 10-8b. As the step input is applied, capacitor C_1 starts charging through $(L_1)/2$. When C_1 gets fully charged, C_2 starts charging. After C_2 has charged, C_3 starts charging, and so on. Thus, the pulse reaches Z_o after all the capacitors have been charged. The time that the total capacitance takes to get fully charged is the delay time, t_d. Assuming that a constant current I_o was drawn from the source E_i when C_1, C_2, C_3, and so on were being charged, then,

$$I_o = \frac{E_i}{Z_i}$$

where:

- I_o is the constant current that flows from the input source while the interconductor capacitance is being charged (A)
- E_i is the amplitude of the input step voltage (V)
- Z_i is the input impedance of the lumped parameter delay line (Ω)

If the load Z_o is made equal to the input impedance Z_i, then the current in the load will also be I_o after the pulse has arrived at the load. Under this condition, there will be a constant current drain equal to I_o from the source, E_i, first charging the interconductor capacitance, then flowing through the load. When such is the case, Z_o is known as the *characteristic impedance* of the delay line. Strictly speaking, the characteristic impedance (Z_o) of a transmission line is equal to its input impedance when the line is infinitely long. The significance of Z_o is that in practice, a delay line is usually terminated in its characteristic impedance to avoid reflections or line oscillations.

For the lumped-parameter delay line shown in Fig. 10-8, it may be shown (see Appendix 2) that:

$$Z_o = \sqrt{\frac{L}{C}} \text{ ohms } (\Omega) \tag{10-10}$$

$$t_d = \sqrt{LC} \text{ seconds (s)} \tag{10-11}$$

and $$f_c = \left(\frac{1.1}{\pi}\right)\left(\frac{1}{\sqrt{LC}}\right) \text{ hertz (Hz)} \qquad (10\text{-}12)$$

where:

Z_o is the characteristic impedance (Ω)
L is the inductance (L)
C is the capacitance (F)
t_d is the delay time (s)
f_c is the upper cutoff frequency (Hz)

Note that the equivalent circuit shown in Fig. 10-8a is a low-pass filter. It attenuates signals that have frequencies above f_c, the upper cutoff frequency.

It has already been mentioned that higher frequency components of a pulse wave control its rise and fall times. Equation (10-12) indicates that a lumped-parameter delay line attenuates all frequency components that are higher than f_c. Therefore, the higher the f_c value of a particular delay line, the better will be the rise and fall times or the steeper will be the edges of the delayed pulse.

Example 10-4

A lumped-parameter delay line (Fig. 10-8a) consists of two inductors of 2 mH each and a capacitor of 100 pF. Find its:

a. Characteristics impedance (Z_o).
b. Delay time (t_d).
c. Cutoff frequency (f_c).

Solution

a.
$$Z_o = \sqrt{\frac{L}{C}} \qquad (10\text{-}10)$$
$$= \sqrt{\frac{2\text{ mH} + 2\text{ mH}}{100 \times 10^{-12}}}$$
$$= \mathbf{6.33\ k\Omega}$$

b.
$$t_d = \sqrt{LC} \qquad (10\text{-}11)$$
$$= \sqrt{4 \times 10^{-3} \times 100 \times 10^{-12}}$$
$$= \sqrt{40 \times 10^{-14}}$$
$$= 6.33 \times 10^{-7}$$
$$= \mathbf{0.633\ \mu s}$$

c. $$f_c = 1.1\frac{1}{\pi}\cdot\frac{1}{\sqrt{LC}} \qquad (10\text{-}12)$$

$$= \frac{1.1}{3.14}\cdot\frac{1}{6.33\times10^{-7}}$$

$$= \mathbf{553.4\ kHz}$$

10-8 Distributed-Parameter Delay Lines

Distributed-parameter delay lines are the actual transmission lines. These are usually either two-conductor transmission lines or coaxial cables. Remember that a transmission line is simply a means of connecting the power source to a load: the electric energy is carried from the source to the load over a transmission line. It is a medium used to carry the electrical energy from one place to another.

It is a common practice to find first the equivalent lumped-parameter delay line for the given distributed-parameter delay line. First equivalent L and C (Fig. 10-8a) are calculated. Then t_d, Z_o and f_c are calculated using Eqs. (10-11), (10-10), and (10-12), respectively.

A two-conductor transmission line is shown in Fig. 10-9a. Its equivalent capacitance and inductance are given by:

$$L = 0.921\log_{10}\left(\frac{2D}{d}\right)\ (\mu\text{H/m}) \qquad (10\text{-}13)$$

$$C = 12.06\left[\frac{k}{\log_{10}\left(\frac{2D}{d}\right)}\right]\ (\text{pF/m}) \qquad (10\text{-}14)$$

where:

D is the distance between the two conductors (center to center)
d is the diameter of each conductor
k is the dielectric constant of the material through which the electric energy travels

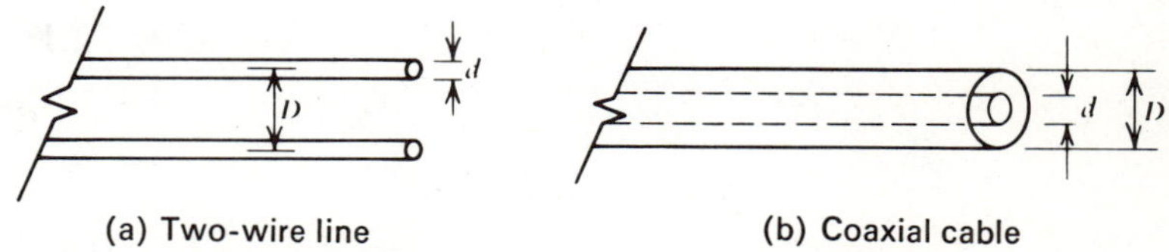

(a) Two-wire line (b) Coaxial cable

Figure 10-9 Distributed-parameter delay lines.

Note that $k = 1$ for air (space) and that the units for D and d must be the same for Eqs. (10-13) and (10-14) to be true.

Substituting these values of L and C [Eqs. (10-13) and (10-14)] in Eqs. (10-11), (10-10), and (10-12), we get:

$$t_d = \sqrt{LC} \tag{10-11}$$

$$= \sqrt{0.91 \log\left(\frac{2D}{d}\right) \times 12.06\left(\frac{k}{\log \frac{2D}{d}}\right)}$$

$$= 3.33 \times 10^{-9}\sqrt{k}\ (\text{s/m}) \tag{10-15}$$

$$Z_o = \sqrt{\frac{L}{C}} \tag{10-10}$$

$$= \sqrt{\frac{0.921 \log\left(\frac{2D}{d}\right)}{12.06\left(\frac{k}{\log 2D}\right)}}$$

$$= \sqrt{\frac{276}{k}\cdot\log\frac{2D}{d}}\ (\Omega) \tag{10-16}$$

$$f_c = \left(\frac{1.1}{\pi}\right)\left(\frac{1}{\sqrt{LC}}\right) \tag{10-12}$$

$$= \left(\frac{1.1}{3.14}\right)\left(\frac{1}{3.33 \times 10^{-9}\sqrt{k}}\right)$$

$$= \frac{1.05 \times 10^{8}}{\sqrt{k}}\ (\text{Hz}) \tag{10-17}$$

A coaxial cable is shown in Fig. 10-9b. Its equivalent capacitance and inductance are given by:

$$C = 24.1 \frac{k}{\log_{10} \frac{D}{d}}\ (\text{pF/m}) \tag{10-18}$$

$$L = 0.46 \log_{10} \frac{D}{d}\ (\mu\text{H/m}) \tag{10-19}$$

where:

C is the equivalent capacitance in pF/m
L is the equivalent inductance in μH/m
k is the dielectric constant of the insulating material through which the electromagnetic waves travel
D is the diameter of the outer conductor
d is the diameter of the inner conductor

Note again that $k = 1$ for air and D and d must have the same units.

Substituting the values for L and C obtained from Eqs. (10-18) and (10-19) in Eqs. (10-10), (10-11), and (10-12), we get:

$$t_d = \sqrt{LC} \tag{10-11}$$

$$= \sqrt{24.1\left(\frac{k}{\log \frac{D}{d}}\right)\left(0.46 \log \frac{D}{d}\right)}$$

$$= 3.33 \times 10^{-9}\sqrt{k} \text{ (s/m)} \tag{10-20}$$

$$Z_o = \sqrt{\frac{L}{C}} \tag{10-10}$$

$$= \sqrt{\frac{24.1[k/(\log D/d)]}{0.46 \log D/d}}$$

$$= 138\left(\frac{\log D/d}{\sqrt{k}}\right) (\Omega) \tag{10-21}$$

$$f_c = \left(\frac{1.1}{\pi}\sqrt{\frac{1}{\sqrt{LC}}}\right) \tag{10-13}$$

$$= \left(\frac{1.1}{3.14}\sqrt{\frac{1}{3.33 \times 10^{-9}\sqrt{k}}}\right)$$

$$= \frac{1.05 \times 10^{8}}{\sqrt{k}} \text{ (Hz)} \tag{10-22}$$

Example 10-5

A two-conductor, 1000-meter-long transmission line has an inductance of 2 mH and a capacitance of 120 pF. Find its:

a. t_d.
b. Z_o.
c. f_c.

Solution

a.

$$t_d = \sqrt{LC} \qquad (10\text{-}11)$$
$$= \sqrt{2 \times 10^{-3} \times 120 \times 10^{-12}}$$
$$= \sqrt{24 \times 10^{-14}}$$
$$= 4.9 \times 10^{-7}$$
$$= \mathbf{0.49\ \mu s}$$

b.

$$Z_o = \sqrt{\frac{L}{C}} \qquad (10\text{-}10)$$
$$= \sqrt{\frac{2 \times 10^{-3}}{120 \times 10^{-12}}}$$
$$= \sqrt{66.67 \times 10^{-6}}$$
$$= \mathbf{8.17\ k\Omega}$$

c.

$$f_c = \left(\frac{1.1}{\pi}\right)\left(\frac{1}{\sqrt{LC}}\right) \qquad (10\text{-}12)$$
$$= \left(\frac{1.1}{3.14}\right)\left(\frac{1}{0.49 \times 10^{-6}}\right)$$
$$= \mathbf{715\ kHz}$$

Example 10-6

A coaxial cable is to be used to accomplish a time delay of 5 μs. The dielectric constant (k) of the insulating material is 5; the diameter of the inner conductor is 0.1 cm and that of the outer conductor is 1.0 cm. Find:

a. Required length of the coaxial cable (delay line).
b. Characteristic impedance of the delay line.
c. Cutoff frequency of the delay line.

Solution

a.

$$t_d = 3.33 \times 10^{-9}\sqrt{k} \qquad (10\text{-}20)$$
$$= 3.33 \times 10^{-9}\sqrt{5}$$
$$= 7.47 \times 10^{-9}\text{, s/m}$$

$$\therefore \quad \text{required length} = \frac{\text{total time delay}}{\text{time delay per meter}}$$
$$= \frac{5 \times 10^{-6}}{7.47 \times 10^{-9}}$$
$$= \mathbf{669\ m}$$

b.

$$Z_o = 138\,\frac{\log D/d}{\sqrt{k}} \qquad (10\text{-}21)$$
$$= \frac{138 \log 1.0/0.1}{\sqrt{5}}$$
$$= \frac{138 \times 1}{2.24}$$
$$= \mathbf{61.6\ ohms}$$

c.

$$f_c = \frac{1.05 \times 10^8}{\sqrt{k}} \qquad (10\text{-}22)$$
$$= \frac{1.05 \times 10^8}{\sqrt{5}}$$
$$= \mathbf{46.9\ mHz}$$

10-9 Electronic Delay Lines

An *electronic delay line* is an electronic circuit that delays the pulse for a fixed amount of time. Any monostable circuit may be used for this purpose. The input pulse is merely used to trigger the monostable device. The output pulse is generated by the monostable circuit itself. The amplitude and duration (t_p) of the output pulse is controlled by the monostable circuits. Thus, a monostable circuit may be used as a pulse delay device as well as a pulse reshaping device.

The most commonly used circuit as an electronic delay line is the monostable multivibrator, which was discussed in Sec. 7-4. Usually, the pulse must be delayed for only one-bit time, which is equal to the pulse width (t_p). For this purpose, an *edge-triggered* monostable multivibrator (MV) may be used. The monostable MV should be triggered by the *trailing* edge of the input pulse.

Figure 10-10a shows a monostable MV being used as a delay circuit. The trailing or negative-going edge (Fig. 10-10c) turns OFF Q_1, the normally ON transistor. It is at this time that the voltage at E_o jumps from V_{CE}(ON) to V_{CC}. How long it stays at V_{CC} depends, basically, upon the time constant C_1R_1 (Sec. 7-4). If t_{p1} is made equal to t_{p2}, and the amplitude of the output pulse is made equal to the amplitude of the input pulse, then the pulse has been delayed by time t_p (one-bit time).

Figure 10-11a shows a level-triggered monostable circuit. This circuit may be used to delay the pulse for less than or equal to the input pulse width, that is, $t_d \leqq t_p$. In the stable state, the output of the monostable MV is at V_{CE}(ON), as shown in Fig. 10-11d. An input pulse applied at E_i charges C_2 with a time constant of R_4C_2. When the voltage across C_2 reaches the turn-ON voltage of Q_3, then, of course, Q_3 turns ON. As Q_3 turns ON, Q_1 turns OFF and Q_2 turns ON. It is at this time that the output voltage jumps to V_{CC}. As explained earlier, the output-pulse width and amplitude may be made equal to the amplitude and width of the input pulse.

The time delay t_d is the time that the capacitor C_2 takes to charge to the turn-ON voltage of Q_3.

Using the general capacitor charge equation, we get:

$$v_C = E_i[1 - e^{-(t/\tau)}] \qquad (3\text{-}4)$$

$$v_{C_2} = V_1[1 - e^{-(t/R_4C_2)}]$$

Assume that the turn-ON voltage for Q_3 is V. Then,

$$V_{ON} = V_1[1 - e^{-(t_d/R_4C_2)}]$$

$$e^{-(t_d/R_4C_2)} = \frac{V_1 - V(\mathrm{ON})}{V_1} \text{ seconds (s)} \qquad (10\text{-}23a)$$

$$t_d = R_4C_2 \ln \frac{V_1}{V_1 - V(\mathrm{ON})} \text{ seconds (s)} \qquad (10\text{-}23b)$$

Note that the time delay circuits shown in Figs. 10-10 and 10-11 are capable of accomplishing time delay as well as adjusting the amplitude and width of the output pulse.

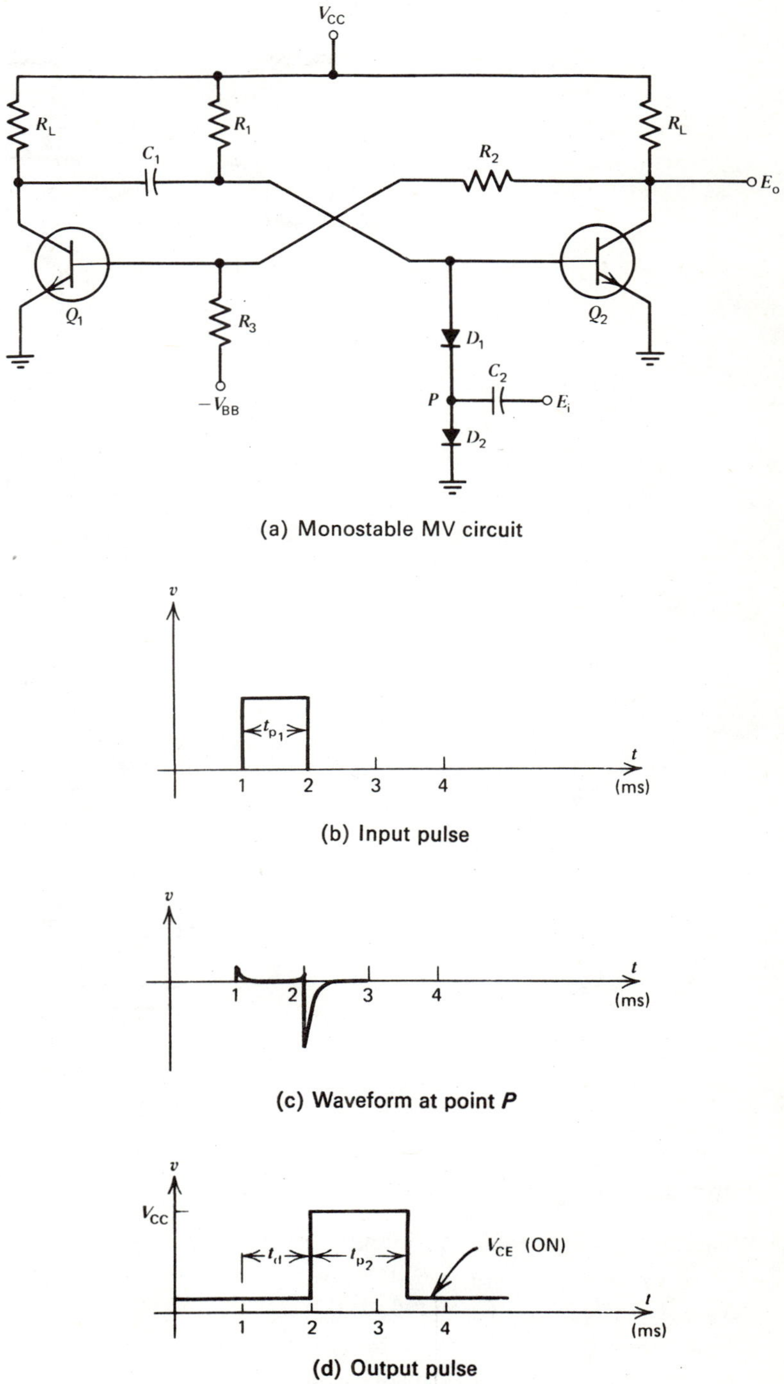

Figure 10-10 Monostable MV as a time-delay device.

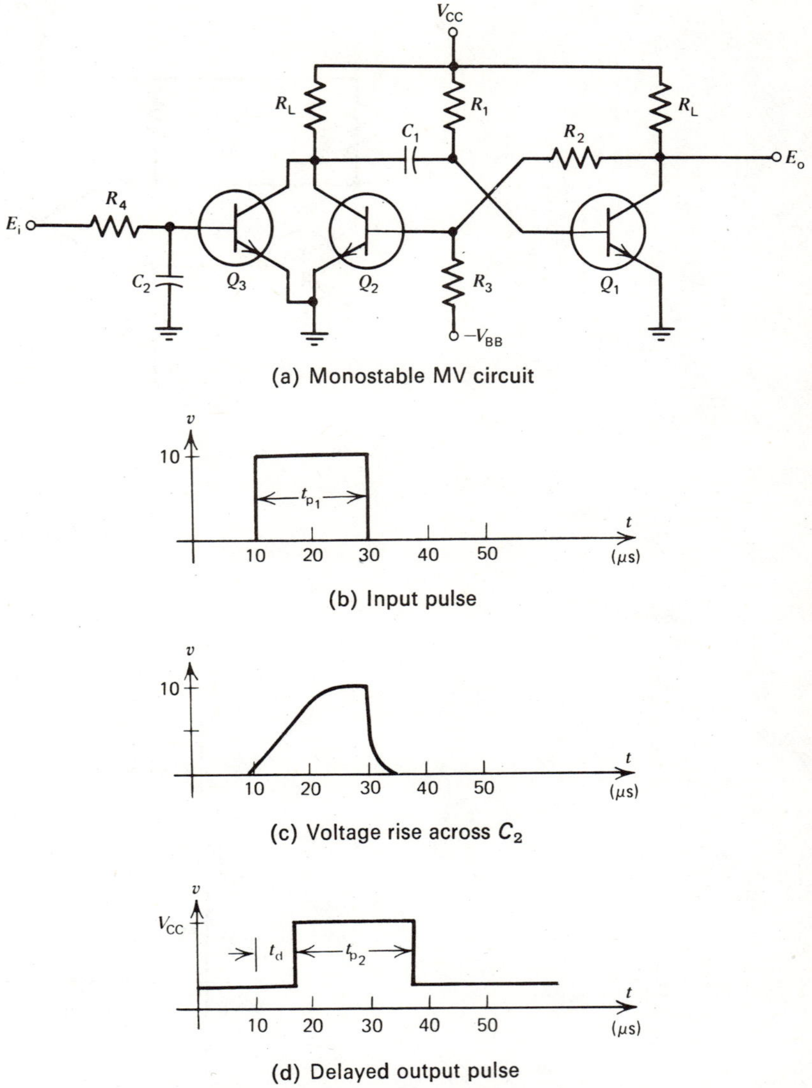

Figure 10-11 Monostable MV as a time-delay device.

Example 10-7

Consider the monostable MV circuit shown in Fig. 10-10a. Assume:

$V_{CC} = 10$ V	$V_{BB} = -10$ V
$R_L = 1.0$ kΩ	$C_1 = 144$ pF
$R_1 = 100$ kΩ	$C_2 = 50$ pF
$R_2 = 5$ kΩ	$D_1 =$ Ideal
$R_3 = 30$ kΩ	$D_2 =$ Ideal

For Q_1 and Q_2, assume $\beta = 100$, $I_C(\max) = 50$ mA, $V_{CE}(\text{ON}) = 0$ V, $V_{BE}(\text{ON}) = 0$ V.

The pulse wave shown in Fig. 10-12a is applied to the input terminal (E_i). Sketch the output waveform appearing at E_o.

Solution

$$E_{o(p\text{-}p)} = V_{CC} - V_{CE}(\text{ON})$$
$$= 10 \text{ V}$$
$$t_p = 0.693\, R_1 C_1 \qquad (7\text{-}8)$$
$$= 0.693 \times 100 \text{ k}\Omega \times 144 \times 10^{-12}$$
$$= \mathbf{10\ \mu s}$$

The voltage waveform appearing across D_2 and the output waveform are drawn in Figs. 10-12b and 10-12c, respectively. Note that the delay time t_d is equal to the pulse width t_p' (10 μs).

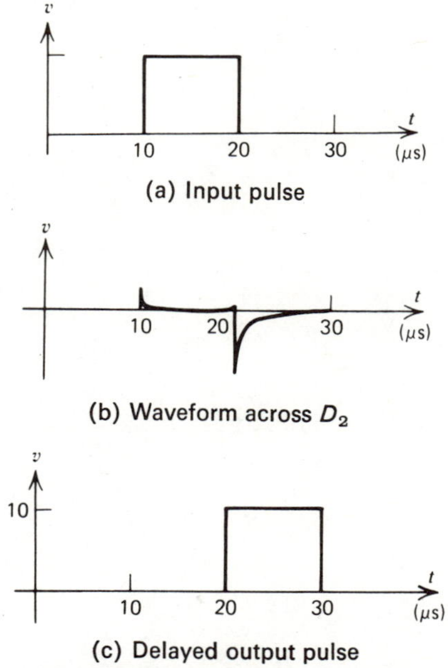

Figure 10-12 Waveforms for Ex. 10-7.

Example 10-8

Design an electronic pulse delay circuit that will delay the pulse wave shown in Fig. 10-12a by 50 μs. Transistors similar to the ones used in Ex. 10-7 are to be used. Assume the turn-ON voltage for Q_3 is 0.5 V.

Solution

The monostable MV circuit that may be used here is shown in Fig. 10-11a. V_{CC}, V_{BB}, R_L, R_1, R_2, R_3, and C_1 are the same as given in Ex. 10-7. Review Sec. 7-5 for the design of the monostable MV circuit itself.

This Example shows the manner in which R_4 and C_2 are chosen such that Q_3 turns ON 50 μs after the input pulse has been applied.

$$I_C(\text{ON}) = \frac{V_{CC} - V_{CE}(\text{ON})}{R_L} \qquad (7\text{-}9)$$

$$= \frac{10\ \text{V}}{1\ \text{k}\Omega}$$

$$= 10\ \text{mA}$$

$$I_B(\text{ON}) = \frac{I_C(\text{ON})}{\beta} \qquad (7\text{-}11)$$

$$= \frac{10\ \text{mA}}{100}$$

$$= 0.1\ \text{mA}$$

$$R_4 = \frac{E_i - V_{ON}}{I_B(\text{ON})}$$

$$= \frac{10 - 0.5}{0.1\ \text{mA}}$$

$$= \mathbf{95\ k\Omega}$$

$$e^{-(t_d/R_4C_2)} = \frac{V_1 - V(\text{ON})}{V_1} \qquad (10\text{-}23)$$

$$= \frac{10 - 0.5}{10}$$

$$= 0.95$$

$$\frac{t_d}{R_4C_2} = 0.0513$$

$$C_2 = \frac{t_d}{0.0513 \times R_4}$$

$$= \frac{50\ \mu\text{s}}{0.0513 \times 95\ \text{k}\Omega}$$

$$= \mathbf{10.3\ nF}$$

10-10 Summary

1. A transformer is a device used mainly either to step up an **ac** voltage or to step down an **ac** voltage. It is also used as a coupling, isolating, and impedance-matching device. It is constructed by magnetically coupling a primary coil to a secondary coil.
2. The input-to-output voltage (or current) ratio of a transformer depends primarily upon its turns ratio.
3. The apparent input power (V_pI_p) is equal to the apparent output power (V_sI_s) for an ideal transformer.
4. A pulse transformer is specially constructed such that its interwinding capacitance is kept to a minimum. A lower interwinding capacitance assures that the high-frequency components of a pulse waveform do not get attenuated.
5. A delay line is a device used to delay a pulse wave for a desired amount of time. There are two types of delay lines, electromagnetic delay lines and electronic delay lines.
6. Transmission lines are used as electromagnetic delay lines. They are based upon the fact that it takes a certain finite time (3×10^8 m/s) for a pulse to travel from one point to another over a transmission line.
7. Two types of electromagnetic delay lines are used, the distributed-parameter type and the lumped-parameter type. Important characteristics of an electromagnetic delay line are its delay time (t_d), characteristic impedance (Z_o) and cutoff frequency (f_c).
8. Electronic delay lines mainly consist of monostable multivibrator circuits. They not only accomplish the desired time delay, but also reshape the pulse.
9. Important equations:

$$a = \frac{N_p}{N_s} \tag{10-1}$$

$$a = V_p/V_s \tag{10-2}$$

$$V_pI_p = V_sI_s = P\,(\text{VA}) \tag{10-3}$$

$$Z_i = a^2R_L\,(\Omega) \tag{10-4}$$

$$t_d = \sqrt{LC}\,(\text{s}) \tag{10-11}$$

$$Z_o = \sqrt{\frac{L}{C}}\,(\Omega) \tag{10-10}$$

$$f_c = \left(\frac{1.1}{\pi}\right)\left(\frac{1}{\sqrt{LC}}\right)(\text{Hz}) \tag{10-12}$$

$$L = 0.921 \log_{10} \frac{2D}{d} \ (\mu\text{H/m}) \tag{10-13}$$

$$C = \frac{12.06\ k}{\log_{10} 2D/d} \ (\text{pF/m}) \tag{10-14}$$

$$C = \frac{24.1\ k}{\log_{10} D/d} \ (\text{pF/m}) \tag{10-18}$$

$$L = 0.46 \log_{10} D/d \ (\mu\text{H/m}) \tag{10-19}$$

$$t_d = 3.33 \times 10^{-9} \sqrt{k} \ (\text{s/m}) \tag{10-20}$$

$$e^{-(t_d/RC)} = \frac{V_1 - V(\text{ON})}{V_1} \tag{10-23}$$

10-11 Glossary

Characteristic impedance. The input impedance of an infinite transmission line.

Cutoff frequency. The upper-half power-point frequency of an electromagnetic delay line.

Delay line. A device, a circuit, or a transmission line used to delay a pulse waveform.

Delay time. The amount of time that a delay line delays the pulse waveform.

Distributed-parameter delay line. A certain length of actual transmission line being used as a delay line.

Electromagnetic delay line. A transmission line being used to delay a pulse waveform.

Electronic delay line. An electronic circuit used to delay a pulse waveform.

Load matching. Making the load equal to the input impedance of the power source.

Lumped-parameter delay line. An electromagnetic delay line simulated by inductor-capacitor combinations.

Magnetic core. A medium used to magnetically link the primary with the secondary of a transformer.

Primary. The coil on the input side of a transformer.

Pulse transformer. A specially designed transformer with an extended high-frequency cutoff point.

Secondary. The coil on the output side of a transformer.

Step-down transformer: A transformer with a transformation ratio (a) less than unity.

Transformation ratio. The ratio between the number of turns at the primary to the number of turns at the secondary.

Transformer. A device consisting of a primary coil magnetically linked to a secondary coil.

10-12 Recapitulation

Return to the Objectives and Self-Evaluation Questions given at the beginning of this chapter and see how well you can answer them. If you cannot answer certain questions satisfactorily, place a check mark (√) next to them and review the appropriate parts of the chapter to find the answers. No attempt should be made to solve the Problems (Sec. 10-13) before answering the Self-Evaluation Questions. Ability to solve the Problems should serve as a test and feedback regarding your understanding of the material presented in this chapter. Answers to the Problems are given on p. 527.

10-13 Problems

10-1 A 120-V, 60-Hz source is connected to a transformer whose transformation ratio is 0.05. Find:
a. The open-circuit output terminal voltage.
b. Whether it is a step-up or a step-down transformer.

10-2 A 10-kΩ load is connected to the transformer of Prob. 10-1. Find:
a. The current in the load.
b. The current flowing in the primary.
c. The power rating of the transformer.

10-3 An amplifier draws 1.5 A when connected to a 12-V, 60-Hz source A 120-V, 60-Hz power source is available. For the transformer that is to be used, find:
a. The transformation ratio.
b. The power rating.

10-4 The pulse waveform shown in Fig. 10-3a is applied to a switching circuit through an ideal pulse transformer whose transformation ratio is 6. Find and sketch the voltage waveform that the switching circuit uses.

10-5 The switching circuit of Prob. 10-4 draws 500 mA when a pulse is present at its input. Find:
a. The power rating of the transformer.
b. The equivalent impedance as seen by the original pulse waveform (at primary).

10-6 A 100-m-long, two-conductor transmission line is being used as a delay line. The relative dielectric constant for the insulating material is 4. Find its delay time.

10-7 What length of the transmission line described in Prob. 10-6 is required to obtain a time delay of 5 μs?

10-8 A lumped-parameter delay line consists of a 2-mH inductor and 100-pF capacitor. Find its:

a. Time delay.
b. Characteristic impedance.
c. Cutoff frequency.

10-9 Repeat Prob. 10-8 with the 100-pF capacitor replaced by a 0.005-μF capacitor.

10-10 Design a lumped-parameter delay line such that its characteristic impedance is 56 kΩ and its delay time is 10 μs.

10-11 Repeat Prob. 10-10 for a desired characteristic impedance of 1 kΩ and time delay of 25 μs.

10-12 A 575-m-long, two-conductor transmission line is being used as a delay line. The two conductors of 0.2-cm diameter each are spaced 0.8 cm apart. The dielectric constant for the insulating material is 6. Find:
a. The delay time.
b. The characteristic impedance.
c. The cutoff frequency.

10-13 Repeat Prob. 10-12 for a 200-m length of the transmission line described therein.

10-14 What length of the transmission line described in Prob. 10-12 is needed to accomplish a time delay of 10 μs?

10-15 A delay line consists of a 150-m-long coaxial cable. The diameter of the inner conductor is 0.1 cm and that of the outer conductor is 0.5 cm. The relative dielectric constant is 4.5 for the insulating material. Find its:
a. Delay time.
b. Characteristic impedance.
c. Cutoff frequency.

10-16 Design a lumped-parameter delay line that is equivalent to 150 m of the coaxial cable described in Prob. 10-15 (find L and C).

10-17 The coaxial cable described in Prob. 10-15 is to be used to obtain a time delay of 5 μs. Find:
a. The required length.
b. The characteristic impedance.
c. The cutoff frequency.

10-18 The pulse waveform shown in Fig. 10-3a is to be delayed for 5 μs (same as pulse width). Design a monostable MV circuit that will accomplish this. Assume the transistor and diodes to be used as ideal with the exception that for the transistor, $\beta = 200$ and $I_C(\text{ON}) = 10$ mA.

10-19 Modify the circuit designed in Prob. 10-18 so that the output pulse is of 0 V to 10 V amplitude and 10 μs duration. The time delay is still 5 μs.

10-20 Repeat Prob. 10-18 for a required time delay of 2 μs. Assume the turn-ON voltage for the triggering transistor to be 0.4 V.

10-14 Experiment

Title Lumped-Parameter Electromagnetic Delay Line

Theory A lumped-parameter electromagnetic delay line consists of an inductance-capacitance combination terminated in its characteristic impedance (Z_o) as shown in Fig. E10-1 (one possible circuit configuration). The important characteristics of a lumped parameter delay line are given by:

$$t_d = \sqrt{LC} \tag{10-11}$$

$$Z_o = \sqrt{\frac{L}{C}} \tag{10-10}$$

$$f_c = \left(\frac{1.1}{\pi}\right)\left(\frac{1}{\sqrt{LC}}\right) \tag{10-12}$$

Equipment Dual-beam oscilloscope: PHILIPS 3230, or equal.
Multifunction generator: EXACT model 126 or equal.
Multimeter: RCA model WV 98C, or equal
Resistors: 470 Ω and 1 kΩ potentiometer.
Capacitors: 0.01 μF and 0.047 μF.
Inductor: 30 mH.

Procedure

1. Connect the lumped-parameter delay line circuit shown in Fig. E10-1.

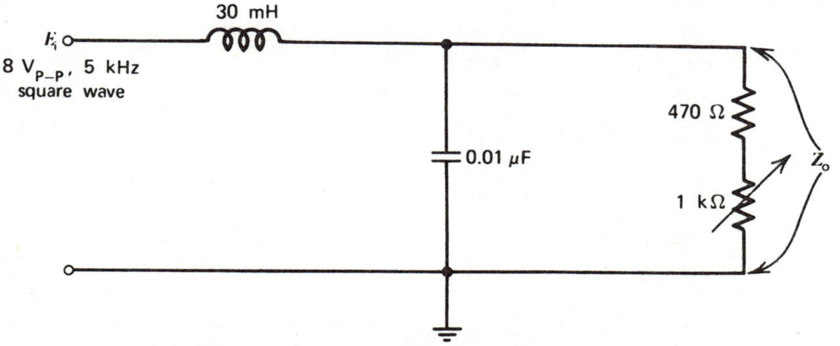

Figure E10-1 A lumped-parameter delay line for Exp. 10.

2. Connect CRO (one beam) across Z_o. Adjust the 1-kΩ potentiometer (Z_o) so that a critically damped waveform appears across Z_o (no overshoot).
3. Observe CRO and draw in Table E10-1 the input as well as the output waveforms. Measure t_d and record it in Table E10-2. Note that t_d is measured from the start of the input pulse to the time when the output pulse reaches 90 percent of its final value.
4. Measure and record in Table E10-2 the ohmic value of Z_o. Note that Z_o is made up of the series combination of 470 Ω and the potentiometer.
5. Change the input to a 8-$V_{p\text{-}p}$ sine wave. Vary the input frequency until the input becomes 5.66 $V_{p\text{-}p}(f_c)$. Make sure that the input stays at 8 $V_{p\text{-}p}$. Record the frequency (f_c) in Table E10-3.
6. Replace the 0.01-μF capacitor with a 0.047-μF capacitor. Repeat Procedure steps 2, 3, 4, and 5. Record your results in the appropriate tables.
7. Disconnect and return all the test equipment to its proper place.

Calculations

1. For the circuit shown in Fig. E10-1, calculate the theoretical values of t_d and Z_o and record the results in Table E10-2, when:
   a. $C = 0.01\ \mu$F.
   b. $C = 0.047\ \mu$F.
2. For the circuit shown in Fig. E10-1, calculate the theoretical values of f_c and record them in Table E10-3, when:
   a. $C = 0.01\ \mu$F.
   b. $C = 0.047\ \mu$F.
3. Calculate the percentage error between the theoretical and experimental values recorded in Tables E10-2 and E10-3. Record your results in the respective Tables.

Tabulation of Results

Table E10-1

Circuit Cap.	Waveform
$C = 0.01\ \mu$F	Input
	Output
$C = 0.047\ \mu$F	Input
	Output

Table E10-2

Circuit Cap.	Parameter	Observed Value	Theoretical Value	Percentage Error
$C = 0.01\ \mu F$	t_d			
	Z_o			
$C = 0.047\ \mu F$	t_d			
	Z_o			

Table E10-3

Circuit Cap.	Parameter	Observed Value	Theoretical Value	Percentage Error
$C = 0.01\ \mu F$	f_c			
$C = 0.047\ \mu F$	f_c			

Questions

1. The output waveforms recorded in Table E10-1 have significant rise and fall times. Explain why.
2. Examine Tables E10-2 and E10-3. How does an increase in the circuit capacitance affect t_d, Z_o, and f_c?

Discussion

1. What conclusions can you draw from the experimental and the theoretical values recorded in Tables E10-2 and E10-3?
2. Account for discrepancies between the theoretical and experimental values recorded in Tables E10-2 and E10-3.
3. Consider the circuit shown in Fig. E10-1. Discuss how t_d, f_c, and Z_o would be affected if the 30 mH coil is replaced by a 10 mH coil.

Chapter 11 Digital Logic Review and IC Circuits

11-1 Objectives

The objectives of this chapter are to:

1. Explain the meaning of the two logic states as used in Boolean algebra.
2. Explain the logic NOT, AND, OR, EXCLUSIVE OR, NOR, and NAND operations.
3. Show how logic expressions are formed, starting with the word description.
4. Show how logic expressions are simplified using either theorems or the Karnaugh map.
5. Show how a logic gate network is realized from the given word description of a problem.
6. Explain the operation of logic gate circuits available in the following IC families:
   - **a.** RTL.
   - **b.** DTL.
   - **c.** TTL.
   - **d.** ECL.
   - **e.** CTL.
   - **f.** HTL.
   - **g.** CMOS.
7. Identify some important characteristics of various logic IC families.

11-2 Self-Evaluation Questions

Watch for the answers to the following questions as you read the chapter. They should help you to understand the material presented. When you have finished the chapter, return to this section and be sure that you can answer all the questions. Then test your understanding by working out solutions to the Problems.

It is strongly recommended that you answer these questions satisfactorily before attempting to solve the Problems (Sec. 11-20).

1. Explain the two-logic-state basis of Boolean algebra and show how it corresponds to an electronic switch.
2. Explain the following logic operations:
   - **a.** NOT.
   - **b.** AND.
   - **c.** OR.
   - **d.** EXCLUSIVE OR.
   - **e.** NOR.
   - **f.** NAND.

3. Show how logic theorems and the Karnaugh map are used to simplify logic expressions.
4. Show how a logic expression is implemented using logic gates.
5. Show how a logic expression is implemented using only either NAND gates or NOR gates.
6. Show a step-by-step procedure used to implement or realize a word problem.
7. Explain the operation of logic gates found in the following families:
   a. RTL.
   b. DTL.
   c. TTL.
   d. ECL.
   e. CTL.
   f. HTL.
   g. CMOS.
8. Compare the important characteristics of the above-named IC families.

11-3 The Two Logic States

Boolean algebra is a specialized branch of mathematics. It is founded on the premise that a statement (variable) is either TRUE or FALSE. If it is not TRUE then it must be FALSE, and vice versa. According to Boolean algebra, a condition or statement is either TRUE or FALSE at any one time. In this bivalued logic there is no other condition. Similarly, a switch is either ON or OFF, either there is a pulse or there is no pulse, either it is day or night, either the car runs or it does not run.

It is common practice to denote TRUE by 1 and FALSE by 0:

$$\text{TRUE} = 1 \quad \text{and} \quad \text{FALSE} = 0$$

Let S represent a switch and say it is TRUE when it is ON and FALSE when it is OFF. In logic algebra, this situation is written in the following manner:

$$S = 1 \text{ (TRUE, switch ON)}$$

and

$$\bar{S} = 0 \text{ (FALSE, switch not ON)}$$

Note that the condition $\bar{S}$ (read as "S not") exists when S is not TRUE. It is the opposite or complement of S.

In actual logic circuits, two voltage levels are assigned to the two logic states, one equal to the TRUE and the other to the FALSE. Assume that:

$$+5\text{ V} = \text{TRUE (1) and } 0\text{ V} = \text{FALSE (0)}$$

Then

$$S = +5\text{ V} \quad \text{and} \quad \bar{S} = 0\text{ V}$$

In any one system that has more than one variable, the same two voltage levels are assigned to the two logic states of each variable. Consider the statement, "The radio (R) comes ON if the switch (S) is turned ON and the antenna (T) is connected." Here, R, S, and T refer to one system. The voltage levels assigned to both of the logic states must be the same for each. S represents a switch (ON–OFF), R represents a radio (ON–OFF), and T represents an antenna (connected–not connected). Then,

$$S = \text{TRUE} = \text{ON} = 1 = +5\text{ V}; \bar{S} = \text{FALSE} = \text{OFF} = 0 = 0\text{ V}$$

$$R = \text{TRUE} = \text{Radio ON} = 1 = +5\text{ V}; \bar{R} = \text{FALSE} = \text{Radio OFF} = 0 = 0\text{ V}$$

and

$$T = \text{TRUE} = \text{Antenna connected} = 1 = +5\text{ V}; \bar{T} = \text{FALSE} = \text{Antenna NOT connected} = 0$$

Example 11-1

Consider the statement, "A car (C) will not run if either there is no fuel (F) or the battery (B) is dead." The two voltage levels for the two logic states are 5 V and 0 V for TRUE and FALSE, respectively. Assign proper voltage levels to each variable.

Solution

C = Car will run = 1 = 5 V; $\bar{C}$ = Car will not run = 0 = 0 V
F = Fuel in tank = 1 = 5 V; $\bar{F}$ = NO fuel in tank = 0 = 0 V
B = Battery NOT dead = 1 = 5 V; $\bar{B}$ = Battery dead = 0 = 0 V

The statement above, "The radio comes ON if the switch is ON and the antenna is connected," may be summarized by an equation:

$$R = S \text{ and } T$$

We shall see later (Sec. 11-7) how to represent this in equation form in Boolean algebra.

11-4 Basic Logic Operations

There are three basic operations used in logic algebra: the NOT, the AND, and the OR. Three additional logic operations—namely, NAND, NOR, and EXCLUSIVE OR—are derived from these three. This will become clearer later on in this Section.

Electronic circuits that are used to perform logic operations are called *logic gates*, or simply *gates*. A NOT gate is used to perform the logic NOT operation. An AND gate performs the logic AND operation, and so on. There are many ways of constructing gates (electronic circuits), as will be seen in the following sections. But remember that an AND gate is an AND gate regardless of its circuit configuration and that it merely performs the logic AND operation. The same is true for all other logic gates. Note again that the same voltage levels are assigned to the TRUE and FALSE situations of various variables in a system.

11-4.1 NOT Gate

The NOT gate is used to *invert* (complement) the state of a variable. It is also known as the *inverter*. The standard symbol of an inverter is shown in Fig. 11-1. The output of a NOT gate is false if the input is true, and vice versa.

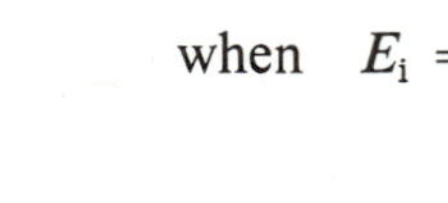

Figure 11-1 NOT gate

The inverter circuit was discussed in Sec. 5-9. Note that:

$$E_o = V_{CC} \qquad \text{when} \quad E_i = V_{CE}(\text{ON})$$

and

$$E_o = V_{CE}(\text{ON}) \quad \text{when} \quad E_i = V_{CC}$$

Example 11-2

In a system in which the logic levels are +12 V and 0 V, two inverters (NOT gates) are connected in series, as shown in Fig. 11-2.

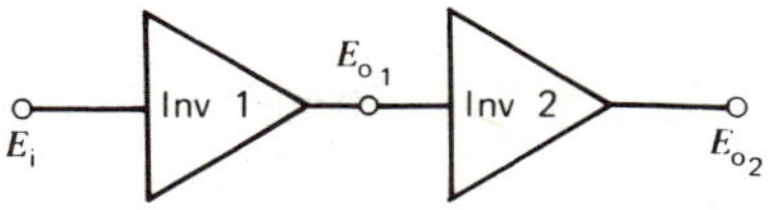

Figure 11-2 Two series-connected NOT gates.

a. Find the output voltage level of inverter #2 if the input to inverter #1 is 12 V.
b. Find the output voltage level of inverter #2 if the input to inverter #1 is 0 V.
c. What happens to the state of a variable if it goes through two series-connected NOT gates.

Solution

a.

$$\text{Input to inverter \#1} = 12\text{ V}$$

$$\therefore \text{Output of inverter \#1} = 0\text{ V}$$

Since the output of inverter #1 is the input to inverter #2,

$$\therefore \text{Output of inverter \#2} = \mathbf{12\ V}$$

b.

$$\text{Input to inverter \#1} = 0\text{ V}$$

$$\therefore \text{Output of inverter \#1} = 12\text{ V}$$

Since the output of inverter #1 is connected to the inverter #2,

$$\therefore \text{Output of inverter \#2} = \mathbf{0\ V}$$

c. It can be seen from the results of parts (a) and (b) of this example that the *state* of the variable emerges the same as the input if it is fed to two series-connected inverters. In fact, it may be concluded that the state of the variable remains the *same* if it is put through any *even*-number groupings of series-connected inverters. The state of the variable will be *complemented* if it is put through any *odd*-number groupings of series-connected inverters.

11-4.2 AND Gate The standard symbol of an AND gate is shown in Fig. 11-3a. A, B, and C are three variable (logic) inputs and E_o is the output. In fact, of course, there can be any practical number of inputs to an AND gate. The function of the AND gate is to perform the logic AND operation, which is:

1. The output is TRUE only if *all* of the inputs are TRUE. In

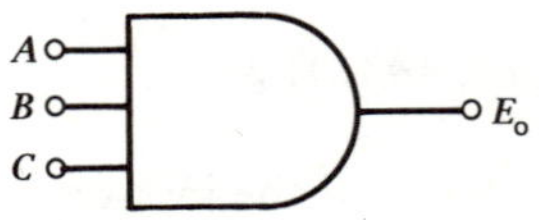

(a) Symbol

A	B	C	E_0
0	0	0	0
0	0	1	0
0	1	0	0
0	1	1	0
1	0	0	0
1	0	1	0
1	1	0	0
1	1	1	1

(b) Truth table

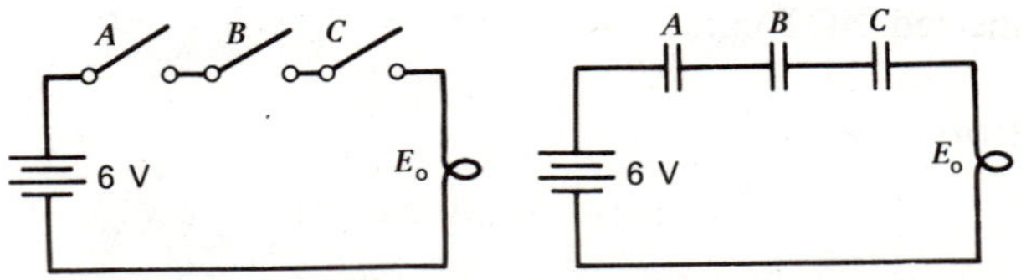

(c) AND gate using toggle switches

(d) AND gate using magnetic relays

Figure 11-3 AND gate.

Fig. 11-3a, E_0 is TRUE only if A is TRUE and B is TRUE and C is TRUE. This is why Fig. 11-3a is called an AND gate.

2. The output is FALSE when one or more than one input is FALSE.

The table shown in Fig. 11-3b tabulates all possible TRUE/FALSE combinations for A, B, and C. The resulting state of the output is shown in column E_0 for each case. The table, thus constructed, is known as a *truth table*. Remember that 1 represents TRUE and 0 represents FALSE.

The AND operation performed by the AND gate shown in Fig. 11-3a is written in logic algebra as:

$$E_0 = A \cdot B \cdot C \tag{11-1a}$$

The dots that appear between $A \cdot B$ and $B \cdot C$ represent the AND operation. These are usually omitted and Eq. 11-1a is written simply as:

$$E_0 = ABC \tag{11-1b}$$

Equation (11-1) means, "E_0 is TRUE if A is TRUE and B is TRUE and C is TRUE." It is usually read as "E_0 equals A **and** B **and** C."

The circuit shown in Fig. 11-3c is an AND gate. Switches A, B, and C represent the three inputs. A lamp is connected at the output (E_o).

Switch ON = TRUE state; Light ON = TRUE state

Switch OFF = FALSE state; Light OFF = FALSE state

It can be seen that the lamp comes ON (TRUE) only when all three switches (A, B, and C) are ON (TRUE). The lamp is not ON when one or more than one switch is OFF (FALSE).

The circuit of Fig. 11-3d also shows an AND gate. Here, the three input switches are the three relays labeled A, B, and C. Note again here that the lamp comes ON *only* when all three relays (A, B, and C) are ON (TRUE).

Example 11-3

Consider the statement, "It rains only when there are clouds in the sky *and* there is also lightning."

a. Assign logic symbols to each variable entity in the statement.

b. Write the equivalent logic equation.

c. Draw the logic gate (symbol) that implements the logic equation.

Solution

a. Rains = R, Clouds = C, Lightning = L

Does not rain = $\bar{R}$, No clouds = $\bar{C}$, No lightning = $\bar{L}$

b. $R = CL$

c.

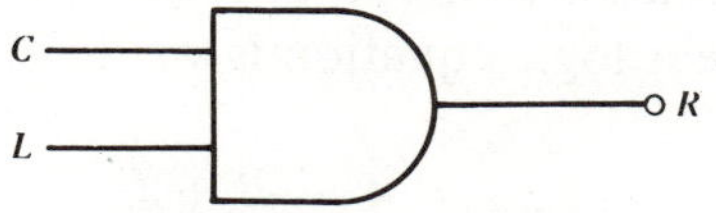

Figure 11-4 AND gate.

11-4.3 OR Gate

The standard symbol for an OR gate is shown in Fig. 11-5a. A, B, and C are three variable inputs and E_o is the output. The function of the OR gate is to perform the logical OR operation, which is:

1. The output is TRUE if one or more than one input is TRUE. In Fig. 11-5a, E_o is TRUE when either A is TRUE or B is TRUE or C is TRUE.
2. The output is FALSE *only* when *all* of the inputs are FALSE.

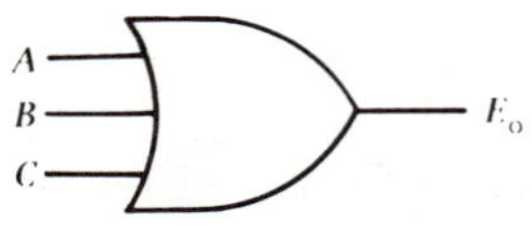

(a) Symbol for OR gate

A	B	C	E_o
0	0	0	0
0	0	1	1
0	1	0	1
0	1	1	1
1	0	0	1
1	0	1	1
1	1	0	1
1	1	1	1

(b) Truth table for OR operation

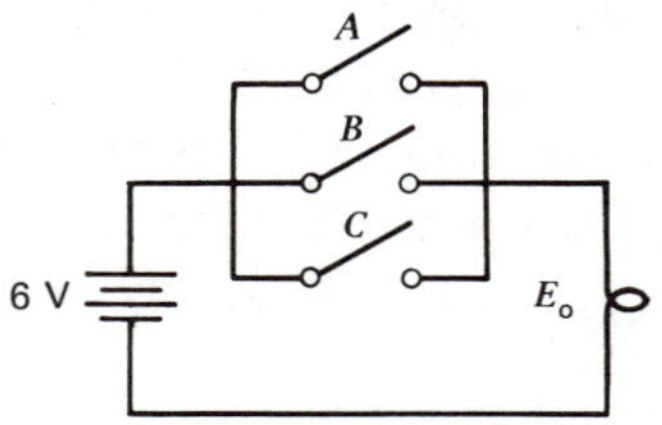

(c) OR gate using toggle switches

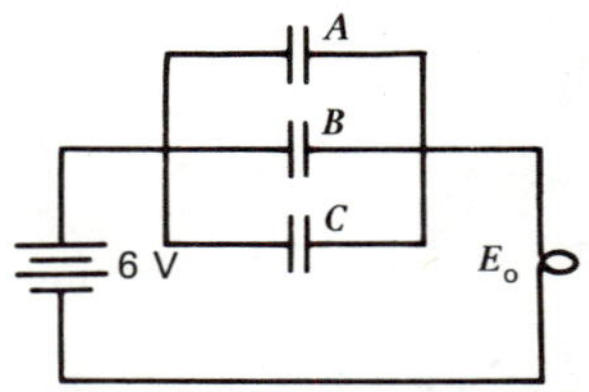

(d) OR gate using magnetic relays

A	B	E_o
0	0	0
0	1	1
1	0	1
1	1	0

(e) Truth table for the EXCLUSIVE OR operation

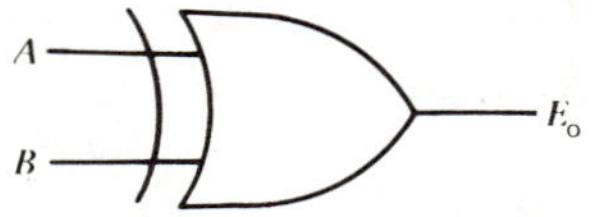

(f) Symbol for an EXCLUSIVE OR gate

Figure 11-5 The OR and the EXCLUSIVE OR operations.

A truth table for a three-input OR operation is shown in Fig. 11-5b. Its equivalent logic equation is written as:

$$E_o = A + B + C \tag{11-2}$$

Equation (11-2) means, "E_o is TRUE when either A is TRUE or B is TRUE or C is TRUE. The "+" (plus) sign represents the OR operation. The equation is usually read as, "E_o equals A **or** B **or** C."

The circuit shown in Fig. 11-5c is an OR gate. The lamp comes ON (TRUE) when either one or two or all three switches are ON. The lamp will be OFF (FALSE) only when all of the three switches are OFF.

An OR gate using relays as input switches is shown in Fig. 11-5d. Note again that the lamp is OFF (FALSE) *only* when all of the relays are open (FALSE). It is ON (TRUE) when one or more than one relay is closed (TRUE).

A close relative of the OR gate is the EXCLUSIVE OR gate. It differs from the OR gate in that the output is FALSE when both of the inputs are TRUE. Its truth table is shown in Fig. 11-5e. Basically, the EXCLUSIVE OR operation is used to identify whether the two inputs are identical or different. Its output is TRUE when the two inputs are different and FALSE when they are the same. The logic expression for an EXCLUSIVE OR operation is:

$$E_o = A\bar{B} + \bar{A}B$$

where A and B are the two inputs and E_o is the output. The standard symbol for an EXCLUSIVE OR gate is shown in Fig. 11-5f.

Example 11-4
Consider the statement, "An alarm turns ON either when there is smoke in the building or there is a fire in the building."
a. Assign logic symbols to each variable entity in the statement.
b. Write the equivalent logic equation.
c. Draw the logic gate (symbol) that implements the logic equation.

Solution
a. Alarm turns ON = A, Smoke = S, Fire = F
Alarm does NOT turn ON = $\bar{A}$, No smoke = $\bar{S}$, No fire = $\bar{F}$
b. $A = S + F$
c.

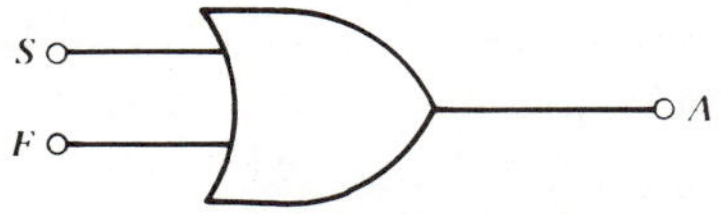

Figure 11-6 OR gate for Ex. 11-4.

11-4.4 NAND and NOR Gates

A NAND gate consists of an AND gate followed by an inverter. It is shown in Fig. 11-7a. A truth table for a 2-input NAND operation is drawn in Fig. 11-7b. The output is FALSE (0) only when both (all) of the inputs are TRUE (1). When one or more than one input is FALSE, the output is TRUE. The equivalent logic equation for this NAND gate is:

$$E_o = \overline{AB} \qquad (11\text{-}3)$$

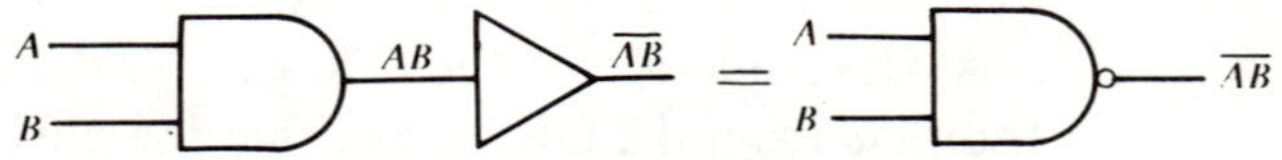

(a) AND-NOT combination for NAND gate

A	B	$A \cdot B$	$\overline{AB}$
0	0	0	1
0	1	0	1
1	0	0	1
1	1	1	0

(b) Truth table

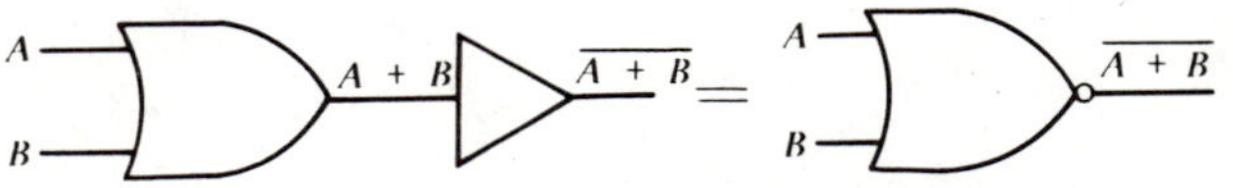

(c) OR-NOT combination for NOR gate

A	B	$A + B$	$\overline{A + B}$
0	0	0	1
0	1	1	0
1	0	1	0
1	1	1	0

(d) Truth table

Figure 11-7 NAND and NOR gates.

A NOR gate consists of an OR gate followed by an inverter, as shown in Fig. 11-7c. The truth table for a 2-input NOR gate is shown in Fig. 11-7d. Note that the output is TRUE only when all of the inputs are FALSE. When one or more than one input is TRUE, the output is FALSE. The equivalent logic equation for the NOR gate is:

$$E_o = \overline{A + B} \tag{11-4}$$

Both NAND and NOR gates are actually derivatives of the three basic logic gates, namely, NOT, AND, and OR. But in practice, the NAND and NOR are the most commonly used gates because of their wide versatility. It is actually possible to implement any logic equation using only NAND gates or only NOR gates (Sec. 11-6). This simplifies the mass production of ICs and brings their cost down considerably.

11-5 Simplification of Logic Expressions

In actual circuit design, each logic expression corresponds to a network of logic gates (Sec. 11-7). Simplifying a logic expression leads to a correspondingly simplified electronic network containing the least number of logic gates.

The simplified expression must be equal to the original expression. The two expressions are equal if they have the same value for every

1	2	3	4	5	6
X	Y	$\overline{Y}$	$X\overline{Y}$	$X\overline{Y} + Y$	$X + Y$
0	0	1	0	0	0
0	1	0	0	1	1
1	0	1	1	1	1
1	1	0	1	1	1

Figure 11-8 Truth table for $X\overline{Y} + Y$ and $X + Y$.

possible combination of the variables. The truth table shown in Fig. 11-8 shows that $X\overline{Y} + Y$ is equal to $X + Y$. In logic algebra, it is written as:

$$X\overline{Y} + Y = X + Y$$

where X and Y are any two variables. Note that columns 5 and 6 in Fig. 11-8 are identical in every respect. Symbols 1 and 0 have the usual connotations of TRUE and FALSE.

The truth-table method as shown in Fig. 11-8 may be used to test the equivalency of any other logic expression or theorem. Some of the logic theorems and laws of Boolean algebra that are used to simplify logic expressions are:

$$X + 0 = X \tag{11-5}$$

$$X + 1 = 1 \tag{11-6}$$

$$X \cdot 1 = X \tag{11-7}$$

$$X \cdot 0 = 0 \tag{11-8}$$

$$X \cdot X = X \tag{11-9}$$

$$X + X = X \tag{11-10}$$

$$X \cdot \overline{X} = 0 \tag{11-11}$$

$$X + \overline{X} = 1 \tag{11-12}$$

$$XY = YX \tag{11-13}$$

$$X + Y = Y + X \tag{11-14}$$

$$(XY)Z = XYZ \tag{11-15}$$

$$(X + Y) + Z = X + Y + Z \tag{11-16}$$

$$X(Y + Z) = XY + XZ \tag{11-17}$$

$$X + YZ = (X + Y)(X + Z) \tag{11-18}$$

$$XY + X\overline{Y} = X \tag{11-19}$$

$$X + XY = X \tag{11-20}$$

$$Y(X + \overline{Y}) = XY \tag{11-21}$$

$$(X + Y)(X + \overline{Y}) = X \tag{11-22}$$

$$X(X + Y) = X \tag{11-23}$$

$$X\overline{Y} + Y = X + Y \tag{11-24}$$

$$XY + YZ + \overline{X}Z = XY + \overline{X}Z \tag{11-25}$$

$$(X + Y)(Y + Z)(\overline{X} + Z) = (X + Y)(\overline{X} + Z) \tag{11-26}$$

$$(X + Y)(\overline{X} + Z) = XZ + \overline{X}Y \tag{11-27}$$

The *inverse* or complement of a Boolean expression is found by applying DeMorgan's law, which is

$$(\overline{X_1 \cdot X_2 \cdot X_3 \cdots X_n}) = \overline{X}_1 + \overline{X}_2 + \overline{X}_3 \cdots + \overline{X}_n \tag{11-28}$$

$$(\overline{X_1 + X_2 + X_3 \cdots + X_n}) = \overline{X}_1\overline{X}_2\overline{X}_3 \cdots \overline{X}_4 \tag{11-29}$$

Example 11-5 shows how the equivalent expressions listed above are used to simplify complicated logic expressions.

Example 11-5

Simplify the following logic expressions:

a. $E_1 = X \cdot 0 + X \cdot 1 + X$

b. $E_2 = AB + A \cdot 0 + B \cdot 1$

c. $E_3 = A\bar{B}C + B + BC + \bar{A}\cdot 0$
d. $E_4 = AB + BC + A\bar{B} + \bar{A}C$
e. $E_5 = A\bar{B}C + ABC\bar{} + \bar{B}C + (\overline{\overline{AB} + \overline{AC}})$
f. $E_6 = \overline{XY\bar{Z}} + (\overline{X + Y}) + \bar{X}(\bar{Y} + Z)$

Solution

a.
$$\begin{aligned} E_1 &= X\cdot 0 + X\cdot 1 + X \\ &= 0 + X + X && \text{(using Eqs. 11-7 and 11-8)} \\ &= \boldsymbol{X} && \text{(using Eq. 11-10)} \end{aligned}$$

b.
$$\begin{aligned} E_2 &= AB + A\cdot 0 + B\cdot 1 \\ &= AB + 0 + B && \text{(using Eqs. 11-7 and 11-8)} \\ &= \boldsymbol{B} && \text{(using Eq. 11-20)} \end{aligned}$$

c.
$$\begin{aligned} E_3 &= A\bar{B}C + B + BC + \bar{A}\cdot 0 \\ &= A\bar{B}C + \underbrace{B + BC} \\ &= A\bar{B}C + B && \text{(using Eq. 11-20)} \\ &= (AC)\bar{B} + B \\ &= \boldsymbol{AC + B} && \text{(using Eq. 11-24)} \end{aligned}$$

d.
$$\begin{aligned} E_4 &= AB + BC + A\bar{B} + \bar{A}C \\ &= \underbrace{AB + BC + \bar{A}C} + A\bar{B} && \text{(rearranging)} \\ &= AB + \bar{A}C + A\bar{B} && \text{(using Eq. 11-25)} \\ &= A + AC && \text{(using Eq. 11-19)} \\ &= \boldsymbol{A + C} && \text{(using Eq. 11-20)} \end{aligned}$$

e.
$$\begin{aligned} E_5 &= A\bar{B}C + (AB\bar{C}) + \bar{B}C + (\overline{\overline{AB}}\cdot\overline{\overline{AC}}) \\ &= A\bar{B}C + AB\bar{C} + \bar{B}C + AB + AC && \text{(double inversion and DeMorgan's law)} \\ &= A\bar{B}C + \bar{B}C + AB + AC \\ &= \bar{B}C + AB + AC \\ &= BA + AC + \bar{B}C && \text{(rearranging)} \\ &= \boldsymbol{BA + \bar{B}C} && \text{(using Eq. 11-25)} \end{aligned}$$

f.
$$\begin{aligned} E_6 &= \overline{XY\bar{Z}} + (\overline{X + Y}) + \bar{X}(\bar{Y} + Z) \\ &= \bar{X} + \bar{Y} + Z + \bar{X}\cdot\bar{Y} + \bar{X}\bar{Y} + \bar{X}Z && \text{(using DeMorgan's law)} \\ &= \bar{X} + \bar{Y} + Z + \bar{X}\cdot\bar{Y} + \bar{X}Z \\ &= \bar{X} + \bar{X}\bar{Y} + \bar{X}Z + \bar{Y} + Z && \text{(rearranging)} \\ &= \bar{X}(1 + \bar{Y} + Z) + \bar{Y} + Z \\ &= \bar{X}(1) + \bar{Y} + Z \\ &= \boldsymbol{\bar{X} + \bar{Y} + Z} \end{aligned}$$

11-5.1 Implementing Logic Expressions Using Logic Gates

Using logic gates to implement a given logic expression is a simple and straightforward process. A logic gate is substituted for each logic operation given in the expression after it has been simplified. The order in which the gates are substituted for follows the general

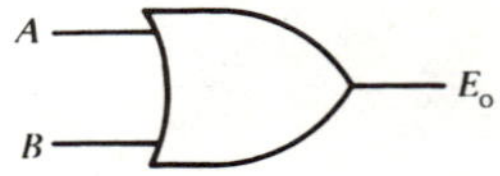

(a) Logic diagram for
$E_o = A + B$

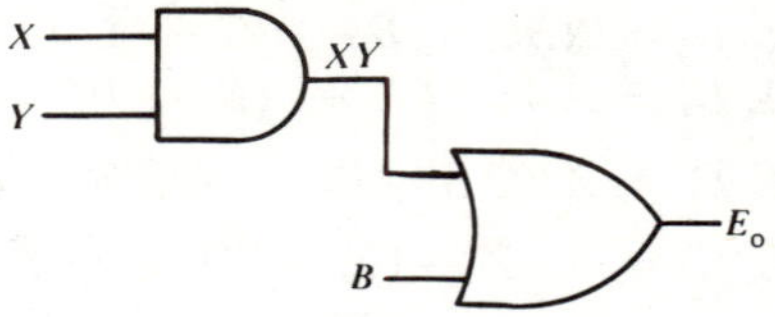

(b) Logic diagram for $E_o = XY + B$

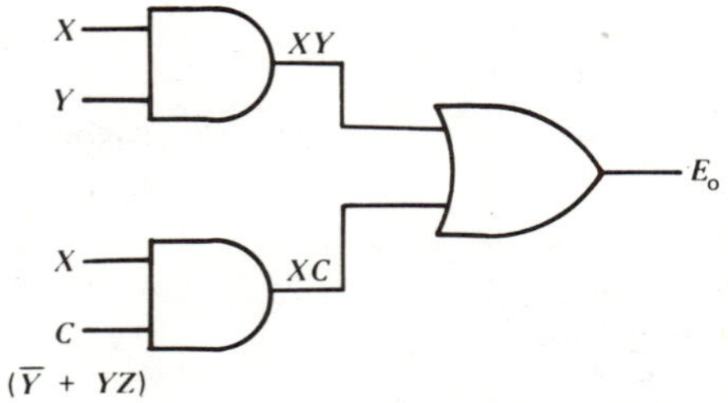

(c) Logic diagram for $E_o = XY + XC$

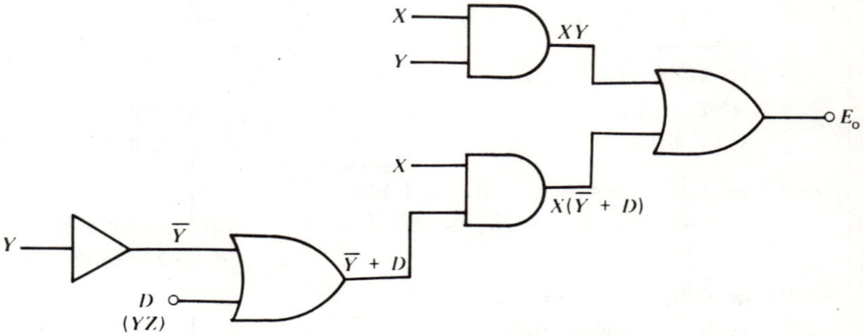

(d) Logic diagram for $E_o = XY + X(\overline{Y} + D)$

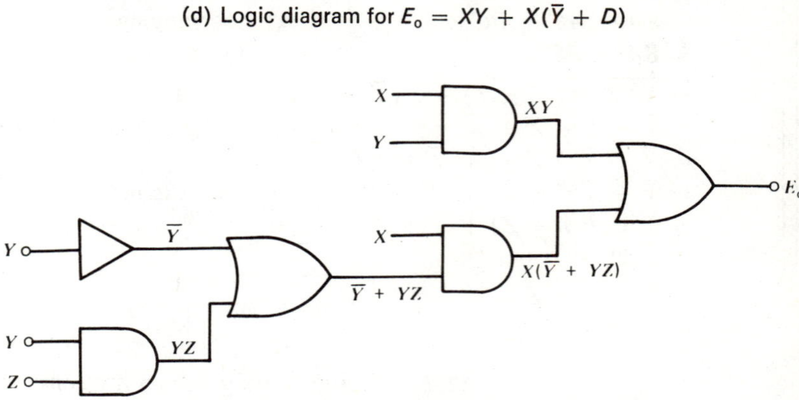

(e) Complete logic diagram for $XY + X(\overline{Y} + YZ)$

Figure 11-9 Step-by-step procedure showing how to implement the logic expression $E_o = XY + X(\overline{Y} + YZ)$.

order of logic operations, which is *parentheses followed by inversion*, AND and OR.

The following step-by-step procedure shows the implementation process. Assume that the expression $E_o = XY + X(\overline{Y} + YZ)$ is to be implemented.

1. This expression may be simplified to the form:

$$E_o = A + B \tag{11-30}$$

where:

$$A = XY \tag{11-31}$$

and

$$B = X(\overline{Y} + YZ) \tag{11-32}$$

The simple OR gate as expressed by Eq. (11-30) is shown in Fig. 11-9a.

2. A is equal to X AND Y (Eq. 11-31). The AND gate representing XY is now added as shown in Fig. 11-9b.

3. B is equal to X AND $(\overline{Y} + YZ)$. The AND gate representing $X(\overline{Y} + YZ)$ is then added as shown in Fig. 11-9c.

4. $\overline{Y}$ OR YZ is implemented next, as shown in Fig. 11-9d.

5. Lastly, an AND gate for YZ is substituted to complete the total gate network as shown in Fig. 11-9e.

Example 11-6

Implement the following Boolean expressions using logic gates:

a. $S = M\overline{N} + P$

b. $E = \overline{A}\overline{B} + \overline{AB} + C$

c. $L = X\overline{Y} + \overline{X}Y + Z(X\overline{Y} + Y\overline{Z})$

Solution

a. See Fig. 11-10a.

b. See Fig. 11-10b.

c. See Fig. 11-10c.

Most logic circuits today contain only either NAND gates or NOR gates. This is possible because a given logic expression may be modified such that it contains only either NAND operations or NOR operations. The process involves the following two steps:

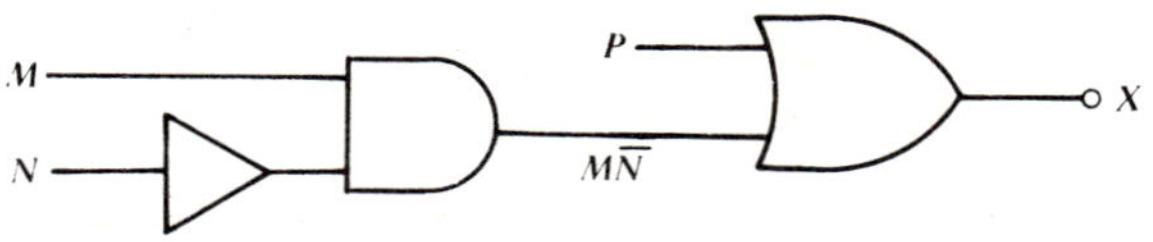

(a) Logic diagram for $X = M\overline{N} + P$

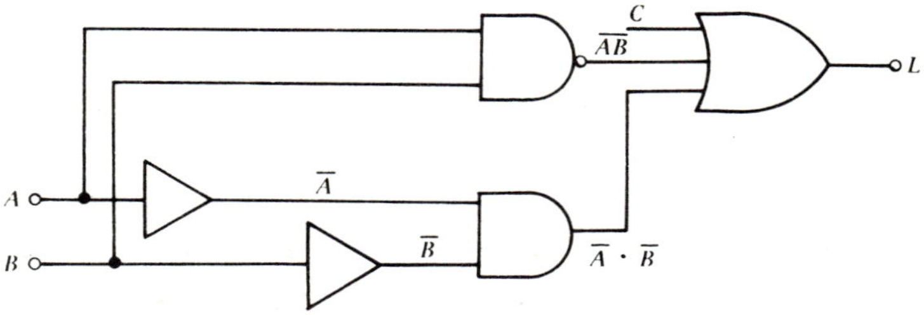

(b) Logic diagram for $L = \overline{A} \cdot \overline{B} + \overline{AB} + C$

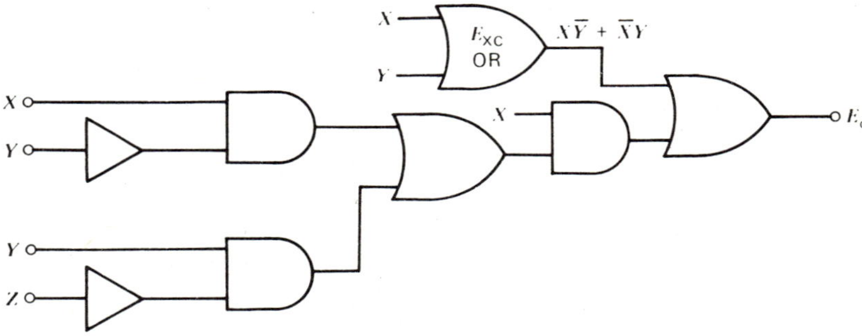

(c) Logic diagram for $E_o = \overline{Y}X + \overline{X}Y + X(X\overline{Y} + Y\overline{Z})$

Figure 11-10 Logic diagrams for Ex. 11-6 (note the EXCLUSIVE OR).

1. Convert the logic expression to either the sum-of-products form or the product-of-sums form. Following are samples of these two forms:

   Sum-of-products form: $L = A\overline{B}C + AC + B\overline{C}$

   Product-of-sums form: $E = (A + \overline{B})(B + C)(\overline{A} + C)$
2. Double invert (invert twice) the expression.

When only NAND gates are to be used, first convert the expression to the sum-of-products form and then double invert it. This process is demonstrated in Ex. 11-7.

Example 11-7

Implement the following logic expression using only NAND gates:

$$S = X\overline{Y}Z + \overline{Y}(Z + \overline{X}) + YZ$$

Solution

1. Convert to sum-of-products form and simplify:

$$\begin{aligned} S &= X\bar{Y}Z + \bar{Y}(Z + \bar{X}) + YZ \\ &= X\bar{Y}Z + \bar{Y}Z + \bar{X}\bar{Y} + YZ \\ &= \bar{X}\bar{Y} + \bar{Y}Z + YZ \end{aligned}$$

2. Double invert.

$$S = \bar{X}\bar{Y} + \bar{Y}Z + YZ$$

$$\bar{S} = (\overline{\bar{X}\bar{Y}})\cdot(\overline{\bar{Y}Z})\cdot(\overline{YZ})$$

$$S = \bar{\bar{S}} = \overline{(\overline{\bar{X}\bar{Y}})\cdot\overline{\bar{Y}Z}\cdot(\overline{YZ})}$$

The above expression contains only NAND gates. It is implemented as shown in Fig. 11-11.

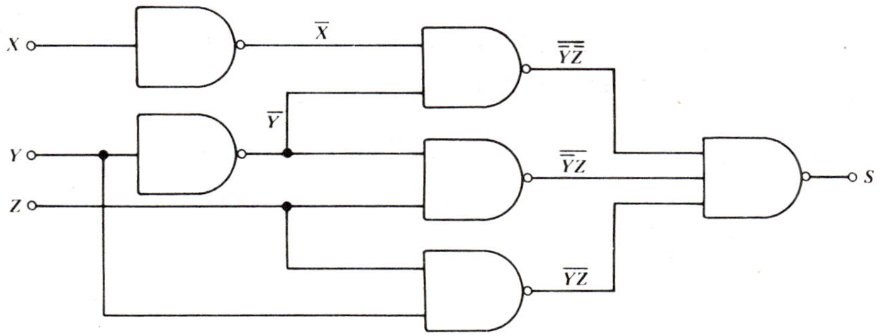

Figure 11-11 Logic diagram for Ex. 11-7.

When only NOR gates are to be used, first convert the expression to the product-of-sums form and then double invert it. The process is demonstrated in Ex. 11-8.

Example 11-8

Implement the following logic expression using only NOR gates:

$$E = AB + \bar{B}C + \bar{A}\bar{B}C + A\bar{B}$$

Solution

1. Simplify the expression (if possible):

$$\begin{aligned} E &= \bar{A}B + \underbrace{\bar{B}C + \bar{A}\bar{B}C}_{} + A\bar{B} \\ &= \bar{A}B + \bar{B}C + A\bar{B} \end{aligned}$$

2. Convert to product-of-sums form:

$$\begin{aligned} E &= \bar{A}B + \bar{B}C + A\bar{B} \\ &= \bar{A}B + \bar{B}(A + C) \\ &= (\bar{A}B + \bar{B})(\bar{A}B + A + C) \qquad \text{(using Eq. 11-18)} \\ &= (\bar{A} + \bar{B})(B + \bar{B})(\bar{A} + A + C)(A + B + C) \\ &= (\bar{A} + \bar{B})(A + B + C) \end{aligned}$$

3. Double invert:

$$\begin{aligned} \bar{E} &= \overline{(\bar{A} + \bar{B})(A + B + C)} \\ &= \overline{(\bar{A} + \bar{B})} + \overline{(A + B + C)} \\ E = \bar{\bar{E}} &= \overline{\overline{(\bar{A} + \bar{B})} + \overline{(A + B + C)}} \end{aligned}$$

The above expression involves only NOR operations. It is implemented as shown in Fig. 11-12. Note that the network contains NOR gates only.

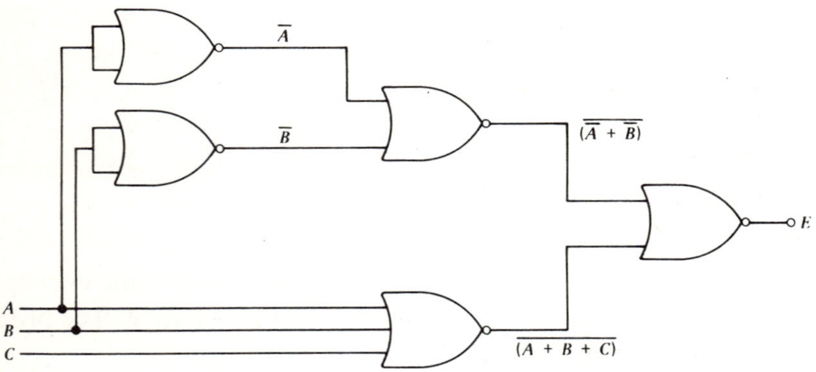

Figure 11-12 Logic diagram for Ex. 11-8.

11-6 Karnaugh Map

The Karnaugh map provides a graphical method for simplifying logic expressions. The map itself is made up of boxes, each of which represents one combination of the variables. The number of boxes required in a particular map depends upon the number of variables in the given expression. Figure 11-13 shows Karnaugh maps for one, two, three, and four variables. Note that a two-variable map requires four (2^2) boxes, a three-variable map requires eight (2^3) boxes, a four-variable map requires sixteen (2^4) boxes, and so on.

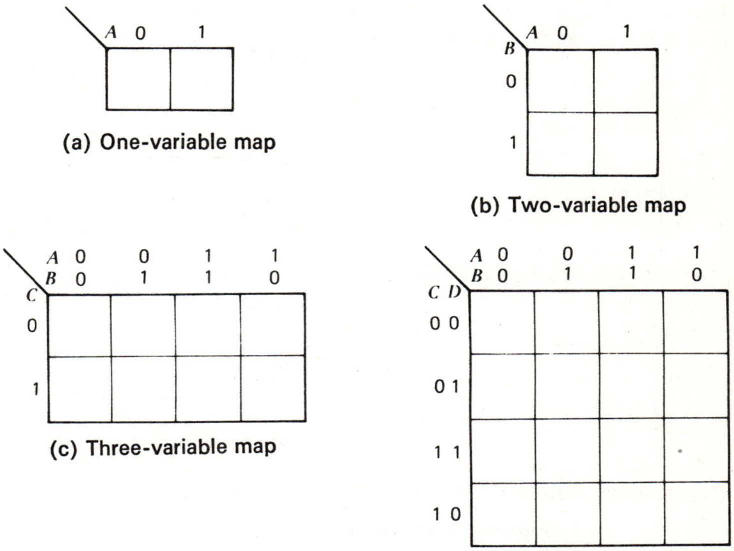

(a) One-variable map

(b) Two-variable map

(c) Three-variable map

(d) Four-variable map

Figure 11-13 One-, two-, three-, and four-variable Karnaugh maps.

Although the Karnaugh map may be used for any number of variables, usually it is only used for two to six variables.

The process of simplifying a logic expression using a Karnaugh map involves the following four steps:

1. Convert the given logic expression to the sum-of-products form, if it is not already in that form (see Sec. 11-5).
2. Draw a Karnaugh map having the appropriate number of boxes.
3. Draw *1*s in boxes that correspond to each product term in the logic expression. Do not draw *1* again if it already has been drawn in a box once.
4. Read the expression off of the map by grouping terms in the simplest way. The idea is to read the maximum number of *1*s in a term. A *1* may be included in one or more than one group (Ex. 11-10).

The logic expression read in step 4 above is the simplified but equivalent version of the given expression. Examples 11-9 and 11-10 demonstrate how the above procedure works.

Example 11-9

Plot the following boolean expressions on Karnaugh maps:

a. $AB + A\bar{B}$

b. $XY + \overline{X}Y + \overline{X}\overline{Y}$
c. $VX\overline{Y}Z + Y\overline{Z} + \overline{X}\overline{Y}Z + X\overline{Y}$

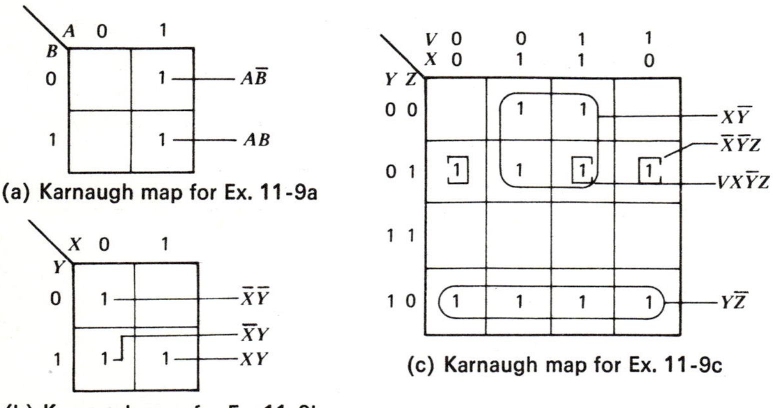

(a) Karnaugh map for Ex. 11-9a

(b) Karnaugh map for Ex. 11-9b

(c) Karnaugh map for Ex. 11-9c

Figure 11-14 Karnaugh maps for Ex. 11-9.

Solution

a. See Fig. 11-14a
b. See Fig. 11-14b
c. See Fig. 11-14c

Example 11-10

Simplify the following Boolean expressions using a Karnaugh map:

a. $E_1 = A(B + \overline{B})$
b. $E_2 = A\overline{B}C + C(\overline{A}B + B) + \overline{A}\overline{B}\overline{C} + \overline{A}\overline{B}$
c. Simplify the expression given in Ex. 11-9c.

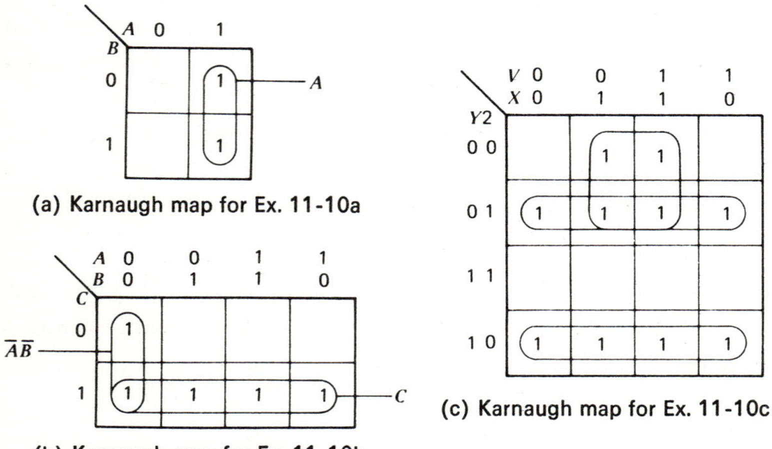

(a) Karnaugh map for Ex. 11-10a

(b) Karnaugh map for Ex. 11-10b

(c) Karnaugh map for Ex. 11-10c

Figure 11-15 Karnaugh maps for Ex. 11-10.

Solution

a. (1) Convert the expression to sum-of-products form:

$$E_1 = A(B + \bar{B})$$
$$= AB + A\bar{B}$$

(2) Draw a two-variable (A and B) Karnaugh map as shown in Fig. 11-15a.

(3) Plot the expression $AB + AB$ in the map as shown in Fig. 11-15a.

(4) Read the map by grouping the maximum number of *1*s in each group. In this case both of the *1*s can be grouped under column A in Fig. 11-15. This is read as:

$$E_1 = A$$

b. (1) Convert the expression to sum-of-products form:

$$E_2 = A\bar{B}C + C(\bar{A}B + B) + \bar{A}\bar{B}C + \bar{A}\bar{B}$$
$$= A\bar{B}C + \bar{A}BC + BC + \bar{A}\bar{B}C + \bar{A}\bar{B}$$

(2) Draw a three-variable map as shown in Fig. 11-15b.

(3) Plot the expression on the map as shown in Fig. 11-15b.

(4) Read the map by looping the maximum number of *1*s in each group. Note that the *1* corresponding to $\bar{A}\bar{B}C$ is common to both of the loops. Loop *1* corresponds to $\bar{A}\bar{B}$ and loop 2 corresponds to C. Therefore,

$$E_2 = \bar{A}\bar{B} + C$$

c. The expression was plotted in Fig. 11-14c. It is replotted in Fig. 11-15c, showing the three *read* loops. Loop 1 corresponds to $Y\bar{Z}$, loop 2 corresponds to $\bar{Y}Z$ and loop 3 corresponds to $X\bar{Y}$. Therefore,

$$\text{Expression} = \text{Loop 1} + \text{Loop 2} + \text{Loop 3}$$
$$= Y\bar{Z} + \bar{Y}Z + X\bar{Y}$$

11-7 Design of Combinational Switching Networks

In practice, a word description of a problem usually is available. Starting with the word description, the design of a combinational switching network involves the following steps:

1. Translate the word problem into a truth table.

2. Assign symbols to various variables in the description.

3. From the truth table, write the equivalent logic equation.
4. Simplify the equation.
5. Implement the logic equation using logic gates.

The following shows a systematic procedure that starts from a word problem and ends up with the required switching network:

The college will hold homecoming *if*:

All the exstudents attend AND the hall is available;

OR

All the 1978 graduates attend AND the dean agrees AND the hall is available.

1. Let:
C = College will hold homecoming (= 1).
$\overline{C}$ = College will not hold homecoming (= 0).
X = All exstudents attend (= 1).
$\overline{X}$ = All exstudents do not attend (= 0).
H = Hall is available (= 1).
$\overline{H}$ = Hall is not available (= 0).
Y = All 1978 graduates attend (= 1).
$\overline{Y}$ = All 1978 graduates do not attend (= 0).
Z = Dean agrees (= 1).
$\overline{Z}$ = Dean does not agree (= 0).

2. The word description is translated into a truth table as shown in Fig. 11-16a. From step 1 above, note that there are four input-variables and one output-variable. The output is marked TRUE (1) when any one of the input conditions are TRUE (1). Rows 13, 14, 15, and 16 are TRUE when the H and X condition is TRUE. Rows 11 and 15 are TRUE when the H, Y, and Z conditions are TRUE.

3. The output is TRUE for rows 11, 13, 14, 15, and 16. Writing the input-variable states for each row:

$$\text{Row 11: } C = H\overline{X}Y\overline{Z}$$

$$\text{Row 13: } C = HX\overline{Y}\overline{Z}$$

$$\text{Row 14: } C = HX\overline{Y}Z$$

$$\text{Row 15: } C = HXY\overline{Z}$$

$$\text{Row 16: } C = HXYZ$$

R. NO	H	X	Y	Z	C
1	0	0	0	0	0
2	0	0	0	1	0
3	0	0	1	0	0
4	0	0	1	1	0
5	0	1	0	0	0
6	0	1	0	1	0
7	0	1	1	0	0
8	0	1	1	1	0
9	1	0	0	0	0
10	1	0	0	1	0
11	1	0	1	0	1
12	1	0	1	1	0
13	1	1	0	0	1
14	1	1	0	1	1
15	1	1	1	0	1
16	1	1	1	1	1

(a) The truth table

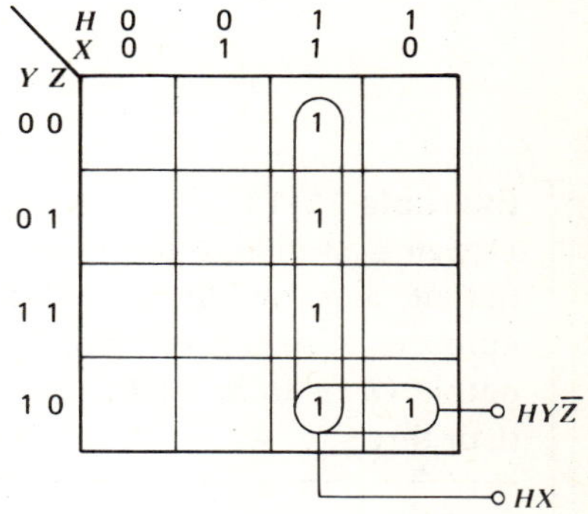

(b) The Karnaugh map

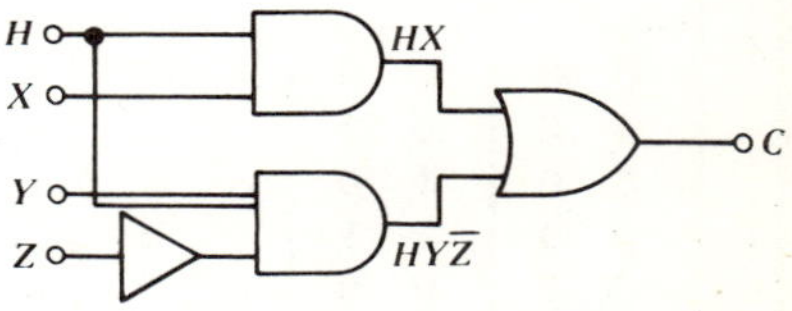

(c) The logic diagram

Figure 11-16 Steps showing the design of a combinational switching network.

C is TRUE when any of the above five logic statements is TRUE. Accordingly, the expression is:

$$C = H\overline{X}Y\overline{Z} + HX\overline{Y}\overline{Z} + HX\overline{Y}Z + HXY\overline{Z} + HXYZ$$

Even though there are *several* ways of writing a logic expression, only *one* is discussed here.

4. Simplify the logic expression:

$$\begin{aligned} C &= H\overline{X}Y\overline{Z} + HX(\overline{Y}\overline{Z} + \overline{Y}Z + Y\overline{Z} + YZ) \\ &= H\overline{X}Y\overline{Z} + HX[\overline{Y}(Z + \overline{Z}) + Y(Z + \overline{Z})] \\ &= H\overline{X}Y\overline{Z} + HX(\overline{Y} + Y) \\ &= H\overline{X}Y\overline{Z} + HX \\ &= H(X + \overline{X}Y\overline{Z}) \\ &= H(X + Y\overline{Z}) \\ &= HX + HY\overline{Z} \end{aligned}$$

See also the Karnaugh map drawn in Fig. 11-16b.

5. A network of logic gates that realizes the above simplified equation is shown in Fig. 11-16c.

Example 11-11

Design a switching network such that it has four inputs and one output, as shown in Fig. 11-17a. N_1 and N_2 represent two binary numbers, each consisting of two bits A B and C D, respectively. The output (T) is to be TRUE only when the sum of $N_1 + N_2$ is greater than three.

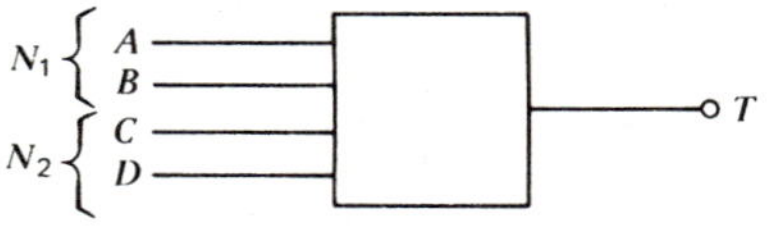

(a) Block diagram

N_1		N_2		Sum	T
A	B	C	D	$N_1 + N_2$	
0	0	0	0	0	
0	0	0	1	1	
0	0	1	0	2	
0	0	1	1	3	
0	1	0	0	1	
0	1	0	1	2	
0	1	1	0	3	
0	1	1	1	4	1
1	0	0	0	2	
1	0	0	1	3	
1	0	1	0	4	1
1	0	1	1	5	1
1	1	0	0	3	
1	1	0	1	4	1
1	1	1	0	5	1
1	1	1	1	6	1

(b) The truth table

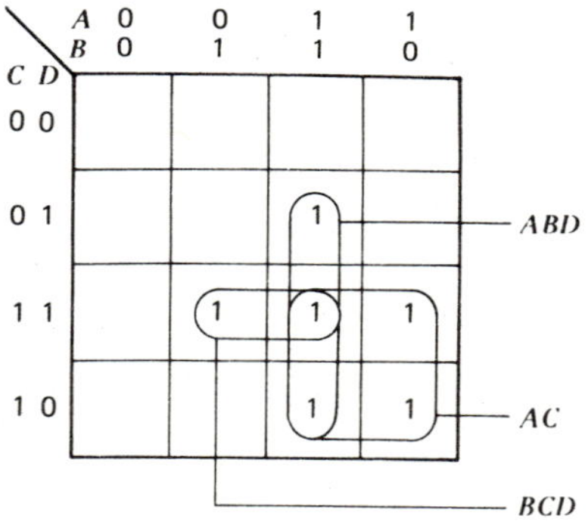

(c) The Karnaugh map

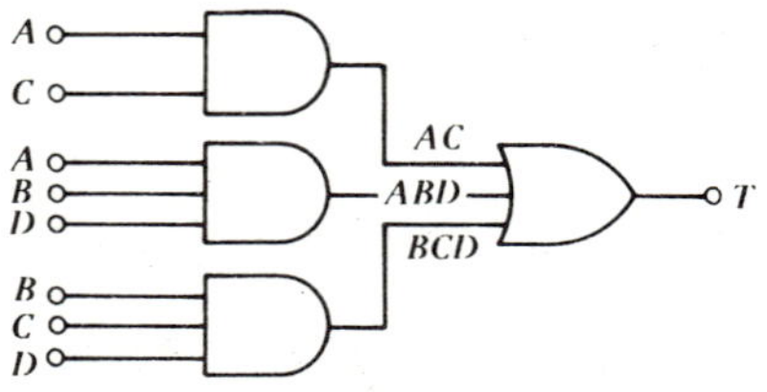

(d) The logic diagram

Figure 11-17 Diagrams for Ex. 11-11.

Solution

1. The truth table is drawn in Fig. 11-17b.
2. The output is TRUE when rows 8, 11, 12, 14, 15, and 16 are TRUE. Consequently, the logic expression is:

$$T = \bar{A}BCD + A\bar{B}C\bar{D} + A\bar{B}CD + AB\bar{C}D + ABC\bar{D} + ABCD$$

3. The Karnaugh map for the above expression is shown in Fig. 11-17c. The simplified expression is:

$$T = AC + ABD + BCD$$

4. The required logic gate switching network is shown in Fig. 11-17d.

11-8 Diode Logic Gates

Diode logic gates use diodes as logic-performing components. The ON/OFF characteristic of a diode is used to accomplish the logic operations.

A two-input diode-AND gate is shown in Fig. 11-18a. A and B are the two inputs. Let us define the two logic states as:

TRUE = 1 = +5 V and FALSE = 0 = 0 V (positive logic)

Also, assume ideal diodes and that $V_{cc} \geqq +5$ V. Note that an ideal diode means that it is equivalent to a closed switch when ON and to an open switch when OFF.
The diode AND circuit operation is:

1. When both A and B are at 0 V, D_1 and D_2 are ON. E_o is at 0 V.
2. When both A and B are at +5 V, D_1 and D_2 are ON. E_o is at +5 V.
3. When A is at 0 V and B is at +5 V, both D_1 and D_2 turn ON immediately after the gate is energized. The forward bias applied across D_1 in this case is higher than that across D_2. Thus, D_1 conducts heavier and turns ON faster than D_2. As D_1 turns ON, 0 V is transferred from A to E_o. Zero volts at E_o reverse-biases D_2 and turns it OFF. E_o remains at 0 V.
4. So we see that 0 V at A and +5 V at B turns D_1 ON and D_2 OFF, resulting in 0 V at E_o.
5. Similarly, when there is +5 V at A and 0 V at B, D_2 is ON and D_1 is OFF, resulting in 0 V at E_o.

The input and output conditions discussed above are tabulated in the truth table of Fig. 11-18b. Note that the output is +5 V only when both (all) inputs are at +5 V. For any other combination of input conditions, E_o is at 0 V. This is exactly how an AND gate operates.

Figure 11-18c shows a 2-input diode-OR gate. It is very similar

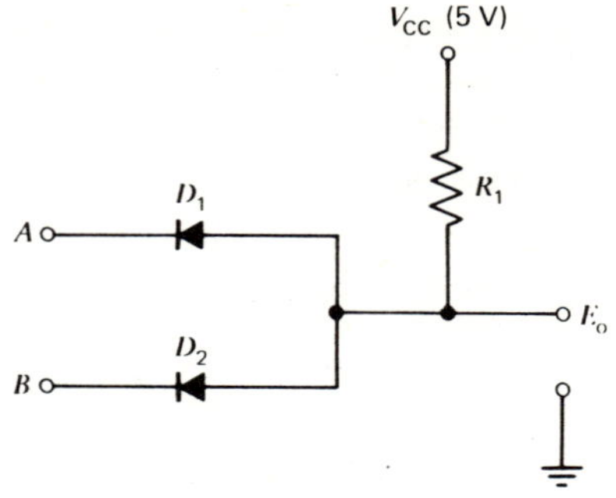

(a) AND gate

A	B	E_o
0	0	0
0	1	0
1	0	0
1	1	1

5 V = 1
0 V = 0

(b) Truth table

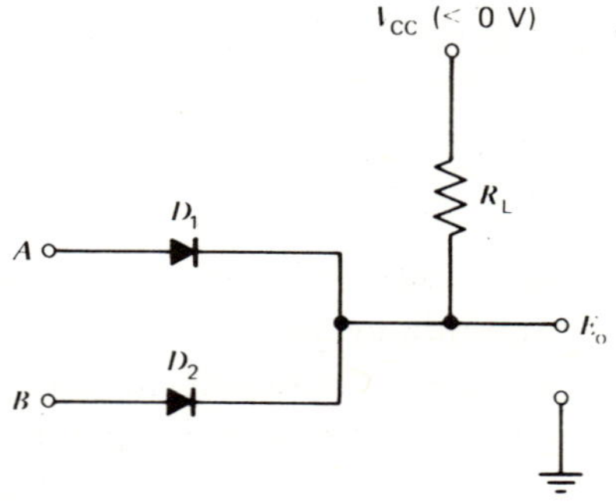

(c) OR gate

A	B	E_o
0	0	0
0	1	1
1	0	1
1	1	1

5 V = 1
0 V = 0

(d) Truth table

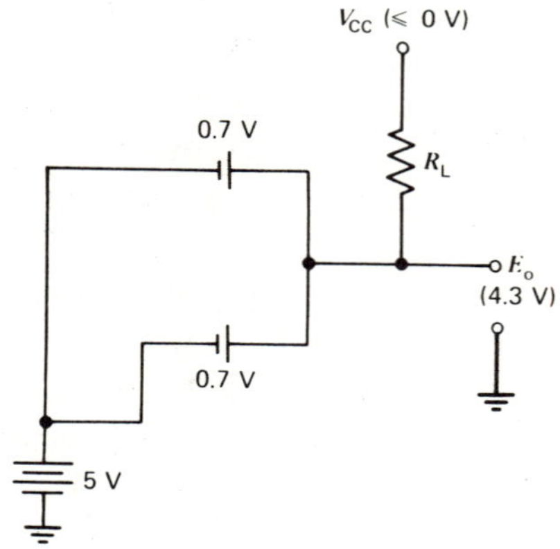

(e) Equivalent circuit when inputs A and B are both at +5 V.

Figure 11-18 Diode gates.

to the AND gate shown in Fig. 11-18a, except that diodes D_1 and D_2 are reversed and $V_{cc} \leqq 0$ V. Again the two logic levels are:

$$\text{TRUE} = 1 = +5\text{ V} \quad \text{and} \quad \text{FALSE} = 0 = 0\text{ V}$$

The diode-OR circuit operation is:

1. When A and B both are +5 V, D_1 and D_2 are ON. E_o is at +5 V.
2. When A and B both are at 0 V, D_1 and D_2 are ON. E_o is at 0 V.
3. When A is at 0 V and B is at +5 V, D_1 turns OFF as D_2 turns ON (conducting heavily), resulting in +5 V at E_o.
4. +5 V at A and 0 V at B turns D_1 ON and D_2 OFF, resulting in +5 V at E_o.

The truth table for the gate of Fig. 11-18c is shown in Fig. 11-18d.

A resistor R_L, shown in both AND and OR gates in Fig. 11-18, is used to limit the current through ON-diodes.

The diode gates discussed above are *not* used commercially because the practical Si diodes, when ON, do have a 0.7 V potential drop across them. Consider again the OR gate of Fig. 11-18c, but this time, assume D_1 and D_2 to be Si diodes. When both A and B are at +5 V, the equivalent circuit shown in Fig. 11-18e applies. E_o is not at +5 V, but rather has degraded to +4.3 V (because of the Si diodes). If such diode gates are cascaded (they usually are), the degradation of the output level accumulates from +4.3 V to +3.6 V to +2.9 V, and so on. This undesirable degradation occurs in diode-OR as well as -AND circuits and makes them unsuitable for practical applications. Another reason for not using discrete-diode gates is the economical and ready availability of TTL gates (Sec. 11-11).

Example 11-12

Consider the two-input OR gate shown in Fig. 11-18c. Assume TRUE = +5 V, FALSE = 0 V, and $V_{cc} = -5$ V. D_1 and D_2 are Si diodes rated at 25 mA each. Calculate the minimum value of R_L required.

Solution

When one or more than one input is at +5 V, the output is at +4.3 V, as shown in Fig. 11-18e. The diode current under these conditions must not exceed 25 mA. The worst-case condition exists when one diode is ON while the other is OFF. For this condition,

$$R_L = \frac{E_o - V_{cc}}{I_{max}}$$

or

$$= \frac{(4.3 + 5)\text{ V}}{25\text{ mA}}$$

$$= \mathbf{372\,\Omega}$$

It is left to the reader to prove that current in each diode is 12.5 mA when both A and B are at +5 V, and 5.8 mA when both A and B are at 0 V.

Example 11-13

The input pulse waveforms shown in Figs. 11-19a and 11-19b are applied to the two inputs (A and B) of the diode-AND gate shown in Fig. 11-18a. Assume ideal diodes.

a. Draw the waveform appearing at E_o.

b. Explain why this gate is sometimes called a *coincidence circuit.*

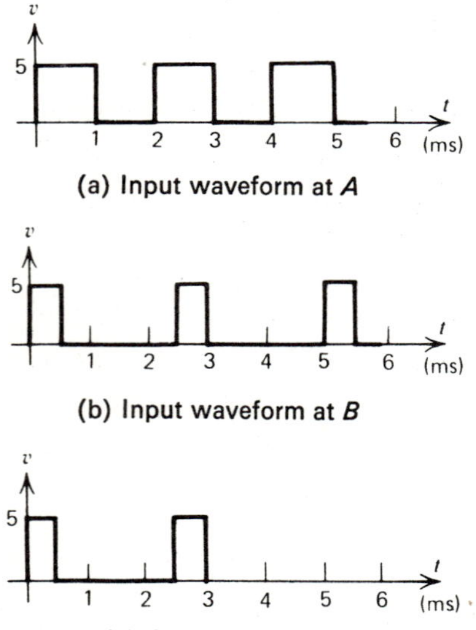

Figure 11-19 Waveforms for Ex. 11-13.

Solution

a. This being an AND gate, the following must be true: $E_o = +5$ V only when $A = +5$ V and $B = +5$ V, and $E_o = 0$ V when either A or B or both are at 0 V. In accordance with this explanation, the output waveform that appears at E_o is shown in Fig. 11-19c.

b. It is also called a coincidence circuit because the output appears only when the two input waves coincide, or appear together at the input.

11-9 Resistor-Transistor Logic (RTL)

There are several so-called commercial families of IC logic gates, which are an improvement over simple diode gates. A family of logic gates is characterized by the constituents of a typical gate circuit. *Resistor-transistor logic* (RTL) gates constitute one such family. All logic gates of this family contain only resistors (R) and transistors (T) in their individual logic (L) circuits.

An RTL NAND gate is shown in Fig. 11-20a. Note that it consists of only resistors and transistors. The two logic levels are:

$$\text{TRUE} = +5\text{ V} \quad \text{and} \quad \text{FALSE} = 0\text{ V}$$

Assume ideal transistors for the description of the operation, given below:

1. R_L is the load resistor used to limit the collector current at saturation. R_A and R_B are the base-current-controlling resistors. Their function is to saturate the corresponding transistor when the input is HIGH (+5 V). The transistor will be OFF when the input is LOW (0 V).

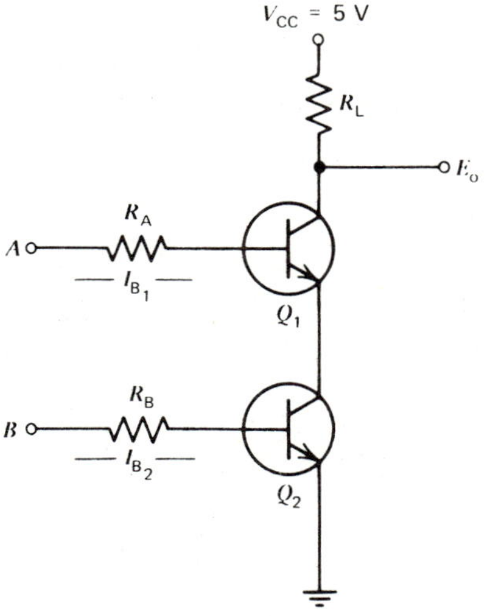

(a) NAND circuit

A	B	$\overline{AB}$
0	0	1
0	1	1
1	0	1
1	1	0

5 V = 1
0 V = 0

(b) Truth table

Figure 11-20 RTL NAND gate.

2. When both of the inputs (A and B) are at 0 V, both Q_1 and Q_2 are OFF (open), and thus E_o is at +5 V. E_o also is at +5 V as long as *either* input A is 0 V (Q_1 open) *or* input B is at 0 V (Q_2 open). This is so because the collector current is blocked as long as either Q_1 is open, or Q_2 is open, or both Q_1 and Q_2 are open.
3. When both A and B are at +5 V, Q_1 and Q_2 both saturate and the saturation collector current flows in R_L. Under this condition E_o is at 0 V.
4. The truth table for this circuit is shown in Fig. 11-20b. It can be seen from the truth table that the circuit is a NAND gate.

Figure 11-21a shows an RTL NOR gate. In this case, the two transistors Q_1 and Q_2 are connected in *parallel.* The circuit operates as follows:

1. When either Q_1 or Q_2 or both Q_1 and Q_2 are ON, E_o is at 0 V.
2. When both Q_1 and Q_2 are OFF, E_o is at +5 V.
3. The truth table for the circuit of Fig. 11–21a is shown in Fig. 11–21b. If +5 V = 1 and 0 V = 0, then the circuit is an RTL NOR gate. Remember that the transistor is ON (saturated) when input is HIGH and OFF when input is LOW.

Commercial logic gates may have any number of inputs. The collector-emitter voltage, V_{CE}(ON), for each *series*-connected transistor becomes a limiting factor on the possible number of inputs. Note that when all the transistors are ON, their combined series-connected V_{CE}(ON) may become significantly higher than 0 V. As a result, appropriate switching may not take place in other

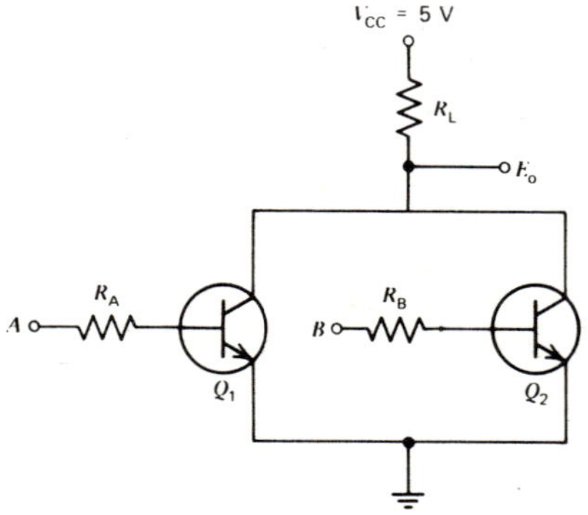

(a) NOR circuit

A	B	$\overline{A+B}$
0	0	1
0	1	0
1	0	0
1	1	0

5 V = 1
0 V = 0

(b) Truth table

Figure 11-21 RTL NOR gate.

related circuits. Therefore, in an RTL NAND gate, transistors with a low V_{CE}(ON) value are used and inputs are limited to some safe number.

In the NOR circuit of Fig. 11-21a, the limiting factor on the number of inputs is the *leakage* current. When all the inputs are LOW, all the transistors are OFF. The total leakage current under this condition is the sum total of all individual parallel-connected leakage currents. The total leakage current flows through R_L and causes a voltage drop, $I_L R_L$, across it. Under this condition, the output voltage, E_o, is:

$$E_o = V_{cc} - I_L R_L \text{ volts (V)} \tag{11-33}$$

where:

V_{cc} is the power supply (V)
I_L is the total leakage current (A)
R_L is the load resistor (Ω)

Example 11-14
Consider the NOR gate circuit shown in Fig. 11-21a. Assume that $R_L = 10\ \text{k}\Omega$ and the leakage current for each transistor is $50\ \mu\text{A}$. Calculate the output voltage level when inputs A and B are both LOW.

Solution
When inputs A and B are both LOW, Q_1 and Q_2 are OFF. Therefore, under this condition, total leakage current flowing through R_L is $100\ \mu\text{A}$.

$$\begin{aligned} E_o &= V_{cc} - I_L R_L && (11\text{-}33) \\ &= 5.0 - 100\ \mu\text{A} \times 10\ \text{k}\Omega \\ &= \mathbf{4.0\ V} \end{aligned}$$

E_o may become significantly lower than 5 V if the combined leakage current becomes sufficient. This of course is undesirable, because a lower voltage level at E_o may not switch other gates. Therefore, in this application, switching transistors of low leakage current are used.

The turn-ON and turn-OFF times of the transistors are improved by connecting capacitors across each input resistor (R_A and R_B). The switching-time-improvement process is similar to the one explained in Sec. 5-11.

11-10 Diode-Transistor Logic (DTL)

Diode-transistor logic is another family of logic gates. A DTL gate is constructed using diodes and transistors as logic elements. The AND/OR operation is accomplished by using diodes (Sec. 11-8), and the NOT operation is accomplished by a transistor inverter.

A DTL NAND gate is shown in Fig. 11-22a, and its associated truth table is shown in Fig. 11-22b. Operation of the diode-AND section was explained in Sec. 11-8, and that of the inverter was explained in Sec. 5-10.

A three-input DTL NOR gate is shown in Fig. 11-22c and its associated truth table is shown in Fig. 11-12d. Again, the operation of the diode-OR gate was explained in Sec. 11-8, and that of the inverter in Sec. 5-10.

A DTL gate switches faster than an RTL gate because the signal passes quickly through the low forward resistance of the diode to the transistor-inverter amplifier. Generally, the diode switching-time is faster than a transistor switching-time, since the latter

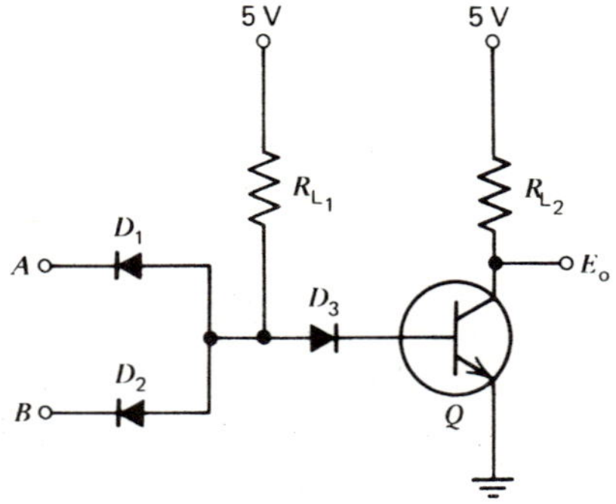

(a) DTL NAND gate

A	B	E_o
0	0	1
0	1	1
1	0	1
1	1	0

5 V = 1
0 V = 0

(b) NAND truth table

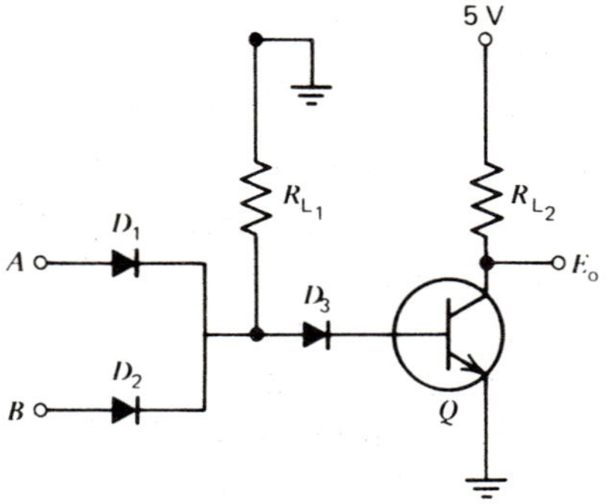

(c) DTL NOR gate

A	B	E_o
0	0	1
0	1	0
1	0	0
1	1	0

5 V = 1
0 V = 0

(d) NOR truth table

Figure 11-22 DTL gates.

must be driven from saturation all the way to cutoff, and vice versa. Techniques to improve the transistor switching time are used, however, in ECL (Sec. 11-12) and CTL (Sec. 11-13).

11-11 Transistor-Transistor Logic (TTL)

The TTL is simply a modification of the DTL. Here, the input diodes of the AND/OR gate are replaced by the emitter of a multi-emitter transistor. The output diode of the DTL gate is replaced by the base-collector (*pn*-junction) diode of the same multi-emitter transistor. To illustrate this point, a three-input diode-AND gate is shown in Fig. 11-23a and its equivalent TTL AND gate is shown in Fig. 11-23b. Note that the four diodes D_1, D_2, D_3, and D_4 have been replaced by a single multi-emitter transistor.

A TTL NAND gate is shown in Fig. 11-24a and its truth table is shown in Fig. 11-24b. The two logic levels are:

$$\text{TRUE} = 1 = +5\text{ V} \quad \text{and} \quad \text{FALSE} = 0 = 0\text{ V}$$

The multi-emitter transistor Q_1 in conjunction with R_{L_1} forms the AND gate and R_{L_2}, R, and Q_2 form the inverter (NOT gate).

Figure 11-25 shows a typical commercially produced two-input TTL NAND gate (7400). A power amplifier stage has been added to the basic NAND circuit. This gives it a higher driving (power) capability. The active pull-up configuration (100 Ω, D_3, Q_3, Q_4) provides low impedance at the output. This results in excellent immunity to **ac** noise. The clamp diodes (D_1 and D_2) at the input limit the overshoot that may otherwise occur.

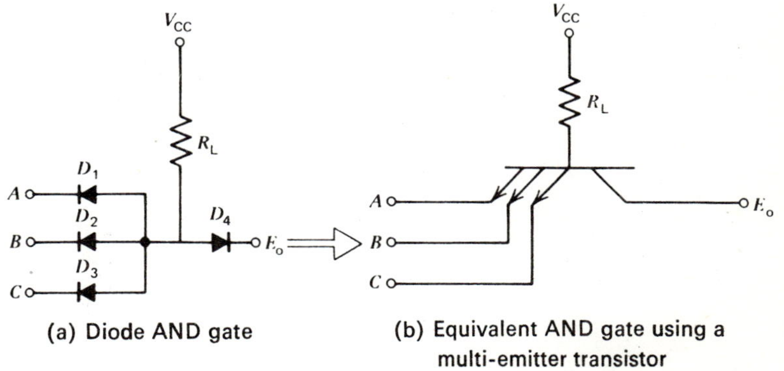

(a) Diode AND gate

(b) Equivalent AND gate using a multi-emitter transistor

Figure 11-23 Diode arrangement and equivalent multi-emitter transistor.

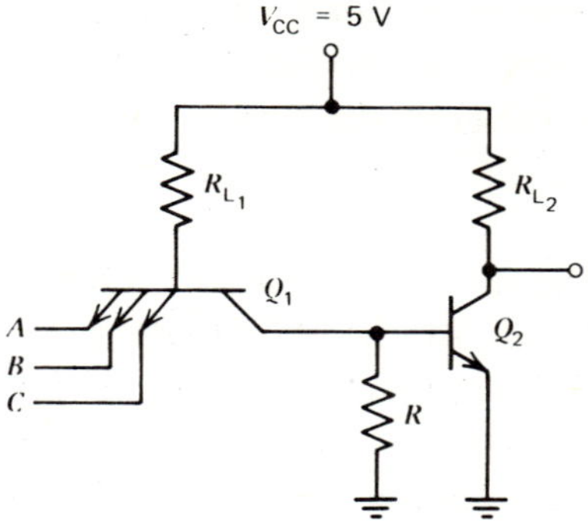

(a) TTL NAND circuit

A	B	C	E_o
0	0	0	1
0	0	1	1
0	1	0	1
0	1	1	1
1	0	0	1
1	0	1	1
1	1	0	1
1	1	1	0

5 V = 1
0 V = 1

(b) Truth table

Figure 11-24 TTL NAND gate.

TTL is the most widely used IC family today. To meet the challenge presented by CMOS (low power-dissipation) and ECL (high switching-speed), improvements have been made in the basic TTL gate described above. The switching speed has been significantly improved by adding a Schottky diode between the base and the collector of the TTL gate. The power dissipation is minimized by choosing high resistances within the gate.

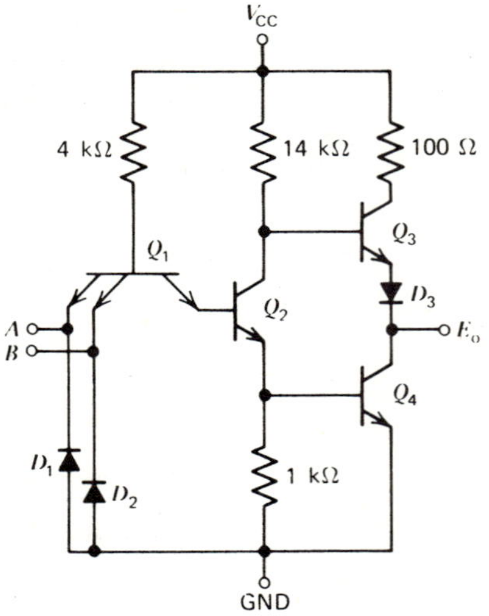

Figure 11-25 Typical IC NAND gate (7400).

11-12 Emitter-Coupled Logic (ECL)

In this family, the emitters of the input logic transistors are coupled to the emitter of a reference transistor. A typical three-input ECL NOR gate is shown in Fig. 11-26a. Q_1, Q_2, and Q_3 are the input-logic transistors. Q_4 is the reference transistor, which in conjunction with R_2 and R_E acts as a constant current source. V_{BB} is the reference voltage. Q_5 and R_3 constitute the power-amplifier output stage.

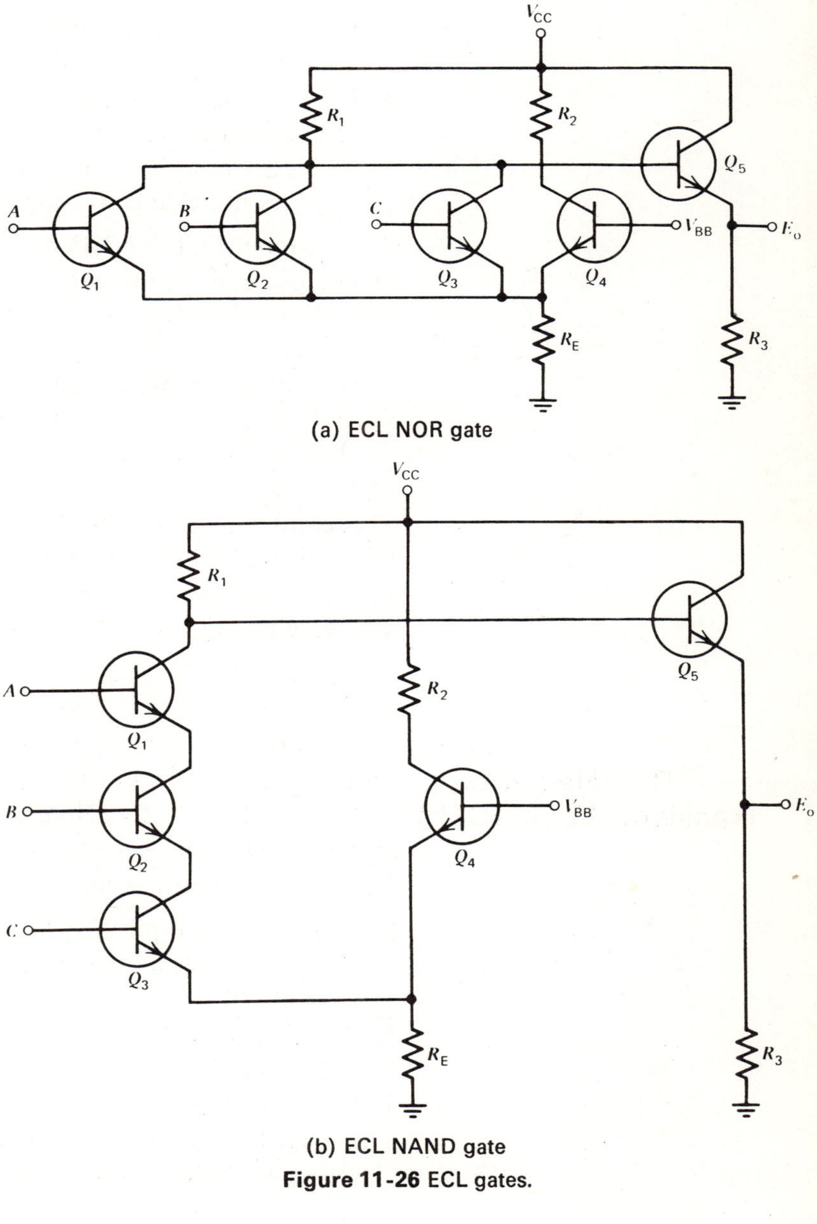

(a) ECL NOR gate

(b) ECL NAND gate

Figure 11-26 ECL gates.

The NOR part of the circuit basically operates in a way similar to the RTL OR gate explained in Sec. 11-9:

1. When one or more than one input is made HIGH (+5 V, which is greater than V_{BB}), the collector current flows through the corresponding transistors.
2. The common collector voltage falls to 0 V. This makes E_o also 0 V (emitter follower).
3. As the current in the input-logic transistors increases, the current in the reference transistor (Q_4) decreases.
4. When all of the inputs are LOW (0 V), Q_1, Q_2, and Q_3 turn OFF and their common collector voltage is +5 V. E_o is also +5 V, because the emitter follower is simply a power amplifier and does not change the logic state.

A three-input ECL NAND gate is shown in Fig. 11-26b. Functions of Q_1, Q_2, and Q_3 are similar to the RTL NAND gate shown in Fig. 11-20a. The functions of Q_4, R_E, Q_5, R_3, and V_{BB} have been explained above.

The emitter coupling used in both the logic stage and the output stage does not allow the transistors to saturate. Thus, the minority carriers do not accumulate at the base-emitter junction (storage time is nonexistent), which results in very fast switching-speeds. The emitter follower at the output provides a low output-impedance, which means that a considerable amount of current may be drawn without loading it. This is a desirable feature, since this gate may be required to drive several other gates.

11-13 Complementary-Transistor Logic (CTL)

Complementary-transistor logic (CTL) is a family that consists of emitter-coupled *pnp* and *npn* transistors. A basic CTL AND gate is shown in Fig. 11-27a. Note that Q_1, Q_2, and Q_3 are *pnp* transistors while Q_4 is an *npn* transistor.

The circuit operates as follows:

1. When all three inputs (A, B, and C) are at 0 V (FALSE), Q_1, Q_2, and Q_3 are ON and the voltage at their common emitter junction (J) is also 0 V.
2. This 0 V, when fed to the common emitter amplifier, keeps Q_4 OFF, which results in 0 V at E_o.
3. When all three inputs are at +5 V (TRUE), Q_1, Q_2, and Q_3 turn OFF, resulting in a potential of +5 V at the junction J as well as at E_o.

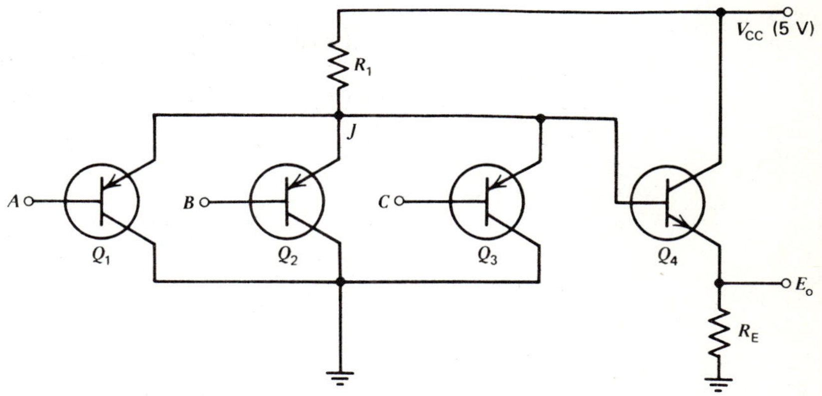

(a) Basic circuit

A	B	C	E_o
0	0	0	0
0	0	1	0
0	1	0	0
0	1	1	0
1	0	0	0
1	0	1	0
1	1	0	0
1	1	1	1

5 V = 1
0 V = 0

(b) Truth table

Figure 11-27 CTL AND gate.

4. When one or more than one input is at 0 V, the corresponding transistors turn ON and the junction voltage as well as E_o becomes 0 V.
5. The above discussion is summarized in the truth table shown in Fig. 11-27b.

In CTL, the transistors never go into saturation when ON. Therefore, their switching time is very fast. The emitter follower at the output ensures that the circuit is able to drive several other gates without degradation of the output voltage because of loading.

11-14 High-Threshold Logic (HTL)

The high-threshold logic (HTL) gates comprise a family of logic IC circuits especially designed to provide immunity against electrical noise. Electrical noise or undesirable signal fluctuations can come either from sources external to the system in question or from within the circuitry itself. These fluctuations may cause erroneous switching of logic circuits. The HTL provides protection against such erroneous switching by providing a higher differential between the two logic-level voltages. Typical HTL logic level voltages are:

$$V_{\mathrm{H}} = 1 = +14\ V \quad \text{and} \quad V_{\mathrm{L}} = 0 = 1\ \mathrm{V}$$

A three-input HTL NAND gate is shown in Fig. 11-28b. A three-input DTL NAND gate is shown in Fig. 11-28a for comparison purposes. Note that the two gates are very similar to each other except for diode D_4. An HTL gate differs from a DTL gate in that:

1. Diode D_4 is a reversed-zener instead of a simple diode, which results in a higher V_{H} switching level.
2. The resistance values of the circuit resistors are a little higher in HTL than in DTL. This reduces the power dissipation in HTL.
3. The magnitude of the supply voltage (V_{cc}) in the HTL is considerably higher than in the DTL.

The circuit operates as follows:

1. When all three inputs (A, B, and C) are HIGH, the voltage at the base of Q_1 is higher than the zener voltage of D_4. Therefore, under this condition,
  a. Q_1 is ON.
  b. D_4 conducts in the reverse direction.

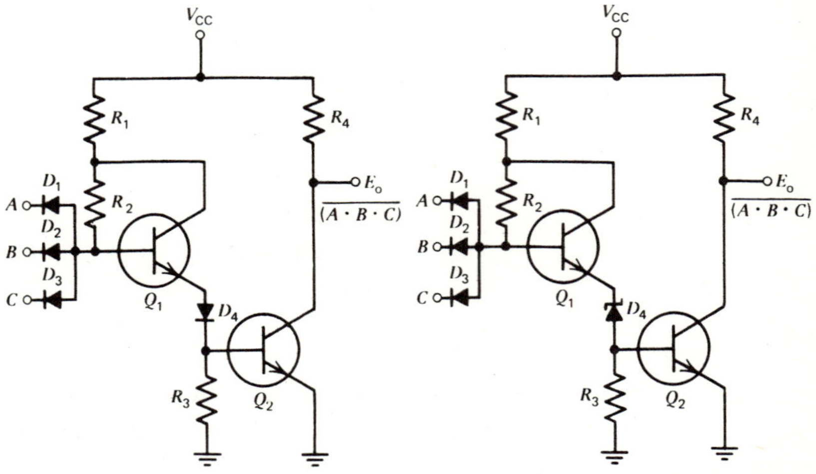

(a) A three-input DTL NAND gate (b) A three-input HTL NAND gate

Figure 11-28 HTL and DTL logic NAND gates.

c. Q_2 is ON.
d. E_o is LOW.

2. When all three inputs are LOW *or* any one of the three inputs is LOW *or* any two of the three inputs are LOW, the base voltage of Q_1 is lower than the zener voltage of D_4. Under any of these conditions,
a. D_4 is OFF.
b. Q_1 and Q_2 are OFF.
c. E_o is HIGH.

11-15 CMOS Logic

The MOSFET switch was introduced in Sec. 5-9 and the CMOS or the COS/MOS inverter was discussed in Sec. 5-10. The complementary or the CMOS-type is the most commonly used circuitry in the fabrication of digital MOSFET ICs. In addition to the inverter, another important building block for the construction of CMOS digital ICs is the *transmission* gate. The basic circuit diagram of a CMOS transmission gate is shown in Fig. 11-29. It operates as follows:

1. When $CL = V_H = V_{DD}$ and $\overline{CL} = V_L = 0$ V, the transmission gate turns ON. Under these conditions, it behaves like a closed switch between the input and output terminals. Depending upon the input voltage level, either the *p*-channel FET is ON or the *n*-channel FET is ON.
a. *p*-channel FET is ON when input is HIGH.
b. *n*-channel FET is ON when input is LOW.

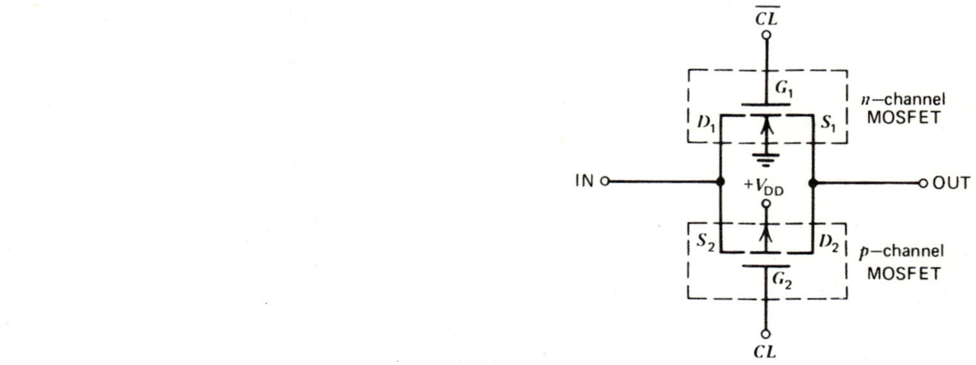

Figure 11-29 A CMOS transmission gate.

2. When $CL = V_L = 0$ V and $\overline{CL} = V_H = V_{DD}$, the transmission gate is OFF. Under these conditions, it behaves like an open switch between the input and output terminals. Regardless of the input voltage level, both the *p*-channel and the *n*-channel FETs remain open. Thus, there is a very high resistance (open switch) between the input and output terminals.

11-15.1 The CMOS Switch A CMOS switch is constructed by connecting an inverter to a transmission gate as shown in Fig. 11-30. The transmission gate:

1. Turns ON (closed switch) when input to the inverter is HIGH.
2. Turns OFF (open switch) when input to the inverter is LOW.

Note that the input to the inverter provides appropriate voltage levels for *CL* and $\overline{CL}$.

All COS/MOS ICs are constructed from two basic building blocks, the *inverter* and the *transmission gate*. The basic inverter is used to construct all NOR and NAND gates. Other, more complicated CMOS IC circuits, such as F/Fs, are formed by combining inverters and transmission gates.

11-15.2 CMOS NOR and CMOS NAND Gates A two-input CMOS NOR gate is shown in Fig. 11-31a. It is basically an inverter consisting of two *n*-channel MOSFETs and two *p*-channel MOSFETs. The *n*-channel MOSFETs are connected in

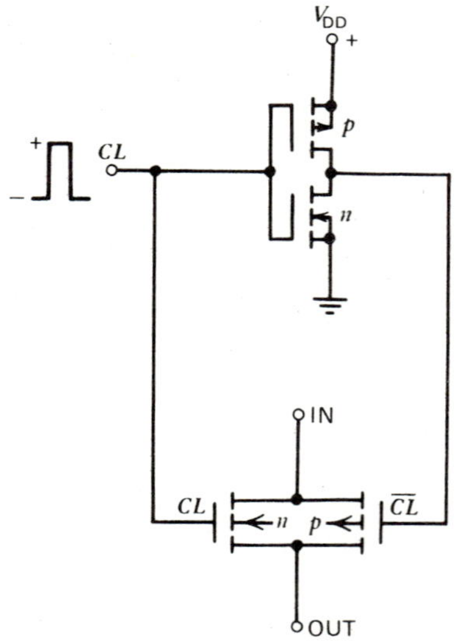

Figure 11-30 Combination of transmission gate and basic inverter to form a switch.

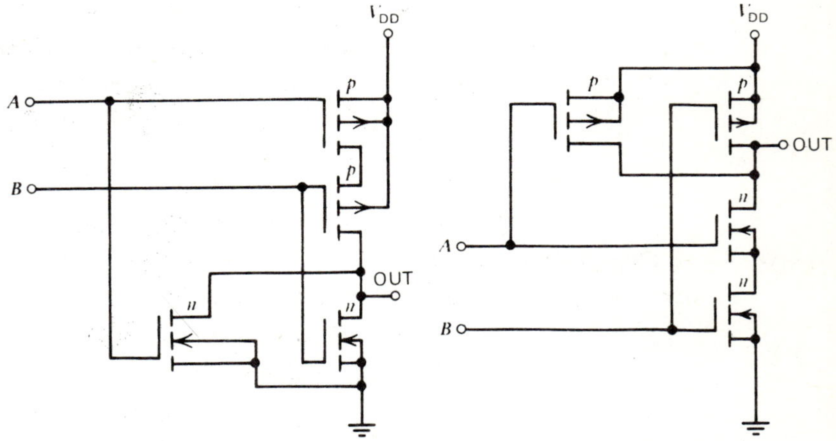

(a) A CMOS NOR gate

(b) A CMOS NAND gate

Figure 11-31 CMOS NOR and NAND gates.

parallel and the two *p*-channel MOSFETs are connected in series. Each input is connected to the gate of one *n*- and one *p*-channel MOSFET. The NOR circuit operates as follows:

1. When either *A* or *B* or both *A* and *B* inputs are HIGH,
  a. The associated *p*-channel MOSFET(s) turns OFF, disconnecting the output from the supply voltage, V_{DD}.
  b. The associated *n*-channel MOSFET(s) turns ON, connecting the output to 0 V.
  c. The output is LOW.
2. When inputs *A* and *B* are both LOW,
  a. Both *p*-channel MOSFETs are ON and the *n*-channel MOSFETs are OFF.
  b. The output is HIGH.

A two-input CMOS NAND gate is shown in Fig. 11-31b. Again, it consists of an inverter having two *p*-channel MOSFETs connected in parallel and two *n*-channel MOSFETs connected in series. Each input is connected to one *p*- and one *n*-channel MOSFET. The NAND circuit operates as follows:

1. When inputs *A* and *B* are both HIGH,
  a. Both *p*-channel MOSFETs are OFF and both *n*-channel MOSFETs are ON.
  b. The ON *n*-channel MOSFETs couple the output to ground.
  c. The output is LOW.

2. When input A or B or both A and B are LOW,
   a. The associated n-channel MOSFET(s) turns OFF and the n-channel MOSFET(s) turns ON.
   b. The output is coupled to V_{DD} and therefore is HIGH.

11-16 Important Characteristics of Logic IC Gates

Today, the TTL IC circuits are the most commonly used circuits in pulse and switching work. Manufacturers usually give quite detailed descriptions of the IC parameters and characteristics. This section discusses *only* the *primary* characteristics of ICs.

1. *Threshold Voltage*. As has been mentioned earlier, two distinct voltage levels are assigned to the two logic states 1 and 0. In fact, though, there is a *certain voltage* range representing the 1 (approximately 2.5 V to 5 V), below which is the 0 (approximately 0 V to 0.8 V) state. The switching action takes place depending upon whether the input is above a minimum of 2.5 V or below 0.8 V. This certain minimum-voltage level at which the circuit switches from one state to the other is called the *threshold voltage* (V_{TH}).

   The threshold voltage is usually given by the manufacturer. If not, one way to approximate it is to simply take the midpoint between the two logic voltage levels. If the two logic levels are 0 V and 5 V, then an approximate threshold voltage is 2.5 V. The IC will not accept voltage below 2.5 V as logical 1. Actual design practice is to choose the logical 1 voltage above 2.5 V and avoid the 1 V to 2.5 V range for any logical level.
2. *Noise Margin*. Noise is any undesired voltage appearing at the input of a logic circuit. A large noise-voltage pulse may inadvertently cause the circuit to switch suddenly and give false information. The difference between the operating voltage level and the threshold voltage is called the *noise margin*.

   Assume that the operating input voltage for the LOW state is 0.8 V and its threshold voltage is 2.5 V. The noise margin is then,

   $$NM = 2.5\text{ V} - 0.8\text{ V} = \mathbf{1.7\text{ V}}$$

   Obviously, the *higher* the noise margin, the less chance there is of false switching as a result of noise.
3. *Propagation Delay*. Regardless of how fast, it takes a certain finite time for a practical logic circuit to switch from one state to another after the input has been applied. Among other things,

the turn-ON or turn-OFF times of a transistor (Sec. 5-11) are the major factors that determine the switching time of a device. The *propagation-delay* time is the total time-delay between the time that the input is applied and the time that the output state gets switched. Note that included in the propagation-delay time is the time that the electric current takes to travel over the connecting wires from input to output (its rate of travel is about 3.33 meters per nanosecond). Typical propagation delays for ICs range from 2 ns to 50 ns.

4. *Fan-In.* A logic IC circuit operates properly so long as the input sink/source current stays below its maximum allowable (rated) value. In practice, the maximum allowable sink/source input current is described in terms of the *maximum allowable inputs*, called the *fan-in* to the IC.

   Assume that fan-in for an IC circuit is 10, with each output rated at 5 mA. Then the maximum allowable input sink/source current is 10 × 5 mA = 50 mA.
5. *Fan-Out.* This is a term used to describe the maximum number of outputs, each rated at certain mA, that an IC logic circuit may have and still operate properly, as in the case of fan-in. Assume that fan-out for a certain IC circuit is 10, with each output rated at 10 mA. Then the maximum allowable output sink/source current is 10 × 10 mA = 100 mA.
6. *Power Dissipation.* The power dissipation of a logic circuit is equal to *one-half* the power that it draws from the supply voltage. It is one-half the power because the logic circuit is assumed to be operating in the 0 state one half of the time and in the 1 state the other half of the time (50 percent duty cycle).

Important characteristics of common IC families are compared in Table 11-1.

Table 11-1

Family	Relative Cost	Propagation Time	Noise Margin	Fan-In	Fan Out	Power Dissipation
RTL	Medium	Medium (12 ns)	Low	Low	Low	Low
DTL	Medium	High (30 ns)	Low	High	Medium	Medium
TTL	Medium	Low (4–12 ns)	High	High	Medium	Medium
HTL	High	Medium (12 ns)	Very high	Medium	Medium	Low
ECL	High	Very low (2 ns)	Medium	Medium	High	High
CTL	High	Low (10 ns)	Medium	Medium	High	High
CMOS	Medium	High (20 ns)	High	High	High	Very low

11-17 Summary

1. Logic or Boolean algebra is based upon the premise that a statement or a condition is either *true* or *false*.
2. There are three basic logic-operations: NOT, AND, and OR. The three derived logic-operations are NAND, NOR, and EXCLUSIVE OR.
3. Logic equations may be written or formed from the given statements or set of conditions.
4. All logic equations may be stated in terms of either only NAND operations or only NOR operations.
5. Electronic circuits used to perform logic operations are called *logic gates*.
6. Logic expressions are simplified using either logic theorems or the Karnaugh map.
7. In practice, a logic expression is first converted to a form in which it contains either only NAND gates or only NOR gates, and then it is realized or implemented by using the appropriate number of gates. The use of only one type of gates keeps the materials cost low.
8. There are several families of logic gates. A family is characterized by the constituents of the typical logic gate. All gates in a family have similar constituent elements. Most common logic families are RTL, DTL, TTL, ECL, CTL, HTL, and CMOS.
9. In addition to the two voltage-logic levels of an IC gate, it is important when designing a logic system to consider its supply voltage, fan-in, fan-out, threshold voltage, noise margin, switching speed, and power dissipation.
10. Important logic theorems:

$$X + YZ = (X + Y)(X + Z) \tag{11-18}$$

$$XY + X\overline{Y} = X \tag{11-19}$$

$$X + XY = X \tag{11-20}$$

$$Y + \overline{Y}X = Y + X \tag{11-24}$$

$$XY + YZ + \overline{X}Z = XY + \overline{X}Z \tag{11-25}$$

$$\overline{(X_1 \cdot X_2 \cdot X_3 \cdots X_n)} = \overline{X}_1 + \overline{X}_2 + \overline{X}_3 + \cdots \overline{X}_n \tag{11-28}$$

$$\overline{(X_1 + X_2 + X_3 \cdots + X_n)} = (\overline{X}_1\overline{X}_2\overline{X}_3 \cdots \overline{X}_n) \tag{11-29}$$

11-18 Glossary

AND gate. A logic gate whose output is TRUE only when all of the inputs are TRUE. The output is FALSE when one or more than one input(s) is (are) FALSE.

CMOS. A family of logic ICs that uses complementary MOSFETs to construct basic logic blocks.

Complementary-transistor logic (CTL). A family of logic gates containing a combination of *npn* and *pnp* transistors.

Diode-transistor logic (DTL). A family of logic gates that contains only diodes and transistors as logic elements.

Emitter-coupled logic (ECL). A family of logic gates in which the emitters of the logic transistors are coupled to the emitter of a reference transistor.

EXCLUSIVE OR. A logic gate whose output is TRUE only when the two inputs are different. The output is FALSE when both of the inputs are the same.

Fan-in. The maximum number of inputs that a given logic gate may have and still operate properly.

Fan-out. The maximum number of outputs that a given logic gate may have and still operate properly.

High-threshold logic (HTL). A family of logic gates having a high noise margin.

Integrated circuit (IC). A complete electronic functional circuit usually fabricated by forming all of its constituent components and connectors on one semiconductor material chip.

Karnaugh map. A graphical method used to simplify logic expressions.

Logic algebra. A branch of mathematics based upon the premise that something is either TRUE or FALSE.

Logic gate. An electronic circuit used to perform a logic operation.

Logic levels. Voltage levels assigned to TRUE and FALSE situations.

Logic states. Names assigned to the two logic situations; that is, the TRUE state and the FALSE state.

NAND gate. A logic gate consisting of an AND gate followed by a NOT gate.

Noise margin. The difference (in volts) between the operating voltage and the threshold voltage.

NOR gate. A logic gate consisting of an OR gate followed by a NOT gate.

NOT gate. A logic gate in which the output is the complement (inverse) of the input.

OR gate. A logic gate whose output is FALSE only when all of the inputs are FALSE, and TRUE if one or more than one input(s) is (are) TRUE.

Resistor-transistor logic (RTL). A family of logic gates consisting of resistors and transistors as logic elements.

Threshold voltage. The voltage level at the input at which the switching takes place.

Transistor-transistor logic (TTL). A family of logic gates consisting of only transistors as logic elements.

11-19 Recapitulation

Return to the Objectives and Self-Evaluation Questions given at the beginning of this chapter and see how well you can answer them. If you cannot answer certain questions satisfactorily, place a check mark (√) next to them and review the appropriate parts of the chapter to find the answers. No attempt should be made to solve the Problems (Sec. 11-20) before answering the Self-Evaluation Questions. Ability to solve the Problems should serve as a test of and feedback regarding your understanding of the material presented in this chapter. Answers to the Problems are given on page 528.

11-20 Problems

11-1 In a certain logic circuit,

$$4.5\text{ V} = 1 = \text{TRUE} \quad \text{and} \quad 0.4\text{ V} = 0 = \text{FALSE}$$

There is an inverter in this logic circuit, the output voltage level of which is measured to be 0.4 V. Find the input voltage level to the inverter at that instant.

11-2 Refer to the logic circuit of Prob. 11-1. If there were four series-connected inverters and the input voltage to the first inverter were 0.4 V, find output-voltage levels of each of the four inverters.

11-3 There is a three-input OR gate in the logic circuit of Prob. 11-1. Two of the inputs are at 4.5 V and the third is at 0.4 V. Find the voltage level at the output.

11-4 There is a four-input AND gate in the logic circuit of Prob. 11-1. Two inputs are at 4.5 V and the remaining two are at 0.4 V. There are three outputs (fan-out). Find the voltage level of each output.

11-5 Repeat Prob. 11-4 for a four-input NAND gate instead of an AND gate.

11-6 Repeat Prob. 11-4 for a four-input NOR gate instead of an AND gate.

11-7 Repeat Prob. 11-3 for a three-input NAND gate instead of an OR gate.

11-8 Input levels to a three-input NOR gate are measured to be 4.5 V, 4.5 V, and 4.5 V. Find the output voltage level (assume the logic levels to be as given in Prob. 11-1).

11-9 Simplify the following logic equations (use theorems):

a. $E_1 = AB + A\bar{B} + B.$

b. $E_2 = XYZ + XY + \bar{X}Z + YZ.$

c. $E_3 = (\bar{X} + \bar{Y})(\bar{\bar{X}}\bar{Y} + X\bar{Y}) + \bar{X}Y.$

d. $E_4 = LM\bar{N}P + \bar{L}MP + M\bar{N}L + \bar{N}LP + \bar{N}L.$

11-10 Simplify the following logic equations using the Karnaugh map:
a. $S_1 = A\bar{B}C + AC + A\bar{B}$.
b. $S_2 = \bar{X}\bar{Y}Z + \bar{X}\bar{Y} + \bar{X}Z + \bar{X}\bar{Y}\bar{Z}$.
c. $L_1 = X\bar{Y}Z + XY\bar{Z} + \bar{X}\bar{Y}Z + XYZ + \bar{X}YZ$.
d. $L_2 = A\bar{B}CD + A\bar{B}\bar{C}D + A\bar{B}\bar{D} + ABC$.

11-11 Convert the following expressions to the sum-of-products form:
a. $E_1 = A(BC + \bar{D}) + A\bar{B}(\bar{B} + \bar{B}C)$.
b. $E_2 = (X + \bar{Y})(\bar{Y} + Z)(X + \bar{Y}Z)$.
c. $E_3 = X\bar{Y}Z + X\bar{Y}(\bar{Y}Z + 0) + X\bar{Z}(1 + \bar{Y}Z + X\bar{Y})$.

11-12 Convert the following expressions to product-of-sums form:
a. $E_1 = A\bar{B} + \bar{A}C$.
b. $E_2 = XY + Z + YZ$.
c. $E_3 = X\bar{Y}Z + \bar{X}Y + XZ$.

11-13 Four chairs—*A*, *B*, *C*, and *D*—are placed in a row. Write a logic equation such that it is TRUE (1) if two or more than two adjacent chairs are occupied. Assume "Chair occupied" = 1 and "Chair not occupied" = 0.

11-14 A switching network has three inputs—*X*, *Y*, and *Z*—and one output, E_o. Write a logic expression such that the output is TRUE when two or more than two inputs are TRUE. Simplify the expression.

11-15 A combinational switching network has four inputs—*A*, *B*, *C*, *D*—and two outputs, *X* and *Y*. The *X* output is TRUE if three or more than three inputs are TRUE. The *Y* input is TRUE if two or less than two inputs are FALSE.
a. Write logic expressions for the *X* and *Y* outputs.
b. Simplify the logic expression for the *Y* output.
c. Implement the simplified logic expression for *Y* using NAND gates only.

11-16 A three-input RTL gate is shown in Fig. 11-32. $A = 12$ V, $B = 0$ V, and $C = 12$ V. Find the output voltage level (E_o).

11-17 Refer to Fig. 11-32. Draw a complete truth table for it and find what kind of a logic gate it is if 12 V = TRUE and 0 V = FALSE.

11-18 If the two voltage-logic levels for the TTL NAND circuit shown in Fig. 11-24 are 0 V and 12 V, what should its supply voltage (V_{CC}) be?

11-19 In a certain logic circuit, the two logic levels are:

$$1.5\text{ V} = 1 \quad \text{and} \quad 0.2\text{ V} = 0$$

Find:
a. Its approximate threshold voltage.
b. Its noise margin.

11-20 The threshold voltage for a certain logic circuit is 1.0 V. The voltage level at the low logic state is 0.4 V. Find its noise margin.

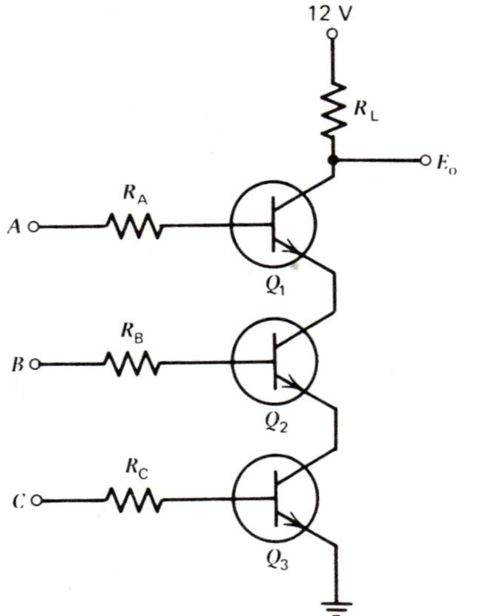

Figure 11-32 RTL gate for Probs. 11-16 and 11-17.

11-21 A maximum of 30 mA may be drawn from a TTL gate without loading it. Each circuit that this gate drives requires 5 mA. Find its fan-out number.

11-22 A TTL NOR gate has a fan-in of 8 with each input requiring 6 mA. Find the total input current that this gate can withstand and still operate properly.

11-23 Refer to Ex. 11-13. If the amplitude of the input waveform being applied at *A* is changed to 0 V and 3 V instead of 0 V and 5 V, how will the output waveform change?

11-21 Experiment

Title **IC Logic Gates**

Theory AND, OR, NOT, NAND, and NOR are five basic logic operations. The electronic circuits used to perform these logic operations are called gates. Only NOT and NAND operations will be used to demonstrate the operation of logic IC gates.

A NOT gate simply inverts the logic state of the input. For a NAND gate:

a. The output is FALSE only when all of the inputs are TRUE.
b. The output is TRUE when one or more than one input is FALSE.

Two TTL ICs, a quad two-input NAND gate and a hex inverter (7400), will be tested in this experiment. The pin-configuration diagrams for the dual-in-line package for these two ICs are shown in Fig. 11-31. Note that in each case, there is more than one individual but identical gate on one IC chip (six NOT gates in the hex inverter). It is suggested that IC bases be used to connect the IC circuits in this experiment.

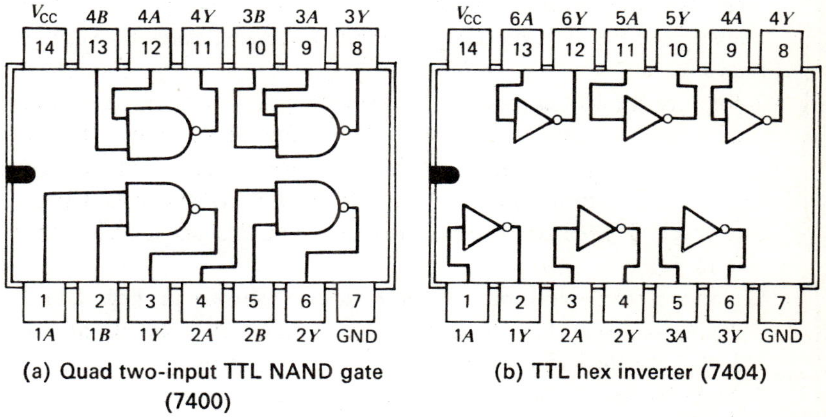

Figure 11-35 Pin-configuration diagram of TTL quad two-input NAND gate (7400) and a hex inverter (7404) in dual-in-line package.

Equipment Power supply: ANATEK model 25–2.00, or equal.
Multimeter: RCA WV98C, or equal.
Hex inverter: 7404, in dual-in-line package.
Quad 2-input NAND gate: 7400, in dual-in-line package.
Two 16-pin dual-in-line type IC bases.
Resistors: 22 Ω, 10 kΩ, 100 kΩ, and 1 kΩ pot.

Procedure

1. Connect one NOT gate of 7404 as shown in Fig. E11-1. Use 4.5 V as the V_{cc}. Note that the numbers in brackets in Fig. E11-1 are the pin numbers.

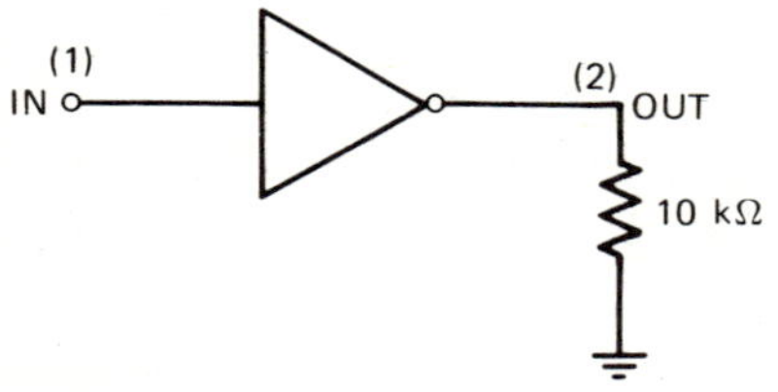

Figure E11-1 One inverter in 7404 for Exp. 11.

2. Apply 0 V (GND) at the input terminal (pin 1). Measure and record in Table E11-1 the output voltage appearing across the 10-kΩ load.
3. Apply 4.5 V at the input terminal. Again measure and record in Table E11-1 the output voltage.
4. Adjust input to 0 V. Slowly increase the input voltage level while continuously monitoring the output voltage. A point will be reached when the output switches from the HIGH to the LOW level. Record in Table E11-2 the input voltage level at this point as V_{TH}.
5. Replace the 10-kΩ load resistance with a series combination of a 22-Ω resistor and a 1 kΩ pot. Apply 0 V at the input. Slowly decrease the pot. resistance while monitoring the output voltage level. The output voltage will begin to drop after a certain point is reached. Measure the pot. resistance at this point and record it in Table E11-3.
6. Connect the two-input NAND gate as shown in Fig. E11-2. Apply 4.5 V at V_{cc}. Note that the numbers in brackets in Fig. E11-2 are the pin numbers.
7. Measure the output voltage when:
   a. $A = 0$ V, $B = 0$ V.
   b. $A = 0$ V, $B = 4.5$ V.

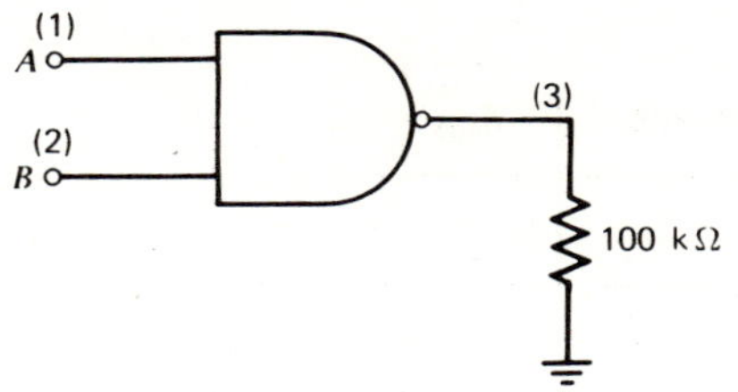

Figure E11-2 One NAND gate in 7400 for Exp. 11.

c. $A = 4.5$ V, $B = 0$ V.
d. $A = 4.5$ V, $B = 4.5$ V.
Record your results in Table E11-4.

8. Connect the output of the two-input NAND gate of Fig. E11-2 to the input of the NOT gate shown in Fig. E11-1. Keep the 10-kΩ load at the output.

9. Repeat Procedure step 7 for the NAND-NOT combination.

10. Disconnect and return all the test equipment to its proper place.

Calculations

1. Calculate the threshold voltage (V_{TH}) as explained in Sec. 11-14. Use the HIGH and LOW values recorded as outputs in Table E11-1 (not 4.5 V and 0 V). Record your results in Table E11-2. Calculate the percentage error between the theoretical and the observed values.

2. Calculate the maximum allowable output current (for one inverter in 7404) from your results as observed in Procedure step 5.

$$I_{max} = \frac{V_H}{22\ \Omega + \text{pot. resistance}}$$

Record your results in Table E11-3.

Tabulation of Results

Table E11-1

Input	Output

Table E11-2

	Observed	Theoretical	Percentage Error
V_{TH}			

Table E11-3

V_H	22 Ω + Pot. res.	I_{max}

Table E11-4

Input *A*	Input *B*	NAND Output Volts	NAND and NOT Output Volts
0	0		
0	4.5		
4.5	0		
4.5	4.5		

Questions

1. Table E11-1 is intended to demonstrate the NOT operation.
   a. Has this been accomplished?
   b. Is this a positive or a negative logic gate?
2. The threshold voltage (V_{TH}) was recorded in Table E11-2. What is its significance in switching circuits?
3. The maximum allowable output-current was recorded in Table E11-3. What is its significance in switching circuits?
4. Table E11-4 is partially intended to demonstrate the NAND operation. Has this been accomplished?
5. The last column in Table E11-4 contains the results of a NAND-NOT combination. Since it is a positive system, what kind of a logic gate does the NAND-NOT combination constitute?

Discussion

1. Compare the experimentally observed results with the theoretically expected ones. Account for discrepancies if there are any.
2. What conclusions do you draw from the results observed in the experiment?
3. Compare the IC gates with the gates that would be made out of discrete components.

Appendix A-1 Derivation of Slope Error Calculations for a Transistor Sweep Generator.

$$v_C = V_{cc} - [V_{cc} - V_{CE}(ON)]e^{-(t/\tau_c)} \tag{8-7a}$$

$$e^{-(t/\tau_c)} = \frac{V_{cc} - v_C}{V_{cc} - V_{CE}(ON)} \tag{A8-1}$$

Differentiating Eq. (8-7a) yields:

$$\frac{dv_c}{dt} = \left[\frac{V_{cc} - V_{CE}(ON)}{\tau_c}\right]e^{-(t/\tau_c)} \text{ (V/s)} \tag{A8-2}$$

Initially, when $t = 0$, the slope is:

$$\left.\frac{dv_c}{dt}\right]_i = \frac{V_{cc} - V_{CE}(ON)}{\tau_c} \text{ (V/s)} \tag{A8-3}$$

And finally, when $t = t_{sp}$, the slope is:

$$\left.\frac{dv_c}{dt}\right]_f = \left[\frac{V_{cc} - V_{CE}(ON)}{\tau_c}\right]e^{-(t_{sp}/\tau_c)}$$

$$= \frac{V_{cc} - V_{CE}(ON)}{\tau_c} \times \frac{V_{cc} - v_C}{V_{cc} - V_{CE}(ON)}$$

$$= \frac{V_{cc} - v_C}{\tau_c} \text{ (V/s)} \tag{A8-4}$$

$$\text{Slope error} = \frac{\text{Initial slope} - \text{Final slope}}{\text{Initial slope}} \times 100$$

Substituting Eqs. (A8-3) and (A8-4), we get:

$$= \frac{v_c - V_{CE}(ON)}{V_{cc} - V_{CE}(ON)} \times 100 \tag{8-8}$$

Appendix A-2 Derivation of Output Voltage Equations of a Bootstrap Sweep Circuit.

Refer to Fig. 8-7b

$$e_o = AV_{th} - AV_{th}e^{-(t/R_{th}C_i)}$$

Differentiating e_o yields its slope:

$$\frac{de_o}{dt} = \frac{AV_{th}e^{-(t/R_{th}C_i)}}{R_{th} \cdot C_i}$$

or

$$= \frac{AV_{th}e^{-[t/R_{th}C(1+A)]}}{R_{th} \cdot C(1 + A)}$$

For a high-gain amplifier, the above expression for the slope reduces to:

$$\frac{de_o}{dt} = \frac{V_{th}}{R_{th} \cdot C} \qquad \left[\begin{array}{l} \text{since } \dfrac{A}{1 + A} = 1 \\ \text{and } e^{-[t/R_{th}C(1+A)]} \cong 1 \end{array}\right]$$

The slope of the output voltage is constant. Further simplifying the above equation yields:

$$de_o = \frac{V_{th}}{R_{th} \cdot C} \cdot dt$$

$$e_o = \frac{V_{th}}{R_{th} \cdot C} \cdot t_{sp}$$

or

$$= \frac{E_i}{RC} \cdot t_{sp} \text{ volts (V)} \qquad (8\text{-}11)$$

It is left as an exercise for the reader to prove that in this case

$$\frac{E_i}{R} = \frac{V_{th}}{R_{th}}$$

Appendix A-3 Derivation of Expression for the Sweep Time in a 555 Timer Sweep-Generator Circuit.

Refer to the circuit shown in Fig. 9-16a.

$$V_{R_E} = V_{cc} - V_{BE} - V_{R_2}$$

$$= V_{cc} - \frac{V_{cc} R_2}{R_1 + R_2} - V_{BE}$$

$$= \frac{V_{cc}R_1 - V_{BE}(R_1 + R_2)}{R_1 + R_2}$$

$$I_E = \frac{V_{R_E}}{R_E}$$

$$= \frac{V_{cc}R_1 - V_{BE}(R_1 + R_2)}{R_E(R_1 + R_2)}$$

$$t_{sp} = \frac{V \cdot C}{I} \text{ (constant current charging a capacitor)}$$

$$= \frac{\frac{2}{3}V_{cc}R_E(R_1 + R_2)C}{R_1V_{cc} - V_{BE}(R_1 + R_2)} \text{ seconds (s)} \qquad \text{(9-26a)}$$

Appendix A-4 Derivations of z_o, t_d, and f_c for an Electromagnetic Delay Line.

In addition to the discussion presented in Sec. 10-7:

$$I_o = \frac{E_i}{Z_i}$$

$$= \frac{E_i}{Z_o} \qquad (Z_i = Z_o)$$

When a capacitor is charged by a constant (**dc**) current, then the voltage buildup across the capacitor is given by:

$$v_c = \frac{I}{C} t \text{ volts (V)} \qquad (1\text{-}8)$$

The capacitor charges to E_i in time t_d by a constant current I_o.

$$E_i = \frac{I_o}{C} \cdot t_d \qquad (\text{A10-1})$$

As the capacitor voltage increases (charges), voltage across the inductance (L) decreases.

$$V_L = L \frac{di}{dt} \qquad (1\text{-}11)$$

$$= L \frac{\Delta I}{\Delta T} \qquad (\text{A10-2})$$

ΔI denotes change in current in the inductor and ΔT denotes the corresponding time elapsed. For the conditions under consideration:

$$V_L = E_i$$

$$\Delta T = t_d$$

$$\Delta I = I_o - 0$$

$$= I_o$$

Equation (A10-2) becomes:

$$E_i = L\frac{I_o}{t_d} \tag{A10-3}$$

Solving Eqs. (A10-1) and (A10-3) yields:

$$\frac{t_d}{C} = \frac{L}{t_d}$$

$$t_d = \sqrt{LC}\text{ seconds (s)} \tag{10-11}$$

Substituting Eq. (10-11) into Eq. (A10-3) yields:

$$E_i = L\frac{I_o}{t_d}$$

or

$$\frac{E_i}{I_o} = \frac{L}{\sqrt{LC}}$$

or

$$Z_o = \sqrt{\frac{L}{C}}\text{ ohms }(\Omega) \tag{10-10}$$

Note that the equivalent circuit shown in Fig. 10-8a is a low-pass filter. Let us find its upper-half power point or the cutoff frequency f_c. Assume:

$$E_i = \text{input voltage}$$

$$E_o = \text{output voltage}$$

$$i = \text{current in the circuit } (L/2 \text{ and } C)$$

$$i = \frac{E_i}{j\left(2\pi f\,\frac{L}{2} - X_c\right)}$$

$$E_o = i(-jX_c)$$

$$= \frac{E_i}{\left(\frac{2\pi fL}{2} - X_c\right)}\cdot(-jX_c)$$

$$\frac{E_o}{E_i} = \frac{-jX_c}{j\left(\frac{2\pi fL}{2} - X_c\right)}$$

At the half-power point:

$$\frac{E_o}{E_i} = 0.707$$

$$0.707 = \frac{-jX_c}{j(\pi f_c L - X_c)}$$

$$0.707\pi f_c L - 0.707X_c = X_c$$

$$0.707\pi f_c L = 1.707X_c$$

since

$$X_c = \frac{1}{2\pi f_c C}$$

$$\therefore f_c = \left(\sqrt{\frac{1.707}{1.414}}\right)\left(\frac{1}{\pi}\right)\left(\frac{1}{\sqrt{LC}}\right)$$

$$= \left(\frac{1.1}{\pi}\right)\left(\frac{1}{\sqrt{LC}}\right) \text{ hertz (Hz)} \qquad (10\text{-}18)$$

Appendix A-5 The Binary Number System.

The decimal number system used in everyday arithmetic today has *ten* different *symbols*, namely, 0, 1, 2, 3, 4, 5, 6, 7, 8, and 9. A decimal number is expressed by using these symbols in a sequence. The position of a symbol or *digit* in the arithmetic expression (number) determines its *weight*. For example:

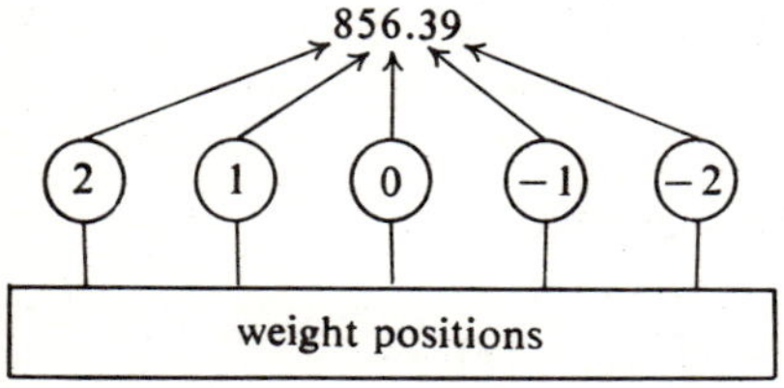

$$= 8 \times 10^{2} + 5 \times 10^{1} + 6 \times 10^{0} + 3 \times 10^{-1} + 9 \times 10^{-2}$$

A *binary* number system uses only *two* symbols, 0 and 1. These symbols, sometimes called *bits*, are arranged in a sequence to express a number. As in the decimal system, each bit in a binary number has some assigned weight in accordance with its position (location) in the number.

The two symbols (0 and 1) used in the binary system are ideally suited for the two logic states (FALSE and TRUE) in Boolean algebra. Usually, a TRUE situation is symbolized by 1 and a FALSE by 0. Also, the two states of a switch (OFF and ON) correspond to the FALSE and TRUE situations in Boolean algebra. Modern logic circuits are the result of the above-mentioned direct correspondence among the binary numbers, Boolean algebra, and the switch.

$$\text{Boolean algebra}\left\{\begin{matrix}\text{TRUE} = 1\\ \text{FALSE} = 0\end{matrix}\right\} \begin{matrix}= \text{ON}\\ = \text{OFF}\end{matrix}\left.\vphantom{\begin{matrix}1\\0\end{matrix}}\right\}\text{switch}$$

$\uparrow$
binary

Other number systems occasionally encountered are the *octal* system (8 symbols) and the *hexadecimal* system (16 symbols). Table A5-1 shows the decimal numbers 0 to 19 and their equivalents in binary, octal, and hexadecimal systems.

Table A5-1

Decimal	Binary	Octal	Hexadecimal	Decimal	Binary	Octal	Hexadecimal
0	0	0	0	10	1010	12	A
1	1	1	1	11	1011	13	B
2	10	2	2	12	1100	14	C
3	11	3	3	13	1101	15	D
4	100	4	4	14	1110	16	E
5	101	5	5	15	1111	17	F
6	110	6	6	16	10000	20	10
7	111	7	7	17	10001	21	11
8	1000	10	8	18	10010	22	12
9	1001	11	9	19	10011	23	13

From the "9" row above, it can be seen that:

$$(9)_{10} = (1001)_2 = (11)_8 = (9)_{16}$$

The subscripts 10, 2, 8 and 16 indicate the number of symbols in that system. The number of symbols in a number system is sometimes called the *radix* or the *base* of that system. The radix of binary is 2, of octal is 8, of decimal is 10, and so on.

Binary-to-Decimal Conversion

As has been shown earlier, each digit in a decimal number system has certain weight in accordance with its location in the number. The same is true for a binary number. Shown below are the binary-to-decimal conversion process and the weights assigned to various bits in a binary number.

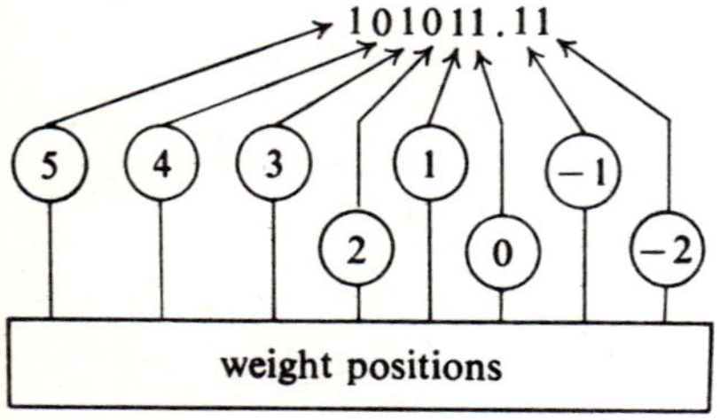

$$= 1 \times 2^5 + 0 \times 2^4 + 1 \times 2^3 + 0 \times 2^2 + 1 \times 2^1 + 1 \times 2^0 + 1 \times 2^{-1} + 1 \times 2^{-2}$$

$$= 32 + 0 + 8 + 0 + 2 + 1 + 0.5 + 0.25$$

$$= 43.75$$

Another way to express this equivalency is:

$$(101011.11)_2 = (43.75)_{10}$$

Decimal-to-Binary Conversion

The decimal-to-binary conversion process is essentially the reverse of the binary-to-decimal conversion process. Assume that $(50.625)_{10}$ is to be converted into its binary equivalent number.

1. Break the integer 50 into powers of 2, as follows:

$32 = 1 \times 2^5$ ←(most significant bit)
$16 = 1 \times 2^4$
$0 = 0 \times 2^3$
$0 = 0 \times 2^2$ $\quad 32 + 16 + 2 = 50 = 110010$
$2 = 1 \times 2^1$
$0 = 0 \times 2^0$ ←(least significant bit)

2. Break the decimal fraction (0.625) into powers of 2, as follows:

$0.5 = 1 \times 2^{-1}$ ←(most significant bit)
$0 = 0 \times 2^{-2}$
$0.125 = 1 \times 2^{-3}$ ←(least significant bit)
$0.5 + 0.125 = 0.625 = 0.101$

Therefore,

$$(50.625)_{10} = (110010.101)_2$$

Another method of converting a decimal number into its binary equivalent is to:

1. Convert the decimal integer by repetitive division by 2 (radix of binary) until the integer quotient is zero.
2. Convert the decimal fraction by repetitive multiplication by 2 (radix of binary) until the fraction is equal to zero.

Let us again convert $(50.625)_{10}$ into its binary equivalent by this method.

1. Repetitive division by 2 (integer).

		Remainder
2	50	
2	25	→0←least significant bit
2	12	→1
2	6	→0
2	3	→0
2	1	→1
2	0	→1←most significant bit

From above,

$$(50)_{10} = (110010)_2$$

2. Repetitive multiplication by 2 (fraction):

$$
\begin{array}{rr}
 & 0.625 \\
 & 2 \\
\hline
\text{most significant bit} \rightarrow 1 \leftarrow & \textcircled{1}.250 \\
 & 2 \\
\hline
0 \leftarrow & \textcircled{0}.500 \\
 & 2 \\
\hline
\text{least significant bit} \rightarrow 1 \leftarrow & \textcircled{1}.000
\end{array}
$$

(Note that each time only the fraction part is multiplied by 2.)
From above,

$$(0.625)_{10} = (0.101)_2$$

Combining the above two results:

$$(50.625)_{10} = (110010.101)_2$$

Answers

Chapter 1 **1-1** $I_{8k} = 0.75$ mA.

1-2 $I_{8k} = 2.5$ mA, $I_{5k} = 1.67$ mA, $I_{10k} = 0.83$ mA.

1-3 $E_i = 33$ V, $V_{8k} = 20$ V, $V_a = 13$ V, $V_{ab} = 8$ V.

1-4 $I_{5k} = 0.86$ mA, $I_{6k} = 0.72$ mA, $I_{3k} = 1.3$ mA, $I_{15k} = 0.263$ mA.

1-5 $Q = 18.8\ \mu C$.

1-6 $V_c = 12.5$ kV.

1-7 $I_C = 22.6 \angle 90^\circ$ mA.

1-8 $I = 0.59 \angle 28^\circ$ mA.

1-9 $C = 531$ pF.

1-10 $X_L = j62.8\ k\Omega$.

1-11 $V_L = 9.4 \angle 38.5^\circ$ V.

1-12 $I_{2k} = 1.17$ mA.

1-13 $V_{6k} = 0.94$ V.

1-14 $I_{2k} = 1.17$ mA.

1-15

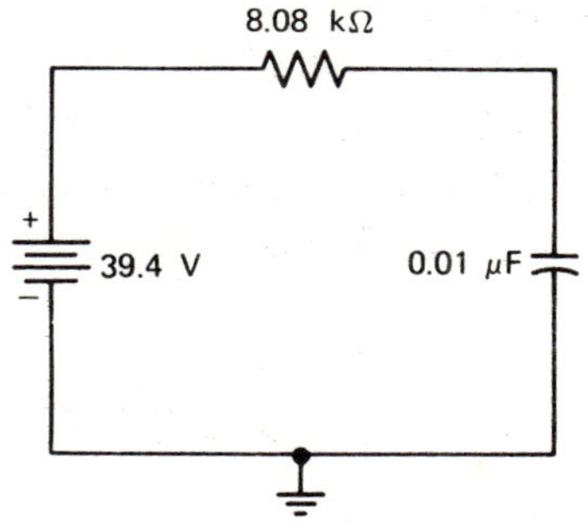

Figure 1-24 Answer to Prob. 1-15.

Chapter 2 **2-1**

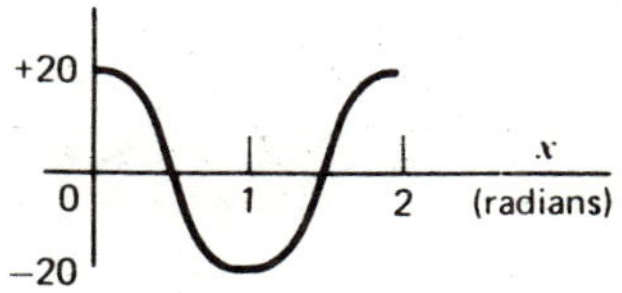

Figure 2-21 Answer to Prob. 1-1.

2-2

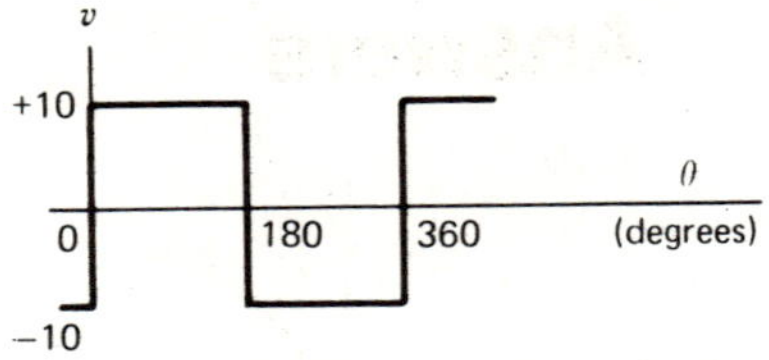

Figure 2-22 Answer to Prob. 1-2.

2-3 $V_{av} = 4$ V.
2-4 $V_{dc} = 5$ V.
2-5 $V_{av} = \dfrac{-4\text{ V}}{5} = -0.8$ V.
2-6 Duty cycle = 20 percent.
2-7 $V_{p\text{-}p} = 14$ V, $t_p = 1$ ms, $t_{rp} = 4$ ms, PRT = 5 ms, PRF = 200 Hz.
2-8 $t_r = 1.5\ \mu$s, $t_f = 1.0\ \mu$s, $t_p = 9.0\ \mu$s.

Chapter 3

3-1 $f_{hc} = 1.59$ kHz.
3-2 $C = 362$ pF.
3-3 $R = 159\ \Omega$.
3-4 $E_o = 7.77$ V.
3-5 $t_{re} = 0.22$ ms.
3-6 Initial slope = 100 kV/s, $t_t = 0.5$ ms, $t_{63\%} = 0.1$ ms.
3-7 Initial slope = 30 V/ms, $t_t = 1.665$ ms, $t_r = 0.733$ ms, $t_{63\%} = 0.333$ ms.

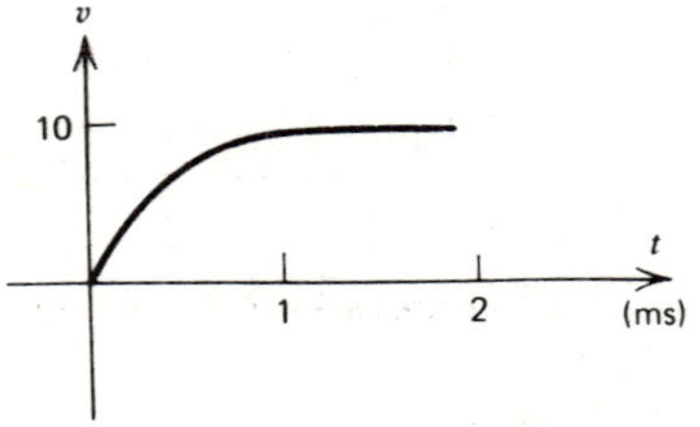

Figure 3-29 Answer to Prob. 3-7.

3-8

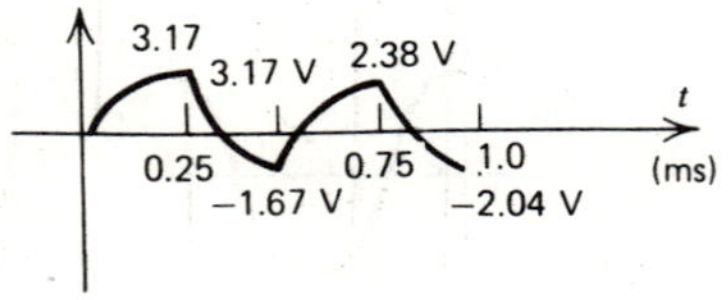

Figure 3-30 Answer to Prob. 3-8.

3-9 $V_{C_{ss}} = \pm 2.15$ V.

3-10

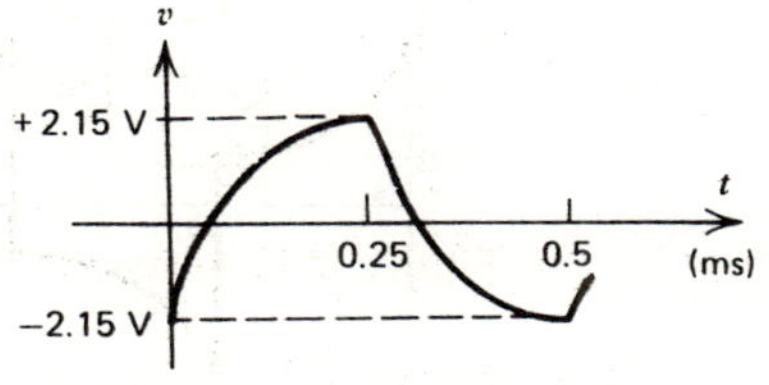

Figure 3-31 Answer to Prob. 3-10.

3-11 Short time constant; output is same as input.

3-12

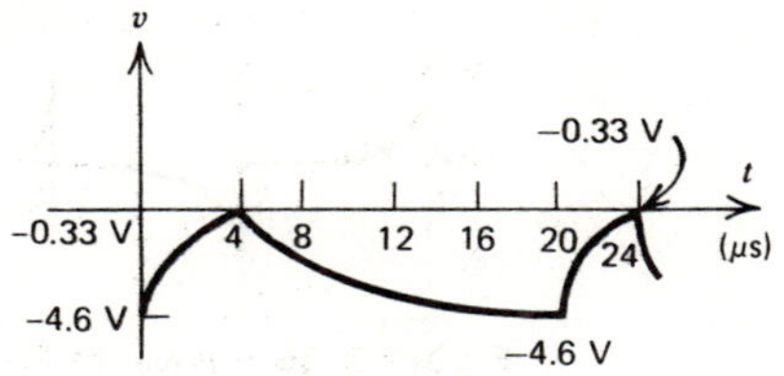

Figure 3-32 Answer to Prob. 3-12.

3-13

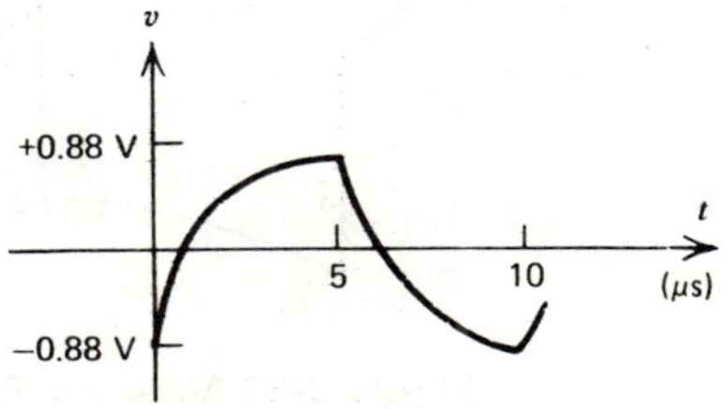

Figure 3-33 Answer to Prob. 3-13.

3-14 $E_o = 17.25$ V.

3-15

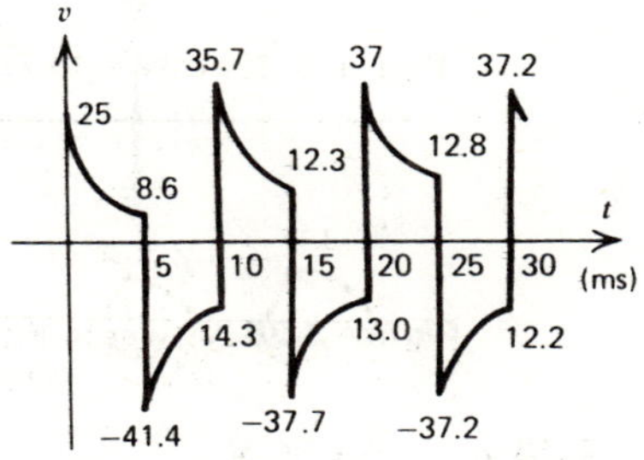

Figure 3-34 Answer to Prob. 3-15.

3-16

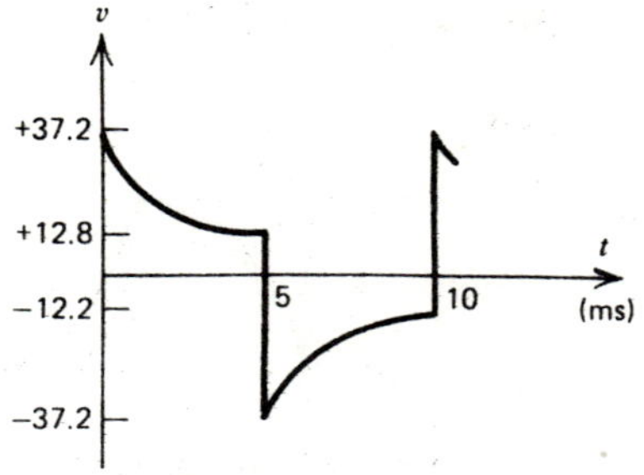

Figure 3-35 Answer to Prob. 3-16.

3-17

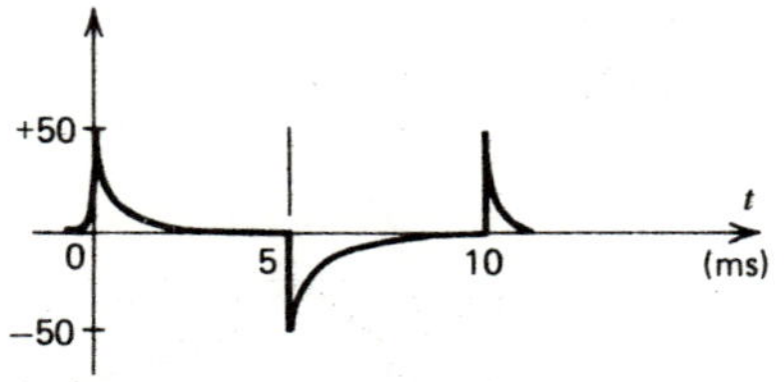

Figure 3-36 Answer to Prob. 3-17 ($t_p > s\tau$).

3-18

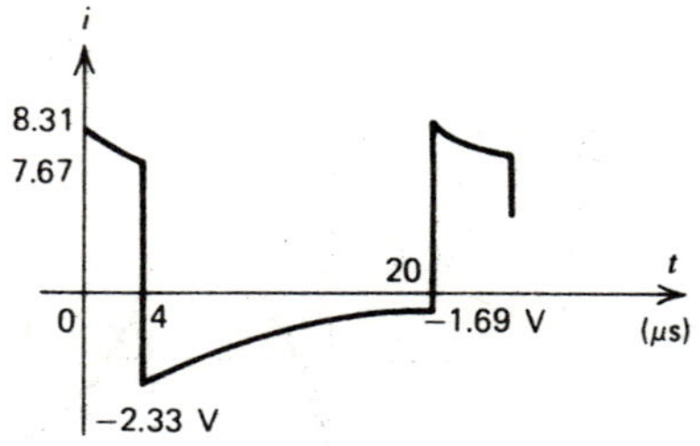

Figure 3-37 Answer to Prob. 3-18.

3-19

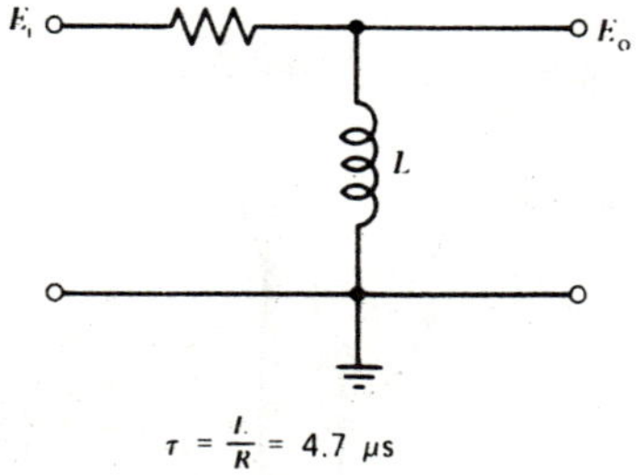

Figure 3-38 Answer to Prob. 3-19.

3-20 $E_o = 34.56$ V.

3-21 Underdamped, $f = 105$ kHz, $R_c = 2.83$ kΩ.

3-22

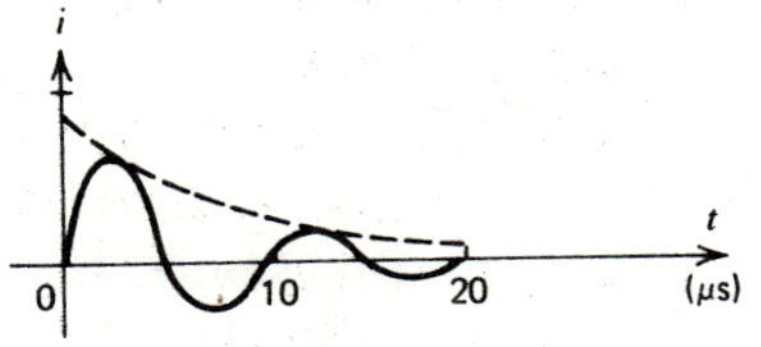

Figure 3-39 Answer to Prob. 3-22.

3-23

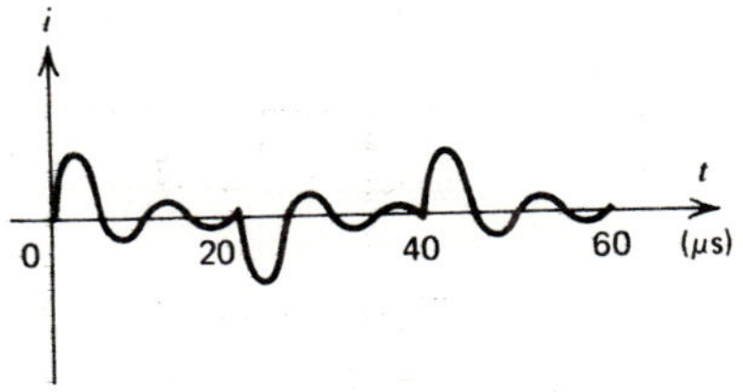

Figure 3-40 Answer to Prob. 3-23.

3-24

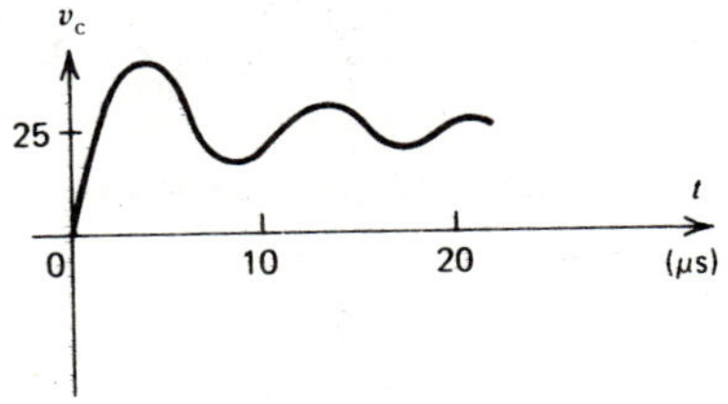

Figure 3-41 Answer to Prob. 3-24.

3-25

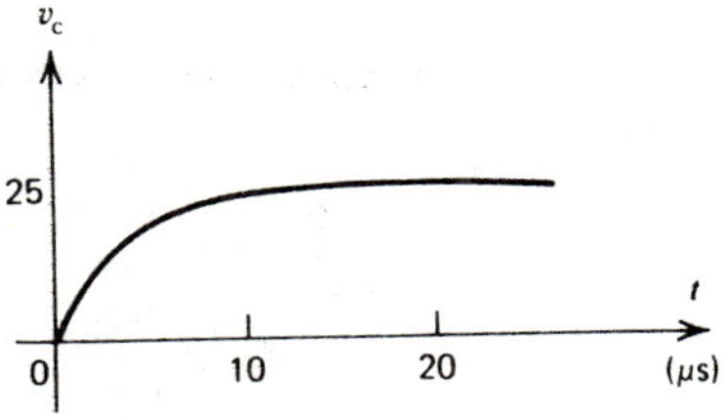

Figure 3-42 Answer to Prob. 3-25.

3-26

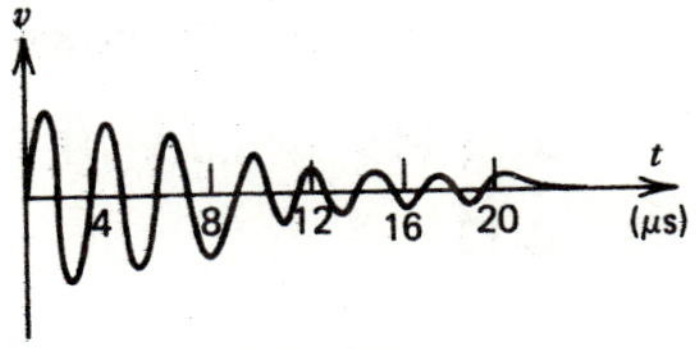

Figure 3-43 Answer to Prob. 3-26.

3-27 $C_1 = 20$ pF.

3-28 Attenuation factor $= 20.05$, $C_1 = 10.5$ pF.

Chapter 4 **4-1** PIV = 40 V, V_K = 0.5 V, R_f = 1.7 Ω, R_r = 20 MΩ.

4-2 2.25 mA.

4-3 The diode current will **a.** increase, **b.** increase, **c.** decrease, **d.** increase, **e.** decrease tremendously (reverse current).

4-4 I_D = 0.286 μA, V_D = 2.86 V.

4-5 R > 560 kΩ.

4-6 0.17 mA, 2.0 μA.

4-7

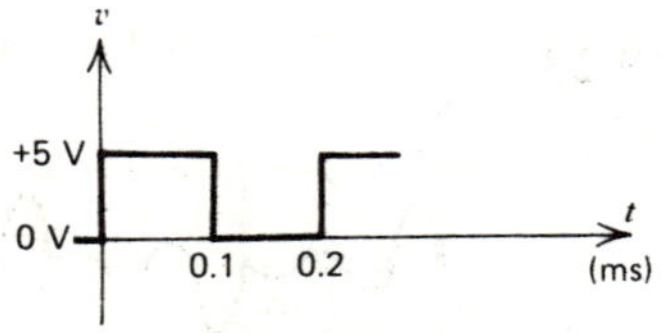

Figure 4-32 Answer to Prob. 4-7.

4-8

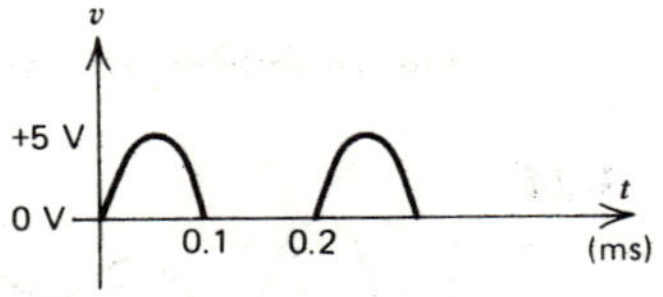

Figure 4-33 Answer to Prob. 4-8.

4-9

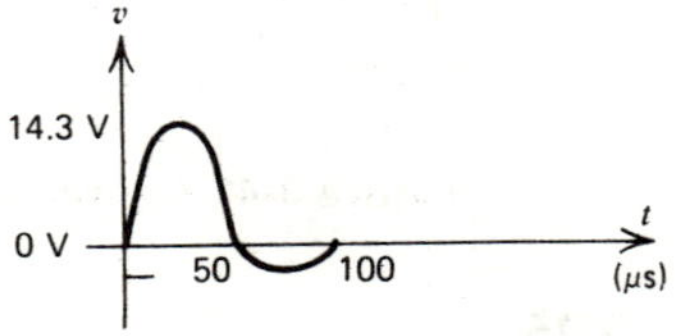

Figure 4-34 Answer to Prob. 4-9.

4-10

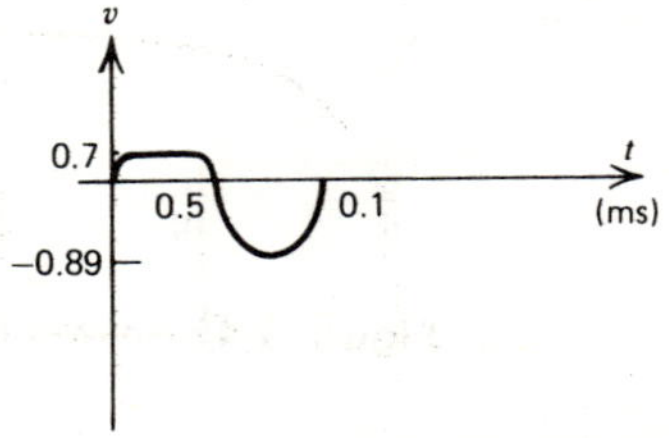

Figure 4-35 Answer to Prob. 4-10.

4-11 PIV = 5.0 V, W = 55 mW.

4-12

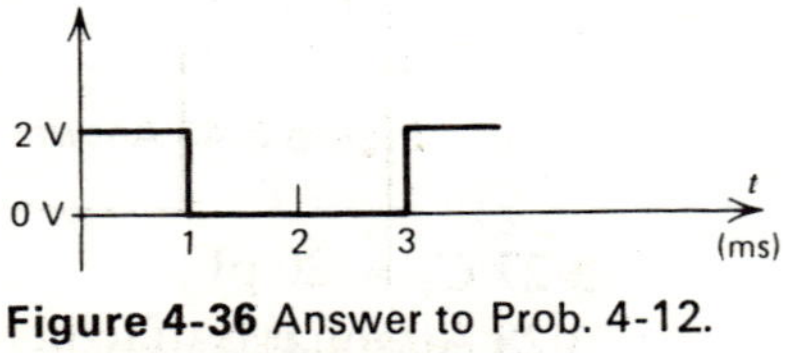

Figure 4-36 Answer to Prob. 4-12.

4-13

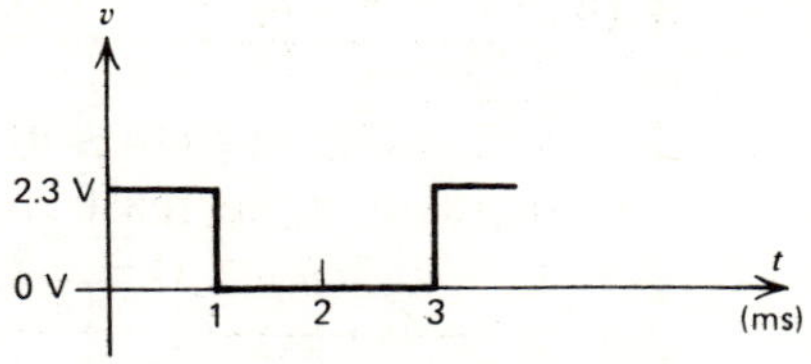

Figure 4-37 Answer to Prob. 4-13.

4-14 Same as Prob. 4-13 above.

4-15

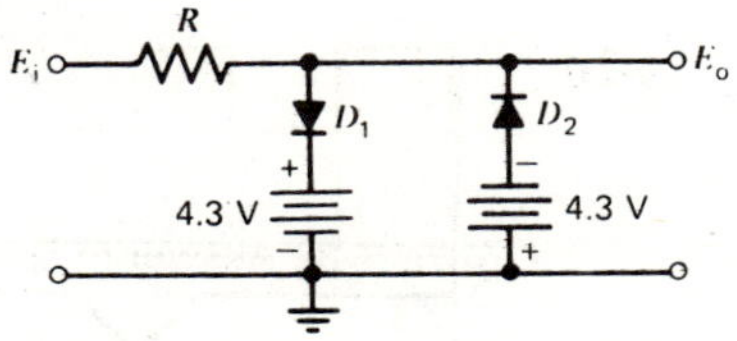

Figure 4-38 Answer to Prob. 4-15.

PIV = 65 V, I_{max} = 65/R. The other diode will still be safe even if one diode opens.

4-16

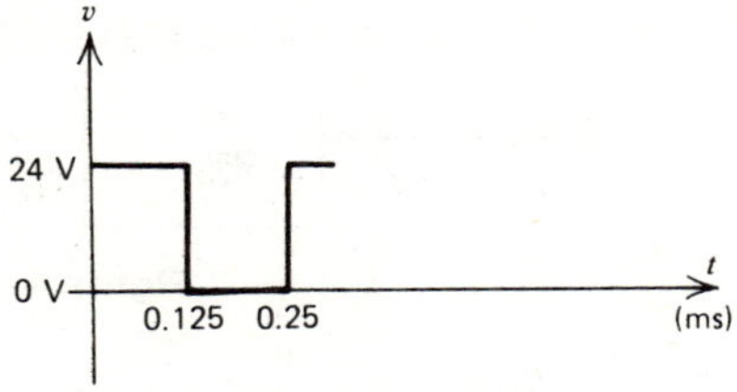

Figure 4-39 Answer to Prob. 4-16.

4-17

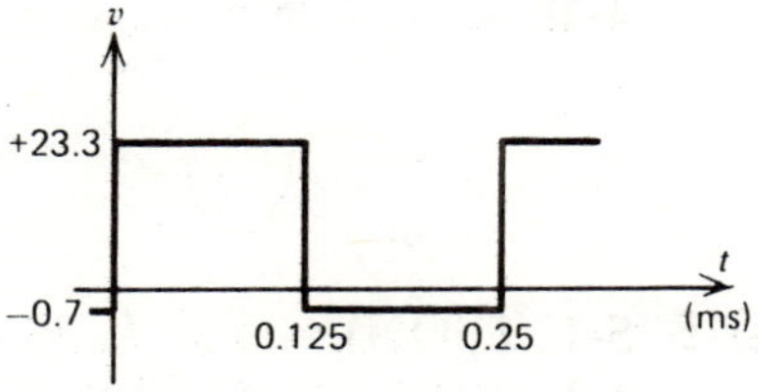

Figure 4-40 Answer to Prob. 4-17.

4-18

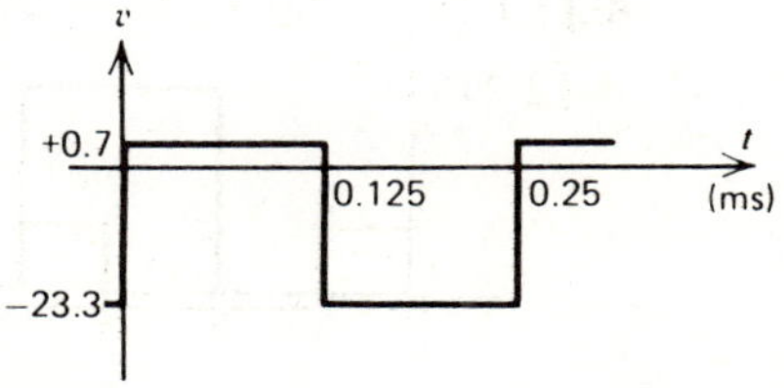

Figure 4-41 Answer to Prob. 4-18.

4-19

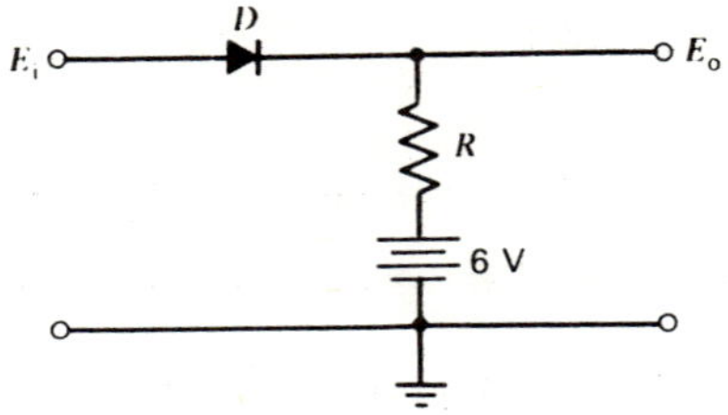

Figure 4-42 Answer to Prob. 4-42.

4-20

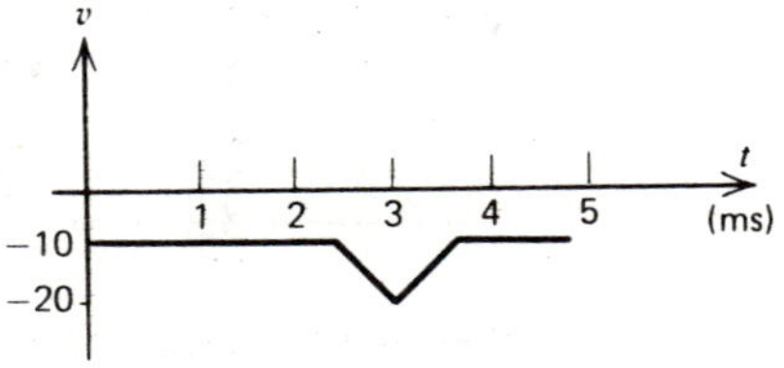

Figure 4-43 Answer to Prob. 4-20.

4-21

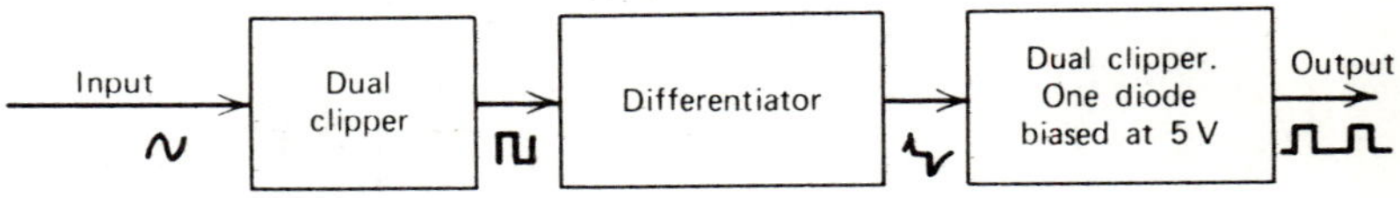

Figure 4-44 Answer to Prob. 4-21.

1. dual clipper; **2.** differentiator with small τ; **3.** negative peak clipper; **4.** biased positive-peak clipper at +5 V **dc.**

4-22 t_{ON} = 1.38 ns, t_{OFF} = 20 ns.

Chapter 5 **5-1** E_i(ON) = 5 V, E_o = 8 V.

5-2 E_i(ON) = 5.652 V, E_o = 10.8 V.

5-3 E_i(ON) = 5.275 V, E_o = 9.2 V.

5-4

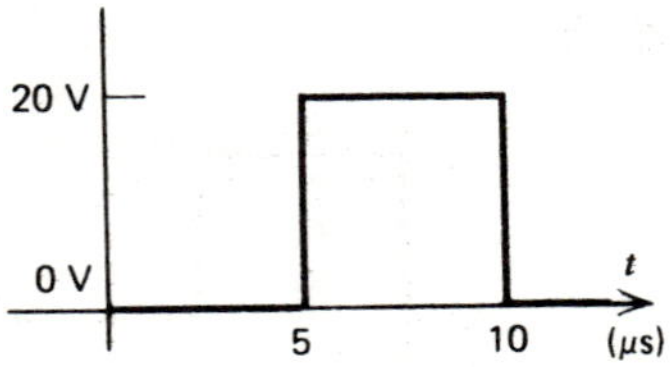

Figure 5-15 Answer to Prob. 5-4.

5-5

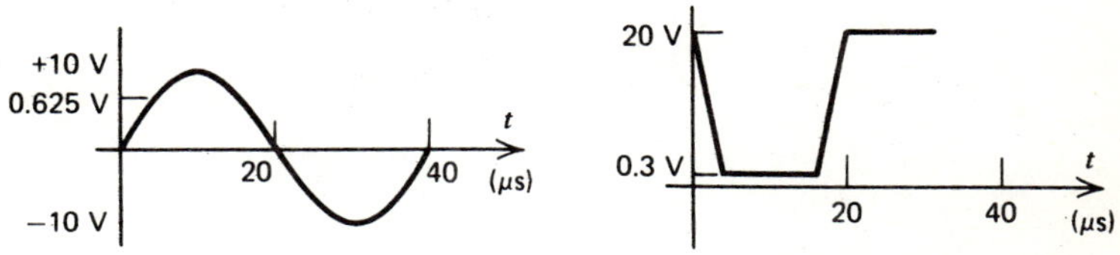

(a) Input waveform (b) Output waveform

Figure 5-16 Answer to Prob. 5-5.

5-6

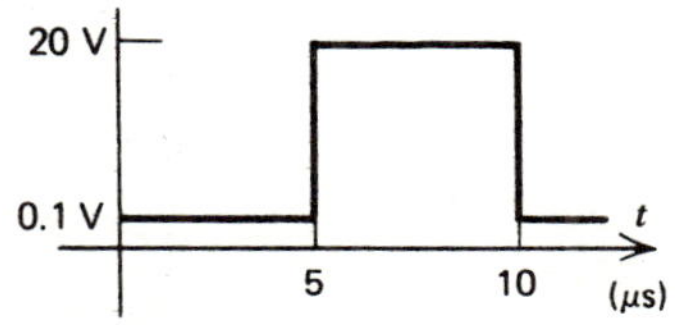

Figure 5-17 Answer to Prob. 5-6.

5-7

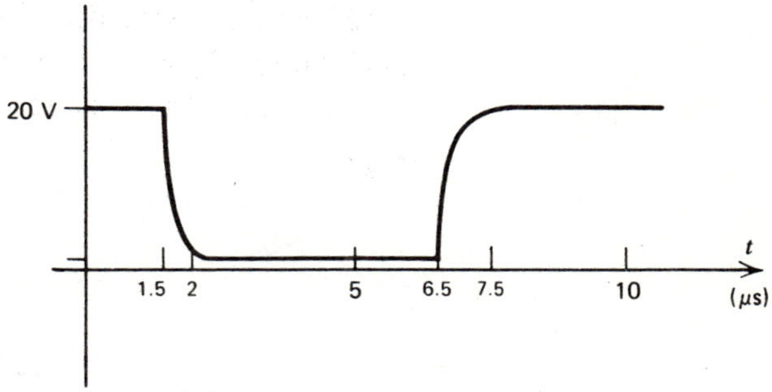

Figure 5-18 Answer to Prob. 5-7.

5-8 12.5 pF $< C_s <$ 40 pF.

5-9

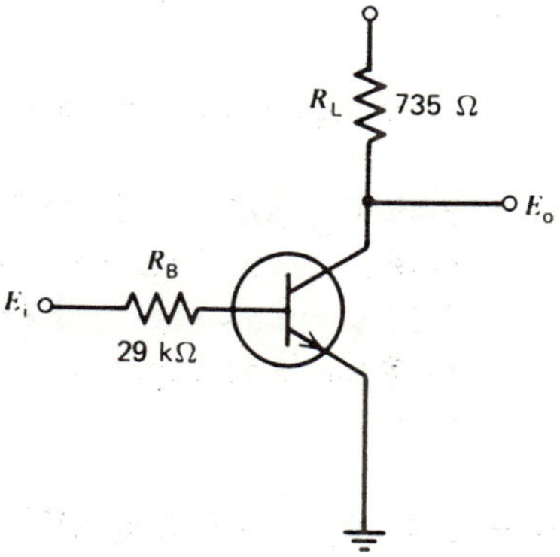

Figure 5-19 Answer to Prob. 5-9.

5-10 See Fig. 5-19. $R_L = 745\ \Omega$; $R_B = 30.3\ k\Omega$.

5-11

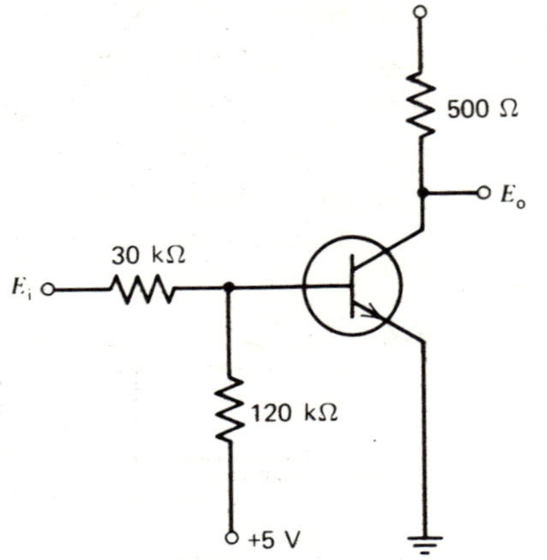

Figure 5-20 Answer to Prob. 5-11.

5-12 The transistor does not saturate in the ON state.
5-13 Make $R_L = 315$, or make $\beta = 158$ (min) for the transistor, or decrease the biasing resistor from 20 kΩ to 14.27 kΩ.

5-14

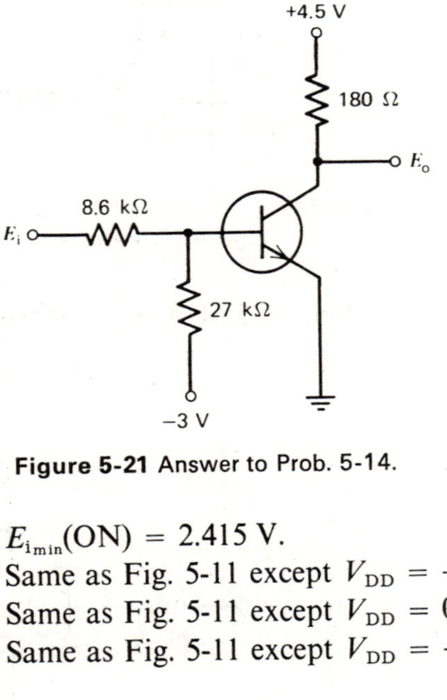

Figure 5-21 Answer to Prob. 5-14.

5-15 $E_{i_{min}}(\text{ON}) = 2.415$ V.
5-16 Same as Fig. 5-11 except $V_{DD} = +5$ V, and $V_{SS} = 0$ V.
5-17 Same as Fig. 5-11 except $V_{DD} = 0$ V and $V_{SS} = -5$ V.
5-18 Same as Fig. 5-11 except $V_{DD} = +5$ V and $V_{SS} = -5$ V.

Chapter 6 **6-1** $R_L = 2\ k\Omega$, $I_B(\text{ON}) = 79.3\ \mu A$.
6-2 $R_1 = 600\ \Omega$, $R_2 = 57.27\ k\Omega$, $R_3 = 630\ k\Omega$.

6-3 $R_1 = 585\ \Omega$, $R_2 = 51.6\ \text{k}\Omega$, $R_3 = 355\ \text{k}\Omega$.
6-4 $R_1 = 680\ \Omega$, $R_2 = 56\ \text{k}\Omega$, $R_3 = 390\ \text{k}\Omega$.
6-5 $R_1 = 1\ \text{k}\Omega$, $R_2 = 139\ \text{k}\Omega$, $R_3 = 1.95\ \text{M}\Omega$.
6-6 $R_1 = 993\ \Omega$, $R_2 = 135\ \text{k}\Omega$, $R_3 = 1.72\ \text{M}\Omega$.
6-7 min $I_c(\text{ON}) = 25.3$ mA, min $\beta = 128.4$.
6-8 Does not work; minimum β required is 131.
6-9 $R_L = 400\ \Omega$, $R_E = 185\ \Omega$, $R_1 = 16.25\ \text{k}\Omega$, $R_2 = 65\ \text{k}\Omega$.
6-10

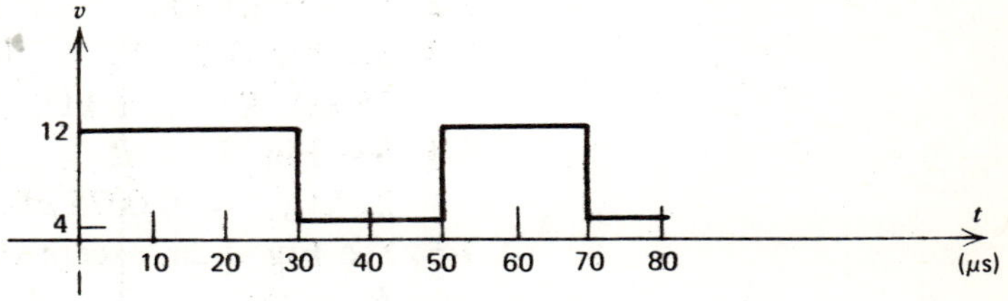

(a) Output waveform at E_{o_1}

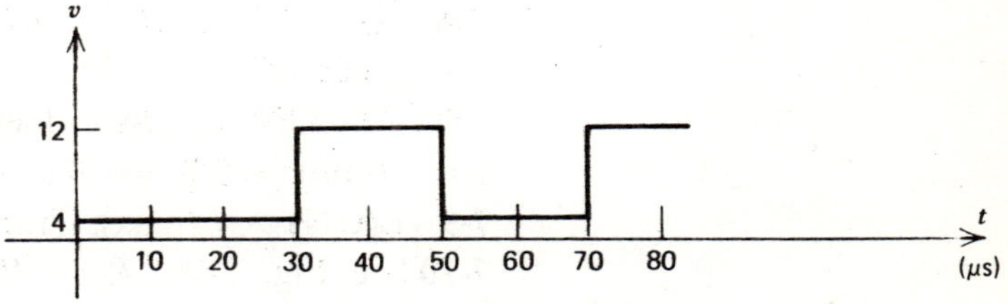

(b) Output waveform at E_{o_2}

Figure 6-32 Output waveforms for Prob. 6-10.

6-11

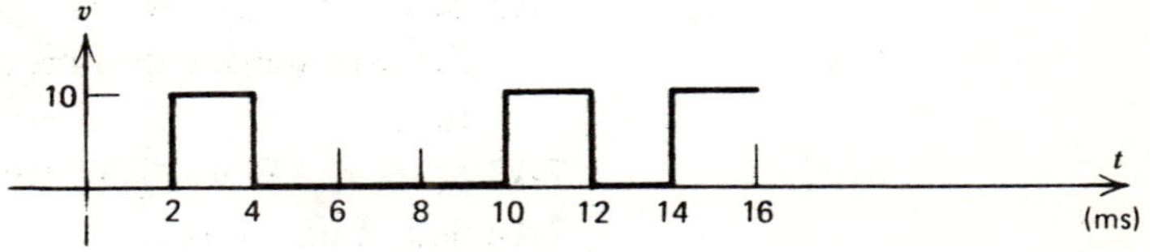

(a) Output waveform at E_{o_1}

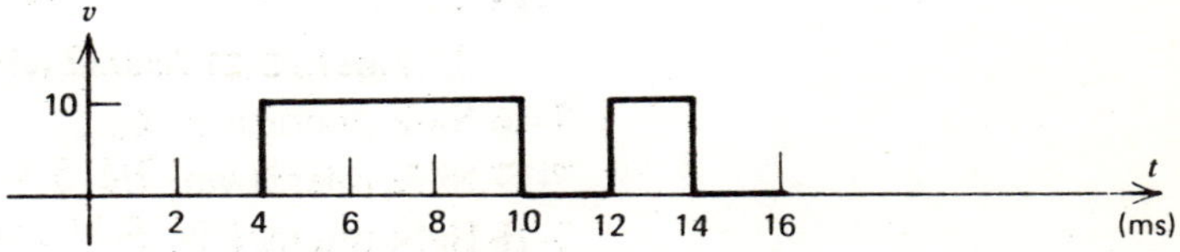

(b) Output waveform at E_{o_2}

Figure 6-33 Output waveforms for Prob. 6-11.

6-12 $f = 2.5$ kHz.
6-13 F/F shown in Fig. 6-23a. $R_D = 19.5\ \text{k}\Omega$, $V_{DD} = +12$ V, $R_1 = 2.1\ \text{k}\Omega$, $V_{GG} = -12$ V, $R_2 = 5.5\ \text{k}\Omega$.

6-14 F/F shown in Fig. 6-23a. $R_D = 950\ \Omega$, $V_{DD} = +10$ V, $R_1 = 2.0\ k\Omega$, $V_{GG} = -12$ V, $R_2 = 5.2\ k\Omega$.

6-15 F/F shown in Fig. 6-24. $V_{DD} = +5$ V.

Chapter 7

7-1 See Fig. 7-2a. $R_L = 200$ ohms, $R_1 = 3.56\ k\Omega$, $R = 4\ k\Omega$, $R_2 = 32\ k\Omega$, $C = 0.00721\ \mu F$.

7-2 $C = 0.018\ \mu F$.

7-3 See Fig. 7-2a. $V_{cc} = 10.2$ V, $R_1 = 3.3\ k\Omega$, $R_L = 200\ \Omega$, $R_2 = 25\ k\Omega$, $R = 3.88\ k\Omega$, $C = 0.00728\ \mu F$.

7-4 See Fig. 7-2a. $R_L = 220\ \Omega$, $R_1 = 2.3\ k\Omega$, $R = 3.9\ k\Omega$, $R_2 = 27\ k\Omega$, $C = 0.0075\ \mu F$.

7-5 See Fig. 7-2a, except that Q_1 and Q_2 are now *pnp* transistors. $R_L = 900\ \Omega$, $R_1 = 60.1\ k\Omega$, $R = 65.9\ k\Omega$, $R_2 = 416.3\ k\Omega$, $C = 0.0258\ \mu F$.

7-6 $R_L = 1\ k\Omega$, $R_1 = 56\ k\Omega$, $R = 68\ k\Omega$, $R_2 = 390\ k\Omega$, $C = 0.025\ \mu F$.

7-7 The ON transistor does not quite saturate.

7-8 $t(\text{min}) = 5.71\ \mu s$.

7-9 $I_c(\text{ON}) > I_c(\text{max})$. Transistors will be destroyed.

7-10 See Fig. 7-16. $R_L = 400\ \Omega$, $V_{cc} = +10$ V, $R = 2\ k\Omega$, $C_1 = 0.6\ \mu F$, $C_2 = 1.2\ \mu F$.

7-11 See Fig. 7-16. $R_L = 470\ \Omega$, $V_{cc} = 10$ V, $C_1 = 0.68\ \mu F$, $C_2 = 1.5\ \mu F$, $R_1 = 21.2\ k\Omega$, $R_2 = 19.2\ k\Omega$.

7-12 See Fig. 7-16. $R_L = 1.5\ k\Omega$, $V_{cc} = 20.1$ V, $C = 68$ pF, $R_1 = 210\ k\Omega$ (series combination), $R_2 = 210\ k\Omega$ (series combination).

7-13 $t_{re} > t_{p_2}$. Does *NOT* function properly.

7-14 See Fig. 7-18. $t_{p_1} = 2.7$ ms, $t_{p_2} = 5.74$ ms, $t_{re_1} = 0.5$ ms, $t_{re} = 235\ \mu s$.

7-15 $V_{cc} = 13$ V, $C_1 = 200$ pF, $R_L = 500\ \Omega$, $C_2 = 200$ pF, $R = 35.2\ k\Omega$, $C_s = 0.032\ \mu F$.

7-16 Safe operation, $E_{o(p\text{-}p)} = 9.524$ V, $t_p = 90.86\ \mu s$, $t_{re} = 2.25\ \mu s$.

7-17 Safe operation, $E_{o(p\text{-}p)} = 9.67$ V, $t_p = 93\ \mu s$, $t_{re} = 3.3\ \mu s$.

7-18 Safe operation, $E_{o(p\text{-}p)} = 9.524$ V, $t_p = 302.8\ \mu s$, $t_{re} = 7.5\ \mu s$.

7-19 $t_p(\text{max}) = 100.5\ \mu s$, $t_p(\text{min}) = 90.86\ \mu s$.

7-20 $V_{DS}(\text{ON}) = 0.52$ V, $E_{o(p\text{-}p)} = 11.48$ V, PRF = 215 Hz.

7-21 $V_{DS}(\text{ON}) = 0.75$ V, $E_{o(p\text{-}p)} = 11.25$ V, PRF = 224 Hz.

7-22 $V_{DS}(\text{ON}) = 0.52$ V, $E_{o(p\text{-}p)} = 11.48$ V, PRF = 281 Hz.

7-23 PRF (max) = 215 Hz, PRF (min) = 172 Hz.

7-24 $C = 40$ nF.

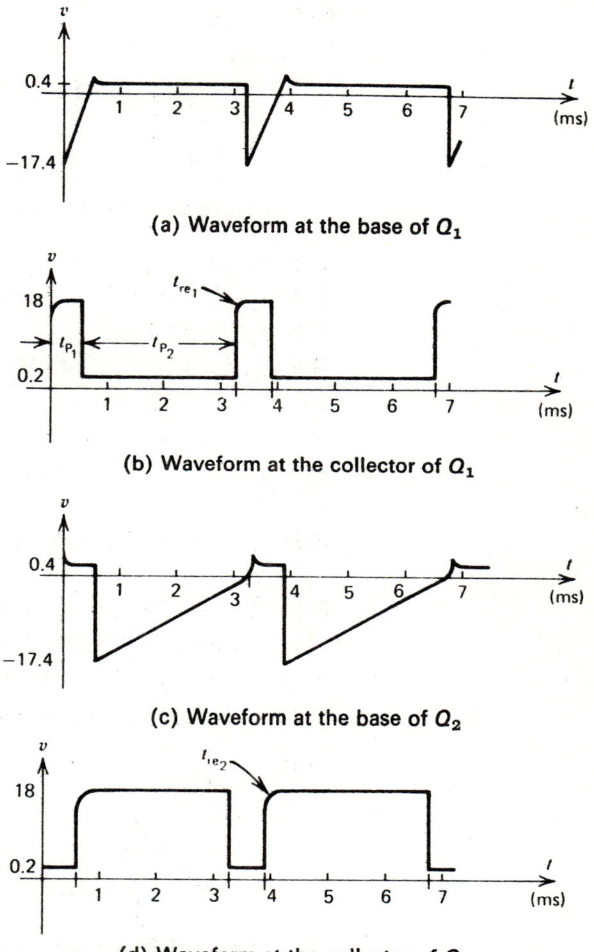

(a) Waveform at the base of Q_1

(b) Waveform at the collector of Q_1

(c) Waveform at the base of Q_2

(d) Waveform at the collector of Q_2

Figure 7-18 Answer to Prob. 7-14.

7-25 $C = 1.6$ nF.
7-26 $t_p = 329\ \mu s$.
7-27 $PRF_{min} = 3.04$ kHz.

Chapter 8

8-1

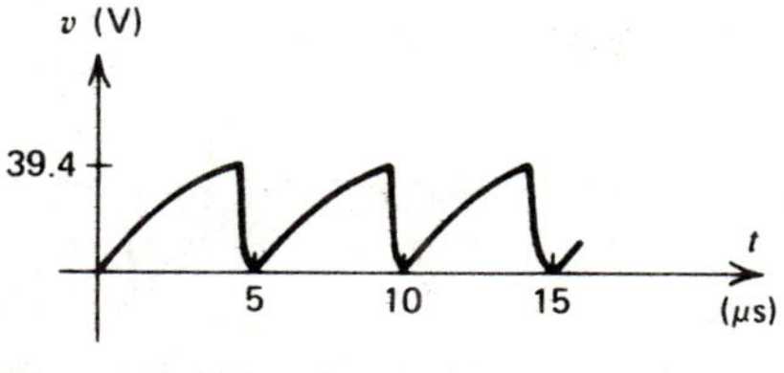

Figure 8-20 Answer to Prob. 8-1.

8-2

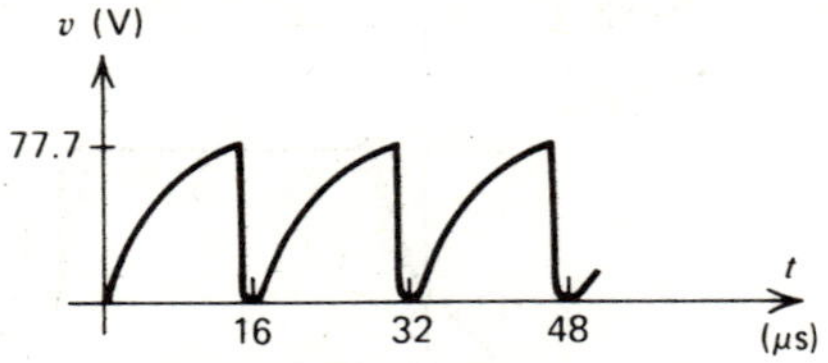

Figure 8-21 Answer to Prob. 8-2.

8-3 Slope error = 39.4%, slope error = 77.7%.
8-4 $I_{min}(ON) < I_H$, ∴ circuit functions satisfactorily.
8-5 $t_{sp} = 34.4\ \mu s$.
8-6 $t_{sp} = 34.4\ \mu s$ to $44.87\ \mu s$.
8-7 $t_{sp} = 1.62$ ms.
8-8 $R = 15\ k\Omega$.
8-9 $I_c(max) = 3.94$ mA, $\beta = 136$, $E_o = 12.23\ V_{p\text{-}p}$, slope error = 62%.
8-10 $I_c(max) = 2.9$ mA, $\beta = 100$, $E_o = 10.11\ V_{p\text{-}p}$, slope error = 51%.
8-11 $R_L = 2.39\ k\Omega$, $R_1 = 89.33\ k\Omega$, $R_2 = 268\ k\Omega$, $C = 0.006\ \mu F$.
8-12 $I_c(max) = 10$ mA, $\beta = 50$, $E_o = 3.70\ V_{p\text{-}p}$.
8-13 $I_c(max) = 13.3$ mA, $\beta = 66.7$, $E_o = 6.67\ V_{p\text{-}p}$.
8-14 $I_c = 1.5$ mA, $E_o = 3.19\ V_{p\text{-}p}$.

8-15

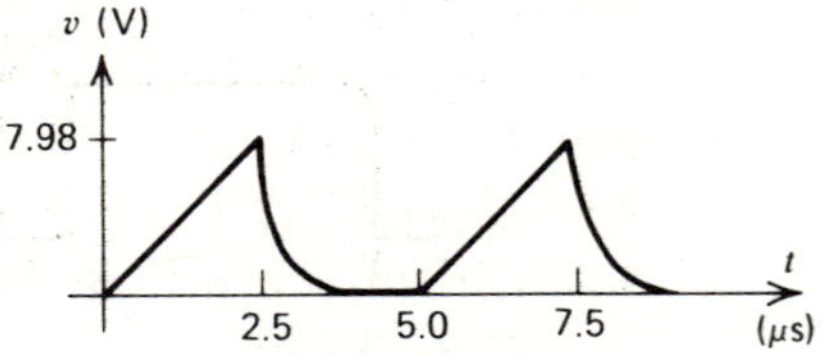

Figure 8-22 Answer to Prob. 8-15.

8-16

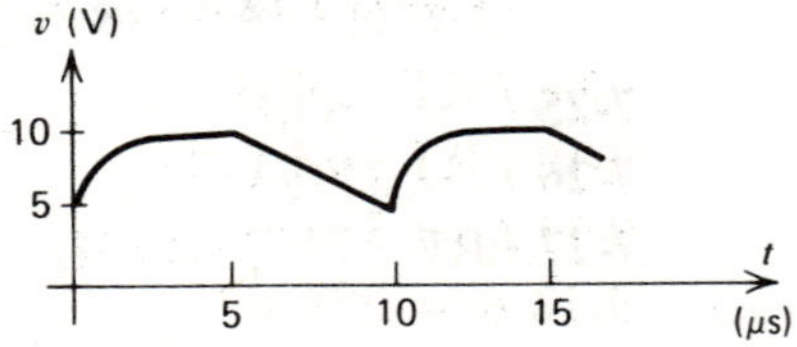

Figure 8-23 Answer to Prob. 8-16.

8-17

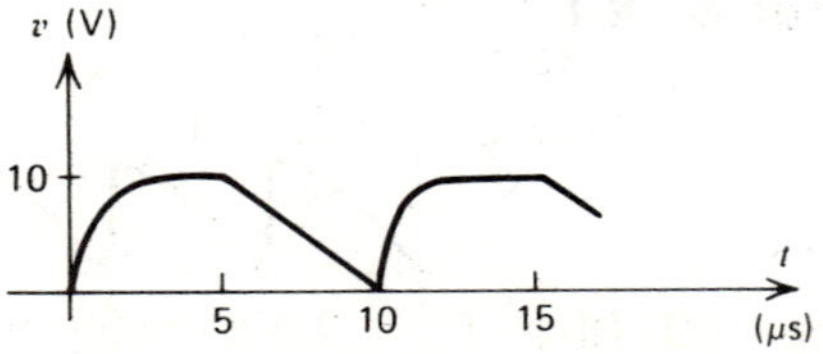

Figure 8-24 Answer to Prob. 8-17.

8-18 $R_E = 2.22\ k\Omega$, $R_1 = 138.9\ k\Omega$, $C = 1.875 \times 10^{-9}F$.

8-19

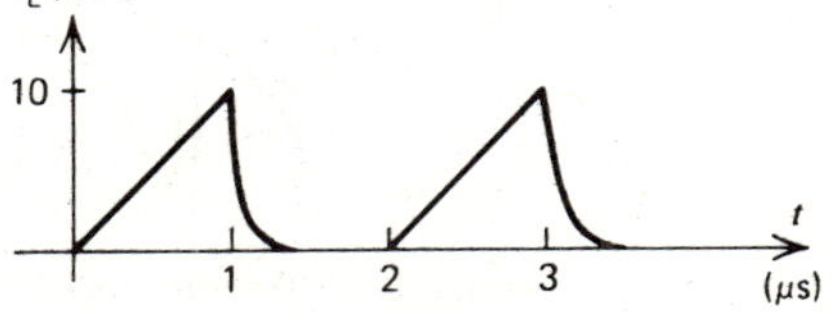

Figure 8-25 Answer to Prob. 8-19.

8-20

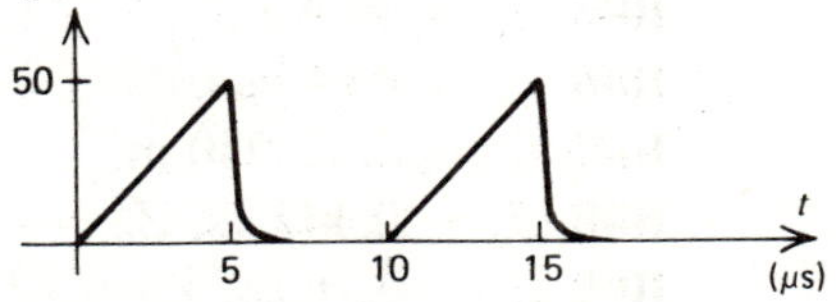

Figure 8-26 Answer to Prob. 8-20.

Chapter 9

9-1 Circuit does work; UTP = 4 V, LTP = 2 V.

9-2 Circuit does work; UTP = 4.6 V, LTP = 2.65 V.

9-3 Circuit shown in Fig. 9-1a; $R_{L_2} = 680\ \Omega$, $R_{L_1} = 48\ k\Omega$, $R_1 = 1.08\ k\Omega$, $R_2 = 144\ k\Omega$, $R_E = 118\ \Omega$, $R_B = 900\ \Omega$.

9-4 Circuit shown in Fig. 9-1a; $R_{L_2} = 696\ \Omega$, $R_1 = 74\ k\Omega$; $R_{L_1} = 1.3\ k\Omega$, $R_2 = 74\ k\Omega$, $R_E = 91\ \Omega$, $R_B = 682.5\ \Omega$.

9-5 $V_p = 15$ V, $R_{B_2} = 2.5\ k\Omega$, $R_{B_1} = 7.5\ k\Omega$, $t_{sp} = 13.84\ \mu s$.

9-6 $V_p = 12$ V, $R_{B_2} = 4\ k\Omega$, $R_{B_1} = 6\ k\Omega$, $t_{sp} = 9.14\ \mu s$.

9-7 Circuit shown in Fig. 9-10a; $V_{BB} = 14$ V, $R_1 = 10.7\ k\Omega$, $C = 390$ pF.

9-8 Circuit shown in Fig. 9-10a; $V_{BB} = 13.33$ V, $R_1 = 28.44\ k\Omega$, $C = 127$ pF.

9-9 PRT = 409 ms.

9-10 PRT = 4.1 ms.

9-11 $t_p = 330\ \mu s$.

9-12 $R = 352\ k\Omega$.

9-13 PRF = 241 kHz.

9-14 PRF = 160 kHz.

9-15 $t_{sp} = 159.7$ ms.

9-16 $C = 7.358\ \mu F$.

Chapter 10

10-1 $V_s = 2.4$ kV, step up.

10-2 $I_s = 240$ mA, $I_p = 4.8$ mA, $P = 576$ W.

10-3 $a = 10$, $P = 18$ W.

10-4

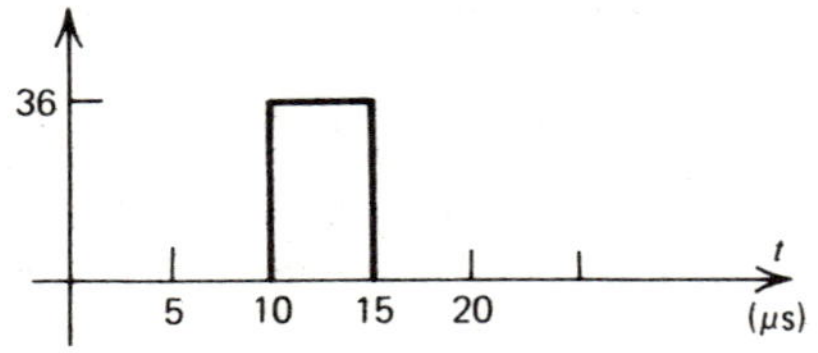

Figure 10-13 Answer to Prob. 10-4.

10-5 $P = 18$ W, $R_{eq} = 2\ \Omega$.
10-6 $t_d = 0.667\ \mu s$.
10-7 Length = 750 m.
10-8 $t_d = 0.447\ \mu s$, $Z_o = 4.47\ k\Omega$, $f_c = 784$ kHz.
10-9 $t_d = 3.16\ \mu s$, $Z_o = 633\ \Omega$, $f_c = 111$ kHz.
10-10 $L = 5.6$ mH, $C = 0.0179\ \mu F$.
10-11 $L = 25$ mH, $C = 0.025\ \mu F$.
10-12 $t_d = 4.7\ \mu s$, $Z_o = 102$ ohms, $f_c = 74.5$ kHz.
10-13 $t_d = 1.6\ \mu s$, $Z_o = 102$ ohms, $f_c = 214$ kHz.
10-14 Length = 1223 m.
10-15 $t_d = 1.06\ \mu s$, $Z_o = 45.5$ ohms, $f_c = 330$ kHz.
10-16 $L = 48\ \mu H$, $C = 0.0233\ \mu F$.
10-17 $l = 714$ m, $Z_o = 45.5$ ohms, $f_c = 69.3$ kHz.
10-18 Circuit is shown in Fig. 10-10a. Assume $V_{BB} = -6$ V, $V_{BE}(OFF) = -1.0$ V. $V_{CC} = 6$ V, $R_L = 600$ ohms, $R_1 = 120\ k\Omega$, $C_1 = 60$ pF, $R_2 = 96\ k\Omega$, $R_3 = 480\ k\Omega$.
10-19 $V_{CC} = 10$ V, $R_L = 1\ k\Omega$, $R_1 = 200\ k\Omega$, $C_1 = 72$ pF, $R_2 = 176\ k\Omega$, $R_3 = 880\ k\Omega$.
10-20 Circuit is drawn in Fig. 10-11a. V_{CC}, V_{BB}, R_L, R_1, R_2, R_3 and C_1 are same as in Ans. 4-18. $R_4 = 112\ k\Omega$, $C_2 = 259$ pF.

Chapter 11

11-1 $E_i = 4.5$ V.
11-2 $E_{o1} = 4.5$ V, $E_{o2} = 0.4$ V, $E_{o3} = 4.5$ V, $E_{o4} = 0.4$ V.
11-3 $E_o = 4.5$ V.
11-4 $E_o = 0.4$ V.
11-5 $E_o = 4.5$ V.
11-6 $E_o = 0.4$ V.
11-7 $E_o = 4.5$ V.
11-8 $E_o = 0.4$ V.
11-9 $E_1 = A + B$, $E_2 = XY + \overline{X}Z$, $E_3 = \overline{X}$, $E_4 = L\overline{N} + \overline{L}MP$.

11-10 See Fig. 11-33a; see Fig. 11-33b; see Fig. 11-33c; see Fig. 11-33d.

C \ AB	00	01	11	10
0				1
1			1	1

(a) Answer to Prob. 11-10a, $S_1 = A\bar{B} + AC$

Z \ XY	00	01	11	10
0			1	
1	1	1	1	1

(c) Answer to Prob. 11-10c, $L_1 = Z + XY$

Z \ XY	00	01	11	10
0	1			
1	1	1		

(b) Answer to Prob. 11-10b, $S_2 = \bar{X}\bar{Y} + \bar{X}Z$

CD \ AB	00	01	11	10
00				1
01				1
11			1	1
10			1	1

(d) Answer to Prob. 11-10d, $L_2 = A\bar{B} + AC$

Figure 11-33 Answer to Prob. 11-10.

11-11 $E_1 = ABC + A\bar{D} + A\bar{B}$, $E_2 = X\bar{Y} + XZ + \bar{Y}Z$, $E_3 = X\bar{Y} + X\bar{Z}$.

11-12 $E_1 = (\bar{A} + \bar{B})(A + C)(\bar{B} + C)$, $E_2 = (X + Z)(Y + Z)$, $E_3 = (X + Y)(Y + Z)(\bar{X} + Z)$.

11-13 $E_o = \bar{A}\bar{B}CD + \bar{A}BC\bar{D} + \bar{A}BCD + A\bar{B}CD + AB\bar{C}\bar{D} + AB\bar{C}D + ABC\bar{D} + ABCD$.

11-14 $E_o = XZ + XY + YZ$.

11-15 $X = \bar{A}BCD + A\bar{B}CD + AB\bar{C}D + ABC\bar{D} + ABCD$, $Y = \bar{A}\bar{B}CD + \bar{A}B\bar{C}D + \bar{A}BC\bar{D} + \bar{A}BCD + A\bar{B}\bar{C}D + A\bar{B}C\bar{D} + A\bar{B}CD + AB\bar{C}\bar{D} + AB\bar{C}D + ABC\bar{D} + ABCD$; $Y = AB + AC + AD + BD + CD$; see Fig. 11-34.

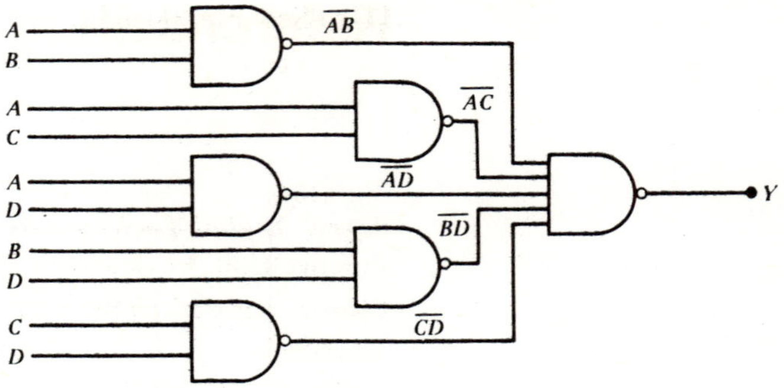

Figure 11-34 Answer to Prob. 11-15c.

11-16 E_o = 12 V.
11-17 NAND gate 0 V = 0, 12 V = 1.

A	B	C	E_o
0	0	0	1
0	0	1	1
0	1	0	1
0	1	1	1
1	0	0	1
1	0	1	1
1	1	0	1
1	1	1	0

11-18 V_{CC} = 12V
11-19 a. V_{TH} = 0.85V
b. NM = 0.65V
11-20 NM = 0.6V
11-21 Fan out = 6
11-22 I = 48mA
11-23 The output [Fig. 11.19c] will now be a 0V-3 pulse wave.

References

Boyce, *Digital Logic and Switching Circuits: Operation and Analysis*, Prentice Hall, Englewood Cliffs, N.J., 1975.

Cirovic, *Basic Electronics: Devices, Circuits and Systems*, Reston, Reston, Va., 1974.

Cutler, *Electronic Circuit Analysis*, Vol. 1, *Passive Networks*, McGraw-Hill, New York, 1960.

Doyle, *Digital Switching and Timing Circuits*, Duxbury, 1976.

Greenfield, *Practical Digital Design Using ICs*, Wiley, New York, 1977.

Hibberd, *Integrated Circuits: A Basic Course for Engineers and Technicians*, McGraw-Hill, New York, 1969.

Jackson, *Introduction to Electric Circuits*, 4th ed., Prentice-Hall, Englewood Cliffs, N.J., 1976.

Leach, *Basic Electric Circuits*, 2nd ed., Wiley, New York, 1976.

Lenk, *Manual for Integrated Circuit Users*, Reston, Reston, Va., 1972.

Malvino, *Transistor Circuit Approximations*, 2nd ed., McGraw-Hill, New York, 1973.

Millman and Taub, *Pulse, Digital and Switching Waveforms*, 2nd ed., McGraw-Hill, New York, 1965.

Mitchell, *Semiconductor Pulse Circuits with Experiments*, Holt, Rinehart, and Winston, New York, 1970.

Nashelsky, *Introduction to Digital Computer Technology*, 2nd ed., Wiley, New York, 1977.

Romanowitz and Puckett, *Introduction to Electronics*, 2nd ed., Wiley, New York, 1976.

Index